**DRILL HALL LIBRARY
MEDWAY**

Edited by
Michael Lämmerhofer and
Wolfram Weckwerth

Metabolomics in Practice

Related Titles

Ekroos, Kim (ed.)

Lipidomics
Technologies and Applications

2012
ISBN: 978-3-527-33098-0

Anzenbacher, P., Zanger, U. M. (eds.)

Metabolism of Drugs and Other Xenobiotics

2012
ISBN: 978-3-527-32903-8

Kalgutkar, A. S., Dalvie, D., Obach, R. S., Smith, D. A.

Reactive Drug Metabolites
Series: Methods and Principles in Medicinal Chemistry

2012
ISBN: 978-3-527-33085-0

Kahl, G.

The Dictionary of Genomics, Transcriptomics and Proteomics
Fourth, greatly enlarged edition

2009
ISBN: 978-3-527-32073-8

Dziuda, D. M.

Data Mining for Genomics and Proteomics
Analysis of Gene and Protein Expression Data

2010
ISBN: 978-0-470-16373-3

Mishra, N. C.

Introduction to Proteomics
Principles and Applications

2010
ISBN: 978-0-471-75402-2

Klipp, E., Liebermeister, W., Wierling, C., Kowald, A., Lehrach, H., Herwig, R.

Systems Biology
A Textbook

2009
ISBN: 978-3-527-31874-2

von Hagen, J. (ed.)

Proteomics Sample Preparation

2008
ISBN: 978-3-527-31796-7

Westermeier, R., Naven, T., Höpker, H.-R.

Proteomics in Practice
A Guide to Successful Experimental Design

2008
ISBN: 978-3-527-31941-1

Edited by Michael Lämmerhofer and Wolfram Weckwerth

Metabolomics in Practice

Successful Strategies to Generate and
Analyze Metabolic Data

WILEY-VCH Verlag GmbH & Co. KGaA

The Editors

Prof. Dr. Michael Lämmerhofer
Eberhard Karls University Tübingen
Pharmazeutisches Institut
Auf der Morgenstelle 8
72076 Tübingen
Germany

Prof. Dr. Wolfram Weckwerth
Universität Wien
Molekulare Systembiologie
Althanstr. 14
1090 Wien
Österreich

All books published by **Wiley-VCH** are carefully produced. Nevertheless, authors, editors, and publisher do not warrant the information contained in these books, including this book, to be free of errors. Readers are advised to keep in mind that statements, data, illustrations, procedural details or other items may inadvertently be inaccurate.

Library of Congress Card No.: applied for

British Library Cataloguing-in-Publication Data
A catalogue record for this book is available from the British Library.

Bibliographic information published by the Deutsche Nationalbibliothek
The Deutsche Nationalbibliothek lists this publication in the Deutsche Nationalbibliografie; detailed bibliographic data are available on the Internet at <http://dnb.d-nb.de>.

© 2013 Wiley-VCH Verlag & Co. KGaA, Boschstr. 12, 69469 Weinheim, Germany

All rights reserved (including those of translation into other languages). No part of this book may be reproduced in any form – by photoprinting, microfilm, or any other means – nor transmitted or translated into a machine language without written permission from the publishers. Registered names, trademarks, etc. used in this book, even when not specifically marked as such, are not to be considered unprotected by law.

Composition Laserwords Private Limited, Chennai, India
Printing and Binding Markono Print Media Pte Ltd, Singapore
Cover Design Bluesea Design, McLeese Lake, Canada

Print ISBN: 978-3-527-33089-8
ePDF ISBN: 978-3-527-65589-2
ePub ISBN: 978-3-527-65588-5
mobi ISBN: 978-3-527-65587-8
oBook ISBN: 978-3-527-65586-1

Printed in Singapore
Printed on acid-free paper

Contents

List of Contributors *XV*
Preface *XXIII*

1 The Sampling and Sample Preparation Problem in Microbial Metabolomics *1*
Walter M. van Gulik, André B. Canelas, Reza M. Seifar, and Joseph J. Heijnen
1.1 Introduction *1*
1.2 Microorganisms and Their Properties *1*
1.3 Sampling Methods *2*
1.3.1 The Need for Rapid Sampling *2*
1.3.2 Sampling Systems *3*
1.4 Quenching *4*
1.4.1 Quenching Procedures and Their Properties *4*
1.4.2 Validation of the Quenching Procedure and Minimization of Metabolite Leakage *5*
1.4.3 Quenching Procedure for Determination of Intracellular Metabolites in the Presence of Extracellular Abundance *6*
1.4.4 Quenching of Bacteria *7*
1.5 Metabolite Extraction *9*
1.5.1 Extraction Methods and Their Properties *9*
1.5.2 Validation of Extraction Methods for Yeast Metabolomics *9*
1.6 Application of ^{13}C-Labeled Internal Standards *13*
1.7 Conclusions *17*
References *18*

2 Tandem Mass Spectrometry Hyphenated with HPLC and UHPLC for Targeted Metabolomics *21*
Gérard Hopfgartner and Emmanuel Varesio
2.1 Introduction *21*
2.2 LC-MS-Based Targeted Metabolomics *22*
2.3 Liquid Chromatography *22*

2.4	Mass Spectrometry 27
2.4.1	Ionization Techniques 27
2.4.2	Mass Analyzers 28
2.5	Sample Preparation 30
2.6	Relative and Absolute Quantification 31
2.7	Applications 32
2.8	Synopsis 34
	References 35

3	**Uncertainty of Measurement in Quantitative Metabolomics** 39
	Raffaele Guerrasio, Christina Haberhauer-Troyer, Stefan Neubauer, Kristaps Klavins, Madeleine Werneth, Gunda Koellensperger, and Stephen Hann
3.1	Introduction 39
3.1.1	MS-Based Techniques in Metabolomics 39
3.1.2	Uncertainty of Measurement in Quantitative Analysis 41
3.1.2.1	Definition 41
3.1.2.2	Uncertainty Calculation According to the Bottom-Up Approach 42
3.2	Uncertainties of Quantitative MS Experiments 48
3.2.1	Uncertainties in Sample Preparation 48
3.2.1.1	Sampling and Sample Preparation in Metabolite Profiling in Fermentations 49
3.2.1.2	Calculation of Sample Preparation Uncertainty for Intracellular Metabolite Quantitation in Yeast: A Practical Example 53
3.2.1.3	LC-MS 59
3.2.2	Uncertainty of Mass Spectrometric Assays (LC-MS and GC-MS Measurements) 61
3.2.2.1	GC-MS 61
3.2.2.2	Calculation of Uncertainty for LC-MS Measurements of Cell Extracts: A Practical Example 63
3.3	Concluding Remarks 66
	Abbreviations 66
	Acknowledgment 67
	References 67

4	**Gas Chromatography and Comprehensive Two-Dimensional Gas Chromatography Hyphenated with Mass Spectrometry for Targeted and Nontargeted Metabolomics** 69
	Song Yang, Jamin C. Hoggard, Mary E. Lidstrom, and Robert E. Synovec
4.1	Introduction and Scope 69
4.2	Sample Preparation for GC-Based Metabolite Profiling 71
4.3	GC–MS and GC × GC–TOFMS Instrumentation for Metabolomics 74
4.4	Data Analysis Strategies and Software 82

4.5	Illustrative Examples and Concluding Remarks 88
	References 89

5	**LC-MS-Based Nontargeted Metabolomics** 93
	Georgios A. Theodoridis, Helen G. Gika, and Ian D. Wilson
5.1	Introduction 93
5.2	LC-MS-Based Untargeted Metabolomics 94
5.2.1	LC Issues 94
5.2.2	Mass Spectrometry 97
5.3	Study Design 98
5.4	Sample Preparation 100
5.5	Analytical Strategies 103
5.6	Data Analysis 104
5.7	Metabolite Identification 107
5.8	Applications 109
5.9	Synopsis 112
	References 113

6	**The Potential of Ultrahigh Resolution MS (FTICR-MS) in Metabolomics** 117
	Franco Moritz, Sara Forcisi, Mourad Harir, Basem Kanawati, Marianna Lucio, Dimitrios Tziotis, and Philippe Schmitt-Kopplin
6.1	Introduction 117
6.2	Metabolomics Technologies 118
6.3	Principles of FTICR-MS 121
6.3.1	Natural Ion Movement Inside an ICR Cell Subjected to Magnetic and Electric Fields 121
6.3.2	Applied Physical Techniques in FTICR-MS 123
6.3.3	Practical Advantages of FTICR-MS 124
6.4	Proceeding in Metabolomics 126
6.4.1	Network Analysis and NetCalc Composition Assignment 126
6.4.2	Statistics on FTICR-MS Datasets 127
6.5	Application Example in Metabolomics Using FTICR-MS Exhaled Breath Condensate 128
6.5.1	The Experiment 128
6.5.2	FT-ICR/MS Measurement 129
6.5.3	Data Preprocessing 129
6.5.4	C–H–N–O–S–P Formula Annotation 130
6.5.5	Statistical Analysis 130
6.5.5.1	Statistical Preprocessing 130
6.5.6	Synthesis of Biochemical Mass Difference Networking and Statistical Results 131
6.6	Conclusion and Remarks 134
	References 134

7	**The Art and Practice of Lipidomics** *137*
	Koen Sandra, Ruben t'Kindt, Lucie Jorge, and Pat Sandra
	Abbreviations *137*
7.1	Introduction *139*
7.2	Lipid Diversity *140*
7.3	Tackling the Lipidome: State-of-the-Art *141*
7.4	LC-MS-Based Lipidomics *146*
7.4.1	Lipid Extraction *146*
7.4.1.1	Biological Fluids and Cellular Material *146*
7.4.1.2	Skin (Stratum Corneum) *148*
7.4.1.3	Solid-Phase Extraction (SPE) *148*
7.4.2	LC–MS(/MS) *151*
7.4.2.1	Retention Time Characteristics *151*
7.4.2.2	Ionization Characteristics *155*
7.4.2.3	Identification of Lipids *156*
7.4.3	Data Processing and Analysis *161*
7.5	GC-MS-Based Lipidomics *165*
7.5.1	Sample Preparation *165*
7.5.2	GC–MS *166*
7.5.3	Data Processing and Analysis *170*
7.6	Conclusion *172*
	References *173*
8	**The Role of CE–MS in Metabolomics** *177*
	Rawi Ramautar, Govert W. Somsen, and Gerhardus J. de Jong
	Abbreviations *177*
8.1	Introduction *177*
8.2	CE–MS *179*
8.2.1	CE Separation Conditions *179*
8.2.2	CE–MS Coupling *180*
8.2.2.1	Interfacing *180*
8.2.2.2	Mass Analyzers *184*
8.3	Sample Pretreatment *185*
8.4	Data Analysis *187*
8.5	Applications *190*
8.5.1	Targeted Approaches *190*
8.5.2	Nontargeted Approaches *202*
8.6	Conclusions and Perspectives *203*
	References *206*
9	**NMR-Based Metabolomics Analysis** *209*
	Andrea Lubbe, Kashif Ali, Robert Verpoorte, and Young Hae Choi
9.1	Introduction *209*
9.2	Platforms for Metabolomics *210*
9.2.1	Mass Spectrometry (MS) *210*

9.2.1.1	Gas Chromatography–Mass Spectrometry (GC-MS)	210
9.2.1.2	Liquid Chromatography–Mass Spectrometry (LC–MS)	211
9.2.1.3	Capillary Electrophoresis–Mass Spectrometry (CE–MS)	211
9.2.1.4	Fourier Transform-Ion Cyclotron Resonance-Mass Spectrometry (FT-ICR-MS)	212
9.2.2	Fourier Transform–Infrared Spectroscopy (FT–IR)	212
9.2.3	Nuclear Magnetic Resonance Spectroscopy: Principles and Techniques	212
9.2.3.1	One-Dimensional Nuclear Magnetic Resonance (^{1}H and ^{13}C NMR)	213
9.2.3.2	J-Resolved Spectroscopy (JRES)	215
9.2.3.3	Correlation Spectroscopy (COSY)	215
9.2.3.4	Total Correlation Spectroscopy (TOCSY)	217
9.2.3.5	Heteronuclear Two-Dimensional Methods	217
9.2.3.6	Combined Two-Dimensional Methods	217
9.3	NMR for Metabolomics	219
9.3.1	Sample Preparation	220
9.3.2	Metabolite Identification	221
9.3.3	Data Analysis: Turning Data into Information, Possibly Knowledge	223
9.3.3.1	Data Preprocessing	223
9.3.3.2	Principal Component Analysis (PCA)	225
9.3.3.3	Partial Least Squares (PLS) Projections to Latent Structures	226
9.3.3.4	Bidirectional Orthogonal-PLS (O2PLS)	227
9.3.3.5	Validation	227
9.4	Applications of NMR-Based Metabolomics	228
9.4.1	Understanding Stress Response	228
9.4.2	Application to Bioactivity Screening	230
9.4.3	Quality Control of Herbal Medicines	231
9.4.4	Chemotaxonomy	232
9.4.5	Agricultural Applications	233
9.5	Future Prospects and Conclusions	233
	References	234
10	**Potential of Microfluidics and Single Cell Analysis in Metabolomics (Micrometabolomics)**	**239**
	Meghan M. Mensack, Ryan E. Holcomb, and Charles S. Henry	
10.1	Introduction	239
10.2	Sample Processing for Metabolomics	240
10.2.1	Solid Phase Extraction	240
10.2.2	Laminar Diffusion	241
10.2.3	Fluidic Pumping for On-Chip Mixing	242
10.3	Microfluidic Separations for Metabolic Analysis	243

10.3.1	Microchip Capillary Electrophoresis	243
10.3.1.1	MCE Systems	243
10.3.1.2	Sample Injection	244
10.3.1.3	Electrophoretic Separations	244
10.3.2	Analyte Detection	245
10.3.2.1	Optical Detection	245
10.3.2.2	Electrochemical Detection	246
10.4	Microfluidics for Cellular Analysis	247
10.4.1	Requirements for Single Cell Metabolomics	247
10.4.2	Types of Microfluidic Instrumentation	248
10.4.3	Biological Questions	249
10.4.3.1	Monitoring Metabolic Response to Stimulation and Cell-to-Cell Signaling	249
10.4.3.2	Pharmacokinetics/Pharmacodynamics	252
10.4.3.3	Clinical Diagnostics	254
10.5	A Look Forward	254
	References	256

11 Data Processing in Metabolomics 261
Age K. Smilde, Margriet M.W.B. Hendriks, Johan A. Westerhuis, and Huub C.J. Hoefsloot

11.1	Introduction and Scope	261
11.2	Characteristics of Metabolomics Data	261
11.2.1	Correlation Structure of Metabolomics Data	261
11.2.2	Informative versus Noninformative Variation	262
11.2.3	Low Samples-to-Variables Ratio	263
11.2.4	Measurement Error	263
11.2.5	Dynamics	263
11.2.6	Nonlinear Relations	264
11.3	Types of Biological Questions Asked	264
11.3.1	Methods Should Follow the Questions	264
11.3.2	Biomarkers	264
11.3.3	Treatment Effects	264
11.3.4	Networks and Mechanistic Insight	265
11.4	Validation	265
11.4.1	Several Levels of Validation	265
11.4.2	Curse of Dimensionality	266
11.4.3	Cross-Validation and Permutations	267
11.5	Overview of Methods	272
11.5.1	Exploratory Analysis	272
11.5.2	ANOVA and Other Univariate Methods	273
11.5.3	Advanced Exploratory Analysis	275
11.5.4	Regression Methods	277
11.5.5	Discriminant Analysis	280
11.5.6	Multilevel Approaches	281

| 11.5.7 | Network Inference *282* |
| | References *283* |

12	**Metabolic Flux Analysis** *285*
	Christoph Wittmann and Jean-Charles Portais
12.1	Introduction *285*
12.2	Prerequisites for Flux Studies *287*
12.2.1	Network Topology and Cellular Composition *287*
12.2.2	Network Formulation and Condensation *287*
12.2.3	Metabolic and Isotopic Steady State *288*
12.2.4	Definition of Isotope Labeling Patterns *289*
12.3	Stoichiometric Flux Analysis *290*
12.4	Labeling Studies Using Isotopes *292*
12.4.1	Radiolabeled Isotopes *293*
12.4.2	Stable Isotopes *295*
12.5	State-of-Art ^{13}C Flux Analysis *296*
12.5.1	Modeling of Carbon Transitions *298*
12.5.2	Experimental Design *299*
12.5.3	Flux Calculation and Statistical Evaluation of Flux Data *300*
12.5.4	Labeling Analysis by Mass Spectrometry *301*
12.5.5	Labeling Analysis by Nuclear Magnetic Resonance Spectroscopy *302*
12.6	Application of Metabolic Flux Analysis *303*
12.6.1	Improvement of Industrial Production Strains *303*
12.6.2	Integration into Systems Biology Approaches *306*
12.7	Recent Advances in the Field *307*
12.7.1	High-Throughput Flux Screening *307*
12.7.2	Flux Dynamics *307*
12.8	Concluding Remarks *308*
	Acknowledgments *308*
	References *308*

13	**Metabolomics: Application in Plant Sciences** *313*
	Gaétan Glauser, Julien Boccard, Jean-Luc Wolfender, and Serge Rudaz
13.1	Introduction *313*
13.2	Sample Preparation *314*
13.2.1	Culture and Harvesting *314*
13.2.2	Storage and Drying *315*
13.2.3	Extraction *315*
13.3	Analytical Methods *316*
13.3.1	NMR-Based Methods *317*
13.3.1.1	Direct NMR Fingerprinting *317*
13.3.1.2	Applications *318*
13.3.1.3	Hyphenation of NMR to Separating Techniques *323*
13.3.1.4	Future Trends *323*
13.3.2	MS-Based Methods *323*

13.3.2.1	Direct MS Methods *324*
13.3.2.2	Hyphenation of MS to Separating Techniques *324*
13.3.2.3	Applications *326*
13.3.3	Combined Approaches *329*
13.4	Metabolite Identification *330*
13.4.1	Interpreting Mass Spectra *330*
13.4.2	Databases *332*
13.5	Structural Elucidation of Novel Metabolites and Validation of Model *334*
13.6	Conclusion and Perspectives *336*
	Acknowledgments *337*
	References *337*

14	**Metabolomics and Its Role in the Study of Mammalian Systems** *345*
	Warwick B. Dunn, Mamas Mamas, and Alexander Heazell
14.1	Introduction – From Early Beginnings *345*
14.2	Hypothesis Generation or Hypothesis-Testing Studies *347*
14.3	Untargeted, Semi-Targeted, and Targeted Analytical Experiments *348*
14.4	Study and Experimental Design *350*
14.5	Sample Types *355*
14.6	Quality Assurance and Quality Control *360*
14.7	Metabolite Annotation and Identification *365*
14.8	Applications *369*
14.8.1	Pregnancy Complications *369*
14.8.2	Cardiovascular Diseases *371*
	Acknowledgments *373*
	References *373*

15	**Metabolomics in Biotechnology (Microbial Metabolomics)** *379*
	Marco Oldiges, Stephan Noack, and Nicole Paczia
15.1	Introduction *379*
15.2	Analytical Methods Applied for Microbial Metabolomics *382*
15.3	Custom-Made Separation for Microbial Metabolomics *384*
15.4	Microbial Metabolomics with Higher Throughput *385*
15.5	Application of Microbial Metabolomics *386*
15.6	Conclusion *388*
	References *388*

16	**Nutritional Metabolomics** *393*
	Hannelore Daniel
16.1	Introduction *393*
16.2	The Metabolome of Human Plasma and Urine: General Considerations *393*
16.2.1	The Plasma Metabolome *394*

16.2.2	The Urinary Metabolome *395*	
16.3	The Food Metabolome and Its Signature in Human Samples *396*	
16.4	The Variability of the Human Metabolome in Health and Disease States *398*	
16.5	The Dynamic Nature of the Metabolome *400*	
16.6	The Future of Nutritional Metabolomics and Research Needs *402*	
	References *403*	

Index *407*

List of Contributors

Kashif Ali
Leiden University
Natural Products Laboratory
Institute of Biology
Einsteinweg 55
2300 RA Leiden
The Netherlands

Julien Boccard
University of Geneva
University of Lausanne
School of Pharmaceutical
Sciences
30, Quai Ansermet
1211 Geneva 4
Switzerland

André B. Canelas
DSM Biotechnology Centre
Alexander Fleminglaan 1
2613 AX Delft
The Netherlands

Hannelore Daniel
Technische Universität München
Molecular Nutrition Unit
Center Institute of Nutrition and
Food Sciences
Gregor Mendel Strasse 2
85350 Freising-Weihenstephan
Germany

Gerhardus J. de Jong
Utrecht University
Biomolecular Analysis
Department of Pharmaceutical
Sciences
Universiteitsweg 99
3584 CG Utrecht
The Netherlands

Warwick B. Dunn
University of Manchester
Manchester Academic
Health Sciences Centre
Centre for Advanced Discovery
and Experimental Therapeutics
(CADET), Central Manchester
NHS Foundation Trust
York Place
Oxford Road
Manchester M13 9WL
UK

and

University of Manchester
Institute of Human Development
Oxford Road
Manchester M13 9PL
UK

Sara Forcisi
Helmholtz-Zentrum
Muenchen-German Research
Center for Environmental Health
Analytical BioGeoChemistry
Ingolstaedter Landstrasse 1
85764 Neuherberg
Germany

Helen G. Gika
Aristotle University of
Thessaloniki
Department of Chemical
Engineering
541 24 Thessaloniki
Greece

Gaétan Glauser
University of Neuchâtel
Laboratory of Fundamental and
Applied Research in Chemical
Ecology
Rue Emile-Argand 11
2000 Neuchâtel
Switzerland

and

University of Neuchâtel
Chemical Analytical Service of the
Swiss Plant Science Web
Rue Emile-Argand 11
2000 Neuchâtel
Switzerland

Raffaele Guerrasio
BOKU–University of Natural
Resources and Life Sciences
Department of Chemistry
Division of Analytical Chemistry
Muthgasse 18
1190 Vienna
Austria

Christina Haberhauer-Troyer
BOKU–University of Natural
Resources and Life Sciences
Department of Chemistry
Division of Analytical Chemistry
Muthgasse 18
1190 Vienna
Austria

Young Hae Choi
Leiden University
Natural Products Laboratory
Institute of Biology
Einsteinweg 55
2300 RA Leiden
The Netherlands

Stephen Hann
BOKU–University of Natural
Resources and Life Sciences
Department of Chemistry
Division of Analytical Chemistry
Muthgasse 18
1190 Vienna
Austria

Mourad Harir
Helmholtz-Zentrum
Muenchen-German Research
Center for Environmental Health
Analytical BioGeoChemistry
Ingolstaedter Landstrasse 1
85764 Neuherberg
Germany

Alexander Heazell
University of Manchester
Institute of Human Development
Oxford Road
Manchester M13 9PL
UK

and

University of Manchester
Maternal and Fetal Health
Research Centre
St Mary's Hospital
Oxford Road
Manchester M13 9WL
UK

Joseph J. Heijnen
Delft University of Technology
Department of Biotechnology
Kluyver Centre for Genomics of
Industrial Fermentation
Julianalaan 67
2628 BC Delft
The Netherlands

Margriet M.W.B. Hendriks
Leiden University
Analytical BioSciences
Leiden Academic Center
for Drug Research
Einsteinweg 55
2300 RA Leiden
The Netherlands

Charles S. Henry
Colorado State University
Department of Chemistry
1872 Campus Delivery
Fort Collins, CO 80523
USA

and

Colorado State University
Department of Chemical and
Biological Engineering
1872 Campus Delivery
Fort Collins, CO 80523
USA

Huub C.J. Hoefsloot
University of Amsterdam
Biosystems Data Analysis
Swammerdam Institute
for Life Sciences
Science Park 904
1098 XH Amsterdam
The Netherlands

Jamin C. Hoggard
University of Washington
Department of Chemistry
Box 351700
Seattle, WA 98195
USA

Ryan E. Holcomb
Colorado State University
Department of Chemistry
1872 Campus Delivery
Fort Collins, CO 80523
USA

and

Colorado State University
Department of Chemical and
Biological Engineering
1370 Campus Delivery
Fort Collins, CO 80523
USA

Gérard Hopfgartner
University of Geneva
University of Lausanne
School of Pharmaceutical
Sciences
Life Sciences Mass Spectrometry
Quai Ernest-Ansermet 30
1211 Geneva 4
Switzerland

Lucie Jorge
Research Institute for
Chromatography
President Kennedypark 26
8500 Kortrijk
Belgium

Basem Kanawati
Helmholtz-Zentrum
Muenchen-German Research
Center for Environmental Health
Analytical BioGeoChemistry
Ingolstaedter Landstrasse 1
85764 Neuherberg
Germany

Ruben t'Kindt
Research Institute for
Chromatography
President Kennedypark 26
8500 Kortrijk
Belgium

Kristaps Klavins
BOKU–University of Natural
Resources and Life Sciences
Department of Chemistry
Division of Analytical Chemistry
Muthgasse 18
1190 Vienna
Austria

Gunda Koellensperger
BOKU–University of Natural
Resources and Life Sciences
Department of Chemistry
Division of Analytical Chemistry
Muthgasse 18
1190 Vienna
Austria

Mary E. Lidstrom
University of Washington
Department of Chemical
Engineering
Box 351202
Seattle, WA 98195
USA

and

University of Washington
Department of Microbiology
Box 355014
Seattle, WA 98195
USA

Andrea Lubbe
Leiden University
Natural Products Laboratory
Institute of Biology
Einsteinweg 55
2300 RA Leiden
The Netherlands

Marianna Lucio
Helmholtz-Zentrum
Muenchen-German Research
Center for Environmental Health
Analytical BioGeoChemistry
Ingolstaedter Landstrasse 1
85764 Neuherberg
Germany

Mamas Mamas
University of Manchester
Manchester Academic
Health Sciences Centre
Centre for Advanced Discovery
and Experimental Therapeutics
(CADET)
Central Manchester NHS
Foundation Trust
York Place
Oxford Road
Manchester M13 9WL
UK

and

University of Manchester
Institute of Human Development
Oxford Road
Manchester M13 9PL
UK

and

Manchester Heart Centre
Manchester Royal Infirmary
Oxford Road
Manchester M13 9WL
UK

Meghan M. Mensack
Colorado State University
Department of Chemistry
1872 Campus Delivery
Fort Collins, CO 80523
USA

Franco Moritz
Helmholtz-Zentrum
Muenchen-German Research
Center for Environmental Health
Analytical BioGeoChemistry
Ingolstaedter Landstrasse 1
85764 Neuherberg
Germany

Stefan Neubauer
BOKU–University of Natural
Resources and Life Sciences
Department of Chemistry
Division of Analytical Chemistry
Muthgasse 18
1190 Vienna
Austria

Stephan Noack
Forschungszentrum Jülich
GmbH (Research Centre Jülich)
Institute of Bio- and Geosciences
(IBG)
IBG-1: Biotechnology
Leo-Brandt-Str.
52428 Jülich
Germany

Marco Oldiges
Forschungszentrum Jülich
GmbH (Research Centre Jülich)
Institute of Bio- and Geosciences
(IBG)
IBG-1: Biotechnology
Leo-Brandt-Str.
52428 Jülich
Germany

Nicole Paczia
Forschungszentrum Jülich
GmbH (Research Centre Jülich)
Institute of Bio- and Geosciences
(IBG)
IBG-1: Biotechnology
Leo-Brandt-Str.
52428 Jülich
Germany

Jean-Charles Portais
Université de Toulouse, INSA,
UPS, INP, LISBP
135 avenue de Rangueil
31077 Toulouse
France

and

INRA, UMR792, Ingénierie des
Systèmes Biologiques
et des Procédés
135 avenue de Rangueil
31077 Toulouse
France

and

CNRS, UMR5504
135 avenue de Rangueil
31077 Toulouse
France

Rawi Ramautar
Utrecht University
Biomolecular Analysis
Department of Pharmaceutical
Sciences
Universiteitsweg 99
3584 CG Utrecht
The Netherlands

and

Analytical Biosciences
Leiden/Amsterdam Center for
Drug Research
Leiden University
The Netherlands

Serge Rudaz
University of Geneva
University of Lausanne
School of Pharmaceutical
Sciences
30, Quai Ansermet
1211 Geneva 4
Switzerland

Koen Sandra
Research Institute for
Chromatography
President Kennedypark 26
8500 Kortrijk
Belgium

Pat Sandra
Research Institute for
Chromatography
President Kennedypark 26
8500 Kortrijk
Belgium

Philippe Schmitt-Kopplin
Helmholtz-Zentrum
Muenchen-German Research
Center for Environmental Health
Analytical BioGeoChemistry
Ingolstaedter Landstrasse 1
85764 Neuherberg
Germany

and

Technische Universität München
Department for
Chemical-Technical Analysis
Research Center Weihenstephan
for Brewing and Food Quality
Chair of Analytical Food
Chemistry
85354 Freising-Weihenstephan
Germany

Reza M. Seifar
Delft University of Technology
Department of Biotechnology
Kluyver Centre for Genomics of
Industrial Fermentation
Julianalaan 67
2628 BC Delft
The Netherlands

Age K. Smilde
University of Amsterdam
Biosystems Data Analysis
Swammerdam Institute
for Life Sciences
Science Park 904
1098 XH Amsterdam
The Netherlands

Govert W. Somsen
Utrecht University
Biomolecular Analysis
Department of Pharmaceutical
Sciences
Universiteitsweg 99
3584 CG Utrecht
The Netherlands

Robert E. Synovec
University of Washington
Department of Chemistry
Box 351700
Seattle, WA 98195
USA

Georgios A. Theodoridis
Aristotle University of
Thessaloniki
Department of Chemistry
541 24 Thessaloniki
Greece

Dimitrios Tziotis
Helmholtz-Zentrum
Muenchen-German Research
Center for Environmental Health
Analytical BioGeoChemistry
Ingolstaedter Landstrasse 1
85764 Neuherberg
Germany

Walter M. van Gulik
Delft University of Technology
Department of Biotechnology
Kluyver Centre for Genomics of
Industrial Fermentation
Julianalaan 67
2628 BC Delft
The Netherlands

Emmanuel Varesio
University of Geneva
University of Lausanne
School of Pharmaceutical
Sciences
Life Sciences Mass Spectrometry
Quai Ernest-Ansermet 30
1211 Geneva 4
Switzerland

Robert Verpoorte
Leiden University
Natural Products Laboratory
Institute of Biology
Einsteinweg 55
2300 RA Leiden
The Netherlands

Madeleine Werneth
BOKU–University of Natural
Resources and Life Sciences
Department of Chemistry
Division of Analytical Chemistry
Muthgasse 18
1190 Vienna
Austria

Johan A. Westerhuis
University of Amsterdam
Biosystems Data Analysis
Swammerdam Institute
for Life Sciences
Science Park 904
1098 XH Amsterdam
The Netherlands

Ian D. Wilson
Biomolecular Medicine
Department of Surgery and
Cancer
Faculty of Medicine
Imperial College
London SW7 2AZ
UK

Christoph Wittmann
Technische Universität
Braunschweig
Biochemical Engineering
Institute
Gauss-Strasse 17
38106 Braunschweig
Germany

Jean-Luc Wolfender
University of Geneva
University of Lausanne
School of Pharmaceutical
Sciences
30, Quai Ansermet
1211 Geneva 4
Switzerland

Song Yang
University of Washington
Department of Chemical
Engineering
Box 351202
Seattle, WA 98195
USA

Preface

Metabolomics is a rapidly growing and expanding subdiscipline in systems biology (Figure 1). The latter attempts to look at biological systems in a holistic manner and tries to analyze the entirety of complex molecular systems in a dynamically comprehensive way. The metabolome is downstream to the genome, transcriptome, and proteome and thus provides a great deal of information on the functional status of a cell or organism. Often it is assumed that the low molecular complement of genes and proteins is easier to analyse, and recent developments of powerful analytical techniques such as nuclear magnetic resonance spectroscopy (NMR) and, in particular, mass spectrometry (MS), hyphenated with separation technologies like liquid chromatography (LC) or gas chromatography (GC) have greatly contributed to the wide popularity of metabolomics technologies in analyzing biological systems. While NMR techniques are still preferentially used by chemists, MS, LC-MS, and GC-MS are more favoured for use by biologists as well. Developments, in particular, in MS technology are extremely rapid and continuous training is needed. Users of metabolomics techniques are seldom specialists in all methodologies and techniques that are needed or are helpful and utilized to address metabolomics questions. This may be related either to separation technologies, mass spectrometry, chemometric data processing or simply to the design of metabolomic studies. Some researchers are experts in a certain application field but are less interested in spending a lot of time to optimize analytical technologies. On the other hand, analytical chemists are sometimes not as familiar with biological background as it would be desirable. Thus, it may occur that one or the other mentioned aspect of significant relevance in metabolomics research is treated more like a black box due to lack of knowledge and user-friendly tools that are currently available. Critical issues may be overlooked and this may lead to suboptimal results. This book aims at presenting current methodologies from an analytical chemist's point of view, with the goal to raise the awareness for strengths, weaknesses, pitfalls, shortcomings, problems of methodologies, and solutions how to overcome them. Usually, methodological shortcomings and problems can only be recognized if suitable quality control measures are implemented in workflows. This book also attempts to draw emphasis on quality assurance and validation of analytical methods in metabolomics research. It should support the idea of implementing method validation as a tool to improve the quality and reliability of the produced

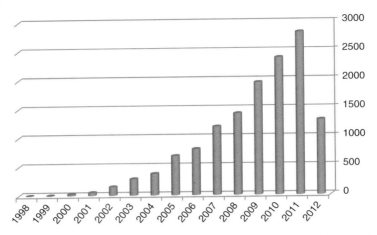

Figure 1 Gain in popularity of metabolomics research as measured by no. of publications with key word "metabolomics" in SciFinder database (as of May 2012).

data, and ascertain that conclusions from metabolomics data are significant. We hope that this book is of significant interest for readers who are new in this field but also for those performing metabolomics analyses for a while.

We have sub-structured the book in two parts. The first part deals with the most common methods utilized in metabolomics and targeted metabolite analysis. The second part focuses on the application of those methods in the major fields of metabolomics.

Inadequate sample preparation can destroy everything in metabolomics research and make results meaningless. To stress their importance, the book starts with a chapter on sampling and sample preparation in microbial metabolomics (Chapter 1). While this chapter specifically focuses on samples from microbial cultivations, it deals with important basic and general ideas that may be of relevance for other samples as well. It is clear that sampling and sample preparation has to be developed specifically for each type of sample, and that questions addressed are certainly different for plant metabolomics, microbial metabolomics, and red biotechnology. Specifics in these fields will be dealt with in the application part. Metabolomics research aims at measuring variances of substances over time, space or populations. The results should not be blurred by excessive analytical variances. It is well known that variances of individual steps in the entire analytical process are additive. Those of early steps in analytical process are usually much larger. This holds, in particular, for sampling and sample preparation. Hence, the importance of this step must not be underestimated and the first chapter emphasises this. The following chapters in the book are dedicated to targeted metabolomics (Chapter 2). This subdiscipline is often thought to be not a real metabolomics approach because only a preselected set of metabolites and not the entirety of the metabolome is analyzed comprehensively. However, we included it in this book because, on the one hand, issues and errors related to analytical methodologies can be better pinpointed by quantitative measurements, for example, ion suppression effects, chemical, isomeric, and isotopic interferences, and so forth. On the other hand, at the end, accurate metabolite quantitation is the final goal of most metabolomics studies. To convey how errors propagate in individual steps to the total uncertainty in the entire metabolomics process, the concept of measurement uncertainty and of uncertainty budgets is presented in the next chapter (Chapter 3). Chapters dealing with methodological aspects of non-targeted metabolomics by GC-MS, GCxGC-MS and LC-MS/MS follow (Chapters 4–6). These techniques are the current workhorses in metabolomics research. Some of these chapters focus more on technological aspects and others on quality assurance issues of analytical measurements. In one chapter, methods employed in lipidomics, a subdiscipline of metabolomics, are described in more detail (Chapter 7). The lipidome is particularly complex and thus it was deemed justified to deal with this topic in more detail. Techniques less broadly used, such as capillary electrophoresis, hyphenated with MS and miniaturized technologies such as microfluidics (micro-metabolomics) as well as NMR based metabolomics follow in the next chapters (Chapters 8–10). All the above techniques, in particular those being utilized in untargeted profiling methods generate a huge body of data. It needs appropriate statistical tools such as multivariate analysis techniques to filter useful information from a noisy background. The most common statistical methods utilized in metabolomics are described in an illustrative manner in the next chapter (Chapter 11). Metabolism is a highly dynamic process with rapid turnover of metabolic intermediates. While most of the above described methods focus on analysis of metabolic samples where the metabolic state is frozen by quenching, metabolic flux analysis deals with the dynamic change of metabolism. Considerations of primary importance in "fluxomics" are described in a separate chapter (Chapter 12). There are certainly many more techniques of relevance in metabolomics research such as imaging methods but are beyond the primary scope of this book.

The second part of the book deals with applications. The application fields covered comprise plant metabolomics (green metabolomics) (Chapter 13), mammalian metabolomics (red metabolomics) (Chapter 14), and metabolomics in biotechnology and food science (white metabolomics) (Chapters 15 and 16). Peculiarities in different application fields are pointed out in these chapters including sample preparation aspects and design considerations of metabolomics experiments and studies. They should make users aware of certain problems, outline which methods are most appropriate and provide an insight to what can be achieved.

The book should give a state-of-the-art picture of the most important techniques written from the analytical chemist's point of view, thus focusing on methodological aspects and technical details, and last but not least quality assurance aspects. We hope that the book is useful and enjoyable to read for both beginners and experts in the field. We greatly appreciate the efforts of all contributing authors and would like to thank them for their excellent articles. Without their valuable contributions this book would not have been possible.

Michael Lämmerhofer
Tübingen, Germany

Wolfram Weckwerth
Vienna, Austria

1
The Sampling and Sample Preparation Problem in Microbial Metabolomics

Walter M. van Gulik, André B. Canelas, Reza M. Seifar, and Joseph J. Heijnen

1.1
Introduction

The reproducibility and accuracy of analysis methods has expanded rapidly over the past years. Hyphenated methods, for example, GC-MS and LC-MS/MS (liquid chromatography-mass spectrometry), are now almost routinely used for (semi-)quantitative analysis of tens to hundreds of metabolites in biological samples. It should be realized, however, that the quality of the applied analytical techniques alone is not sufficient to obtain meaningful results. Other important aspects, which are often overlooked, are the methods that are applied to withdraw the samples from a microbial cultivation and the subsequent sample processing procedures. Immediately after withdrawal of the samples, the metabolism should be arrested to preserve the metabolic snapshot represented by the sample by preventing further metabolic conversions. Subsequently, an appropriate extraction method should be applied, which guarantees complete and unbiased release of all metabolites from the cells without significant degradation and inactivation of enzymatic activity. Finally, an accurate and high throughput analytical platform is required for metabolite quantification. In this chapter, we focus on the essential requirements of the sampling and sample processing procedures as well as the methods to validate if these requirements are met in practice.

1.2
Microorganisms and Their Properties

Microorganisms show very large differences in properties, for example, size, structure, degree of complexity, heterogeneity, physiology, nutrient requirements, and so on. These aspects should be considered beforehand because they can be important for the choice of the sampling and sample processing methods to be used. They also determine the meaning of the obtained results, for example, the interpretation of metabolite measurements in cell extracts of prokaryotic microorganisms, which contain only one compartment (cytosol), is more straightforward compared

Metabolomics in Practice: Successful Strategies to Generate and Analyze Metabolic Data, First Edition.
Edited by Michael Lämmerhofer and Wolfram Weckwerth.
© 2013 Wiley-VCH Verlag GmbH & Co. KGaA. Published 2013 by Wiley-VCH Verlag GmbH & Co. KGaA.

to measurements in cell extracts of eukaryotes. As eukaryotic (micro)organisms contain several compartments (cytosol, mitochondria, peroxisomes, etc.) in which the concentrations of metabolites can be very different, the interpretation of the measurements, which are in fact whole cell averages, can be problematic if the metabolites quantified are present in more than one cellular compartment.

1.3
Sampling Methods

1.3.1
The Need for Rapid Sampling

As quantitative metabolomics aims at obtaining accurate snapshots of intracellular metabolite levels, the withdrawal of samples from the microbial culture under study should be sufficiently fast to prevent significant changes in metabolite levels arising from ongoing metabolic activity. This is particularly important if samples are taken from cultures for which one of the medium components is the growth-limiting nutrient (e.g., chemostat or fed-batch cultivations). If the sampling procedure requires too much time, the small amount of growth-limiting nutrient present in the sample will be rapidly exhausted, which will result in changes in metabolic fluxes and consequently in changes in intracellular metabolite levels. Similarly, if samples are taken from aerobic cultivations, the amount of oxygen dissolved in the liquid phase will be rapidly exhausted because of the consumption of oxygen by the cells combined with the very limited solubility of oxygen in water.

It should be realized that the speed at which the ongoing metabolic activity of the cells in the culture sample occurs, and thus the time of exhaustion of the limiting nutrient and/or the available oxygen in the liquid phase, directly depends on the biomass density and the growth rate of the culture from which the sample is withdrawn. For relatively high density and fast growing cultures, this could happen in a few seconds. Considering the pool sizes of intracellular metabolites, which range from 10^{-3} to 10^2 μmol g^{-1} dry cell mass [1–3] and the fluxes through the metabolic network, the apparent turnover times of metabolite pools range from fractions of a second to tens of minutes [4]. This implies that to obtain meaningful results a dedicated rapid sampling system is required, whereby the residence time of the sampled culture broth in the sampling system is in the subsecond range.

Furthermore, the metabolic activity of the cells should be arrested as soon as the sample enters the sampling vial. This has generally been accomplished by introducing the culture broth with sufficient velocity (i.e., as a liquid jet) into a dedicated quenching solution (Section 1.4) such that it is instantaneously mixed. A common approach is to use a cold aqueous solvent mixture, for example, 60% aqueous methanol, to cool down the sample below $-20\,^\circ$C, thereby minimizing the enzymatic activity. It has been shown that with this method effective quenching of metabolism is obtained only if the mixing of the sample with the cold aqueous quenching liquid is instantaneous [5]. Another approach is to combine quenching with extraction, by sampling directly into the extraction solution (e.g., perchloric acid) or by integrating extraction in the sampling mechanism itself [6].

1.3.2
Sampling Systems

An overview of different rapid sampling systems, both manually operated and fully automated, has been provided by Schaedel and Lara [7]. Apart from being sufficiently fast, other important requirements for rapid sampling systems are that the amount of sample withdrawn is reproducible; the dead volume is negligible (to prevent that stagnant liquid residing in the system is mixed up with the sample); and especially for sampling during highly dynamic experiments, the time between subsequent samples is sufficiently small to fully capture the dynamics.

A convenient manually operated rapid sampling system for withdrawal of broth samples from bench scale bioreactors has been described by Lange et al. [8]. This system consists of a sampling port, inserted in the wall of the bioreactor, with an internal diameter of 1 mm, which is connected to a tube adapter. Sampling is started by removing the dead volume by flushing into the waste. Subsequently, the sample tube is evacuated and directly thereafter, the sample is withdrawn from the bioreactor, facilitated by the vacuum in the sample vial and a slight overpressure in the bioreactor. The liquid flows and evacuation of the sample tube are controlled by electromagnetic pinch valves operated by a timer, allowing the sample volume to be precisely adjusted, that is, with a standard deviation of less than 2%. The authors reported that with this system, they could withdraw samples of 1 ml from a bioreactor, operated at an overpressure of 0.3 bar, within 0.7 s. The residence time of the sample in the system was below 100 ms. However, with this system, the sampling frequency could not be increased much above 1 sample per 5 s because of the many manual handlings that had to be performed. Therefore Schaefer et al. [9] developed a completely automated sampling device, whereby the sampling vials were fixed in transport racks, which were moved by a step engine underneath a continuous jet of culture broth, with a flow rate of $3.3\ ml\ s^{-1}$, from a stirred tank bioreactor. In this way, each sampling vial, which contained a cold quenching solution, could be filled within 220 ms resulting in a sampling rate of approximately 4.5 samples per second. This automated sampling device was applied to investigate the intracellular metabolite dynamics of glycolysis in Escherichia coli after rapid glucose addition to a glucose-limited steady state culture. A disadvantage of this approach is the large amount of broth that is withdrawn, requiring a relatively large bioreactor.

A different approach to integrated sampling, quenching, and extraction from a bioreactor culture has been published by Schaub et al. [6]. Hereby, short time heating of the sample is used as the procedure to quench all metabolic activity, at the same time extracting the metabolites from the cells. This is achieved by using a helical coil heat exchanger that allowed continuous withdrawal of the sample from a bioreactor, whereby the broth was rapidly heated to 95 °C. The helical geometry was chosen to enhance radial mixing, thus improving the plug-flow characteristics of the system. After extraction, the cell debris can be removed by filtration. This sampling device allows withdrawing approximately 5 samples per second. The method has been applied to the analysis of the growth-rate-dependent *in vivo* dynamics of glycolysis in *E. coli* [10].

1.4
Quenching

1.4.1
Quenching Procedures and Their Properties

Different sampling techniques can be combined with various quenching procedures. An essential property of a quenching procedure is that it should accomplish that all metabolic activity is arrested in the moment of sampling. This can be achieved in various ways, for example, by injecting the sample in a solution with an extreme pH (highly acidic or alkaline), by fast heating and subsequent cooling of the sample (see above), by fast cooling to a temperature below $-20\,°C$ or by combining different approaches. Rapid cooling can be achieved by injecting the sample in liquid nitrogen or into a cold quenching solution. Examples are cold perchloric acid ($-20\,°C$), 1 M alcoholic KOH 50% (v/v) methanol at $-20\,°C$ or cold aqueous 60% (v/v) methanol at $-40\,°C$.

An important aspect in deciding which quenching method should be used for a particular case is whether the complete broth sample could be extracted (i.e., cells with the surrounding medium) or whether the cells should be separated from the surrounding medium before extraction. Hereby, it should be realized that in common laboratory-scale bioreactor cultivations, the volume of the surrounding medium is 2 orders of magnitude larger than the total cell volume.

This implies that even low concentrations of metabolites outside the cells could compromise the quantification of the intracellular metabolite levels, if the metabolite measurements are carried out in extracts from total broth samples. This is illustrated by the results of metabolite measurements carried out in glucose-limited chemostats of *E. coli*, of which the results are shown in Table 1.1. As shown in Table 1.1, the measured metabolite concentrations in the culture filtrate are 2–3 orders of magnitude lower than the intracellular concentrations. Nevertheless, if the intracellular and extracellular levels are expressed as amounts per gram biomass dry weight (DW) present, it appears that for some metabolites (in this example malate and pyruvate) the amount present in the extracellular medium can be quite significant (around 50%). Thus, if metabolite measurements are carried out in extracts of total broth samples, the levels of certain metabolites will be significantly overestimated.

In order to separate the cells from the surrounding medium before the metabolite extraction step, a quenching procedure must be used, during which the cell membrane integrity is retained. This clearly excludes quenching procedures employing

Table 1.1 Example of extracellular and intracellular metabolite levels measured in a glucose-limited chemostat of *Escherichia coli* K12 MG1655 [3]. To obtain intracellular concentrations a cellular volume of 2.5 ml g^{-1} of dry weight was used.

	Expressed as concentrations		Expressed as amounts per gram of dry weight		
	Culture filtrate (μM)	Cells (μM)	Culture filtrate (μmol g^{-1}DW)	Cells (μmol g^{-1}DW)	Out/in (−)
Glucose-6P	1.3	568	0.13	1.42	0.09
Pyruvate	5.2	300	0.49	0.75	0.65
Malate	4.0	376	0.38	0.94	0.40
Alanine	2.3	536	0.22	1.34	0.16
Glutamate	16.5	29 880	1.56	74.7	0.02

extremely alkaline or acidic solutions, or short time heating is used to arrest metabolic activity. Rapid freezing in liquid nitrogen is also thought to compromise cell integrity because of the formation of ice crystals.

The most commonly used procedure that enables separation of the cells from the cultivation medium is cold methanol quenching [11], combined with cold centrifugation. This method has been used, for example, in combination with a manual rapid sampling system [8], whereby 1 ml of sample is immediately injected in 5 ml of aqueous 60% methanol at −40 °C. After a cold centrifugation step, whereby the temperature of the quenched sample is kept below −20 °C, and an optional washing step, the obtained cell pellet can be extracted for quantification of intracellular metabolites, without interference of extracellular compounds. It should be noted here that this procedure can only be applied successfully if the metabolites are contained within the cells and do not leak out into the quenching/washing solution.

1.4.2
Validation of the Quenching Procedure and Minimization of Metabolite Leakage

From metabolite measurements carried out in the supernatant after cold centrifugation of the quenched sample, de Koning and van Dam [11] concluded that metabolite leakage during cold methanol quenching of yeast was insignificant. This was later confirmed by other authors [8, 12–14]. Contrary to this, Villas-Bôas *et al.* [15] reported significant leakage of several metabolite classes, especially organic acids and a few amino acids from yeast cells after cold methanol quenching.

Later on a systematic study was performed to trace the fate of a large number of metabolites during cold methanol quenching of yeast culture samples [1]. The authors carried out metabolite measurements in all different sample fractions, that is, total broth, culture filtrate, cell pellet, and quenching solutions. To obtain accurate quantification of metabolites in all sample fractions, isotope dilution mass spectrometry (IDMS) was applied (Section 1.6). This approach allowed them to check the mass balance for each metabolite, and in this way verify the consistency of the measurements, that is, the total amount of each metabolite, as measured in a total broth sample, should be quantitatively found back in the different sample fractions (cell pellet, culture filtrate, and quenching/washing solution). From the obtained results, the authors concluded that quenching of *Saccharomyces cerevisiae* samples in aqueous 60% methanol at $-40\,°C$ induced metabolite leakage to such an extent that for most metabolites the levels are underestimated by at least twofold. By varying the methanol content and quenching temperature, they found the optimal condition for yeast quenching and reported that metabolite leakage is entirely prevented by quenching in 100% methanol at a temperature $\leq -40\,°C$, with a sample/quenching solution ratio $\leq 1:5$.

The same approach has been used to quantify metabolite leakage during cold methanol quenching, and subsequently optimize the procedure, for other eukaryotic microorganisms. de Jonge et al. [16] found that quenching of *Penicillium chrysogenum* in 60% aqueous methanol at $-40\,°C$ also resulted in significant metabolite leakage into the quenching solution. In contrast to what was observed for *S. cerevisiae*, increasing the methanol content of the quenching liquid resulted in increased instead of decreased metabolite leakage. The authors reported that metabolite leakage was minimal if mycelium was quenched in 40% aqueous methanol at a temperature of $-20\,°C$. For cold methanol quenching of *Pichia pastoris*, it was observed that the methanol content of the quenching solution was not very critical, and for most metabolites measured, leakage was acceptable for methanol contents between 40 and 100% and quenching temperatures between -27 and $-40\,°C$ [17]. However, from a critical evaluation, it was found that quenching in 60% methanol at a temperature of $-27\,°C$ resulted in the lowest metabolite leakage.

The above findings illustrate that apparently no single condition exists, which is suitable for the quenching of different eukaryotic microorganisms, possibly because of differences in membrane composition. Consequently, it seems necessary to validate and optimize the quenching procedure for each different organism. Although modifications might be required, cold methanol quenching appears to be applicable for quantitative metabolomics of different eukaryotic microorganisms and can thus be used to separate the cells from the surrounding medium to avoid interference with metabolites present therein.

1.4.3
Quenching Procedure for Determination of Intracellular Metabolites in the Presence of Extracellular Abundance

In some cases, it is important to quantify levels of intracellular metabolites that are present in high concentrations in the surrounding medium. Typical examples are excreted products such as antibiotics or organic acids by high producing industrial strains. In this case, proper quantification of the intracellular levels requires complete removal of the amounts present outside the cells and thus efficient removal of the extracellular medium before metabolite extraction. This could be attempted by increasing the number of washing steps after cold methanol quenching and centrifugation.

However, for efficient removal of highly abundant extracellular metabolites, repeated steps of resuspension of the quenched cell pellet in a cold washing solution and subsequent cold centrifugation would be required because of the carryover of remaining medium after decantation. This is not only a laborious procedure but could also lead to increased metabolite leakage because of the prolonged exposure to the cold methanol containing quenching/washing solution or to partial cell disruption due to the repeated centrifugation steps. For this reason, Douma et al. [18] compared the efficiency of cold centrifugation combined with several washing steps with a procedure whereby cold methanol quenching was combined with rapid cold filtration and subsequent filtration-based washing. They applied these procedures to quantify the metabolites involved in the penicillin biosynthesis pathway, in a high producing strain of P. chrysogenum. For several of these metabolites, for example, the side-chain precursor phenylacetic acid (PAA), the end product penicillin G (penG), and several by-products, the amounts present in the extracellular medium are 3–4 orders of magnitude higher than the amounts present inside the cells.

In the centrifugation-based method, removal of the extracellular metabolites from the cells was achieved by means of multiple centrifugation and resuspension steps with the cold quenching solution. The cold filtration method was found to be highly superior to the centrifugation method to determine intracellular amounts of metabolites related to penG biosynthesis and indeed allowed the quantification of compounds of which the extracellular amounts were 3–4 orders of magnitude higher than the intracellular amounts.

1.4.4
Quenching of Bacteria

Also for bacteria, cold methanol quenching has been applied. Bolten and coworkers [19] reported that cold methanol quenching of different gram-positive and gram-negative bacteria resulted in a loss of more than 60% of most metabolites from the cells.

From a thorough validation study of cold methanol quenching for E. coli [3], whereby metabolite levels were measured in different sample fractions, it was found that a major part of the intracellular metabolites were lost from the cells and ended up in the quenching solution. The authors observed no positive effect of buffering or increasing the ionic strength of the quenching solution. Therefore, they proposed, an alternative to measure the metabolite levels in the total broth sample as well as in the culture supernatant, whereafter the intracellular levels can be obtained by subtraction. This method was successfully applied to quantify intracellular metabolite levels of E. coli grown in an aerobic glucose-limited chemostat at a dilution rate of $0.1\ h^{-1}$ [3] and during short term highly dynamic conditions following a glucose pulse [20, 21].

1 The Sampling and Sample Preparation Problem in Microbial Metabolomics

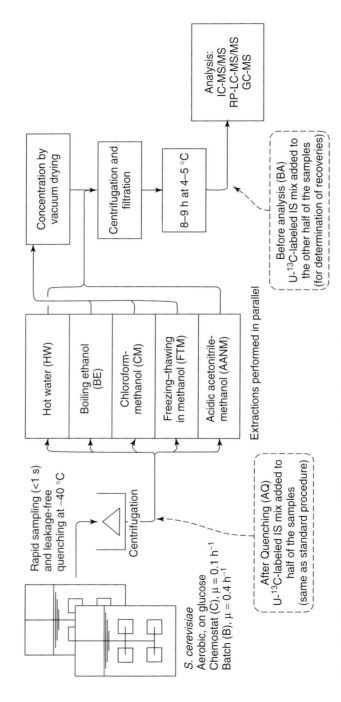

Figure 1.1 Schematic representation of the experimental procedure used for the quantitative evaluation of the applicability of five different extraction procedures for yeast metabolomics. (Source: Figure from Canelas et al. [39].) IC, ion-chromatography; RP-LC, reversed-phase LC; ion-pair LC.

1.5
Metabolite Extraction

1.5.1
Extraction Methods and Their Properties

After obtaining a quenched and washed cell pellet, the metabolites contained within the cells should be extracted completely and quantitatively from them, to allow meaningful quantification. Quite a number of different extraction methods have been described in literature and are based on heating, extreme pH, organic solvents and mechanical disruption, or combinations of these. In earlier works, mainly perchloric acid extraction [22, 23] or hot extraction methods have been applied, for example, short-term boiling in water [24, 25] or aqueous ethanol [26, 27]. More recently, milder methods have been advocated, for example, freeze thaw cycles in cold methanol [28], cold chloroform-methanol (CM) extraction [11], and acidic acetonitrile-methanol (AANM) extraction [29].

Essential requirements that a proper extraction method should fulfill are completeness of extraction, that is, all metabolites should be completely released from the cells; complete denaturation of all enzymes present to prevent further interconversion of metabolites during subsequent sample processing steps; and at the same time, no significant degradation and/or chemical conversion. Finally, the obtained extract should be compatible with the analysis techniques, which will be used for quantification.

Several authors have attempted to evaluate different methods for the extraction of metabolites from different organisms. A summary is given in Table 1.2. It can be seen from this table that the methods that are considered "good" or "best" by some authors are considered as "poor" by others. Clearly, the completeness of extraction might well be different for different microorganisms; however, in some cases, different methods are recommended for the same organism, for example, for *E. coli* (Table 1.2). Important requirements such as the denaturation of enzyme activity and the absence of metabolite degradation should, in principle, not be dependent on the organism used.

1.5.2
Validation of Extraction Methods for Yeast Metabolomics

With the aim to identify the most appropriate extraction method for yeast metabolomics, Canelas *et al.* [39] carried out a quantitative evaluation of the essential requirements, that is, completeness of extraction, enzyme denaturation, and absence of metabolite degradation for five different methods, namely, extraction with hot water (HW), boiling ethanol (BE), CM, freezing-thawing in methanol (FTM), and AANM. The procedure followed is shown schematically in Figure 1.1.

It should be noted that completeness of extraction can never be guaranteed because the exact metabolite contents of the cells cannot be known beforehand.

Table 1.2 Survey of comparisons of extraction procedures for intracellular metabolite analysis in microorganisms.

Source	Microorganism(s)	PCA	KOH	HW	BE	CM	FTM	AANM	Notes
Bagnara and Finch [30]	E. coli	+++	−	+	−	−	−	−	Quantitative, TLC, six metabolites (nucleotides), no recoveries. Tested one other method (TCA).
Lundin and Thore [31]	Five bacteria species	++	+++	+++	+++	−	−	−	Quantitative, enzymatic analysis, three metabolites (ATP, ADP, AMP), recoveries by spiking. Tested six other methods.
Larsson and Olsson [32]	Four algae species	+++	−	++	++	−	−	−	Quantitative, enzymatic analysis, three metabolites (ATP, ADP, AMP), no recoveries. Tested three other methods.
Dekoning and Vandam [11]	S. cerevisiae	+++	−	−	−	+++	−	−	Quantitative, enzymatic analysis, up to 13 metabolites (data not shown), recoveries by spiking (only CM)
Gonzalez et al. [12]	S. cerevisiae	++	−	−	+++	−	−	−	Quantitative, enzymatic analysis, six metabolites, recoveries by spiking
Hajjaj et al. [33]	Monascusruber	++	++	−	+++	−	−	−	Quantitative, enzymatic analysis, two to six metabolites, recoveries by spiking
Hans et al. [13]	S. cerevisiae	−	−	+++	+++	−	−	−	Quantitative, HPLC, 17 metabolites (amino acids), recoveries by spiking (only BE)
Maharjan and Ferenci [28]	E. coli	+	+	−	++	+	++	−	2D-TLC, semiquantitative for efficacies (total extract intensity and relative intensities for 13 metabolites), no recoveries. Tested one other method (hot methanol)

1.5 Metabolite Extraction

Reference	Organism								Notes
Jernejc [34]	Aspergillus niger	+++	+++	—	++	—	—	—	Quantitative, enzymatic analysis of three metabolites (organic acids), recoveries assessed by measuring the stability of analytical grade standards during extraction. GC-MS, qualitative for efficacies (number of peaks detected), quantitative for recoveries, 27 metabolites, by spiking
Villas-Bôas et al. [15]	S. cerevisiae	+	+	—	+	++	++	—	Qualitative, ESI-MS (richness and reproducibility of mass spectra), no recoveries
Wang et al. [35]	E. coli	—	++	—	+	—	+++	—	Quantitative, enzymatic analysis, seven metabolites, recoveries by standard additions
Hiller et al. [36]	E. coli	+	—	+++	++	—	—	—	LC-MS, semiquantitative for efficacies (peak heights for nearly 100 metabolites) and recoveries, by spiking (peak heights for 12 metabolites). Tested two other methods (and some variations).
Rabinowitz and Kimball [29]	E. coli	—	—	—	—	+	+	+++	Quantitative, enzymatic analysis, three metabolites, recoveries by spiking (only FTM). Tested one other method (chloroform–water)
Faijes et al. [37]	Lactobacillus plantarum	+++	—	—	+++	++	+++	—	GC-MS, qualitative (number of peaks detected) and semiquantitative (relative abundance of 21 metabolites) for efficacies, no recoveries
Winder et al. [38]	E. coli	+	+	—	+++	+++	+++	—	

[a]Legend: +, poor/bad; ++, fair; +++, good; underlined, best among the methods tested. Some variations in extraction time, buffers, solvent concentrations and temperatures are considered within the same method.
[b]PCA, perchloric acid; KOH, potassium hydroxide; HW, hot water; BE, boiling ethanol; CM, chloroform–methanol; FTM, freeze-thawing in methanol; AATNM, acidic acetonitrile-methanol; TCA, trichloroacetic acid.
Source: Table from Canelas et al. [39].

The only way to determine this is to compare how a number of different extraction methods perform with respect to the amounts of metabolites obtained, under the assumption that at least one of the methods will achieve complete extraction.

The second essential property, denaturation of enzymatic activity, can be evaluated by monitoring whether any interconversion of metabolites takes place after the extract has been obtained, that is, during the subsequent sample processing steps. The third requirement, absence of metabolite degradation, can be evaluated by quantifying the metabolite losses during extraction.

Canelas et al. [39] evaluated both metabolite conversion as well as degradation by comparing the obtained metabolite levels after extraction of the same sample in the presence of fully U-^{13}C metabolite mixture with the levels obtained after extraction in the absence of the U-^{13}C mixture (see also the Section 1.6). In the latter case, the U-^{13}C mixture was added after extraction, thus allowing analysis with IDMS. With this procedure, each metabolite is quantified relative to a U-^{13}C-labeled standard, thus correcting for analytical artifacts, for example, sample matrix effects. If the U-^{13}C-labeled standard mix is added before extraction, partial metabolite degradation can also be effectively corrected for because the ^{12}C and U-^{13}C labeled forms of each metabolite are chemically identical and therefore their ratio will not change if partial degradation occurs [40, 41]. This procedure effectively corrects for partial degradation of metabolites during the extraction procedure.

By comparing the metabolite levels in cell extracts obtained in the presence and absence of the U-^{13}C standard mix, the overall process recovery can be quantified for each metabolite analyzed.

$$\text{recovery}_{x,i} = \frac{[x]_i^{BA}}{[x]_i^{AQ}} \tag{1.1}$$

where the superscript BA indicates that the U-^{13}C standard mix was added before analysis, that is, extraction was carried out in the absence of internal standards (ISs), and AQ indicates that the standard mix was added directly after quenching, that is, the extraction was carried out in the presence of ISs.

This recovery includes nonspecific losses sample handling, metabolite degradation as well as (inter)conversion. Furthermore, to take into account the time that the samples have to stay in a cooled auto sampler before analysis, the extracts were incubated for 8–9 h at a temperature of 4–5 °C. If only nonspecific losses would occur, a very similar recovery for all metabolites with a value above 90% would be expected. Significantly lower recoveries would be an indication for metabolite degradation, whereas if recoveries well above as well as well below 100% would occur, this would strongly indicate the occurrence of interconversion of metabolites because of the remaining enzymatic activity.

Canelas et al. determined the overall process recoveries for the extraction methods shown in Figure 1.1 for 44 metabolites in cell samples of *S. cerevisiae* grown in two different ways, namely in glucose-limited chemostats and in unlimited batch cultures on glucose [39]. The results are shown in Figure 1.2. It can be seen from this figure that the overall process recoveries for the first three extraction methods (HW, BE, and CM) are close to one for the majority of the quantified metabolites, whereby BE shows the flattest profile. For the other two methods (FTM and AANM), however, large deviations were observed. In case of the FTM method, recoveries were found which were either much higher or much lower than one. This shows that interconversion of metabolites has occurred. It should be noted that the FTM method (freeze-thawing cycles in 50% (v/v) aqueous methanol) is a mild extraction method that will cause cell disruption and release of metabolites but is unlikely to result in complete denaturation of enzymes.

The AANM method did not result in recoveries above 100%, and hence no indications were obtained for interconversion of metabolites. However, especially for the large metabolites (i.e., on the basis of a larger mw), the recoveries were poor (Figure 1.2). The reason for this is not clear. Also, this method seems to be relatively mild because it is carried out at a low temperature and the time of exposure to acidic conditions is not extremely long (45–60 min). From the observation that especially polar charged metabolites exhibit low recoveries, Canelas et al. [39] suggested that these could have been caused by poor solubility of these compounds in the cold solvent-rich mixture. If this results in precipitation, a part would be lost during the centrifugation and filtration steps of the sample processing procedure. For further details on the evaluation of these five extraction methods for yeast metabolomics, we refer to the paper of Canelas et al. [39]. The final conclusion of these authors was that, at least for yeast, from the five methods tested, BE and CM should be considered as the most reliable methods in terms of completeness of extraction, prevention of metabolite conversion, and metabolite stability.

1.6
Application of ^{13}C-Labeled Internal Standards

Finally, the determination of the metabolites in the obtained cell extracts can be performed with different analytical techniques. Before the application of mass spectrometry for quantification of compounds became common practice, mainly enzymatic assays and chromatographic methods, for example, high-performance liquid chromatography (HPLC), were applied. The advantage of enzymatic assays is their specificity. Their disadvantage is that for each metabolite a different assay has to be used, which makes the analysis of a large number of metabolites very laborious while the amount of sample required is relatively large.

At present, high-throughput hyphenated mass spectrometric techniques, whereby liquid- or gas-chromatographic separation is coupled to mass spectrometric detection, are routinely used for the measurement of small molecules. These methods allow the simultaneous identification and quantification of a large number of metabolites with high selectivity, adequate sensitivity, and minimal sample use (e.g., 5–10 µl for the quantification of several tens of compounds versus 100 µl or more for the enzymatic assay of a single metabolite).

In particular, liquid chromatography/electrospray ionization mass spectrometry (LC–ESI-MS) is nowadays widely used for the quantification of phosphorylated carbon compounds, which are intermediates of primary metabolic pathways [42,

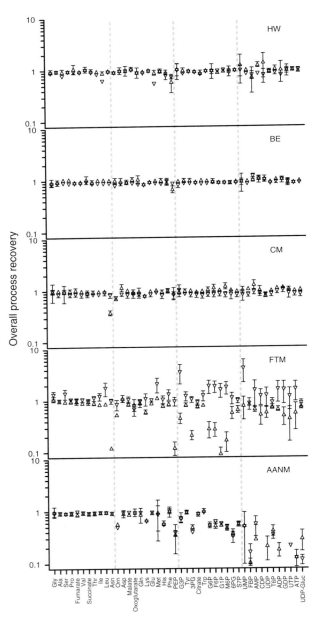

Figure 1.2 Overall process recoveries (calculated from Eq. (1.1)) for 44 metabolites analyzed, in the order of increasing molecular weight, for each of the extraction methods, under two growth conditions (glucose limited chemostat and batch cultivation). Data are averages and standard deviations of duplicate samples each analyzed twice. Legend: ∇, chemostat; Δ, batch. Dashed vertical lines are for guidance only. (Source: Figure from Canelas et al. [39].)

43]. However, the performance of LC–ESI-MS can be compromised by ion suppression effects [44] that result in a changing signal response of the analyte, owing to changes in the sample matrix. The composition of the sample matrix is influenced by, for example, the applied quenching and extraction procedures, the species of microorganism used, and the composition of the cultivation medium.

This drawback of LC–ESI-MS quantification can be tackled by applying the isotope dilution technique, whereby for each analyte a stable isotope analog (i.e., isotopolog) is used as an IS in the MS analysis [45]. This technique has found a large variety of applications, as reviewed by Baillie [46]. Because of the high physical-chemical similarities between the labeled IS and the analyte, degradation during sample preparation, instrument drift, and ion suppression effects in LC–ESI-MS can be compensated for. This technique has been widely applied to different research areas, for example, pharmacological, medical studies [46–48]. Also, for microbial metabolomics, it has been shown that the application of IDMS is indispensible to accomplish accurate and reproducible quantification of metabolites [40, 41, 49, 50].

To apply this method, an isotopolog has to be included as IS for each metabolite to be analyzed. Because these isotopologs are not commercially available for most metabolites, they must be produced by cultivating microorganisms on 100% U-^{13}C labeled medium and subsequent extraction to obtain a cell extract containing a 100% U-^{13}C labeled metabolite mixture. If this cell extract is added to each sample before metabolite extraction, partial degradation of metabolites, losses during sample handling as well as instrument drift, and ion suppression effects during LC-MS analysis can be effectively corrected for.

A suitable procedure for the production of the ^{13}C extract is (small-scale) fed-batch cultivation [41]. The advantage of this cultivation method is that the growth rate of the culture can be set to a desired value by controlling the medium feed rate. Cultivation of cells at a certain growth rate can be relevant, for example, to obtain certain desired metabolite levels. Examples are producing strains for which product formation only occurs within a particular range of growth rates (e.g., a high producing strain of P. chrysogenum for which the rate of penicillin production, and consequently the intracellular concentrations of the intermediates of the biosynthesis pathway, is highest at a specific growth rate of around 0.03 h^{-1} [51]). Another example is baker's yeast (S. cerevisiae) that should be cultivated below the so-called critical growth rate ($\sim$0.3 h^{-1}) because otherwise massive formation of ethanol occurs, resulting in a much lower biomass yield on the supplied U-^{13}C glucose and thus much lower concentrations of U-^{13}C labeled metabolites in the obtained cell extract.

Harvesting of the U-^{13}C-labeled cells can be performed by quenching portions of, for example, 100 ml of broth in 500 ml of cold ($-40\,^\circ$C) aqueous methanol followed by extraction in 75% aqueous BE, as described for the production of U-^{13}C yeast extract [41].

Metabolite measurement with IDMS, whereby U-^{13}C labeled cell extract is used as IS, requires the addition of an amount of ^{13}C extract to the quenched cell samples before extraction. To enable proper quantification, a calibration line has to be made, whereby a fixed amount of the ^{13}C extract is added to different dilutions of an unlabeled standard mixture of the metabolites to be quantified. LC-MS analysis of the calibration mixtures yields a calibration line, whereby for each compound the $^{12}C/^{13}C$ peak area ratio is plotted against the (known) concentration of the unlabeled metabolite in the different dilutions of the standard mixtures. If the volume fraction of the ^{13}C extract in the calibration mixture is identical to the volume ratio of the ^{13}C extract added to each samples versus the volume of the final cell extract, calibration line can be directly applied to convert the measured $^{12}C/^{13}C$ peak area ratio's to the concentrations of unlabeled metabolites in the samples.

It should be noted that partial degradation of metabolites during extraction and/or matrix and ion suppression effects will not result in changes in the $^{12}C/^{13}C$ metabolite ratios because chemical degradation, matrix and ion suppression effects of labeled and unlabeled metabolite analogs should be the same (isotopologs are chemically identical). Also, volume losses due to sample handling will not affect the $^{12}C/^{13}C$ ratio. Therefore, these issues will not affect the outcome of the quantification. Examples of calibration lines for the adenosine nucleotides AMP, ADP, and ATP are shown in Figure 1.3. It can be clearly seen from this figure that the constructed calibration lines based on peak areas of unlabeled standards were

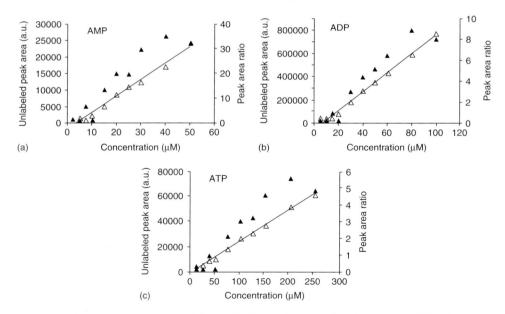

Figure 1.3 Comparison of obtained calibration curves based on measured peak areas of unlabeled nucleotide standards (closed triangles) and peak area ratios of unlabeled to labeled internal standards (open triangles) for monophosphate (AMP), diphosphate (ADP), and triphosphate (ATP) adenosine nucleotides. a.u., arbitrary units. (Source: Figure from Seifar et al. [52].)

highly nonlinear compared with the calibration lines based on the ID approach. This shows that the application of U-^{13}C-labeled isotopologs as ISs improves the quality of MS-based metabolome quantification by correcting for partial losses during sample preparation as well as through improvement of the precision of the LC-EI-MS based analytics. It should be realized that enzymatic (inter)conversions of metabolites in the presence of ^{13}C IS mix will result in changes in ^{12}C/^{13}C ratios of the metabolites involved, and thus the application of this standard mix cannot correct for this kind of conversions. The applied extraction procedure should therefore result in complete enzyme denaturation to prevent enzymatic (inter)conversions to occur.

1.7 Conclusions

To obtain meaningful snapshots of the microbial metabolome, rapid sampling and instantaneous quenching of all metabolic activity to prevent further interconversion of metabolites is essential.

An important aspect is the separation of the cells from the surrounding medium, before the extraction procedure, as in many cases intracellular metabolites are also present outside the cells. Although the metabolite concentrations in the extracellular medium are usually low, the much larger volume fraction of the medium in a culture sample, compared to the volume occupied by the cells, could easily result in overestimation of metabolite levels. To avoid this, a quenching procedure should be applied, which does not induce metabolite leakage from the cells and thus allows separation of cells and surrounding liquid after quenching without causing metabolite losses.

So far, no single condition has been found that is suitable for the quenching of different eukaryotic microorganisms; possibly, because of differences in membrane composition. Consequently, it seems necessary to validate and optimize the quenching procedure for each different organism. Although modifications might be required, cold methanol quenching appears to be applicable for quantitative metabolomics of different eukaryotic microorganisms and can thus be used to separate the cells from the surrounding medium to avoid interference with metabolites present therein.

In contrast to this, no suitable quenching procedure has been reported, which allows removal of extracellular metabolites in case of bacterial cultures. Although cold methanol quenching can be applied to bacteria, subsequent cold centrifugation and washing will result in significant loss of metabolites, because of substantial diffusion from the cells into the cold methanol solution, during and after quenching.

The second step after obtaining the cell samples is an extraction procedure to release the metabolites from the cells. Essential requirements that a proper extraction method should fulfill are completeness of extraction, that is, all metabolites should be completely released from the cells; complete denaturation of all enzymes present to prevent further interconversion of metabolites during subsequent sample processing steps; and at the same time no significant degradation and/or chemical conversion. Several methods have been described in the literature, which do not fulfill these requirements. At present, hot aqueous ethanol boiling seems to be the most common procedure, while HW boiling and CM extraction yield comparable results. These three methods are common in that the enzymes present in the cell samples are effectively denaturated, and therefore enzymatic (inter)conversion of metabolites does not occur.

Metabolite quantification is mostly accomplished with MS-based methods (mainly LC- and GC-MS). This allows the application of IDMS, using stable isotope labeling, for example, ^{13}C, to improve the reproducibility and accuracy of the measurements.

References

1. Canelas, A.B., Ras, C., ten Pierick, A., van Dam, J.C., Heijnen, J.J., and van Gulik, W.M. (2008) *Metabolomics*, **4**, 226–239.
2. Nasution, U., van Gulik, W.M., Ras, C., Proell, A., and Heijnen, J.J. (2008) *Metab. Eng.*, **10**, 10–23.
3. Taymaz-Nikerel, H., de Mey, M., Ras, C., ten Pierick, A., Seifar, R.M., van Dam, J.C., Heijnen, J.J., and van Gulik, W.M. (2009) *Anal. Biochem.*, **386**, 9–19.
4. Canelas, A.B., Ras, C., ten Pierick, A., van Gulik, W.M., and Heijnen, J.J. (2011) *Metab. Eng.*, **13**, 294–306.
5. Wellerdiek, M., Winterhoff, D., Reule, W., Brandner, J., and Oldiges, M. (2009) *Bioprocess. Biosyst. Eng.*, **32**, 581–592.
6. Schaub, J., Schiesling, C., Reuss, M., and Dauner, M. (2006) *Biotechnol. Progr.*, **22**, 1434–1442.
7. Schaedel, F. and Franco-Lara, E. (2009) *Appl. Microbiol. Biotechnol.*, **83**, 199–208.
8. Lange, H.C., Eman, M., van Zuijlen, G., Visser, D., van Dam, J.C., Frank, J., de Mattos, M.J.T., and Heijnen, J.J. (2001) *Biotechnol. Bioeng.*, **75**, 406–415.
9. Schaefer, U., Boos, W., Takors, R., and Weuster-Botz, D. (1999) *Anal. Biochem.*, **270**, 88–96.
10. Schaub, J., Mauch, A., and Reuss, M. (2008) *Biotechnol. Bioeng.*, **99**, 1170–1185.
11. de Koning, W. and van Dam, K. (1992) *Anal. Biochem.*, **204**, 118–123.
12. Gonzalez, B., Francois, J., and Renaud, M. (1997) *Yeast*, **13**, 1347–1355.
13. Hans, M.A., Heinzle, E., and Wittmann, C. (2001) *Appl. Microbiol. Biotechnol.*, **56**, 776–779.
14. Loret, M.O., Pedersen, L., and François, J. (2007) *Yeast*, **24**, 47–60.
15. Villas-Bôas, S.G., Højer-Pedersen, J., Åkesson, M., Smedsgaard, J., and Nielsen, J. (2005) *Yeast*, **22**, 1155–1169.
16. de Jonge, L.P., Douma, R.D., Heijnen, J.J., and van Gulik, W.M. (2011) *Metabolomics*, **8**, 727–735.
17. Carnicer, M., Canelas, A.B., van Pierick, A., Zeng, Z., van Dam, J.C., Albiol, J., Ferrer, P., Heijnen, J.J., and van Gulik, W.M. (2012). *Metabolomics*, **8**, 284–298.
18. Douma, R.D., de Jonge, L.P., Jonker, C.T.H., Seifar, R.M., Heijnen, J.J., and van Gulik, W.M. (2010) *Biotechnol. Bioeng.*, **107**, 105–115.
19. Bolten, C.J., Kiefer, P., Letisse, F., Portais, J.C., and Wittmann, C. (2007) *Anal. Chem.*, **79**, 3843–3849.
20. de Mey, M., Taymaz-Nikerel, H., Baart, G., Waegeman, H., Maertens, J., Heijnen, J.J., and van Gulik, W.M. (2010) *Metab. Eng.*, **12**, 477–487.
21. Taymaz-Nikerel, H., van Gulik, W.M., and Heijnen, J.J. (2011) *Metab. Eng.*, **13**, 307–318.
22. Hancock, R. (1958) *Biochim. Biophys. Acta*, **28**, 402–412.
23. Hommes, F.A. (1964) *Arch. Biochem. Biophys.*, **108**, 36–46.

24. Gale, E.F. (1947) *J. Gen. Microbiol.*, **1**, 53–76.
25. Work, E. (1949) *Biochim. Biophys. Acta*, **3**, 400–411.
26. Bent, K.J. and Morton, A.G. (1964) *Biochem. J.*, **92**, 260.
27. Fuerst, R. and Wagner, R.P. (1957) *Arch. Biochem. Biophys.*, **70**, 311–326.
28. Maharjan, R.P. and Ferenci, T. (2003) *Anal. Biochem.*, **313**, 145–154.
29. Rabinowitz, J.D. and Kimball, E. (2007) *Anal. Chem.*, **79**, 6167–6173.
30. Bagnara, A.S. and Finch, L.R. (1972) *Anal. Biochem.*, **45**, 24–34.
31. Lundin, A. and Thore, A. (1975) *Appl. Microbiol.*, **30**, 713–721.
32. Larsson, C.M. and Olsson, T. (1979) *Plant Cell Physiol.*, **20**, 145–155.
33. Hajjaj, H., Blanc, P.J., Goma, G., and Francois, J. (1998) *FEMS Microbiol. Lett.*, **164**, 195–200.
34. Jernejc, K. (2004) *Acta Chim. Slov.*, **51**, 567–578.
35. Wang, Q.Z., Yang, Y.D., Chen, X., and Zhao, X.M. (2006) *Chin. J. Anal. Chem.*, **34**, 1295–1298.
36. Hiller, J., Franco-Lara, E., and Weuster-Botz, D. (2007) *Biotechnol. Lett.*, **29**, 1169–1178.
37. Faijes, M., Mars, A.E., and Smid, E.J. (2007) *Microb. Cell Fact.*, **6**, 27.
38. Winder, C.L., Dunn, W.B., Schuler, S., Broadhurst, D., Jarvis, R., Stephens, G.M., and Goodacre, R. (2008) *Anal. Chem.*, **80**, 2939–2948.
39. Canelas, A.B., ten Pierick, A., Ras, C., Seifar, R.M., van Dam, J.C., van Gulik, W.M., and Heijnen, J.J. (2009) *Anal. Chem.*, **81**, 7379–7389.
40. Mashego, M.R., Wu, L., van Dam, J.C., Ras, C., Vinke, J.L., van Winden, W.A., van Gulik, W.M., and Heijnen, J.J. (2004) *Biotechnol. Bioeng.*, **85**, 620–628.
41. Wu, L., Mashego, M.R., van Dam, J.C., Proell, A.M., Vinke, J.L., Ras, C., van Winden, W.A., van Gulik, W.M., and Heijnen, J.J. (2005) *Anal. Biochem.*, **336**, 164–171.
42. Buchholz, A., Takors, R., and Wandrey, C. (2001) *Anal. Biochem.*, **295**, 129–137.
43. van Dam, J.C., Eman, M.R., Frank, J., Lange, H.C., van Dedem, G.W.K., and Heijnen, S.J. (2002) *Anal. Chim. Acta*, **460**, 209–218.
44. Annesley, T.M. (2003) *Clin. Chem.*, **49**, 1041–1044.
45. Sojo, L.E., Lum, G., and Chee, P. (2003) *Analyst*, **128**, 51–54.
46. Baillie, T.A. (1981) *Pharmacol. Rev.*, **33**, 81–132.
47. Bajad, S. and Shulaev, V. (2007) *Trends Anal. Chem.*, **26**, 625–636.
48. Ciccimaro, E. and Blair, I.A. (2010) *Bioanalysis*, **2**, 311–341.
49. Buescher, J.M., Moco, S., Sauer, U., and Zamboni, N. (2010) *Anal. Chem.*, **82**, 4403–4412.
50. Vielhauer, O., Zakhartsev, M., Horn, T., Takors, R., and Reuss, M. (2011) *J. Chromatogr. B*, **879**, 3859–3870.
51. van Gulik, W.M., de Laat, W.T.A.M., Vinke, J.L., and Heijnen, J.J. (2000) *Biotechnol. Bioeng.*, **68**, 602–618.
52. Seifar, R.M., Ras, C., van Dam, J.C., van Gulik, W.M., Heijnen, J.J., and van Winden, W.A. (2009) *Anal. Biochem.*, **388**, 213–219.

2
Tandem Mass Spectrometry Hyphenated with HPLC and UHPLC for Targeted Metabolomics

Gérard Hopfgartner and Emmanuel Varesio

2.1
Introduction

The understanding of diseases onset, toxicity of chemicals, the effect of lifestyle over life span, the pharmacology of new pharmaceuticals would greatly be enhanced if one could identify and quantify metabolites involved in the various biological processes and use them as a biological readout (biomarkers). The monitoring at a molecular level of endogenous and exogenous metabolites present in biological systems requires sophisticated analytical techniques and can be performed either in a nontargeted or targeted way. While nontargeted metabolomics relies on a more holistic approach of "all one can measure," targeted metabolomics is based on the identification and quantification of selected metabolites with known characteristics, which are substrates of an enzyme, direct products of a protein, or members of a pathway. As most living species such as humans, animals, or plants are exposed to chemicals, pharmaceuticals, drugs, pollutants, or food additives, such compounds are often present in their intact form or as metabolites in biological matrices and are becoming part of the metabolome.

With the introduction of atmospheric pressure ionization and in particular electrospray ionization (ESI), liquid chromatography combined to mass spectrometric detection has become an essential tool for quantitative targeted analysis in various fields such as pharmaceutical, environmental, toxicology, biology, and clinical chemistry and is also becoming a working horse in -omics techniques. While in bioanalysis LC-MS (mass spectrometry) has been extensively used over the past two decades in the selected reaction monitoring (SRM) mode to measure single analyte concentrations, metabolomics studies require multicomponents assays. Furthermore, with continuous technological improvements in mass spectrometric performance regarding sensitivity, resolving power in chromatography, data handling, or with multidimensional approaches, the simultaneous analysis of hundreds of analytes in the low picogram range has become reality.

Metabolomics in Practice: Successful Strategies to Generate and Analyze Metabolic Data, First Edition.
Edited by Michael Lämmerhofer and Wolfram Weckwerth.
© 2013 Wiley-VCH Verlag GmbH & Co. KGaA. Published 2013 by Wiley-VCH Verlag GmbH & Co. KGaA.

2.2
LC-MS-Based Targeted Metabolomics

The human metabolome database (version 2.5) contains over 7900 metabolites [1] covering various classes of compounds such as lipids, carbohydrates, amino acids, biogenic amines, nucleotides, vitamins, organic acids, lipids, and steroids. In human serum 4229 confirmed and highly probable metabolites have been reported with their concentrations range [2]. Interestingly, more than 3000 compounds are lipids. Theoretically, one could develop analytical assays for all classes of compounds; however, there are several points to consider, such as the nature of the analyte, the dynamic range of the metabolites, the sample volume, and the analytical throughput as well as precision and accuracy of the assay. As illustrated in Figure 2.1, the dynamic range of plasma/serum metabolites covers almost 9 orders of magnitude, which is analytically challenging. As full scan mass spectrometric analysis has a dynamic range of about 3–4 orders of magnitude, it becomes obvious that many metabolites, which have also different MS response factors, cannot be covered by nontargeted metabolomic analysis. Hyphenation between separation sciences such as gas chromatography, capillary electrophoresis, or liquid chromatography and MS has become so powerful mainly because of their orthogonality. Considering the analysis on a single platform, LC-MS has been found to be the best hyphenation among the various combinations because of its flexibility and robustness [3]. On the other hand, regarding the chemical diversity of the metabolites liquid chromatography can be a confining factor because of relatively long analysis time and several different retention mechanisms, which are mandatory to analyze polar versus nonpolar metabolites. As MS can be considered as a separation technique that differentiates analytes based on their mass-to-charge ratio (m/z), some authors have proposed infusion-based or flow–injection-based (FIA-MS) metabolomics, where basically no analyte is lost [4, 5]. FIA-MS can be considered as a high-throughput technique, but isomeric or isobaric metabolites will be measured as a sum and not as individual metabolites that can jeopardize the biological readout. Besides the classical LC-SRM/MS quantitative approach, novel chromatographic LC materials as well as high-resolution MS are opening innovative possibilities in targeted metabolomics. The problem of how many metabolites can be detected by generic LC-MS/MS is nicely illustrated in Figure 2.2 showing the overlap of serum metabolites detected by five global analytical methods. Surprisingly, only less than 3% of all analytes were detected by LC-ESI-MS/MS [2].

2.3
Liquid Chromatography

Unlike gas chromatography, high-performance liquid chromatography (HPLC) is not limited to the analysis of volatile thermostable compounds. Different classes of analytes, for example, from elements to large biomolecules, can be analyzed by HPLC including hydrophilic, lipophilic, neutral, acidic, or basic compounds.

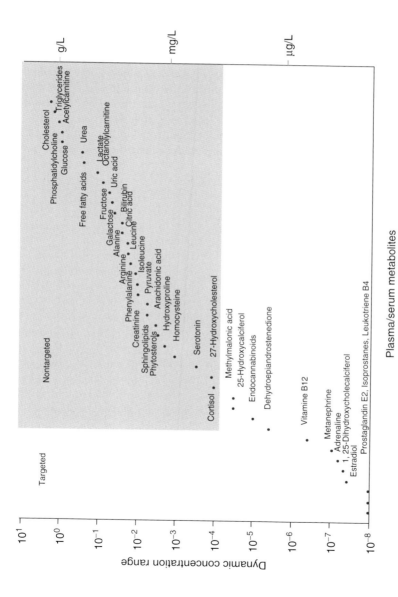

Figure 2.1 Dynamic range of plasma/serum metabolites. (Source: Adapted with permission from Ref. [6].)

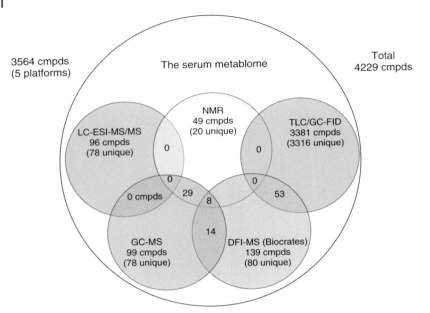

Figure 2.2 Venn diagram showing the overlap of serum metabolites detected by global nuclear magnetic resonance (NMR), gas chromatography–mass spectrometry (GC–MS), liquid chromatography/gas chromatography with flame ionization detection (LC/GC-FID), LC-ESI-MS/MS, and MS/MS methods compared to the detectable serum metabolome. (Source: From Ref. [2].)

The choice of the stationary phase that is most suited for the separation is directly related to the analyte's chemical properties. HPLC comprises different modes of separation including reversed-phase, normal phase (NP), hydrophilic interaction, ion exchange, size exclusion, and chiral chromatography. The most commonly used stationary phases in liquid chromatography are reversed-phase such as C_{18}-based materials. While several types of detection can be used with HPLC (e.g., UV or fluorescence spectroscopy, electrochemistry), MS based on triple quadrupole (QqQ), ion trap, or high-resolution MS instruments remains the detector of choice in metabolomic studies. When performing non-targeted metabolomics, the HPLC separation power and selectivity are important to separate as many compounds as possible in a single sample and multidimensional LC approaches are growing of interest [7, 8]. The LC parameters that need to be optimized for targeted metabolomics are some what different; in general, the focus is to quantify simultaneously several hundreds of analytes and based on their structures several important physicochemical parameters for sample preparation or chromatography can easily be determined. Owing to their poor retention on C_{18} reversed-phase columns, polar metabolites are challenging to analyze and some alternative separation mechanisms are available including (i) hydrophilic interaction liquid chromatography (HILIC) [9], (ii) aqueous normal phase chromatography (ANP) [10], (iii) ion pairing chromatography (IP-LC) [11], (iv) reversed-phase pentafluorophenylpropyl (PFPP) columns, and

(v) mixed modes columns. Compared to conventional size columns packed with 3.5 or 5.0 μm particles, the introduction of sub-2.0 μm particles [12, 13] or core-shell materials [14] has greatly improved the peak capacity and shortens the analysis time. As most of these columns can be operated at high linear velocities, typically 1 and 2 mm i.d. columns should be selected to match the optimal performance of the current MS ionization interfaces.

HILIC is generally reported as a variant of NP chromatography because the chromatographic separation is usually performed on a polar stationary phase such as silica, amino, diol, or cyano [15, 16]. While in NP-HPLC an apolar mobile phase is typically used, mobile phase similar to reversed phase RP-HPLC are used in HILIC [17]. HILIC combines several separation mechanisms (ion chromatography, NP-HPLC, and RP-HPLC) and seems to be particularly well suited for the separation of uncharged highly hydrophilic and amphiphilic analytes. Unfortunately, the reported drawbacks remain, such as the slow re-equilibration time in gradient elution mode, column bleeding, and degradation. Recently, chromatographic packing material has emerged where hydride groups (Si-H) replace 95% of the silanols (Si-OH) on the surface [10]. The less polar surface reduces the attraction of the stationary phase for water, allowing the operation of the columns in ANP mode.

Good chromatographic separation on reversed-phase chromatography of negatively charged compounds such as nucleotides, sugar phosphates, and carboxylic acids can also be achieved with anionic ion pairing agents (tributyl or hexylamine) [11, 18], while cationic analytes such as amines or amino acids can be efficiently separated using volatile perfluorocarboxylic acids such as heptafluorobutyric acid [19]. However, ion pairing agents can induce signal suppression and pollute the LC lines or the ionization interface where the effect becomes only visible when switching from negative to positive mode or vice versa. Reversed-phase PFPP columns offer an attractive complement to HILIC and ion pairing approaches for the analysis of polar metabolites. Lv *et al.* [20] investigated four different PFPP columns for targeted metabolomics. A total of 112 hydrophilic metabolites could be measured by SRM within 8 min of running time to obtain a metabolite profile from the various biological samples.

As one would like to measure as many compounds as possible in a single sample, the use of several HPLC systems are becoming mandatory and various combinations have been described [21]. Lu *et al.* [22] reported a dual chromatographic method where HILIC chromatography was used with positive ion mode detection, while tributylamine ion pairing agent was used in conjunction with reversed-phase chromatography and a detection in the negative mode. The platform was successfully applied to the quantitative analysis of 250 water-soluble metabolites. More complex multiplexing has been described by Wei *et al.* [7] to assay 205 analytes (divided into three subgroups: amino acids, sugars, and nucleotides and organic acids) in a single analysis and is depicted in Figure 2.3. The platform uses three gradient pumps, one injector with three injection loops, and three different types of analytical columns (i.e., a Synergi Polar RP column, a Luna Phenyl-Hexyl column, and an Atlantis T3 ODB column). A custom program was

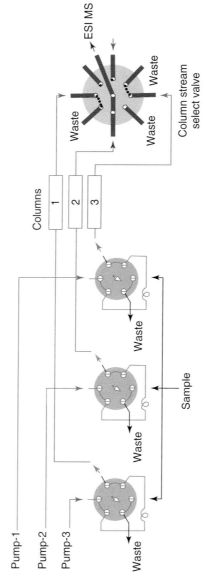

Figure 2.3 Multiplexed LC-SRM/MS Method Setup based on three chromatographic systems and one MS detector: (1) reversed-phase on a Luna Phenyl-Hexyl column to separate aminoacids, and nucleic bases, (2) ion pairing on an Atlantis T3 OBD column to separate nucleotides, sugar, sugar phosphates, and sugar alcohols, and (3) a Synergi Polar-RP to separate organic acids and some nucleotides. (Adapted with permission from Ref. [7].)

developed to introduce the sample in three injectors in a sequential way. Detection is performed in the selection reaction monitoring either in positive or negative mode covering a dynamic range of 8 orders of magnitude. The cycle time of the analysis is of 9.5 min.

The metabolites that are isomers cannot in most cases be distinguished by MS because they have the same mass-to-charge ratio and similar fragmentation pattern. Therefore, isomers need baseline HPLC separation to be quantified properly. A representative example, which illustrates the challenge of separation sciences for targeted metabolomics, is phosphorylated carbohydrates that are not suited for reversed-phase separation but were successfully analyzed by using mixed mode anion exchangers [23].

The chemical diversity of metabolites calls for multidimensional and orthogonal separation mechanisms. One point that should always be kept in mind is the sample throughput and therefore the analysis time. The major goal of targeted metabolomics is to analyze large cohorts of samples, which needs to be done in a timely manner and cost-effective way. To allow accurate quantitative analysis on QqQ instruments, special attention should be given to analytes cross-talk (see below), as well as to isomers coelution. As in most cases, the target analytes are well characterized, and sample preparation based on the intrinsic characteristics of the analytes is almost mandatory.

2.4
Mass Spectrometry

2.4.1
Ionization Techniques

While for untargeted metabolomics various ionization techniques should be considered including electron impact (EI), ESI, atmospheric pressure chemical ionization (APCI), atmospheric pressure photoionization (APPI), or matrix-assisted laser desorption (MALDI) [24] for targeted metabolomics, one could select the ionization type according to the intrinsic nature of the analytes. ESI remains the technique of choice in LC-MS-based targeted metabolomics studies. While the ionization process is instrument-independent, different ion current or ion current ratio will be measured for the various metabolites when performing the analysis on different instruments. Several reasons can count for that, such as ion sampling geometry or source operation conditions. Looking at the interfaces from the different MS vendors (e.g., AB Sciex, Agilent Technologies, BrukerDaltonics, Shimadzu, Thermo Fisher, and Waters), one can rapidly realize that the interface design changes significantly from one instrument to another. The interface design and the jet expansion into the vacuum will affect for a given analyte the ion distribution, as, for example, the ion ratio between the protonated molecule, the ammonium, and the sodium adduct. Most of electrospray interfaces depending on the LC flow rate operate in pneumatically assisted electrospray mode with the application of heat to enhance the ionization process. The application of heat can be detrimental to certain analytes and needs to be investigated in details. In atmospheric pressure ionization, the presence of coeluting compounds can significantly affect the response factor of the analyte of interest. This phenomenon is reported as matrix-dependent signal suppression or enhancement and is considered as one of the major drawbacks in quantitative LC-MS analyses. In most cases, the use of an internal standard (IS) such as a coeluting structural analog or the isotopically labeled version of the analyte allows to overcome this issue.

2.4.2
Mass Analyzers

Since their introduction in the late 1970s, QqQ mass spectrometers have become the working horses for quantitative LC-MS analyses based on the SRM mode [25]. Often this mode where the precursor is selected in the first quadrupole (Q1) and a specific fragment in the third quadrupole (Q3) is also referred to as multiple reaction monitoring (MRM) mode. With SRM, a transition needs to be selected for each analyte as well as the optimum collision energy (CE), which is generally instrument specific. Quadrupoles are typically operated at unit mass resolution. SRM is a sensitive and selective mode, however, sometimes coeluting analytes can generate a signal unrelated to the analyte, which is referred to as cross-talk. Cross-talk can occur from isobaric compounds, adducts, or fragments of different analytes and can become critical for quantitative analysis when chromatographic resolution is minimized. Modern QqQ have SRM dwell times as low as 1 ms allowing several hundreds of transitions to be monitored simultaneously. The dwell times (e.g., 5 vs 100 ms) do not affect the signal intensity, but lower dwell times generate more noise that can affect the limit of quantification (LOQ) of the assay.

In MS, the *duty cycle* can be defined as the time needed for acquiring data for a single experiment. When several experiments are considered, the sum of the duty cycles is the analysis cycle time. In LC-MS, at least 12 data points need to be collected over the LC peak for accurate end precise quantitative analysis. Therefore, the MS acquisition duty cycle (i.e., the analysis cycle time) needs to match the specificity of the chromatography. With ultrahigh pressure liquid chromatography (UHPLC) or fast chromatography (e.g., with core-shell columns), fast acquiring MS are mandatory to avoid jeopardizing the chromatographic performance. Figure 2.4 shows the LC-MS extracted ion current trace at m/z 380.1 of seven isobaric metabolites. At the base, the LC peak is of about 6 s wide. The present MS acquisition comprises five different MS experiments with different duty cycles and a total analysis cycle time of 400 ms. For each of the five experiments, only one data point will be generated every 400 ms even if the duty cycle is of only a few milliseconds as for the last experiment.

Time-scheduled SRM is a concept based on LC retention time to optimize the number of analytes and dwell times [26] using time windows. Ion trap instruments are not very well suited for targeted analysis in metabolomics because of their too

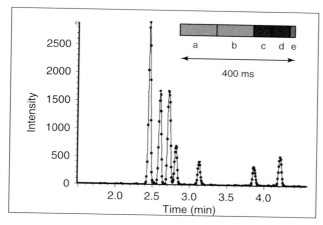

Figure 2.4 Extracted ion current trace of m/z 380.1 for the first experiment (a). The UHPLC-MS analysis consisted of five different looped experiments (a–e) with different duty cycles and a total MS analysis cycle time of 400 ms. Each dot represents one data point.

low duty cycle in MS/MS mode. On the other hand, QqQ linear ion traps are of interest because of their ability to operate in the SRM mode as well as to obtain, with similar sensitivities, enhanced product ion spectra for confirmatory analysis or for method development [27].

Recent instrumental improvements in high-resolution MS offers interesting new possibilities for quantitative analysis either in full scan or in MS/MS mode. In practice, there are two categories of mass spectrometers that can achieve high resolution with good mass accuracies (<5 ppm): the time-of-flight (TOF) geometry MS and the Fourier transform mass spectrometry (FT-MS) including the Fourier transform ion cyclotron resonance (FT-ICR) and the Orbitrap. The difference regarding resolving power between these instruments is the duty cycle, which is only critical in MS/MS mode. With a FT-MS instrument such as the Orbitrap, working at a resolving power of 10 000 requires a duty cycle of 100 m, and a resolution power of 100 000 is achieved at the expense of a duty cycle of 1–1.5 s [28]. In the case of a TOF geometry instrument (such as the Triple TOF 5600), the resolving power is not dependent on the duty cycle of the MS and the system can acquire data as fast as 10 ms with resolving power of 20 000–30 000 either in MS or MS/MS mode [29]. Compared to QqQ approaches, high resolution quantification requires less tuning but the major benefit is the combination of simultaneously acquired qualitative and quantitative analyses (QUAL/QUAN). Lu et al. [30] reported the ability to measure known and unanticipated metabolites for cellular extracts from Baker's yeast using a standalone Exactive Orbitrap and found its performance comparable to modern QqQ.

Ion mobility mass spectrometry (IMS) is a gas phase process where ions are separated by their different mobility in a buffer gas under the influence of an electric field. Several flavors of IMS have been described [31] in which analytes are separated by their size-to-charge ratio based on the analyte's cross-section, its charge state, and interactions with the buffer gas. This technique is particularly suited for isomers separation, and IMS remains a promising front-end high throughput technique. While IMS provides additional separation selectivity, its current peak capacity is significantly lower than for gas chromatography (GC) or HPLC; IMS can neither smooth nor minimize the matrix effects occurring during the ionization process. Only a few applications using IMS have been described for metabolomic investigations [32], but it will certainly play a major role in the future in targeted metabolomics as an additional separation dimension to HPLC [33].

2.5
Sample Preparation

Study design and sample collection are critical steps before analysis and should be carefully considered [34]. Two major points need to be investigated: (i) inactivation of metabolism and (ii) sample integrity during storage [35]. In biological matrices such as cells, urine, plasma, and tissues, many analytes are degrading overtime and their stability should be known or investigated under various conditions. Besides many others, organic solvents quenching or the use of liquid nitrogen are two largely used methods for stopping the metabolism. To further enhance sample stability, pH adjustment or the addition of stabilizers has also been considered. Ideally, no sample preparation would be the best choice, but, as for many analytical techniques, LC-MS suffers from limited dynamic range (3–5 orders of magnitude) and from matrix effects caused by coeluting compounds. Assay robustness or the ability to analyze hundreds of samples on a daily basis is also key and therefore a good sample preparation for the target analytes is highly desirable to guarantee reliable data. Besides the stabilization of the sample and removal of interferences, sample preparation also allows to concentrate metabolites and to extend the dynamic range of the metabolites covered. Depending on the nature of the sample, that is, liquid or solid, various strategies can be considered including: direct injection, liquid–liquid extraction (LLE), solid-phase extraction (SPE), supercritical fluid extraction (SFE), accelerated solvent extraction (ASE), microwave-assisted extraction (MWE), protein precipitation (PP), and membrane methods such as dialysis or ultracentrifugation [4]. Gika et al. [36] have reviewed various sample preparation before LC-MS-based metabolomics for blood-derived samples. In targeted metabolomics, to obtain time profiles on single subject can be of importance. As traditional sample collection may suffer from various errors and issues with storage and transport, beside the traditional dialysis method, solid phase microextraction (SPME) has emerged as an attractive alternative for direct sample collection and preparation [37]. SPME is a solvent-free sample preparation technique in which the analyte is isolated and enriched in a single step. It consists generally of a fiber coated with an absorbing material. The device can be used to collect either analytes from breath or from blood.

2.6
Relative and Absolute Quantification

Relative quantification of the analyte can be quite straightforward. In general, area ratios are determined between the analyte and an added reference compound not genuinely present in the sample. However, several points need to be carefully considered, such as the MS response of the analyte that should be linear, taking into consideration that in most cases 3 orders of magnitude of linearity are common with atmospheric pressure ionization. The reference compound should ideally coelute with the analyte to compensate for matrix effects. Structural analogs of the analyte or better an isotopically labeled version of the analyte with the incorporation of stable isotopes such as $[^2H]$, $[^{13}C]$, or $[^{15}N]$ can be used successfully as reference compound. As their response factor ratio (analyte/reference compound) can be determined, they can also be used for single point quantitative estimation of the analyte of interest.

LC-MS-based absolute quantification in the SRM mode is widely applied to the analysis of pharmaceuticals and their metabolites in biological matrices [38] where method validation and sample analysis follow international guidelines [39]. High-resolution MS is also more and more employed for quantitative analysis [40]. A basic requirement is the availability of certified reference material. One of the first steps is to define a LOQ and a dynamic range of the assay. In bioanalysis, LC-SRM/MS quantification is mostly performed by building a calibration curve by spiking the analyte into the same matrix as the study samples (e.g., plasma or urine). Most of LC-MS-based assays use an IS at a fixed concentration to compensate for losses in sample preparation or during analysis, and the calibration curves are built based on the area or height of analyte to IS ratio. Stable isotope labeled internal standard (SIL-IS) are the best choice because they normalize for matrix effects occurring in the ionization process. The purity of the SIL-IS is critical because it will affect the LOQ as well as the dynamic range. In general, $[^{13}C]$ or $[^{15}N]$ labeling is preferred to $[^2H]$ labeling because of the risk of deuterium exchange. Before running study samples, the assays are validated according to international recommendations, and precision and accuracy are determined over a certain concentration range based on quality control (QC) samples. For targeted metabolomics assays, metabolites are generally endogenous analytes and are present at various concentration levels in the samples; consequently, the preparation of calibration samples is almost impossible. Owing to the fact that matrix effects can significantly jeopardize the quantification, the preparation of the calibration sample is critical. Several alternatives can be considered to prepare calibration and QC samples: (i) the use of a surrogate matrix (alternative species, buffers, proteins solution, etc.), (ii) the removal (stripping)of the analyte of interest from the matrix, or (iii) the use of the standard addition method. The standard addition method that relies on the serial addition of standard to the sample could also be envisaged but is rarely used for large sets of samples because it is time consuming. A method to quantify an analyte present in a high background level in matrix such as metabolites has been described by Li *et al.* using a "surrogate" analyte approach

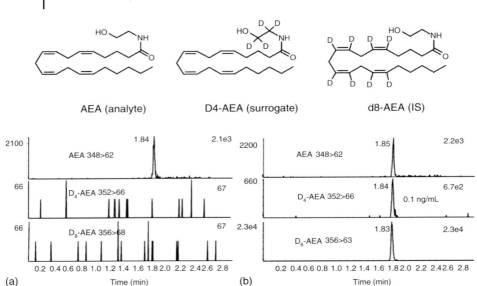

Figure 2.5 Surrogate analyte concept: (a) plasma sample with specific signal of AEA and (b) plasma sample spiked with surrogate analyte (d4-AEA) at 0.1 ng ml^{-1} and IS (d8-AEA). (Source: Adapted with permission from Ref. [42].)

[41]. The principle is quite simple and specific to mass spectrometric detection and is based on the replacement of the analyte by a stable isotope-labeled standard (surrogate) that differs at least by 2 Da to generate the calibration samples in the matrix of interest. The surrogate will keep the same properties as the compound of interest and will only differ by its mass. The IS will then be either a structural analog or a secondary stable isotope labeled standard. Ethanolamides (fatty acid amides) are potential biomarkers to determine target engagement for fatty acid amide hydrolase. Jian et al. [42] described an assay for the quantification of anandamide (AEA), oleoylethanolamide (OEA), and palmitoyletahnolamide (PEA) in human plasma using D4-AEA, D4-OEA, and $^{13}C_2$-PEA as "surrogate analytes." Figure 2.5a shows in the SRM transition m/z 348 → 62 the presence of AEA. The use of surrogate analyte D4-AEA to construct the calibration curve (Figure 2.5b SRM trace m/z 352 → 66) allows building a calibration curve with a LOQ of 0.1 ng ml^{-1} for AEA. In this example, the coelution of the three analytes is almost complete, which is the best case to minimize matrix effects.

2.7
Applications

Owing to the challenge to analyze the complete metabolome with a single method, holistic metabolomics approaches are required to identify biomarkers specific to a disease, to determine the effect of a drug, or to evaluate environmental exposure. However, most of the untargeted LC approaches that require long gradient are not quantitative and do not allow high throughput analysis as required for biomarker validation, drug discovery or drug development, and population studies or integrated health care. Accurate and precise quantification of biomarkers is essential for the biological readout. Therefore, targeted metabolomics focuses on the quantification of defined sets of structurally known and biochemically annotated metabolites. Targeted LC-MS assays are generally developed for single or for several hundreds of metabolites. The application of tandem MS for the screening of amino acids and carnitines has already been introduced in the 1990s for the detection of inborn errors [43]. The application of targeted metabolomics has expanded significantly over the last years in the field of clinical chemistry for inherited metabolic diseases, cancer studies, diabetes, and coronary heart disease [6]. Several reviews have extensively summarized the published work around targeted metabolomics [21, 44], and the most commonly analyzed metabolites are summarized in Table 2.1.

Owing to the high dynamic range and the chemical diversity of the metabolites present in biological matrices, no single method will be able to measure in a single chromatographic run all the metabolome. The quantification performance of the analytical methods is also variable depending on the fit of purpose. Some of them are extensively validated according to the regulatory guidelines that require quite complex method development and validation steps, but others are referred to as *semiquantitative methods* with no minimum validation. One should avoid using the term *semiquantitative analysis*, as it mimics the half-full or half-empty glass story. Precision and accuracy based on QC samples should always be defined

Table 2.1 Selected references of targeted LC-MS assays described in the literature.

Analytes	References
Amino acids	[45–51]
Anionic and aromatic metabolites	[52]
Acylcarnitines	[43]
Bile acids	[53]
Biogenic amines	[54]
Carboxylic acids	[55]
Carbohydrates	[56]
Eicosanoids	[57]
Endocannabinoids	[42]
Glycerophospholipids	[58–62]
Long-chain acyl-coenzymes A	[63]
Nucleosides	[64]
Steroids	[65, 66]
Vitamins	[67]

and monitored when analyzing study samples. The acceptance criteria may vary significantly depending on the fit of purpose of the analysis.

Various commercial LC-MS based platforms (e.g., Biocrates, Metabolon, and Metanomics) have been developed in the past and are contributing to the expansion of the technique [68]. Also various commercial kits for LC-SRM/MS analysis of metabolites are becoming available such as the SteroIDQ from Biocrates [65]. The ready-to-use kit covers calibrator set, QCs, IS, and buffer additives for the analysis of 80 samples and enables the analysis of 16 steroid hormones (in human serum). The aim of the assay is to allow standardized LC-SRM/MS analysis of a comprehensive steroid hormone profile in clinical diagnostics.

For absolute quantification, in the SRM mode or high-resolution MS, the use of a reference compound is mandatory, whereas for relative quantification method development can be based either on metabolomics MS libraries or on experimental data. To be able to fully exploit the datasets generated by metabolomics experiments, the development of user-friendly and publicly accessible databases becomes mandatory [69].

Ideally, in targeted SRM experiments each precursor → product ion transition should be unique for each metabolite. Owing to the diversity of the metabolites and their numerous isomers, several signals can often be observed in one transition and metabolite cross-talk can be an issue. The signal can occur from a different metabolite eluting at different or similar retention time. Selectivity of the assay can be jeopardized, and the use of a confirmatory transition can help to monitor the selectivity of the assay. Depending on the number of monitored analytes, the combination of information-dependent acquisition (IDA) on a QqQ linear ion trap using SRM as survey scan and an enhanced product ion as dependent scan can also be envisaged [70]. With multicomponents targeted analysis, data processing can become challenging. Beside the commercially available quantification software, more and more open access tools are becoming available. For example, Wong *et al.* [71] have developed the Metabolite Mass Spectrometry Analysis Tool (MMSAT) available on the web and using mzXML input format for comparing multiple SRM experiments across multiple samples to enable quantitative comparison of each metabolite between samples.

2.8
Synopsis

LC-SRM/MS-based analysis was initially applied to develop quantitative assays to follow the pharmacokinetics (PK) of drugs and their metabolites. It was further applied to quantify low molecular weight compounds in various fields such as environmental, food science, and clinical chemistry. More recently, LC-SRM/MS has emerged as a key technique for proteomics where it has been used for the quantification of large sets of peptides as surrogates of proteins [26]. At present, LC-MS-based metabolomics is performed in two steps: a nontargeted and a targeted analysis. The first step focuses on the identification of metabolites that are up- or down-regulated in a relatively limited set of samples. In the second step, quantitative assays are developed for potential biomarkers that can be measured in large sets of samples. While QqQs remain the most sensitive and reliable MS instruments for quantitative analysis, they are limited for qualitative analysis. As for PK analysis, accurate and precise quantification of biomarkers is of importance to follow their time course in subjects. Ideally, one would like to perform qualitative and quantitative analysis (QUAL/QUAN) in the same analysis. Moreover, postacquisition relative quantification would be of great interest. The introduction of fast acquisition high-resolution mass spectrometers has opened new possibilities in QUAL/QUAN approaches, which would also be beneficial for metabolomics studies. Even more sample preparation and multidimensional LC separation combined with ion mobility will be required to tackle efficiently the metabolome [72]. It turns out to be that the analytical challenge is not to measure single analytes but to perform multicompounds analysis. They are still major analytical issues to address regarding analytes stability, sample preparation, separation, detection, and data processing. In addition, it is key to centralize as an open resource all the relevant analytical information regarding metabolites quantification (e.g., compounds libraries, analytes stability, MS/MS spectra, assays repository, chromatographic retention factor libraries, etc.).

References

1. Wishart, D.S., Tzur, D., Knox, C., Eisner, R., Guo, A.C., Young, N., Cheng, D., Jewell, K., Arndt, D., Sawhney, S., Fung, C., Nikolai, L., Lewis, M., Coutouly, M.A., Forsythe, I., Tang, P., Shrivastava, S., Jeroncic, K., Stothard, P., Amegbey, G., Block, D., Hau, D.D., Wagner, J., Miniaci, J., Clements, M., Gebremedhin, M., Guo, N., Zhang, Y., Duggan, G.E., MacInnis, G.D., Weljie, A.M., Dowlatabadi, R., Bamforth, F., Clive, D., Greiner, R., Li, L., Marrie, T., Sykes, B.D., Vogel, H.J., and Querengesser, L. (2007) *Nucleic Acids Res.*, **35**, D521–D526.
2. Psychogios, N., Hau, D.D., Peng, J., Guo, A.C., Mandal, R., Bouatra, S., Sinelnikov, I., Krishnamurthy, R., Eisner, R., Gautam, B., Young, N., Xia, J., Knox, C., Dong, E., Huang, P., Hollander, Z., Pedersen, T.L., Smith, S.R., Bamforth, F., Greiner, R., McManus, B., Newman, J.W., Goodfriend, T., and Wishart, D.S. (2011) *PLoS ONE*, **6** (2), e16957.
3. Büscher, J.M., Czernik, D., Ewald, J.C., Sauer, U., and Zamboni, N. (2009) *Anal. Chem.*, **81** (6), 2135–2143.
4. Dettmer, K., Aronov, P.A., and Hammock, B.D. (2007) *Mass Spectrom. Rev.*, **26** (1), 51–78.
5. Lin, L., Yu, Q.A., Yan, X.M., Hang, W., Zheng, J.X., Xing, J.C., and Huang, B.L. (2010) *Analyst*, **135** (11), 2970–2978.
6. Becker, S., Kortz, L., Helmschrodt, C., Thiery, J., and Ceglarek, U. (2012) *J. Chromatogr. B*, **883-884**, 68–75.
7. Wei, R., Li, G.D., and Seymour, A.B. (2010) *Anal. Chem.*, **82** (13), 5527–5533.
8. Guo, K., Peng, J., Zhou, R.K., and Li, L. (2011) *J. Chromatogr. A*, **1218** (23), 3689–3694.
9. Buszewski, B. and Noga, S. (2012) *Anal. Bioanal. Chem.*, **402** (1), 231–247.
10. Pesek, J.J. and Matyska, M.T. (2005) *J. Sep. Sci.*, **28** (15), 1845–1854.
11. Coulier, L., Bas, R., Jespersen, S., Verheij, E., van der Werf, M.J., and Hankemeier, T. (2006) *Anal. Chem.*, **78** (18), 6573–6582.

12. Unger, K.K., Skudas, R., and Schulte, M.M. (2008) *J. Chromatogr. A*, **1184** (1-2), 393–415.
13. Wu, N.J. and Thompson, R. (2006) *J. Liquid Chromatogr. Related Technol.*, **29** (7-8), 949–988.
14. Gritti, F., Leonardis, I., Abia, J., and Guiochon, G. (2010) *J. Chromatogr. A*, **1217** (24), 3819–3843.
15. Hemstrom, P. and Irgum, K. (2006) *J. Sep. Sci.*, **29** (12), 1784–1821.
16. Cubbon, S., Antonio, C., Wilson, J., and Thomas-Oates, J. (2010) *Mass Spectrom. Rev.*, **29** (5), 671–684.
17. Jandera, P. (2011) *Anal. Chim. Acta*, **692** (1-2), 1–25.
18. Luo, B., Groenke, K., Takors, R., Wandrey, C., and Oldiges, M. (2007) *J. Chromatogr. A*, **1147** (2), 153–164.
19. Kaspar, H., Dettmer, K., Gronwald, W., and Oefner, P.J. (2009) *Anal. Bioanal. Chem.*, **393** (2), 445–452.
20. Lv, H.T., Palacios, G., Hartil, K., and Kurland, I.J. (2011) *J. Proteome Res.*, **10** (4), 2104–2112.
21. Xiao, J.F., Zhou, B., and Ressom, H.W. (2012) *Trends Anal. Chem.*, **32**, 1–14.
22. Lu, W., Bennett, B.D., and Rabinowitz, J.D. (2008) *J. Chromatogr. B*, **871** (2), 236–242.
23. Hinterwirth, H., Laemmerhofer, M., Preinerstorfer, B., Gargano, A., Reischl, R., Bicker, W., Trapp, O., Brecker, L., and Lindner, W. (2010) *J. Sep. Sci.*, **33** (21), 3273–3282.
24. Nordstrom, A., Want, E., Northen, T., Lehtio, J., and Siuzdak, G. (2008) *Anal. Chem.*, **80** (2), 421–429.
25. Yost, R.A. and Enke, C.G. (1978) *J. Am. Chem. Soc.*, **100** (7), 2274–2275.
26. Gallien, S., Duriez, E., and Domon, B. (2011) *J. Mass Spectrom.*, **46** (3), 298–312.
27. Hopfgartner, G., Varesio, E., Tschappat, V., Grivet, C., Bourgogne, E., and Leuthold, L.A. (2004) *J. Mass Spectrom.*, **39** (8), 845–855.
28. Makarov, A. and Scigelova, M. (2010) *J. Chromatogr. A*, **1217** (25), 3938–3945.
29. Hopfgartner, G., Tonoli, D., and Varesio, E. (2011) *Anal. Bioanal. Chem.*, **402** (8), 2587–2596.
30. Lu, W.Y., Clasquin, M.F., Melamud, E., Amador-Noguez, D., Caudy, A.A., and Rabinowitz, J.D. (2010) *Anal. Chem.*, **82** (8), 3212–3221.
31. Kanu, A.B., Dwivedi, P., Tam, M., Matz, L., and Hill, H.H. (2008) *J. Mass Spectrom.*, **43** (1), 1–22.
32. Dwivedi, P., Schultz, A.J., and Hill, H.H. (2010) *Int. J. Mass Spectrom.*, **298** (1-3), 78–90.
33. Varesio, E., Le Blanc, J.C., and Hopfgartner, G. (2011) *Anal. Bioanal. Chem.*, **402** (8), 2555–2564.
34. Alvarez-Sanchez, B., Priego-Capote, F., and de Castro, M.D.L. (2010) *Trends Anal. Chem.*, **29** (2), 111–119.
35. Alvarez-Sanchez, B., Priego-Capote, F., and de Castro, M.D.L. (2010) *Trends Anal. Chem.*, **29** (2), 120–127.
36. Gika, H. and Theodoridis, G. (2011) *Bioanalysis*, **3** (14), 1647–1661.
37. Ouyang, G., Vuckovic, D., and Pawliszyn, J. (2011) *Chem. Rev.*, **111** (4), 2784–2814.
38. Hopfgartner, G. and Bourgogne, E. (2003) *Mass Spectrom. Rev.*, **22** (3), 195–214.
39. Bioanalytical Method Validation, US Department of Health and Human, Food and Drug Administration. Guidance for Industry (2001), Services, FDA, Center for Drug Evaluation and Research, Rockville, MD.
40. Zhang, N.R., Yu, S., Tiller, P., Yeh, S., Mahan, E., and Emary, W.B. (2009) *Rapid Commun. Mass Spectrom.*, **23** (7), 1085–1094.
41. Li, W. and Cohen, L.H. (2003) *Anal. Chem.*, **75** (21), 5854–5859.
42. Jian, W.Y., Edom, R., Weng, N.D., Zannikos, P., Zhang, Z.M., and Wang, H. (2010) *J. Chromatogr. B-Anal. Technol. Biomed. Life Sci.*, **878** (20), 1687–1699.
43. Shushan, B. (2010) *Mass Spectrom. Rev.*, **29** (6), 930–944.
44. Roux, A., Lison, D., Junot, C., and Heilier, J.F. (2011) *Clin. Biochem.*, **44** (1), 119–135.
45. Piraud, M., Vianey-Saban, C., Petritis, K., Elfakir, C., Steghens, J.P., and Bouchu, D. (2005) *Rapid Commun. Mass Spectrom.*, **19** (12), 1587–1602.
46. Armstrong, M., Jonscher, K., and Reisdorph, N.A. (2007) *Rapid Commun. Mass Spectrom.*, **21** (16), 2717–2726.

47. Kaspar, H., Dettmer, K., Chan, Q., Daniels, S., Nimkar, S., Daviglus, M.L., Stamler, J., Elliott, P., and Oefner, P.J. (2009) *J. Chromatogr. B*, **877** (20-21), 1838–1846.
48. Waterval, W.A.H., Scheijen, J.L.J.M., Ortmans-Ploemen, M.M.J.C., Habets-van der Poel, C.D., and Bierau, J. (2009) *Clin. Chim. Acta*, **407** (1-2), 36–42.
49. Harder, U., Koletzko, B., and Peissner, W. (2011) *J. Chromatogr. B*, **879** (7-8), 495–504.
50. Piraud, M., Ruet, S., Boyer, S., Acquaviva, C., Clerc-Renaud, P., Cheillan, D., and Vianey-Saban, C. (2011) *Methods Mol. Biol. (Clifton, N.J.)*, **708**, 25–53.
51. Guo, K., Ji, C., and Li, L. (2007) *Anal. Chem.*, **79** (22), 8631–8638.
52. Buescher, J.M., Moco, S., Sauer, U., and Zamboni, N. (2010) *Anal. Chem.*, **82** (11), 4403–4412.
53. Xiang, X., Han, Y., Neuvonen, M., Laitila, J., Neuvonen, P.J., and Niemi, M. (2010) *J. Chromatogr. B*, **878** (1), 51–60.
54. de Jong, W.H., de Vries, E.G., and Kema, I.P. (2011) *Clin. Biochem.*, **44** (1), 95–103.
55. Johnson, D.W. (2005) *Clin. Biochem.*, **38** (4), 351–361.
56. Jannasch, A., Sedlak, M., and Adamec, J. (2011) *Methods Mol. Biol.*, **708**, 159–171.
57. Ferreiro-Vera, C., Mata-Granados, J.M., Priego-Capote, F., Quesada-Gómez, J.M., and Luque De Castro, M.D. (2011) *Anal. Bioanal. Chem.*, **399** (3), 1093–1103.
58. Xia, Y.Q. and Jemal, M. (2009) *Rapid Commun. Mass Spectrom.*, **23** (14), 2125–2138.
59. Zhao, Z. and Xu, Y. (2009) *J. Chromatogr. B*, **877** (29), 3739–3742.
60. Cutignano, A., Chiuminatto, U., Petruzziello, F., Vella, F.M., and Fontana, A. (2010) *Prostaglandins Other Lipid Mediat.*, **93** (1-2), 25–29.
61. Kasumov, T., Huang, H., Chung, Y.M., Zhang, R., McCullough, A.J., and Kirwan, J.P. (2010) *Anal. Biochem.*, **401** (1), 154–161.
62. Lee, J.Y., Min, H.K., and Moon, M.H. (2011) *Anal. Bioanal. Chem.*, **400** (9), 2953–2961.
63. Blachnio-Zabielska, A.U., Koutsari, C., and Jensen, M.D. (2011) *Rapid Commun. Mass Spectrom.*, **25** (15), 2223–2230.
64. Hsu, W.Y., Lin, W.D., Tsai, Y.H., Lin, C.T., Wang, H.C., Jeng, L.B., Lee, C.C., Lin, Y.C., Lai, C.C., and Tsai, F.J. (2011) *Clin. Chim. Acta*, **412** (19-20), 1861–1866.
65. Koal, T., Schmiederer, D., Pham-Tuan, H., Rohring, C., and Rauh, M. (2012) *J. Steroid Biochem. Mol. Biol.*, **129** (3-5), 129–138.
66. Ceglarek, U., Kortz, L., Leichtle, A., Fiedler, G.M., Kratzsch, J., and Thiery, J. (2009) *Clin. Chim. Acta*, **401** (1-2), 114–118.
67. Adamec, J., Jannasch, A., Huang, J.J., Hohman, E., Fleet, J.C., Peacock, M., Ferruzzi, M.G., Martin, B., and Weaver, C.M. (2011) *J. Sep. Sci.*, **34** (1), 11–20.
68. Suhre, K., Meisinger, C., Doring, A., Altmaier, E., Belcredi, P., Gieger, C., Chang, D., Milburn, M.V., Gall, W.E., Weinberger, K.M., Mewes, H.W., de Angelis, M.H., Wichmann, H.E., Kronenberg, F., Adamski, J., and Illig, T., (2010) *Plos One*, **5** (11), e13953.
69. Go, E. (2010) *J. Neuroimmune Pharmacol.*, **5** (1), 18–30.
70. Thomas, A., Hopfgartner, G., Giroud, C., and Staub, C. (2009) *Rapid Commun. Mass Spectrom.*, **23** (5), 629–638.
71. Wong, J.W.H., Abuhusain, H.J., McDonald, K.L., and Don, A.S. (2012) *Anal. Chem.*, **84** (1), 470–474.
72. Varesio, E., Le Blanc, J.C., and Hopfgartner, G. (2012) *Anal. Bioanal. Chem.*, **402** (8), 2555–2564.

3
Uncertainty of Measurement in Quantitative Metabolomics

Raffaele Guerrasio, Christina Haberhauer-Troyer, Stefan Neubauer, Kristaps Klavins, Madeleine Werneth, Gunda Koellensperger, and Stephan Hann

3.1
Introduction

3.1.1
MS-Based Techniques in Metabolomics

Metabolomics, in its truest sense, refers to the global untargeted and unbiased analysis – including both identification and quantification – of the complete metabolome of a given biological system [1–5]. This is to date impossible because analytical platforms that are available nowadays, despite important and innovative developments, still lack overall and nonbiased applicability [3–5]. This chapter focuses on the implementation of the concept of measurement uncertainty (MU) in quantitative metabolite analysis. Accordingly, this introduction highlights mass spectrometry (MS) in the context of metabolic profiling, which is the quantification of selected metabolites [3] usually based on existing knowledge of a certain metabolic pathway enabling the detailed optimization of the analytical procedure [5, 6].

The two separation techniques, which are most commonly combined with MS-based detection for metabolite analysis, are gas chromatography (GC) and liquid chromatography (LC). Both, LC-MS and GC-MS offer a range of advantages, but also suffer from some drawbacks [3–7]. GC is known for its high separation efficiency. In addition, electron impact (EI) ionization allows for reliable identity confirmation, stable (matrix-robust) ionization efficiency, and low-cost operation. Besides EI, chemical ionization (CI) offers the advantages of soft ionization techniques, such as a lower degree of fragmentation and the consequent production of intact molecular ions for further fragmentation via collision-induced dissociation (CID). The most apparent disadvantage of GC is its limited applicability to nonvolatile or thermally labile compounds, demanding for tedious derivatization procedures. On the one hand, derivatization increases sample preparation time as well as the complexity and the variance of the analytical procedure. However, derivatization also enhances selectivity (by selective reaction) and often also

Metabolomics in Practice: Successful Strategies to Generate and Analyze Metabolic Data, First Edition.
Edited by Michael Lämmerhofer and Wolfram Weckwerth.
© 2013 Wiley-VCH Verlag GmbH & Co. KGaA. Published 2013 by Wiley-VCH Verlag GmbH & Co. KGaA.

the stability of the analytes. Therefore, GC is preferentially used for targeted metabolomics or the screening for selected substance classes determined by the selected derivatization reaction. High-performances liquid chromatography (HPLC) provides various separation mechanisms and is therefore applicable to a wide range of analytes. The restrictions of HPLC mainly arise from its combination with ESI (electrospray ionization) – or APCI-MS (atmospheric pressure chemical ionization), which offer only limited flexibility in terms of eluent composition and often suffer from matrix-dependent suppression effects during compound ionization.

Concerning mass analyzers, a range of different methodologies are available with highly diverse capabilities [7]. When combined with soft ionization techniques such as CI, ESI, APCI, or atmospheric pressure photoionization (APPI), single MS devices such as quadrupole or time-of-flight (TOF) instruments lack fragmentation via CID and are therefore limited regarding selectivity and signal-to-noise (S/N) ratios. Compared to quadrupole mass analyzers, TOF instruments offer higher scan speed, higher S/N ratios, and higher selectivity due to high mass resolution. In addition, high mass accuracy allows for the determination of molecular formulas.

Regarding the setup of multidimensional MS, one has to distinguish between MS^n in space and MS^n in time. The former is generally performed as MS/MS (tandem) experiments, most commonly using triple quadrupole or Q-TOF systems. Both produce fragment ions via CID and can be used for structural analysis and identity confirmation. Moreover, isolation of selected fragment ions allows for high selectivity and background reduction in quantitative analysis. The latter MS^n technology, MS^n in time, is realized using trapping systems – by common quadrupole ion traps, linear ion traps, or more advanced Fourier-transformation-based platforms such as FT-ICR-MS instruments or LIT-Orbitraps. Most frequently, these systems are used for metabolite identity confirmation. They provide highly detailed fragmentation information – and in the case of FT-ICR-MS and LIT-Orbitrap, a very high level of mass accuracy and resolution. However, their scan speed limits the applicability of ultrafast chromatography with narrow peaks. In conclusion, all analytical GC- and LC-MS-systems described earlier are platforms suitable for targeted metabolomics, and, thus, the quantitative analysis of small metabolites, which are highly differing in concentration as well as their chemical properties and which are present in matrix-rich samples. The complementary use of these mass spectrometric techniques allows to benefit from high separation efficiency (100 000 theoretical plates and more in GC), large dynamic range (e.g., 6 orders of magnitude for triple quadrupole devices), and high S/N ratios, thanks to background reduction either by selective fragmentation or high mass resolution. In addition, all mass spectrometric techniques offer the advantage of error compensation by the use of isotopically labeled internal standards. Still, none of these systems can be regarded as nonbiased or a Jack of all trades. Therefore, the calculation of total combined uncertainty is inevitable to gain fundamental understanding of the sources contributing to the overall uncertainty of quantitative results obtained with MS-based analysis.

3.1.2
Uncertainty of Measurement in Quantitative Analysis

3.1.2.1 **Definition**
The definition of the term *uncertainty of measurement* is given by the International Vocabulary of Basic and General Terms in Metrology (VIM) as follows [8]:

> "A nonnegative parameter characterizing the dispersion of the quantity values being attributed to a measurand, based on the information used."

Uncertainty expresses the intrinsic reliability of the results that is often erroneously identified with repeatability precision. It is a new approach, which gives more information as traditional strategies based on the assessment of MU via repetitive measurements. As a matter of fact, repeating measurements does not improve the reliability of results, as it does not consider all possible sources of uncertainty. The third edition of *Guide for Measurement Uncertainty* (GUM), *Quantifying uncertainty in analytical measurement* [9] and the VIM define a common way of estimating uncertainty and provide a common glossary for measurement vocabulary; the most important definitions are given below.

Measurand is a particular quantity subject to measurement (e.g., concentration of histidine in the cellular extract of *pichia pastoris*). The *analyte* is the compound or species, which is measured (histidine). The MU of a measurand can be expressed as the standard deviation, the range, or the width of a confidence interval. MU is an intrinsic property of every measurement result demonstrating the metrological quality of a measurement. Moreover, it improves the knowledge about the measurement procedure and allows identifying major uncertainty contributors, which should finally lead to an improvement of the overall procedure. As a matter of fact, correct determination of MU is a prerequisite for the reliable interpretation of comparative results. *Accuracy* is the closeness of the agreement between the result of a measurement and the true value of the measurand quantity. *Precision* can be expressed as repeatability, intermediate precision – also obtained from repetitive measurements – or reproducibility from interlaboratory studies.

Different approaches can be used to calculate the uncertainty of a system: the GUM provides a very detailed bottom-up approach, in which all possible sources of uncertainty are identified and assessed via expression of a standard uncertainty – this definition is discussed in Section 1.2.3.1 -and summed up with the most appropriate mathematical model to estimate the combined standard uncertainty of the method. More specifically, the GUM is only calculating uncertainty derived from statistical fluctuations of the system (also *random uncertainties*) because it assumes that the measurand fulfills the requirement of traceability. According to the VIM, the traceability is [8]:

> "The property of a measurement result whereby the result can be related to a reference through a documented unbroken chain of calibrations, each contributing to the measurement uncertainty."

Another widely applied MU method is called *top-down approach* in which data recovered by quality assurance and quality control (QA/QC) studies are used to estimate the uncertainty contribution of a process. Validation parameters such as trueness, precision, robustness, and interlaboratory and intralaboratory comparisons are often involved in top-down MU approaches [10]. The advantages and disadvantages of the two strategies have been extensively reviewed [10–15], and the big difference reseeds in the diverging principles they address in the QA/QC. While the bottom-up MU calculation follows a detailed protocol, which allows identifying the sources of random uncertainty and propagating them through the measurand model, the top-down is more prone to assess – and correct – the bias affecting the method, omitting or minimizing the study of random uncertainty. The accuracy of the bottom-up approach in identifying uncertainties has a salty price: the complexity of the MU calculation protocol – already elaborated in the case of a simple process as calculation of a standard concentration – grows exponentially in difficulty with the complexity of the mathematical model and of the analytical platform and hence the respective mathematical model. The top-down approach being based on QA/QC protocols has the main objective for minimization of, the distance between the measured value of the measurand and its true value. In this approach the random fluctuations that affect the measurand are not a priority and parameters such as robustness, reproducibility, and intermediate-precision become the most valuable tools to validate the method.

In conclusion, it appears clear that the combination of top-down and bottom-up protocols leads to the most comprehensive MU evaluations possible because the understanding of random as well as systematic influences is the only way to obtain reliable and accurate results.

3.1.2.2 Uncertainty Calculation According to the Bottom-Up Approach

Many authors assess the difficulty of performing MU calculations of complex analytical procedures with the bottom-up approach [11, 14, 15]. Nevertheless, it is still the most widely used approach, as it gives a global overview of all sources of uncertainty along with their relative contributions (uncertainty budget), and can be used to efficiently reduce the total combined uncertainty of an analytical procedure.

The third edition of *GUM Eurachem/CITAC* [9] illustrates the procedure, in which all the uncertainties affecting the studied method have to be identified, converted into standard uncertainties, and summed up through the appropriate mathematical model in order to calculate the total combined uncertainty. According to the bottom-up approach, the overall uncertainty of a measurand y is called *combined standard uncertainty* $(u_c(y))$. In case of independent variables, $u_c(y)$ is the square root of the sum of squares (RSS) of the individual uncertainty sources. x_1 to x_n as shown in Eq. (3.1)

$$u_c(y(x_1, x_2, \ldots x_n)) = \sqrt{\Sigma_{i=1,n} u(y, x_i)^2} \qquad (3.1)$$

$$u_c(y(x_1, x_2, \ldots x_n)) = \sqrt{\Sigma_{i=1,n} c_i^2 u(x_i)^2} \tag{3.2}$$

Equation (3.2) is a more general version of Eq. (3.1) and introduces the sensitivity coefficient c_i, that is the partial derivatives of y with respect to x_i [$c_i = \partial y/\partial x_i$].

The *combined standard uncertainty* $u_c(y)$ is then multiplied by the coverage factor k to obtain the *expanded uncertainty* [U]. U is the range in which the measurand true value resides with a defined statistical probability. The coverage factor k is usually a value between 2 and 3 that in a normal distribution corresponds to a level of confidence between 95 and 99.7%, respectively.

The bottom-up protocol for calculation of U is given below. However, it is important to mention that a universal protocol does not exist, because the involved uncertainty sources change significantly depending on the type of measurand (e.g., concentration of a standard solution, biomass in a microbial fermentation, and product purity in a downstream process) and the employed analytical platform.

Definition of the Measurand The selection of a measurand has to be unambiguous regarding what is being measured and also regarding the mathematical model equation, which is linked to the measurand value. A typical example of a rational or method-independent measurand is the mass isotopomer distribution of a metabolite in flux experiments. The mass isotopomer distribution, - a vector that assigns to the number of different isotope-containing species of a metabolic compound, the corresponding relative abundance - is a result, which theoretically has the same value regardless of the analytical platform used (GC-MS, LC-MS, or NMR). A classical-method-dependent measurand, however, is the concentration of a metabolite in a cellular extract, which is mainly determined by the techniques involved in sample preparation. Since, the extraction step is often focused on the optimum recovery of a selected pool of compounds (amino acids, nucleotides, lipids, proteins, DNA, etc.), it is often defined as method dependent. The most frequently stated example of a classical-method-dependent measurand is the determination of extractable fat. Translated to metabolomics, the evaluation of changes in the concentration profiles of one and the same organism under different cultivation conditions are can be regarded as method dependent measurands, accepting the bias which might be introduced by the extraction method.

Model Equation After defining the measurand, a mathematical equation describing the measurand model has to be built. Each parameter or variable in this equation, which is necessary for the calculation of the measurand value, is a source of uncertainty. In addition, further factors can be included in order to take into account environmental or empirical effects as well as repeatability, parameters, which might not be physically associated to any variable in the initial measurand model. This is often applied for introducing sample inhomogeneity as a possible source of uncertainty.

3 Uncertainty of Measurement in Quantitative Metabolomics

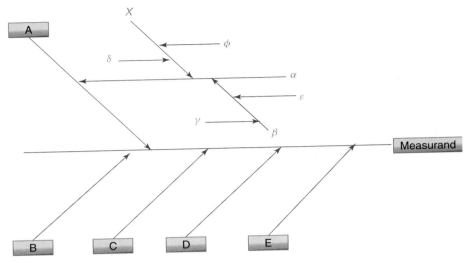

Figure 3.1 Example of a cause-and-effect diagram (CaED).

Identification of Uncertainty Sources The widely used Ishikawa diagram (also called bonefish or cause-and-effect diagram (CaED), shown in Figure 3.1) is a simple and straightforward approach to identify, define, and group uncertainty sources as well as to illustrate how they relate to each other. The CaED construction is a QA tool with a long tradition and has been well described [9, 16].

The CaED is composed of a main branch that leads straight to the measurand result and secondary branches that contribute to the main branch. The main branch represents the final outcome of the measurand, while for each input quantity of the general measurand equation, a secondary branch is added. The tertiary branches represent the uncertainty sources that influence the respective input quantities of the model equation. Once the CaED is drawn, it may be necessary to group certain sources of uncertainty in additional secondary branches. The model equation might need to be additionally modified via multiplication by factors, which do not explicitly show up in the equation used for calculating the measurand result, but whose contribution to overall uncertainty seems substantial.

A more detailed protocol for uncertainty source evaluation is given by the Eurachem/CITAC guide in the appendix D [9].

Uncertainty Source Quantification Detailed instructions of how to identify and characterize single uncertainty source are out of the scope of this chapter. The reader is referred to the comprehensive *Eurachem/CITAC* guide [9].

Briefly, the main criteria used for identification and quantification of uncertainties are listed as follows:

- Experimental variation of input variables (assessment of uncertainty effects by modifying one variable at a time in order to test the range in which the uncertainty budget can be applied).
- Comparison of measurement results with certificates of e.g certified reference materials, tuning solutions, and quality controls or internal standards (traceability).
- Modeling of theoretical principles (in case of chemical reactions or processes that show dependence from, e.g., temperature or time). It is necessary to include into the model equation a dependence from this effect or assess the uncertainty with determinate constraints proper of the resulted experiment. It is also possible to characterize the system uncertainty in a range of conditions in which the model equation behaviors are known (mathematical approximations).
- Judgment based on experience. The MU evaluation is neither matter of mathematical or routine-based assessment. It is something for which experience, detailed knowledge, and commonsense are needed. The operator as well has a great importance as uncertainty sources himself and analyst of the dataset.

Combined Standard Uncertainty Calculation The GUM divides uncertainty contributions into two groups, type A and type B. Type A uncertainties can be obtained via several direct repetitive observations of the measurand. When a Gaussian dispersion of data is assumed, the standard uncertainty of repetitive measurements corresponds to the standard deviation of the mean value. Type B uncertainties, on the other hand, are usually based on experience that does not originate from direct observations but from other sources such as literature, certificated reference materials, manufacturing information, or handbooks. A detailed protocol on how to transform uncertainty contribution type B into standard uncertainties is listed in section 8.1 of the GUM.

Once the uncertainty sources have been characterized, quantified, and expressed as standard uncertainties, they are combined following the concept of error propagation using Eq. (3.2). The choice of the model to calculate the *combined standard uncertainty* ($u_c(y)$) is a critical point, for which a technical debate is ongoing [9, 11, 12]. Section 8 of the GUM provides several versions of Eq. (3.2) that can be used in case if the measurand model equation fulfills particular requirements. The truth is that for each individual procedure, a specific mathematical model has to be developed in order to calculate the *total combined uncertainty*. It is also true that the mathematical model is just a tool to create a relationship between the uncertainty contributions and the measurand; a possibility to simplify the *total combined uncertainty* equation is to combine just the major uncertainty sources and leave out the negligible ones.

Calculation of the Value of the Measurand and the Associated Expanded Uncertainty
The expanded uncertainty [U] is calculated by multiplying $u_c(y)$ by the chosen coverage factor k. In order to choose the appropriate k value according to the resulting uncertainty, different factors (measurand peculiarities, instrument performances,

Table 3.1 The Kragten spreadsheet for uncertainty propagation.

		A	B	C	D
1	→ Input quantities →	A	B	C	D
2	Standard uncertainty →	α	β	χ	δ
	Values of input quantities ↓				
3	A	$a+\alpha$	a	A	A
4	B	b	$b+\beta$	B	B
5	C	c	c	$c+\chi$	C
6	D	d	d	D	$d+\delta$
7	$M = F(A,B,C,D)$	$M = f(A+\alpha, B, C, D)$	$M = f(A, B+\beta, C, D)$	$M = f(A, B, C+\chi, D)$	$M = f(A, B, C, D+\delta)$
8	—	$= M - M\alpha$	$= M - M\beta$	$= M - M\chi$	$= M - M\delta$
9	$\mathrm{DIF}_{\mathrm{sum}} = \mathrm{DIF}_\alpha + \mathrm{DIF}_\beta + \mathrm{DIF}_\chi + \mathrm{DIF}_\delta$	$\mathrm{DIF}_\alpha{}^2 = (M - M\alpha)^2$	$\mathrm{DIF}_\beta{}^2 = (M - M\beta)^2$	$\mathrm{DIF}_\chi{}^2 = (M - M\chi)^2$	$\mathrm{DIF}_\delta{}^2 = (M - M\delta)^2$
10	% Uncertainty contribution	$= \frac{(M-M\alpha)^2}{DIF_{sum}}$	$= \frac{(M-M\beta)^2}{DIF_{sum}}$	$= \frac{(M-M\chi)^2}{DIF_{sum}}$	$= \frac{(M-M\delta)^2}{DIF_{sum}}$
11	$1 < n < 3$	← k	—	—	—
12	$k \times \mathrm{SQRT}(\mathrm{DIF}_{\mathrm{sum}})$	← $k \times u_c(y)$	—	—	—
13	$100 \times k \times u_c(y)$	← Relative uncertainty (%)	—	—	—

validation studies, number of replicates, etc.) should be considered as shown in the subchapter 8.3 of the GUM. In most cases $k = 2$, corresponding to a coverage of 95% (two standard deviations), is appropriate.

Analysis of Uncertainty Contributions (Budget of Total Combined Uncertainties)
When applying the Kragten [9, 17] approach via spreadsheet or by using dedicated software [18], it is possible to evaluate and assess the weight of the different uncertainty contributions with respect to the total combined uncertainty. In the original paper of the 1994, Kragten gives a detailed and comprehensive explanation for setting up such a spreadsheet, which is described in the following text (Table 3.1).

The spreadsheet shows as many columns as the number of uncertainty sources +2. In row 1 and column 1, all input quantities of the model equation are listed. Column 2 lists the values of all input quantities, which are used to calculate the value of the measurand. In row 2, the standard uncertainties attributed to each of the input quantities are entered. Rows 3–6 list the input quantities reported in the measurand general equation. In these rows the corresponding standard uncertainties are systematically added to the input quantity value (A with α, B with β, etc.). In row 7, the model equation used to calculate the measurand M as a function of the input quantities A, B, C, and D is given. In row 8, the difference between the measurand, to which the *standard uncertainty* is added, and the nonmodified measurand is reported. Row 9 reports the squares of the differences recovered in row 8. The sum of the squared difference is calculated subsequently in order to calculate the *combined standard uncertainty* of the measurand. Row 10 reports the uncertainty distance (from row 8) divided by the square root of the sum of all the deviations. It is the value (it can be expressed as a percentage) that clarifies the single uncertainty source contribution. In row 11, the coverage factor is specified, while in row 12, the total combined uncertainty (U) is calculated by multiplying ($u_c(y)$) by k. In row 13, the relative uncertainty is calculated as percentage by dividing ($u_c(y)$) by the measurand and normalizing by 100.

It is important to keep in mind that the Kragten approach is based on the assumption that the measurand value M changes linearly within the range $M \pm M_{(\alpha,\beta,\chi,\delta)}$. Thanks to the versatility of spreadsheets, it is easily possible to validate this assumption by subtracting the standard uncertainties instead of adding them. As a rule of thumb, linearity should not be a problem if the interval $M \pm M_{(\alpha,\beta,\chi,\delta)}$ is relatively small. This demands high accuracy when quantifying uncertainties, even if generally the $M \pm M_{(\alpha,\beta,\chi,\delta)}$ is a parameter depending exclusively on the experimental settings used for measurand analysis. On the other hand, if the interval $M \pm M_{(\alpha,\beta,\chi,\delta)}$ is large, the Kragten approach may not be valid, and for the calculation of ($u_c(y)$) other approaches are needed.

Reporting and Documentation of Uncertainty Uncertainty estimation is a procedure used in many scientific contexts and performed by many operators in order to standardize and assess measurement performances (e.g., EN ISO 17025 demands for a procedure to determine and assess uncertainty of measurement). Uncertainty estimations need a high level of comprehensive documentation in order to make the MU calculation traceable for analysts and operators, which will perform corresponding studies in the future. A description of the methods and protocols used to calculate the measurand results and a list of all uncertainty sources with full documentation of their identification and quantification have to be stated. In addition, conversions from uncertainty to *standard uncertainty* have to be described, and all other parameters used for calculation of combined uncertainty (e.g., covariance used in uncertainty budgeting) need to be stated.

3.2
Uncertainties of Quantitative MS Experiments

A major breakthrough in biotechnology was the implementation of cell factories, for example, for the production of enzymes, pharmaceuticals, antibiotics, food additives, or bulk chemicals. Recently, with the introduction of the metabolomics' method portfolio, cell factory design advanced to a new stage enabling well directed metabolic engineering. One important tool in this context of optimizing cell factories concerns quantification of intracellular metabolite levels, often referred to as *metabolic profiling*. This approach is based on targeted MS analysis, for example, the quantitative determination of selected metabolites sharing the same metabolic pathway or chemical properties. Typical metabolites are intermediates from central carbon metabolism, purine and pyrimidine nucleotides, and amino acids. The aim of quantitative metabolite profiling is to obtain unbiased *in vivo* snapshots of the metabolic state of the investigated biological system.

3.2.1
Uncertainties in Sample Preparation

Accurate quantification of intracellular metabolites by MS-based strategies relies on the application of dedicated sampling and sample preparation procedures. More specifically, three operations, that is, cell sampling, quenching of cellular metabolism and extraction of metabolites have to be fulfilled. Owing to the high reactivity of a plethora of metabolites inside cells, rapid sampling and cell quenching techniques are absolutely required. Evidently, these multiple step procedures preceding MS analysis are highly critical and potentially affected by all kind of error sources (systematic, random, or spurious). The identification of the main uncertainty sources and the calculation of the combined standard uncertainties are powerful tools to assess the quality of analytical protocols and indicate the most relevant steps for further improvements.

As mentioned in the previous section, the first step of uncertainty budgeting is describing the whole analytical process in a model, which details each step of the analytical process. In the following, the different steps and approaches of MS-based metabolite profiling in fermentations are explained, and the critical aspects for quantitative analysis are discussed.

3.2.1.1 Sampling and Sample Preparation in Metabolite Profiling in Fermentations

Up to date different dedicated devices for rapid sampling of fermentations have been developed. These devices are predominantly in-house developments and not commercially available. Variations in the sample volume, in the maximal number of samples per run, and in the driving force for the liquid flow (e.g., overpressure, vacuum, or and peristaltic pump) result in a wide range of designs. The main performance parameters of these constructions are the sampling time and the reproducibility of sampling. A comprehensive overview of sampling devices has already been compiled [19]. Sampling is followed immediately (within seconds or even subseconds after sampling) by quenching. Quenching aims at instantly stopping the enzymatic activity of the sample. Most frequently, this is accomplished by an abrupt change of sample temperature, for example, to low temperatures (e.g., $< -20\,°C$). In the case of whole broth analysis, combining quenching and extraction in a single step, the sample is treated at elevated temperatures (e.g., $>80\,°C$) [20], within seconds after sampling. The time the cells take from the reactor until quenching (the sampling time) is extremely critical and has to be adapted according to the analyzed metabolites and their enzymatic turnover times [21]. Turnover of intermediates of the central metabolic pathway, for example, is in the range of seconds, whereas for amino acids, turnover is in the range of minutes [21], which renders sampling time by far less critical in amino acid analysis. Quick separation of the cells from the quenching solution and rapid washing are necessary to avoid leakage of intracellular metabolites, which occurs when metabolites diffuse out of the cell into the quenching solution. Hence quenching conditions have to be thoroughly optimized for the investigated organism with respect to leakage. The separation of the biomass from supernatant after quenching and washing of the biomass before extraction are also very critical steps with regard to accurate quantification.

The choice of adequate quenching and extraction methods, which can also be combined in a single step, depends on the investigated metabolites as well as on the investigated sample. The investigation of tissues (e.g., leaves, muscles) relies on harsh quenching and extraction such as freezing with liquid nitrogen and grinding with a pestle. Biotechnological samples, derived from microorganisms grown in suspension in bioreactors, are often quenched by sampling the culture broth directly into a cold quenching solution [22]. The range of investigated microorganisms spans gram-negative and gram-positive bacteria, yeasts, and filamentous fungi. Fragile mammalian cells require again specialized techniques, which consider the lack of cell walls and the fact that these cells are often grown adherently [23, 24]. All these cells differ in their physiological background, in particular, in the individual cell membrane composition and the existence of a cell wall, as well as, for example, their surface-to-volume ratio. Hence, various quenching and extraction protocols need to be established depending on the cell type, to which they are applied. After separation of the quenched cells from the supernatant either by centrifugation or filtration and washing the biomass, the cells are subjected to extraction, which is besides rapid sampling and quenching one of the most critical steps. During extraction, the cell walls are chemically and/or thermally permeabilized or mechanically disrupted to exhaustively extract the relevant metabolites into the liquid phase.

In this context, it is important to distinguish extraction efficiency, also called extraction efficacy and extraction recovery. The extraction efficiency is the ability of a certain extraction method to release metabolites from the cells and can – owing to the lack of reference materials – only be obtained by direct comparison of extraction methods based on different physical and chemical principles [25–27]. The extraction recovery, on the other hand, is obtained from spiking experiments, for example, by spiking isotopically labeled metabolites to the sample before and after extraction [25]. Extraction recovery takes into account, for example, losses of solvent, losses due to adsorption or degradation, or interconversion of metabolites due to the presence of not fully inactivated enzymes, but does not consider incomplete extraction. To prevent enzymatic and nonenzymatic decay after quenching, sample handling and storage are generally conducted at subambient temperatures. Thus comprehensive sample preparation protocols have recently been developed and are currently in use in order to make cultivated microbes accessible to LC-MS or GC-MS techniques for metabolite quantification. For detailed information, the reader is referred to comprehensive reviews on this area [20, 21].

Table 3.2 gives an overview of some recent articles on methods for quenching and extraction of microbial samples and a summary of the reported standard uncertainties derived from biological replicates. For yeasts, for example, the prominent technique for quenching is sampling into cold methanol solution [22, 25, 28–30]. For cell extraction, cold (subambient temperatures) as well as hot extraction techniques (e.g., boiling ethanol) are commonly used [22, 25, 28].

As pointed out in Section 3.1, the total combined uncertainty of a quantitative value is not revealed by repeating experiments. In Table 3.2, the reported and experimentally assessed standard uncertainties do not reflect contributions of leakage, extraction efficiency, storage stability, and sometimes also extraction recovery to the total combined uncertainty. We think, however, that it is an absolute demand assessing the uncertainty budget in this complex quantification task. Only in this way, it can be evaluated whether differences in measured metabolite concentration levels are significant. Moreover, the basic calculations offer the tools needed to decide which step is a major contribution to overall uncertainty and bias and hence enable well-directed measures for improving sample preparation protocols. Despite these facts, metrological considerations are rarely discussed in the field. In row 2 and 3 of Table 3.2, different approaches of sample preparation are compared in terms of repeatability. Two procedures aiming at the determination of intracellular metabolites were studied thereby specifically addressing the leakage problem: (i) a differential technique where intracellular metabolite levels are obtained from metabolite measurement in total broth and extracellular liquid and (ii) direct measurement of intracellular metabolite levels [22, 25]. As can be seen in

3.2 Uncertainties of Quantitative MS Experiments | 51

Table 3.2 Uncertainties reported in selected recent publications for common techniques in quantitative microbial metabolomics.

Microorganism	Sampled amount	Quenching	Extraction	Measured metabolites	Intracellular metabolite levels	Reported uncertainty	References
Pichia pastoris	0.63 ± 0.01 g	60% (v/v) MeOH, $-27\,°C$	75% (v/v) EtOH, $95\,°C$	37 (central carbon metabolites, amino acids)	Direct measurement: $0.06-200$ µmol · g_{CDW}^{-1} differential: $0.05-208$ µmol · g_{CDW}^{-1}	Repeatability (n not stated direct measurement: $0.4-26\%$ RSD[a] differential: $1.3-80\%$ RSD[a]	Carnicer et al. [28]
Saccharomyces cerevisiae	1.2 ± 0.6 g	100% (v/v) MeOH, $-40\,°C$	Comparison of five extraction methods	44 (central carbon metabolites, nucleotides, and amino acids)	chemostat: $0.1-122$ µmol · g_{CDW}^{-1} batch: $0.06-55$ µmol · g_{CDW}^{-1}	Repeatability of four samples analyzed twice $0.9-20\%$ RSD[a] (chemostat) $2.9-31\%$ RSD[a] (batch)	Canelas et al. [25]
Saccharomyces cerevisiae	1 ± 0.05	100% (v/v) MeOH, $-40\,°C$	75% (v/v) EtOH, $95\,°C$	34 (central carbon metabolites, amino acids)	$0.113-174$ µmol · g_{CDW}^{-1}	Repeatability of at least two samples analyzed twice $<0.3-19\%$ RSD[a]	Canelas et al. [22]

(continued overleaf)

Table 3.2 (Continued)

Microorganism	Sampled amount	Quenching	Extraction	Measured metabolites	Intracellular metabolite levels	Reported uncertainty	References
Escherichia coli	sampled cell volume is calculated	combined quenching and extraction ACN/MeOH/H_2O = 40 : 20 : 20 (v/v/v), 0.1% (v/v) HCOOH, $-20\,°C$		>100 metabolites	Amino acids: 1.82×10^{-5} – 9.60×10^{-2} mol·l^{-1}	Amino acids: 2.7–113% RSU^b	Bennet et al. [31]
Penicillium chrysogenum	1 ± 0.05 ml, quantified by weighing	60% (v/v) MeOH, $-40\,°C$	75% (v/v) EtOH, 95 °C	15 (central carbon metabolites, adenine nucleotides)	IDMS: 0.14–6.25 µmol·g_{CDW}^{-1} no IDMS: 0.14–4.92 µmol·g_{CDW}^{-1}	Repeatability of eight samples 3.9–14% RSD^a (IDMS) 10–29% RSD^a (no IDMS)	Nasution et al. [29]
Corynebact glutamicum	5 ml	Omitted; fast filtration	150 µm amino-butyrate, 15 min, 100 °C	11 amino acids	0.3–385 µmol·g_{CDW}^{-1}	Repeatability (n not stated 3.4–68% RSD^a	Wittmann et al. [30]

[a] Relative standard deviations were calculated from reported standard deviations.
[b] Relative standard uncertainties were calculated from reported 95% confidence intervals of the concentration.

row 2 and 3 of Table 3.2, higher relative standard deviations (RSDs) are obtained for the differential technique, which are due to the introduction of additional sources of error. Nasution et al. [29] evaluated the effect of internal standardization by adding a fully ^{13}C labeled yeast extract as internal standard before metabolite extraction. His data (row 7) demonstrate the usefulness of internal standardization not only in terms of bias correction (partial degradation, nonlinearity of the LC-MS method) but also in terms of improved standard uncertainties.

In the following sections, an exemplary uncertainty budget calculation is given for a typical metabolite quantification procedure.

3.2.1.2 Calculation of Sample Preparation Uncertainty for Intracellular Metabolite Quantitation in Yeast: A Practical Example

General Sample Workflow for Quantification of Intracellular Metabolites Each step in sample preparation contributes to the MU (to a different degree). In this practical example, the identification of uncertainty sources is based on the workflow for sample preparation as an alternative approach to CaEDs.

The common workflow for the analysis of intracellular metabolites consists of six major steps as shown in Figure 3.2: (i) sampling from the bioreactor, (ii) quenching of enzymatic activity, (iii) separation of biomass and supernatant and washing of the biomass, (iv) metabolite extraction from the cell, (v) removal of organic solvent, and (vi) reconstitution, dilution, or enrichment. Deviations from this general workflow are common. Further steps might need to be included, such as the addition of isotopically labeled internal standards before extraction (as is shown in Figure 3.2) or derivatization before GC-MS analysis.

The addition of internal standards is a common approach in quantitative LC-MS analysis in order to overcome errors from matrix-dependent ionization efficiency. If the internal standard is added before extraction, however, certain systematic errors stemming from extraction and sample preparation steps following extraction are also corrected. The metabolite concentrations in the reconstituted extracts, which finally undergo LC-MS or GC-MS analysis, are usually linked to the amount of sampled biomass, to cell dry weight (CDW), protein concentration, or cell

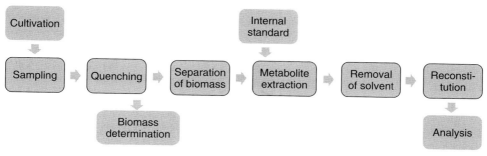

Figure 3.2 Example of a sample workflow.

number. Steps such as gravimetric CDW determination (of the culture broth or of the quenched cells) or determination of protein content hence also need to be considered as sources of uncertainty, which contribute to the combined standard uncertainty. Removal of the extraction solvent and subsequent reconstitution with other solvents are additional steps that might aim to a solvent change before LC-MS or GC-MS analysis or to preconcentration of the extract.

Sample Preparation Protocol for the Uncertainty Estimation The quantitation of an intracellular metabolite of the central carbon metabolism is in the focus of this uncertainty estimation. The metabolite is extracted from yeast cells cultivated under defined growth conditions. The cells are sampled, quenched, separated from the supernatant and washed immediately. The cell pellet is then spiked with an internal standard solution containing the same metabolite enriched with ^{13}C atoms. Extraction is performed, the supernatant is separated from the cell debris, and the extraction solvent containing the compound is evaporated to complete dryness of the sample. Before LC-MS/MS analysis, the dried sample is reconstituted with water and a dilution step is performed. The CDW of the quenched cell suspension is determined gravimetrically.

Derivation of the Model Equation and Definition of the Measurand All sources of uncertainty for the sampling protocol described earlier are defined and quantified following the bottom-up approach. A model equation is developed, which considers all uncertainty sources in sample preparation but excludes those of the subsequent analytical measurement techniques. It illustrates the reduction of combined uncertainty by the use of isotopically labeled internal standard during sample preparation.

Equation (3.3) describes how the concentration of a metabolite in the organic cell extract, c_{Extr} (µmol × l^{-1}) is given by the intracellular concentration of the metabolite (c_{Cell}) under investigation as well as sample preparation parameters. The latter are extraction efficiency (E_{extr}), which is defined as the extractable fraction of the respective metabolite, and its extraction recovery R_{extr}. In addition, the actually obtained concentration c_{Extr} depends on the volume of quenched cell suspension, which is sampled (V_{sample}, culture broth + quenching solution), the CDW ($g_{CDW} \times l^{-1}$) of the quenched sample volume, the volume of internal standard (V_{IS}), and the volume of the organic extraction solvent (V_{Extr}).

The factor L introduced in Eq. (3.3) takes into account uncertainties due to leakage during quenching and washing. Incomplete extraction, degradation, and interconversion of metabolites during extraction are taken into account by the extraction efficiency E_{extr} and the extraction recovery R_{extr}. Determination of the CDW introduces uncertainties linked to the volumetric and gravimetric measurements, while all volumetric dosages of sample, extraction solvent and volume of internal standard introduce uncertainties, which are related to volumetric measurements.

$$c_{Extr} = c_{Cell} \times E_{extr} \times CDW \times \frac{V_{sample}}{(V_{Extr} + V_{IS})} \times L \times R_{extr} \qquad (3.3)$$

where

c_{Extr} = concentration of the compound in the cell extract ($\mu mol \times l^{-1}$)
c_{Cell} = intracellular concentration of the metabolite ($\mu mol \times g_{CDW}^{-1}$)
E_{extr} = extraction efficiency
CDW = cell dry weight of quenched cell suspension ($g_{CDW} \times l^{-1}$)
V_{sample} = volume of quenched cell suspension, which is sampled (liter)
V_{Extr} = volume of solvent used for extraction (liter)
V_{IS} = volume of internal standard (liter)
L = leakage factor
R_{extr} = extraction recovery.

A minor loss of extract results from the separation of the extract from the cell debris by centrifugation and decanting. The volume being recovered, $V_{Extr\ recov}$, is subsequently evaporated, and the dried extract is redissolved in a defined volume of water $V_{reconst}$. The reconstituted extract is further diluted to reach the concentration levels suitable for LC-MS or GC-MS measurements (dilution factor d). The concentration of the metabolite in the solution, which is finally subjected to analysis, c_{final} ($\mu mol \times l^{-1}$) is hence described by Eq. (3.4). Compound losses during evaporation are expressed by the recovery factor R_{evap}.

$$c_{final} = c_{Extr} \times \frac{V_{Extr\ recov}}{V_{reconst}} \times R_{evap} \times d \qquad (3.4)$$

where

c_{final} = final concentration of the metabolite in the measurement solution ($\mu mol \times l^{-1}$)
$V_{Extr\ recov}$ = recovered volume of extract after decanting (liter)
$V_{reconst}$ = volume of sample solution after reconstitution of dried extract (liter)
R_{evap} = recovery of the drying step
d = dilution factor.

The combination of Eqs. (3.3) and (3.4) gives the final Eq. (3.5) for tracking the metabolite concentration from the cell to the solution, which is finally measured.

$$c_{final} = c_{Cell} \times E_{extr} \times CDW \times \frac{V_{sample} \times V_{Extr\ recov}}{(V_{Extr} + V_{IS}) \times V_{reconst}} \times L \times R_{extr} \times R_{evap} \times d \qquad (3.5)$$

An analogous equation (Eq. (3.6)) can be developed for the final concentration of the isotopically labeled IS in the vial ($c_{final\ IS}$) in $\mu mol \times l^{-1}$ in which n_{IS} is the amount of internal standard (μmol) spiked to the sample before extraction

$$c_{final\ IS} = n_{IS} \times \frac{V_{Extr\ recov}}{(V_{Extr} + V_{IS}) \times V_{reconst}} \times R_{extr} \times R_{evap} \times d \qquad (3.6)$$

where

$c_{vial\ IS}$ = concentration of internal standard in the sample vial ($\mu mol \times l^{-1}$)
n_{IS} = amount of internal standard spiked before extraction (μmol)

The internal standardization allows for correction of the fluctuation of compound concentration that occurs after spiking. This can easily be demonstrated by dividing

Eq. (3.5) by Eq. (3.6)

$$\frac{c_{final}}{c_{final\ IS}} = \frac{c_{Cell} \times E_{extr} \times CDW \times \frac{V_{sample} \times V_{Extr\ recov}}{(V_{Extr}+V_{IS}) \times V_{reconst}} \times L \times R_{extr} \times R_{evap} \times d}{n_{IS} \times \frac{V_{Extr\ recov}}{(V_{Extr}+V_{IS}) \times V_{reconst}} \times R_{extr} \times R_{evap} \times d} \tag{3.7}$$

$$\frac{c_{final}}{c_{final\ IS}} = c_{Cell} \times E_{extr} \times CDW \times \frac{V_{sample}}{n_{IS}} \times L \tag{3.8}$$

The resulting Eq. (3.8) shows that the ratio $c_{final}/c_{final\ IS}$ becomes independent of many uncertainty sources (volume changes, recovery of evaporation, etc.), which do effect both c_{final} and $c_{final\ IS}$. Sources of uncertainty that cannot be compensated by the use of internal standards are (i) source sconnected to the determination of the biomass sampled (here CDW of the quenched solution and V_{sample}), (ii) leakage during quenching and washing, (iii) uncertainties introduced with the addition of the internal standard (uncertainty connected with the volumetric dosage), and (iv) extraction efficiency.

In the calculation carried out above, it is assumed that the extraction recovery R_{extr} is the same for the *intracellular* metabolite and for the isotopically labeled standard, which is spiked *onto* the sample.

Finally, the repeatability factor K_{rep} is included resulting in Eq. (3.7). This factor groups the uncertainty deriving from repetitions of the sample preparation procedure (repeatability of the sample preparation procedure). K_{rep} is generally affected by fluctuations such as temperature (e.g., temperature of quenching and extraction), time (e.g., time of quenching, washing, and extraction), or composition of the extraction solvent and does not include repeatability of LC-MS measurements. In addition, the intracellular concentration and the extraction efficiency are grouped in a single factor ($c_{released} = c_{cell} \times E_{extr}$), with $c_{released}$, being the intracellular concentration of the metabolite released by extraction.

$$\frac{c_{final}}{c_{final\ IS}} = c_{released} \times CDW \times \frac{V_{sample}}{n_{IS}} \times L \times K_{rep} \tag{3.9}$$

where

K_{rep} = repeatability factor for sample preparation

$c_{released}$ = intracellular concentration of the metabolite released by extraction ($\mu mol \times g_{CDW}^{-1}$).

Equation (3.9) is the final model equation of the measurand. It expresses the ratio of the metabolite concentration to the internal standard concentration $c_{final}/c_{final\ IS}$ in the solution, which is subjected to analysis.

Uncertainty Estimation For the subsequent calculation of the combined standard uncertainty, all identified sources of uncertainty, which are present in the model equation, are characterized and quantified (Table 3.3). The example concerns the quantification of proline extracted from *Pichia pastoris* according to the optimized procedures published by Carnicer et al. [28] (quenching of *P. pastoris*) and Canelas et al. (extraction) [25].

3.2 Uncertainties of Quantitative MS Experiments

Table 3.3 Characterization and quantification of uncertainty sources for sample preparation.

Input quantities	Values	Unit	RSD	Uncertainty	Unit	Type	Distribution	Standard uncertainty
$c_{released}$	23.10	$\mu mol \cdot g_{CDW}^{-1}$	5[a]	1.155	$\mu mol \cdot g_{CDW}^{-1}$	A	Normal	1.155
CDW	4.50	$g_{CDW} \cdot l^{-1}$	2.20[b]	0.099	$g_{CDW} \cdot l^{-1}$	A	Normal	0.0990
V_{sample}	0.002	l	—	2.0×10^{-5}[c]	l	B	Triangular	8.16×10^{-6}
n_{IS}	0.200	μmol	—	0.005[c]	μmol	B	Triangular	0.0020
L	1.00[d]	—	—	0.02[d]	—	B	Rectangular	0.012
K_{rep}	1.00	—	5.0	0.05[e]	—	A	Normal	0.050

[a] Note: repeatability of extraction is not included in this parameter, instead it is included in K_{rep}.
[b] On the basis of the repeatability of the CDW determination, $n = 5$.
[c] Uncertainty derived from the manufacturers specification of the pipette.
[d] Experiments in our laboratory addressing leakage showed that its contribution is negligible if quenching is optimized and filtration is performed instantaneously. Leakage of proline was found to be 2%.
[e] $n = 3$.

Calculation of Combined Uncertainty and Expanded Uncertainty Combined uncertainty and expanded uncertainty are calculated based on the data given in Table 3.3 using the Kragten approach (Table 3.4).

Figure 3.3 shows that the major contribution to uncertainty stems from two input quantities, K_{rep} and $c_{released}$ (>40% each). The uncertainty in $c_{released}$ is solely attributed to the extraction efficiency E_{extr}, as the fluctuations in intracellular concentration are included in K_{rep}. Extraction efficiency is known to be a very critical parameter, and although its uncertainty is set to a value as low as 5% for this example, it does form a major contribution to combined uncertainty.

Table 3.4 The Kragten spreadsheet for sample preparation uncertainty propagation.

Input quantities →		$c_{released}$	CDW	V_{sample}	n_{IS}	L	K_{rep}
Standard unc →		1.155	0.099	8.16×10^{-6}	0.002	0.012	0.050
$c_{released}$	23.10	24.25	23.10	23.10	23.10	23.10	23.10
CDW	4.50	4.500	4.599	4.500	4.500	4.500	4.500
V_{sample}	0.002	0.002	0.002	0.002	0.002	0.002	0.002
n_{IS}	0.200	0.200	0.200	0.200	0.202	0.200	0.200
L	1.00	1.00	1.00	1.00	1.00	1.01	1.00
K_{rep}	1.00	1.00	1.00	1.00	1.00	1.00	1.05
$c_{final}/c_{final\,IS}$	1.04	1.0915	1.0624	1.0437	1.0290	1.0515	1.0915
Dev →	—	0.052	0.023	0.004	−0.011	0.012	0.052
Sum Dev² →	0.0062	0.0787	← Total combined uncertainty				
K →	2	0.16	← Expanded uncertainty				
		15%	← Relative uncertainty				

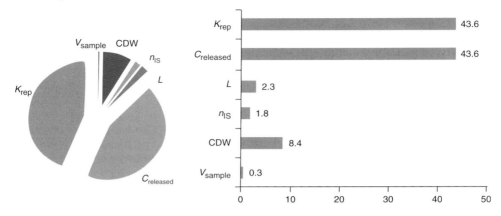

Figure 3.3 Relative contribution to combined standard uncertainty.

It has to be pointed out, however, that these calculations are based on the uncertainties of thoroughly optimized conditions and that the amino acid proline, which was chosen as example of this uncertainty calculation, is (i) not prone to leakage and is (ii) easy to extract, which is underlined by the range of diverse methods, which all show the same extraction efficiency [25].

For metabolites, which are more difficult to extract, the influence of extraction efficiency to total combined uncertainty can be expected to be the major contribution to uncertainty. The relevance of extraction efficiency in metabolomics has recently been underlined by Duportet *et al.* [32]. On the basis of quantitative metabolite data derived with different extraction methods, this study clearly shows how the biological interpretation of the results is impacted by the extraction method used. In the uncertainty estimation described earlier, the contribution of leakage to total combined uncertainty was shown to be negligible. Nevertheless, leakage does become a major contribution if quenching and washing conditions are not thoroughly optimized.

Uncertainty increases even further if metabolites are prone to interconversion, which is, for example, the case for the redox pair glutathione (GSH) and glutathione disulfide (GSSG). Glutathione is easily oxidized to GSSG. As intracellular GSH concentration levels exceed GSSG levels by a factor of 20 and more in microbial cells, GSSG levels can be altered significantly, even if only a small fraction of glutathione is oxidized. Derivatization of GSH is the most common approach to prevent this interconversion reaction. It is frequently overlooked, however, that derivatization is often not quantitative and oxidation of residual GSH still has the potential to significantly increase GSSG levels. The uncertainty that arises from this interconversion hence contributes to total combined uncertainty and needs to be included in uncertainty estimations for GSSG analysis.

Reporting Uncertainty

$$\text{Measurand}: c_{\text{final}}/c_{\text{final IS}} = 1.04 \frac{\mu\text{mol}_{\text{metabolite}} \times l^{-1}}{\mu\text{mol}_{\text{IS}} \times l^{-1}}$$

$$\text{Expanded uncertainty } (k=2): U = 0.16 \frac{\mu\text{mol}_{\text{metabolite}} \times l^{-1}}{\mu\text{mol}_{\text{IS}} \times l^{-1}}$$

3.2.1.3 LC-MS

LC-MS is one of the most widely used analytical techniques in both targeted and untargeted metabolite analysis. Usually, LC-MS systems consist of many different components that can affect analysis and hence influence results. To evaluate the performance of a given LC-MS method, the calculation or estimation of its uncertainty is a powerful tool. Usually, analytical performance is evaluated by carrying out method validation in terms of intermediate precision, accuracy, and/or robustness. These data, however, give little or even no information about the contribution of analytical instrumentation to overall uncertainty.

Various sources of uncertainty arise from both HPLC and MS systems. Major sources in HPLC systems are variations of pumps, ovens, injectors leading to fluctuations of mobile phase composition, column temperature, and injection volumes, respectively. Mobile phase composition and flow rate are identified as the most important parameters affecting retention times and peak widths in gradient elution [33]. The imperfect functioning of the pump or electronics can cause random and/or systematic deviations from preset mobile phase composition. The capacity factor is significantly affected by the changes in the mobile phase therefore small changes in mobile phase composition leads to large deviation of the retention time. Compared to manual mixing of the mobile phase, online mixing is recommended in order to minimize variations in mobile phase composition. Manual preparation can be more accurate in the case of isocratic elution, if the mobile phase contains less than 10% of any one solvent and especially if a low flow rate is being used, Moreover the separation factors of acidic or basic components are also significantly affected by the pH of the mobile phase.

In partition and adsorption chromatography, column temperature is known to affect solute diffusion and mass transfer as well as mobile phase viscosity in such a way that the retention times usually decrease - or fluctuate - as the temperature increases. It is estimated that, in order to keep the retention time stable within $\pm 1\%$, it is necessary to maintain the temperature stable within $\pm 0.35\,^\circ\text{C}$ [33]. However, small changes in retention time are critical in terms of compound identification while peak areas are in most cases not affected significantly.

Another source of variability is the injection system. The inherent variability in the position of the syringe plunger position occurs whenever it is moved by the autosampler. This can lead to variation in the injection volume and hence in the corresponding peak areas. Therefore the appropriate volume of the syringe needs to be selected in order to reduce injection systems contribution to the uncertainty of the measured peak area. It is therefore important to select a syringe of appropriate volume; otherwise the injection system may contribute significantly to the uncertainty in measured peak areas. Two measures to mitigate or eliminate this source of uncertainty are full loop injection or the implementation of an internal standard, ideally an isotopically labeled standard.

Mass spectral overlap is another source of uncertainty in LC-MS. When using internal standardization, as described earlier, mass spectral overlap between the isotopically labeled standard and the analyte may occur depending on the isotopic purity of the internal standard in use. This has also to be kept in mind, when using uniformly ^{13}C (U^{13}C) labeled cell extracts as internal standards [34], which rapidly becomes more and more widespread. The approach currently being followed is to improve the isotopic purity of the ^{13}C standard by preventing unnecessary incorporation of ^{12}C, when cultivating U^{13}C-labeled cells for standard production (omission of unlabeled nonessential vitamins, removal of ^{12}CO$_2$ from the air [35]). Nevertheless, this uncertainty source has to be critically evaluated especially for metabolites with a low number of carbon atoms, which are more prone to a mass spectral overlap of metabolite and impurities of the ^{13}C-labeled internal standard.

The integration process contributes to uncertainty of measured peak areas; baseline noise and the identification of the beginning and end of the peak are examples of the possible sources of variations linked to this factor. These two factors that affect accuracy and precision of peak area measurement are interrelated because the background noise blurs the base of the peak making it difficult for the integrator to identify the beginning and end of the peak. The relative error of the integrated area because of noise is inversely proportional to the signal-to-noise ratio: the higher the noise, the poorer the precision of the measured area. This error will be systematic, if the integration parameters are set incorrectly. Correct setting of integration parameters is especially important for incompletely resolved peaks. The vertical drop method, which is commonly used to integrate incompletely resolved peaks, is a major source of uncertainty [33, 36].

The estimation of the uncertainty of the peak area includes contributions from all uncertainty sources that arise from instrumentation. In the case of concentration sensitive detectors such as mass spectrometers coupled to LC by electrospray ionization, the peak area of the analyte (A) depends on the detector sensitivity (S_c), amount of the injected analyte (m), and mobile phase velocity (F) according to the following equation (Eq. (3.10)):

$$A = \frac{S_c \times m}{F} \tag{3.10}$$

The amount of the injected analyte depends on the repeatability of the injection volume, cross contamination, and temperature effects on the injection volume. The mobile phase velocity F depends on the variation of the mobile phase flow rate, mobile phase composition, and column temperature. Detector sensitivity S_c depends on the repeatability and drift of readings as well as the repeatability and precision of detector settings [37].

According to Leito et al. [37], the expression of the uncertainty of peak area u(A) includes several components that can be grouped and expressed as follows:

for standards,

$$u(A_{std}) = \sqrt{u(A_{std-rep})^2 + u(A_{std-drift})^2 + u(A_{std-integr})^2} \quad (3.11)$$

or alternatively for samples,

$$u(A_{sample}) = \sqrt{\begin{array}{l} u(A_{sample-rep})^2 + u(A_{sample-drift})^2 \\ + u(A_{sample-integr})^2 + u(A_{sample-nonlin})^2 \end{array}} \quad (3.12)$$

The repeatability component $u(A_{rep})$ groups the repeatability of the analytical platform including injection, variation in mobile phase flow rate, and composition and the detector.

The drift component $u(A_{drift})$ sums up the temporal drift of system parameters including the temperature drift of the system and drift of the mass accuracy of the mass spectrometer.

The nonlinearity component $u(A_{nonlin})$ is considered exclusively for $u(A_{sample})$. This component is generated when a linear calibration is used to describe a slightly nonlinear behavior. This uncertainty component is sometimes fairly important but is often neglected. Nonlinearity of the calibration curve is a systematic effect that, in principle, is sufficiently corrected by the use of multiple point calibrations. Therefore, this component only needs to be considered when a one point calibration is used [37].

3.2.2
Uncertainty of Mass Spectrometric Assays (LC-MS and GC-MS Measurements)

3.2.2.1 GC-MS

GC-MS can be regarded as key technology in metabolomics and fluxomics because of its suitability for a wide range of compounds, high separation efficiency, and the reproducibility of EI ionization. In terms of uncertainty, we would like to briefly discuss three substantial differences between GC and LC. (i) Derivatization is a prerequisite for GC analysis of polar compounds; (ii) injection includes vaporization of the sample for the transfer of the sample onto the column and is a crucial step in GC; and (iii) the mass spectra obtained with EI ionization are more complex, as EI is a hard ionization technique.

For GC-MS experiments, polar analytes need to be derivatized in order to convert them to nonpolar, volatile, and stable derivatives. Currently, trimethylsilylation with prior methoximation is the technique that is most widely used [38]. Trimethylsilylation replaces active hydrogen atoms in hydroxyl, carboxylic, amine, and thiol groups by the trimethylsilyl group, while methoximation converts ketone groups into the more stable and less polar methoxyamino groups.

Stability of the analytes might be significantly improved by derivatization (which is, for example, the case for sulfur compounds containing thiol groups), as well as selectivity and sensitivity. There are, however, several drawbacks of derivatization, which all contribute to increased uncertainty, incomplete derivatization, matrix-dependent derivatization kinetics, formation of more than one derivative from a single metabolite, formation of identical derivatives from different analytes (e.g., degradation of arginine in citrulline and ornithine), and artifact formation [39]. Automated just in time online derivatization has the potential to improve precision of repeated sample analyses by standardizing derivatization parameters such as derivatization time. However, it cannot compensate bias caused by these sources of systematic error.

Sample injection is another crucial step in GC, as the sample needs to be vaporized. Many factors (mode of injection, injector temperature, selection of the liner, sample volume, carrier gas flow rate, head pressure, contamination, or insufficient inertness of the liner, etc.) influence the transfer of the analyte onto the column. Sources of uncertainty for gas chromatographic analysis in general including a comprehensive section on injection have been reviewed by Barwick [33].

The sample matrix in metabolomics is highly concentrated and complex causing buildup of debris in the liner of the injection port. In addition, many metabolites are thermolabile. Injection systems with automated liner exchange and programmed temperature injection hence considerably expand the applicability of GC-MS in metabolomics [40]. As in LC-MS, uncertainty of injection can generally be improved by the use of isotopically labeled standards.

Currently, EI ionization is used nearly exclusively in metabolomics. While this technique has several advantages (such as highly reproducible spectra, which can be searched in databases [41]), it has to be kept in mind that EI is a hard ionization technique leading to extensive fragmentation of the analyte. Hence only low-intensity ions might be available for selective detection [42], and mass spectra are more complex, which in turn leads to mass spectral overlaps being more likely than with soft ionization techniques. With the advent of MS/–MS technology for GC, this uncertainty source can be expected to be of less relevance. The model equation (Eqs. (3.16) and (3.20)) developed for uncertainty estimation of LC-MS measurements also hold true for GC-MS analysis but has to be expanded by two additional factors, which cover the uncertainty of derivatization and injection, in particular.

Under well-optimized conditions and using internal standards, uncertainty of injection contributes little to total combined uncertainty. The uncertainty contribution of derivatization, however, needs to be carefully evaluated. Owing to the complexity of derivatization, this sample preparation step can be the major source of uncertainty depending on the type of analyte under investigation. This is especially true for metabolites, which form more than one derivative from a single metabolite. Kanani and Klapa [43] strikingly showed in a recent study how the standard uncertainty of derivatization time can be decreased by a factor of approximately 10 for these analytes. The approach is characterized by a data evaluation strategy based on weighted summation of the peak areas of all derivatives, instead of data evaluation of a single derivative. The study clearly shows how effective uncertainty can be improved by focusing on critical steps in the analytical procedure.

3.2.2.2 Calculation of Uncertainty for LC-MS Measurements of Cell Extracts: A Practical Example

In the following section, uncertainty calculation is performed for a metabolite from the central carbon metabolism. The parameters for LC-MS/MS analysis are optimized in order to obtain optimal peak shape, high sensitivity, and a sufficiently high data acquisition rate. The quantification of the compound is performed using linear regression (six calibration levels). Internal standardization with isotopically labeled standards is used leading to an area ratio for each standard and sample.

All the calculations are carried out as described in the Chapter 1.

Measurand Definition LC-MS-based quantification of intracellular metabolites addresses the determination of the concentration of metabolites in the cell extract. The concentration of the cell extract subjected to analysis is later on linked with factors such as the amount of sample, extraction recovery, sample dilution, or sample concentration in order to obtain a final result of the metabolite in the investigated sample, for example, micromoles metabolite per gram cell dry weight (μmol g_{CDW}^{-1}). For this uncertainty estimation of the LC-MS analysis, however, these uncertainty sources are not taken into account and the *measurand* is defined as the concentration of the studied metabolite in the extract (c_i). Both uncertainty sources, that is, sample preparation and LC-MS measurement, will be finally combined in order to give the total combined uncertainty of the entire procedure (Section 3.1.2.2).

Model Equation Sample peak area and the parameters of the calibration curve are used to calculate the analyte concentration in the sample extract. If there is linear dependence between the analyte concentrations $C_1 \ldots C_n$ and the respective analytical signal (e.g., peak area) $A_1 \ldots A_n$, the regression line equation can be written as follows:

$$A_i = C_i \times B_1 + B_0 \tag{3.13}$$

where B_0 and B_1 are intercept and slope, respectively. Using this equation, the concentration can be expressed as follows:

$$C_i = \frac{(A_i - B_0)}{B_1} \tag{3.14}$$

When using internal standards, however, these peak areas, the ratio between the area of the analyte (A_{sample}) and the internal standard (A_{IS}) is used for data evaluation.

$$R_s = \frac{A_{sample}}{A_{IS}} \tag{3.15}$$

Combining Eqs. (3.12) and (3.13), the final definition of the measurand is derived as follows:

$$C_i = \frac{R_s - B_0}{B_1} \qquad (3.16)$$

Uncertainty Estimation for Components In this example, we estimate the uncertainty of the different input quantities from experimental observations. To estimate the uncertainty of peak area, values obtained from repeated sample injections are used. The standard deviation (Eq. (3.17)) is used as uncertainty for experimentally obtained values because normal distribution can be assumed for these data.

$$s = \sqrt{\frac{\sum_{i=1}^{n}(x_i - x)^2}{n-1}} \qquad (3.17)$$

Accordingly, the intercept (B_0) and slope (B_1) values are calculated using Eqs. (3.17) and (3.18) from repetitive calibrations ($n = 3$). Standard deviations of the area ratios R_{Stdi} of each standard i representing the repeatability precision obtained within a 24 h sequence are calculated and converted to standard uncertainties. Average values are used to calculate the final result. The calibration standards were obtained via stepwise dilution (dilution of solid standard followed by three intermediate dilutions). The limits of the weighting step was set to 0.1% (standard uncertainty calculated for rectangular distribution), and the uncertainty of each pipetted volume was set to 0.1% (standard uncertainty calculated for triangular distribution).

$$B_0 = \frac{R \times \Sigma_i C_i^2 - C \times \Sigma_i C_i \times R_{Stdi}}{\Sigma_i C_i^2 - n \times \overline{C^2}} \qquad (3.18)$$

$$B_1 = \frac{\Sigma_i C_i \times R_{Stdi} - n \times R \times C}{\Sigma_i C_i^2 - n \times \overline{C^2}} \qquad (3.19)$$

Calculation of Total Combined Uncertainty and Uncertainty Budget The total combined uncertainty of the final result is calculated after estimation of the uncertainties of each component of the mathematical model. For calculations of the combined standard uncertainty, the Kragten approach was performed using the GUM Workbench pro [18] (www.metrodata.de). Results are given in the Tables 3.5 and 3.6.

The results clearly indicate that the weighting and pipetting steps are not contributing to the uncertainty of LC-MS quantification. As a matter of fact, the contribution of all those steps accounts to 0.2% of the total uncertainty. The repeatability precision of the area ratio in the sample and the uncertainty of the measured ratios of the standards are the major contributors.

Finally, we have integrated the uncertainty obtained for the sample preparation procedure described in Section 3.2.1.2 by introducing a factor S (Eq. (3.20)). This factor was set to 1 with a given standard uncertainty of 7.6%.

$$C_i = \frac{S \times R_S - B_0}{B_1} \qquad (3.20)$$

Table 3.5 Uncertainty budget for LC-MS analysis of proline.

Input quantity	Contribution
Area ratio std 1 (R_{Std1})	0.60%
Area ratio std 2 (R_{Std2})	2.20%
Area ratio std 4 (R_{Std4})	10.40%
Area ratio std 5 (R_{Std5})	25.60%
Area ratio std 6 (R_{Std6})	16.10%
Area ratio sample (R_S)	44.90%
Analyte concentration	1.89 µmol l^{-1}
Expanded uncertainty ($K = 2$)	0.13 µmol l^{-1}

Table 3.6 Uncertainty budget for sample preparation and LC-MS analysis of proline.

Input quantity	Contribution
Area ratio std 4 (R_{Std4})	1.10%
Area ratio std 5 (R_{Std5})	2.70%
Area ratio std 6 (R_{Std6})	1.70%
Area ratio sample (R_S)	4.70%
Sample preparation (S)	89.60%
Analyte concentration	1.89 µmol l^{-1}
Expanded uncertainty ($K = 2$)	0.37 µmol l^{-1}

Calculation of the total combined uncertainty resulted in the values above listed.

As can be readily observed, the major contribution to the total combined uncertainty is derived from the sample preparation. Increasing the precision of MS detection thereby reducing the uncertainty of calibration and MS determination of the analyte will not significantly affect the uncertainty of the quantification task. Moreover, it has to be mentioned that in the case of proline, the excellent extraction efficiency and hence excellent extraction repeatability represents an ideal case. Consequently, for other analytes this contribution might be even higher.

3.3
Concluding Remarks

Accurate quantitative metabolite profiling in cell factories by mass spectrometric assays is a highly challenging endeavor. The assessment of uncertainty budgets, which is neither an experimental task nor a purely mathematical operation, is an important analytical exercise in this field. All possible input values contributing finally to a total combined uncertainty are critically evaluated and considered when setting up a model equation. We could show that despite the fact that isotopically enriched standards were added to the samples at a very early stage of the analytical process and hence compensate for a multitude of factors, sample preparation was the major limitation for obtaining low uncertainties associated with the metabolite concentration values. As a matter of fact, this conclusion is also valid for untargeted strategies as employed in metabolite fingerprinting. Although no quantitative values or concentrations are determined in these assays, data is finally evaluated in a comparative way and the significance of two sets of data is strongly linked to their expanded uncertainty.

Abbreviations

APCI	Atmospheric pressure chemical ionization
CaED	Cause-and-effect diagram
CDW	Cell dry weight
CID	Collision-induced dissociation
CRM	Certified reference material
ESI	Electrospray ionization
FT-ICR	Fourier transform ion cyclotron resonance
GC	Gas chromatography
GSH/GSSG	Reduced/oxidized glutathione
GUM	Guide for uncertainty measurement
HPLC	High-performance liquid chromatography
ISO	International organization for standardization
IUPAC	International Union of Pure and Applied Chemistry
K	Coverage factor
LC	Liquid chromatography
LIT	Linear ion trap
LOC	Level of confidence
MS	Mass spectrometry
MU	Measurement uncertainty
QA	Quality assurance
QC	Quality control
RSD	relative standard deviation
RSS	Square root of the sum of the square
SD	Standard deviation
St.unc.	Standard uncertainty
TOF	Time-of-flight

U	Expanded measurement uncertainty
$u_c(y)$	Combined standard uncertainty
VIM	International Vocabulary of Basic and General Terms in Metrology

Acknowledgment

This work has been supported by the Federal Ministry of Economy, Family and Youth (BMWFJ), the Federal Ministry of Traffic, Innovation and Technology (bmvit), the Styrian Business Promotion Agency SFG, the Standortagentur Tirol and ZIT – Technology Agency of the City of Vienna through the COMET-Funding Program managed by the Austrian Research Promotion Agency FFG. EQ BOKU VIBT GmbH is acknowledged for providing LC-MS/MS and GC-MS/MS instrumentation

References

1. Fiehn, O. (2001) *Comp. Funct. Genomics*, **2**, 155–168.
2. Fiehn, O. (2002) *Plant Mol. Biol.*, **48**, 155–171.
3. Villas-Bôas, S.G., Mas, S., Åkesson, M., Smedsgaard, J., and Nielsen, J. (2005) *Mass Spectrom. Rev.*, **24**, 613–646.
4. Dettmer, K., Aronov, P.A., and Hammock, B.D. (2007) *Mass Spectrom. Rev.*, **26**, 51–58.
5. Allwood, J.W., Ellis, D.I., and Goodacre, R. (2008) *Physiol. Plant.*, **132**, 117–135.
6. Isaaq, H.J., Van, Q.N., Waybright, T.J., Muschik, G.M., and Veenstra, T.D. (2009) *J. Sep. Sci.*, **32**, 2183–2199.
7. Allwood, J.W. and Goodacre, R. (2010) *Phytochem. Anal.*, **21**, 33–47.
8. Barwick, V.J. and Prichard, E. (eds) (2011) *Eurachem Guide: Terminology in Analytical Measurement – Introduction to VIM 3*, ISBN: 978-0-948926-29-7. www.eurachem.org.
9. Ellison, S.L.R., Williams, A. (eds) (2012) *Eurachem/CITAC guide: Quantifying Uncertainty in Analytical Measurement*, Third edition, ISBN 978-0-948926-30-3. www.eurachem.org..
10. Hund, E., Massart, D.L., and Smeyer-Verbeke, J. (2001) *Trends Anal. Chem.*, **20** (8), 394–406.
11. Rozet E., Marini R.D., Ziemons E., Dewé W., Rudaz S., Boulanger B., and Hubert P. (2011) *Trends Anal. Chem.*, **30** (5), 797–806.
12. Rozet E., Rudaz S., Marini R.D., Ziemons E., Boulanger B., and Hubert P. (2011) *Anal. Chim. Acta*, **702**, 160–171.
13. Vanatta, L.E. and Coleman, D.E. (2007) *J. Chromatogr. A*, **1158**, 47–60.
14. White, G.H. (2008) *Clin. Biochem. Rev.*, **20**, 53–60.
15. Meyer, V.R. (2007) *J. Chromatogr. A*, **1158**, 15–24.
16. Ellison, S.L.R. and Barwick, V.J. (1998) *Accredit. Qual. Assur.*, **3**, 101–105.
17. Kragten, J. (1994) *Analyst*, **119**, 2161–2165.
18. Losinger, W.C. (2004) *Am. Stat.*, **38** (2), 165–167.
19. Schädel, F. and Franco-Lara, E. (2009) *Appl. Microbiol. Biotechnol.*, **83**, 199–208.
20. Mashego, M.R., Rumbold, K., de Mey, M., Vandamme, E., Soetaert, W., and Heijnen, J. (2007) *Biotechnol. Lett.*, **29**, 1–16.
21. van Gulik, W. (2010) *Curr. Opin. Biotechnol.*, **21**, 27–34.
22. Canelas, A.B., Ras, C., ten Pierick, A., van Dam, J.C., Heijnen, J.J., and van Gulik, W.M. (2008) *Metabolomics*, **4**, 226–239.
23. Sellick, C.A., Hansen, R., Maqsood, A.R., Dunn, W.B., Stephens, G.M., Goodacre, R., and Dickson, A.J. (2009) *Anal. Chem.*, **81**, 174–183.

24. Dettmer, K., Nürnberger, N., Kaspar, H., Gruber, M.A., Almstetter, M.F., and Oefner, P.J. (2011) *Anal. Bioanal. Chem.*, **399**, 1127–1139.
25. Canelas, A.B., ten Pierick, A., Ras, C., Seifar, R.M., van Dam, J.C., van Gulik, W.M., and Heijnen, J.J. (2009) *Anal. Chem.*, **81**, 7379–7389.
26. Sellick, C.A., Knight, D., Croxford, A.S., Maqsood, A.R., Stephens, G.M., Goodacre, R., and Dickson, A.J. (2010) *Metabolomics*, **6**, 427–438.
27. Ritter, J.B., Genzel, Y., and Reichl, U. (2008) *Anal. Biochem.*, **373**, 349–369.
28. Carnicer, M., Canelas, A.B., ten Pierick, A., Zeng, Z., van Dam, J., Albiol, J., Ferrer, P., Heijnen, J.J., and van Gulik, W. (2011) *Metabolomics*, doi: 10.1007/s11306-011-0308-1
29. Nasution, U., van Gulik, W.M., Kleijn, R.J., van Winden, W.A., Proell, A., and Heijnen, J.J. (2006) *Biotechnol. Bioeng.*, **94**, 159–166.
30. Wittmann, C., Krömer, J.O., Kiefer, P., Binz, T., and Heinzle, E. (2004) *Anal. Biochem.*, **327**, 135–139.
31. Bennet, B.D., Kimball, E.H., Gao, M., Osterhout, R., Van Dien, S.J., and Rabinowitz, J.D. (2009) *Nat. Chem. Biol.*, **5**, 593–599.
32. Duportet, X., MereschiAggio, R.B., Carneiro, S., and Villas-Boâs, S.G. (2011) *Metabolomics*, doi: 10.1007/s11306-011-0324-1
33. Barwick, V.J. (1999) *J. Chromatogr. A*, **949**, 13–33.
34. Mashego, M.R., Wu, L., van Dam, J.C., Ras, C., Vinke, J.L., van Winden, W.A., van Gulik, W.M., and Heijnen, J.J. (2004) *Biotechnol. Bioeng.*, **85** (6), 620–628.
35. Wu, L., Mashego, M.R., van Dam, J.C., Proell, A.M., Vinke, J.L., Ras, C., van Winden, W.A., van Gulik, W.M., and Heijnen, J.J. (2005) *Anal. Biochem.*, **336**, 164–171.
36. Norman D. (1998) *Chromatographic Integration Methods*, 2nd edn, Royal Society of Chemistry.
37. Leito, S., Mölder, K., Künnapas, A., Herodes, K., and Leito, I. (2006) *J. Chromatogr. A*, **1121**, 55–63.
38. Dunn, W.B. (2008) *Phys. Biol.*, **5**, 011001 (24 pp), doi: 10.1088/1478-3975/5/1/011001
39. Little, J.L. (1999) *J. Chromatogr. A*, **844** (1-2), 1–22.
40. Denkert, C., Budczies, J., Kind, T., Weichert, W., Tablack, P., Sehouli, J., Niesporek, S., Dominique Könsgen, D., Dietel, M., and Fiehn, O. (2006) *Cancer Res.*, **66** (22), 10795–10804.
41. Tohge, T. and Fernie, A.R. (2009) *Phytochemistry*, **70**, 450–456.
42. Radim, Š., Hajšlová, J., Kocourek, V., and Tichá, J. (2004) *Anal. Chim. Acta*, **520**, 245–255.
43. Kanani, H. and Klapa, M.I. (2007) *Metab. Eng.*, **9**, 39–51.

4
Gas Chromatography and Comprehensive Two-Dimensional Gas Chromatography Hyphenated with Mass Spectrometry for Targeted and Nontargeted Metabolomics

Song Yang, Jamin C. Hoggard, Mary E. Lidstrom, and Robert E. Synovec

4.1
Introduction and Scope

Researchers in the metabolomics field aim to identify and quantify low molar mass molecules found in a cell that support general cell life functions and participate in metabolic reactions [1–3]. This chemical information is then used to gain a better understanding of the metabolic pathways under investigation [4, 5]. To this end, researchers strive to develop and provide instrumental platforms and methods for analyzing metabolites present in a sample using comprehensive analytical techniques. The various techniques include UV, IR, FT-IR, and FT-ICR spectroscopies [6], nuclear magnetic resonance (NMR) [7, 8], liquid chromatography (LC) [9], capillary electrophoresis [10], and LC or gas chromatography (GC) combined with mass spectrometry (LC-MS and GC-MS) [5, 11–16]. Recently, comprehensive two-dimensional (2D) GC coupled to time-of-flight mass spectrometry (GC × GC–TOFMS) has emerged as a powerful chemical analysis tool to study complex samples [17–26], which are found in metabolomics investigations [22–30]. Many studies comparing GC × GC–TOFMS to one-dimensional (1D) GC–MS have demonstrated the superior advantages of GC × GC–TOFMS because of increased chromatographic resolution, peak capacity, and detection sensitivity. Nonetheless, all instrumental platforms presently available are challenged to meet the demanding criteria for current and emerging metabolomics studies: sufficiently broad coverage of the metabolome of the species investigated, unambiguous metabolite identification, good detection sensitivity, reliability, speed, cost, and ease of downstream data analysis. Correspondingly, more than one technology is often applied in concert in a cross-platform approach [31–35]. However, GC-based metabolomics analyses are very popular because many of these criteria are adequately met.

The goals and challenges for any metabolomics investigation are diverse. From the chemical perspective, it is important to know what researchers wish to learn about the biochemical system under investigation, and hence, what information needs to be obtained from the analytical data. From the technical perspective it is important to identify the experimental design, sample preparation, analytical

Metabolomics in Practice: Successful Strategies to Generate and Analyze Metabolic Data, First Edition.
Edited by Michael Lämmerhofer and Wolfram Weckwerth.
© 2013 Wiley-VCH Verlag GmbH & Co. KGaA. Published 2013 by Wiley-VCH Verlag GmbH & Co. KGaA.

measurement, and data analysis challenges that must be addressed to provide the desired information leading to an enlightened outcome [36]. In this chapter, we focus on GC–MS and GC × GC–TOFMS based instrumental platforms to better understand how to optimally apply these techniques and to assess the role these play in the field of metabolomics.

Broadly viewed, metabolomics studies can be categorized into three groups, targeted metabolomics, nontargeted metabolomics or a combination of targeted and nontargeted approaches. For targeted metabolomics, the analyst has predefined a list of metabolites of interest, standard solutions of the metabolites are generally available for quantification, and separation and detection protocols have often been previously worked out and are readily followed (e.g., most clinical lab applications). This is generally the case for hypothesis-driven research, in which the knowledge of the biochemical pathway indicates which metabolites are of interest for the study in hand, the specific metabolites are again targeted. An important area that often uses a targeted approach is "fluxomics," in which an isotopically labeled substrate is fed to the organism under investigation, and the uptake of the isotopic label is measured temporally in order to elucidate valuable biological pathway information [37–40]. The disadvantage of a targeted analysis is that information regarding nontargeted metabolites is unlikely to be discovered, unless by chance. In reality, many targeted studies aim toward discovery, with the constraint that only a priori selected metabolites are measured. For the purpose of this chapter, however, a discovery-based approach is defined and described in which the experimental design involves initial measurement of nontargeted metabolites. The analyst must design the experiments so as to facilitate the discovery of initially nontargeted metabolites that change from one experimental condition to the next, that is, upregulated or downregulated in concentration [26, 41] or temporally changing concentrations [42, 43]. For example, the biochemical question might be "what metabolites are upregulated or downregulated in a particular growth medium of yeast when comparing wildtype and mutant strains?" The experimental design would require growing the two yeast strains under the same conditions, preparing the samples for GC-based analysis, and then collecting and analyzing the data to find the upregulated or downregulated metabolites. These basic steps, in the context of targeted and nontargeted discovery-based metabolomics, are described in more detail later in this chapter, primarily in the data analysis section.

Many important challenges must be addressed to advance the field of GC-based metabolomics with MS detection. The metabolites must be made amenable to gas phase analysis at relatively high temperature (separations employing temperature programming from ∼50 to 350 °C). The metabolites must be derivatized to make them amenable to high temperature gas phase analysis with robust reagent and/or isotopic labeling chemistry [44–46], to facilitate quantitative measurements. Indeed, an important challenge in metabolomics is the lofty goal of unambiguously identifying and determining absolute concentration levels for all metabolites in a biological system. Many GC–MS and GC × GC–TOFMS studies still provide relative concentrations instead of absolute concentrations. Robust approaches to obtain absolute quantification are an ongoing challenge. For the GC and MS instrumental components, a set of very basic design considerations must be addressed in order to achieve superior chromatographic resolution, high peak capacity, fast separations, and low limits of detection (LODs). The challenges and trade-offs associated with addressing these instrumental considerations are described. Recent advances have been made in optimizing the performance of GC and MS instrumentation that strive to provide optimal data, combined with retention time and mass spectral libraries for positive metabolite identification [14, 47]. Finally, once the data has been collected, depending on the experimental design, the next step is data analysis to transform the experimental data to biologically relevant information [48–56].

In this chapter, we begin by describing sample preparation methods and associated challenges. We then cover GC–MS and GC × GC–TOFMS instrumentation for metabolomics. Next, the data analysis challenges and methods are described. Throughout these three sections, both targeted and nontargeted experimental design approaches are integrated. Finally, illustrative examples and concluding remarks are provided that point toward the future needs and challenges for GC-MS-based metabolomics.

4.2
Sample Preparation for GC-Based Metabolite Profiling

Development of robust and reliable sample preparation methods for metabolite profiling is a critical aspect of metabolomics for GC–MS and GC × GC–TOFMS instrumental methods [57]. Sample preparation methods are generally of two types, either involving sampling of the headspace vapor above a given (liquid or solid) sample [46] or more commonly, the isolation of metabolites directly from cells, fluids, plants, and so on [44, 58, 59]. For the first sample preparation type, a sample preparation method referred to as solid phase microextraction (SPME) is gaining wide acceptance because it is an elegant sampling approach providing both simplicity and excellent performance [46]. A SPME fiber coating is exposed to the headspace and sufficiently volatile metabolites adsorb or absorb to the coating. The SPME fiber is then inserted into the inlet of the GC whereby the metabolites desorb to form the injected sample pulse for the chromatographic separation. SPME readily provides a means to survey, and in many instances to quantify, metabolites in complex samples without generally requiring derivatization chemistry or requiring significant sample cleanup. The potential benefits of SPME for volatile metabolomics, such as the capture of short-lived and unstable metabolites, have been demonstrated for the detection of plant pathogenic infections, the evaluation of food quality, and the recognition of human diseases [60, 61]. Another example benefitting from the use of SPME involved collecting the vapor headspace of volatile or semivolatile metabolites above cacoa beans [62]. These reports indicate that a combination of different analytical techniques is often needed to sufficiently survey a broad range of unbiased metabolomic pathways.

As most common sample types are initially liquid or solid phase, a majority of the applications of GC-based metabolomics focus on isolation of metabolites directly from cells, or biologic fluids. An example of this common format of preparation would be sampling the metabolome of yeast cells. For this second sample preparation type, most of the metabolites may not be sufficiently volatile nor thermally stable for GC-based analysis. Three major steps must be carefully performed to analyze the metabolites in this case: quenching of metabolism inside the cells; extraction of all or the most abundant metabolites from the cells; and derivatization of polar functional groups that are thermally labile at the temperature or insufficiently volatile such that the derivatized metabolite products are thermally stable and made volatile for GC analysis. A preponderance of effort has been devoted to developing quenching and extraction protocols for mammalian cells, blood plasma, plants and microbes [44, 58, 59, 63]. Different sample types usually require the use of slightly different preparation protocols in order to optimize the conditions. Indeed, application of processing methods in the microbial biology field is growing in importance, providing insight into the metabolic pathways that direct metabolic engineering for producing many valuable primary metabolites. Below we use a microbial system as an example to demonstrate the sample quenching and extraction steps, along with the inherent challenges that must be met to achieve useful sample preparation results.

As the turnover rate of the microbial metabolites involved in central metabolism is in the range of seconds to minutes, rapid quenching for stopping the enzyme reactions and for keeping the cell membrane intact is essential to obtain an accurate snapshot of the metabolome. Cold-methanol/water mixture ($< -40\,°C$) is a widely applied protocol for yeast, and it has been further adapted for other microbes [64, 65]. However, as there are significant chemical and physical differences between the cell membrane structures of eukaryotic and prokaryotic organisms, this protocol has been shown to lead to severe leakage (>60%) of the metabolites (e.g., many amino acids and carboxylic acids) in both gram-positive bacteria (e.g., *Bacillus subtilis* and *Corynebacterium glutamicum*) and gram-negative bacteria (e.g., *Escherichia coli* and *Gluconobacter oxydans*) [63]. Many other methods including cold-glycerol/saline solution, cold-ethanol/sodium-chloride solution, fast filtration protocols, and an integrated sampling procedure have been developed to optimize quenching either in batch culture or bioreactor culture [63, 66–68]. These protocols can reduce the loss of metabolites tested in the specific microbes, but some of the protocols have other potential shortcomings. For example, it may be difficult to sufficiently remove the glycerol from the samples after quenching. It is also challenging to suitably apply filtration for dynamic experiments or steady-state measurements for a flux analysis.

Metabolites can be extracted from microbial cells using different protocols, such as those based on the use of boiling ethanol, cold methanol or acetonitrile (either alone or in combination with an aqueous buffer), freeze-thaw cycles in methanol, and percholoric acid [64]. It has been recently demonstrated that a boiling ethanol solution can achieve the best performance for both hydrophilic and hydrophobic metabolites simultaneously, in terms of efficacy and metabolite recoveries [69]. As is also the case with the challenge of trying to cover the metabolome via a single instrumental platform, no single standardized sample preparation method has been identified, which is able to cover the entire metabolome for literally any sample type. Future targeted and untargeted metabolomics investigations will still need to develop standardized protocols to provide a sufficiently broad coverage of the metabolites extracted. Notably, no matter which extraction method is used, introduction of an internal standard (IS) is usually essential for accurately providing relative or absolute quantification. An IS can be natural (^{12}C) or isotopically labeled (^{13}C or ^{15}N), and can be added at different stages of sample preparation and analysis to monitor either the recovery of each preparation step or the performance of the subsequent GC–MS or GC × GC–TOFMS analysis. For example, an IS added before the derivatization step can be used to obtain derivatization efficiency and correct for variance. An IS added before sample injection can correct any variances produced during running the samples on the instrument. By adding the IS, especially a global ^{13}C-labeled IS before the sample preparation, *in situ* corrections can be made for most variations (e.g., decomposition or interconversion of labile metabolites) arising from sample preparation except for the release of metabolites from the cells [65, 70]. However, the increased sample complexity that results from addition of the ^{13}C-labeled IS also produces challenges for the data analysis step. Essentially, signal deconvolution of the GC–MS and/or GC × GC–TOFMS data in order to identify and quantify metabolites of interest is made more demanding. In this regard, the richness of ion fragmentation information in the data provides both advantages and disadvantages depending on the degree of chromatographic and spectral overlap in truly complex samples. The subject of addressing the data analysis challenges are discussed in detail later in this chapter.

Compared to using LC-MS instrumentation, most metabolites analyzed by GC–MS and GC × GC–TOFMS instrumentation require chemical derivatization to reduce the polarity of the functional groups, which facilitates the GC separation and protects thermally labile metabolites (the second sample preparation type). The efficacy of a derivatization procedure can depend on a number of factors including reaction conditions such as temperature and time, the nature of the derivatization reagent and the solvent(s), and the amount of the derivatization reagent. The most commonly applied chemical derivatization reactions include the following: methoximation of ketone group and trimethylsilyl (TMS) or *tert*-butyldimethylsilyl (TBDMS) modification of hydroxyl, carboxyl, and primary amine groups. Silylation reagents and the resulting derivatives are generally moisture sensitive, which require the derivatized samples to be totally dried and tightly sealed to prevent inadvertent reaction by water vapor. TBDMS derivatives are approximately 10^4 times more stable to hydrolysis than TMS derivatives. Taking into account this benefit, differential derivatization can be carried out by using isotopically light and heavy (D6) forms of TMDMS reagent because the possibility of isotopomer scrambling is diminished during the TMDMS derivatization [39]. This differential isotope coding method was an effective means to simultaneously compare the concentration of metabolites between two samples. Moreover, owing to the paramount importance of ^{13}C flux analysis in metabolic engineering, TBDMS derivatization is more widely used for amino acids and carboxylic acids because it yields abundant fragment ions (m/z M-57, where M is the molecular ion of metabolites), which is convenient for ^{13}C-labeling pattern assignments [44, 71]. It is noteworthy that any derivatization chemistry may produce artifacts, challenging both targeted and untargeted metabolomics investigations and requiring accurate identification and quantification [71]. Moreover, current derivatization chemistry protocols can add additional peaks to the chromatogram as some metabolites can be derivatized in different ways (e.g., glycine-2TMS and glycine-3TMS) [72]. In addition, it is possible that two different metabolites can generate the same or very similar derivatized molecules (e.g., arginine-TMS and ornithine-TMS; 3-OH butyric acid-TMS; and 3-OH isobutyric acid-TMS) and have coeluting retention times, challenging identification and perhaps requiring better chromatographic separation conditions, that is, requiring more chromatographic resolution [71, 73].

4.3
GC–MS and GC × GC–TOFMS Instrumentation for Metabolomics

GC provides a large degree of chemical selectivity along the retention time axis. Analytes (i.e., metabolites) are separated as they flow through a GC column using a suitable carrier gas (e.g., hydrogen, helium, or nitrogen) via a dynamic equilibrium of the metabolite concentration profile (peak) between the stationary phase in the column (generally a film on the inside wall of the capillary column) and the carrier gas flowing through the column, as a function of the oven temperature that houses the column(s), and other instrument components (primarily the injector and detector). While manual injection can be practiced, in the metabolomics field, it is often deemed necessary to use automated sample injection (and in some laboratories automated sample preparation) to achieve suitable reproducibility and to reduce the need for human interaction with sample preparation and with the instrument. The detector for GC-based metabolomics is commonly MS. Temperature programming of the GC oven is essential to optimize separation performance (e.g., ~50–350 °C range). When temperature programming is properly applied, the analyst can achieve peak widths for all peaks throughout the separation that have a width similar to that of the narrowest peak, which is the unretained peak leaving the column at the dead time.

With modern instrumentation, the GC columns are almost always of the open tubular capillary variety with an inside diameter (i.d.) of 100, 180, or 250 μm and the stationary phase of suitable thickness coated on the inside of the capillary wall. Regarding the stationary phase thickness, there is a trade-off between GC peak broadening and the injected sample mass. A thicker phase allows for more injected sample mass (so the column is not "overloaded"); however, each individual peak is broadened because of slower mass transfer of the metabolite between the stationary phase and the carrier gas mobile phase. The most common stationary phases are polymer based, but there are also other phase types such as ionic liquid and cyclodextrin phases. Polymer stationary phases typically applied in metabolomics studies include dimethyl polysiloxane (e.g., BP-1, ZB-1), 5% diphenyl/95% dimethyl polysiloxane (e.g., DB-5, SPB-5), and trifluoropropylmethyl polysiloxane (e.g., Rtx-200, AT-210). A typical nonpolar column would be a BP-1, while a polar column would be a Rtx-200. These two columns are a good combination with GC × GC. All stationary phases have maximum temperature restrictions to avoid or minimize stationary phase "bleed" out of the column, and also most stationary phases do not perform well, or for any length of time, if the sample injected contains any significant amount of water. Hence, for metabolomics studies, injected samples must be essentially free of water in order not to degrade the stationary phase, which ultimately degrades separation performance. Recently, an ionic liquid phase column has been demonstrated to not be limited by this water constraint (SLB-IL59). The primary requirements, and hence limitations, for use of GC are that the analytes (metabolites) must be sufficiently volatile and thermally stable to be amenable to the gas phase separation process. This is why chemical derivatization is generally required for sample preparation to analyze derivatized metabolites from a variety of compound classes including amino acids, organic acids, sugars, sugar phosphates, purines, fatty acids, and fragments of Coenzyme A metabolites. Metabolites more suited to LC methods and not GC include, but are not limited to, phospholipids, acylcarnitines, nucleotides, and Coenzyme A metabolites (e.g., malonyl-CoA).

If a univariate detector is used with 1D-GC, for example, flame ionization detection (FID), GC-FID produces a 1D chromatogram, which is a data vector for each sample injection, with a first order data structure. Generally, a GC-FID instrument does not provide sufficient chemical selectivity and analyte identification power to be useful for metabolomics, as only the retention time matching is provided. However, GC-FID or any other GC-univariate detector combination may be useful for chemical fingerprinting studies. When coupled with MS detection, the widely used GC–MS instrument is obtained and the added chemical selectivity from the MS dimension renders the instrument extremely powerful for metabolomics studies because the combination of having both a retention time and a mass spectrum to identify metabolites is provided [13, 14, 47]. The mass spectra should be obtained at a suitable duty cycle (described later) as metabolite peaks elute from the GC column. The resulting data structure for GC–MS is 2D, and represented as a matrix, and is referred to as a *second order data structure*. An example of a GC–MS chromatogram (total ion current or TIC) for a metabolite sample is shown in Figure 4.1, with the mass spectrum for the peak indicated in the inset of the figure. The malate peak is identified by the mass spectrum match with a library and also can be simultaneously matched to the retention time if retention times for metabolites of interest are also in the library [47]. As described in more detail constraints of using the GC and MS in tandem exist that must be considered and addressed, if the analyst is to optimally glean the most information from their samples in a metabolomics study.

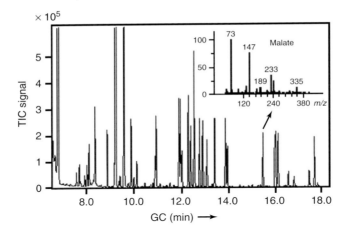

Figure 4.1 A GC–MS chromatogram of a metabolite sample, showing the mass spectrum for the malate peak that has been used, along with the retention time, to identify the metabolite. The sample was derivatized in two steps. The first step was methoximation using methoxyamine hydrochloride (25 mg ml^{-1}), and the second step was trimethylsilylation using TMS reagent (BSTFA/TMCS, 99 : 1). The column was an Rtx-5MS (30 m × 0.25 mm × 0.5 μm film).

The relatively recent emergence of the GC × GC–TOFMS instrument (~10 years ago), has significantly benefited the metabolomics field. The GC × GC–TOFMS instrument can selectively and sensitively determine a large number of compounds in a single analysis of a very complex sample. The resulting data structure for a single sample run is third order, that is, a 2D separation combined with a MS dimension. This instrument separates the greatest number of metabolite species with excellent sensitivity using relatively small sample sizes. The instrument is relatively rapid and inexpensive to operate, and the separations are reproducible. The data obtained is incredibly rich with information. With GC × GC–TOFMS, a 2D separation is produced with the first separation dimension often a nonpolar stationary phase column and a separation time of 30–60 min. The second separation dimension is performed in real time as effluent (carrier gas and derivatized metabolites) leaves the first column, with the second column separation run time taking ~1–5 s (referred to as the *modulation period*). Species leaving the first column should have ~three to four modulation periods to maintain the comprehensiveness of the 2D separation (i.e., to preserve the quantitative integrity and resolving power of the first dimension separation) [74, 75]. For example, a first column peak (a metabolite concentration profile), that is 8 s wide (four standard deviation peak width), would suggest the use of a 2 s modulation period to provide a sufficiently comprehensive separation. In the second dimension, a polar stationary phase column is used, providing a complementary separation relative to the first column (i.e., a high degree of informational orthogonality between the two columns is applied [76]), hence the species that are not separated on the first dimension at a given retention time have the opportunity to be separated on the second dimension column. The

TOFMS then detects the metabolites as they leave the second column, with each species producing a chromatographic peak in the second dimension with a width of ~100 ms. The mass spectral scan rate for a typical TOFMS collects spectra at a rate of 500 Hz (2 ms per spectrum) over a m/z range of 5–1000, so there are ~50 spectra per peak before averaging. For most GC-based metabolomics studies, however, a m/z range of 40–600 m/z is sufficient. Generally, one will average spectra to obtain ~10–20 spectra per peak width before software analysis. Thus, GC × GC–TOFMS provides a considerable amount of data for a given complex sample (e.g., typically ~300–400 MB per sample run), and it has become clear that there are also significant challenges to readily glean useful information from this large volume of data. Indeed, the data analysis challenges are amplified when one considers the experimental design for a discovery-based study, for example, when it may be necessary to run many samples from at least two sample classes and to collect a minimum of three replicate runs for each sample. For such discovery-based studies using GC × GC–TOFMS, it is common to produce 50–100 or more sample runs with the instrument. Software development to analyze the GC × GC–TOFMS data must keep pace with the applications that researchers have in mind for this powerful technology. Historically, in the early stages of GC × GC–TOFMS based metabolomics studies, reliable and user-friendly data processing, alignment, and analysis tools were not available. At present, a number of software tools have been developed and made available, although the process of transforming raw data into biological knowledge would still benefit from further automation efforts. There is an ongoing challenge to reduce the computational demand of handling GC × GC–TOFMS data, especially for high throughput analysis involving the study of many samples, including replicates of each sample. The software is based on applying the appropriate chemometric multivariate data analysis tools. Chemometrics is generally described as the development and use of mathematical techniques to extract useful information from data acquired through chemical analyses [55, 77, 78]. Thus, it is imperative that the most recent advances in the chemometrics field be adapted in user-friendly software for the metabolomics analyses of GC × GC–TOFMS data. More about this issue is discussed in the data analysis section.

The overall goals for applying GC–MS and GC × GC–TOFMS to metabolomics studies can be summarized as follows. The analyst should seek to chromatographically separate as many metabolites as possible and to do so in the shortest separation run time as possible. A quantitative metric that defines this first goal is to optimize the peak capacity of the separation. The peak capacity is the number of evenly spaced peaks that can be separated at unit resolution over a specified run time that contains peaks (basically the time window between the dead time and the last eluting peak). An example of the number of observed peaks and the peak capacity achieved for the GC × GC–TOFMS separation of a metabolite sample is shown in the TIC chromatograms in Figure 4.2a,b, respectively, for the same separation. In Figure 4.2a, showing GC × GC–TOFMS data, peak finding software found ~800 peaks, whereas in Figure 4.2b, the peak finding software found only ~200 peaks for the same separation in a GC–MS

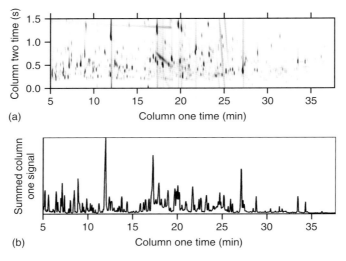

Figure 4.2 (a) A GC × GC–TOFMS chromatogram (total ion current, TIC) of a derivatized metabolite sample. (b) The GC–MS equivalent of (a) formed by summing the second dimension TIC signal onto the first dimension time axis. The first dimension column was a 250 μm i.d. 20 m Rtx-5MS (0.5 μm film), and the second dimension column was a 180 μm i.d. 2 m Rtx-200 (0.2 μm film). The carrier gas was helium at a 1 ml min^{-1} flow rate.

mode, obtained by summing the second dimension signal onto the first dimension time axis, resulting in a separation functionally equivalent to a GC–MS separation. The peak widths are ∼5 s on the first dimension column and on average about 90 ms on the second dimension column. On the basis of a run time of 35 min (less than the actual time due to the solvent delay and a small additional reduction for the increasing peak width for the isothermal hold at the end), the peak capacity is ∼420 for the GC–MS version of the separation (Figure 4.2b,) and ∼7000 for the GC × GC–TOFMS separation (Figure 4.2a). The MS must provide a scan rate with sufficient scans across the peak widths in either the lone separation dimension for GC–MS or the second dimension separation of GC × GC–TOFMS to be able to optimize the peak capacity of the separation.

It is generally not prudent to design the first dimension separation in GC × GC to purposely function at reduced performance (higher peak widths) relative to GC–MS even though this is commonly (and often inadvertently) practiced. Concurrently, the second goal for applying GC–MS and GC × GC–TOFMS is to achieve low LODs for trace analysis [79, 80], which is facilitated by keeping the GC peaks narrow (minimizing dilution of the detected concentration relative to the injected concentration) and by using a sensitive, selective, and low noise detector such as a quadrupole mass spectrometry (qMS) or a TOFMS. By doing so with a short separation run time, high throughput conditions can be achieved, which facilitate running more samples per time. Finally, the key goal is to provide informative data that can be readily analyzed by user-friendly software and available databases, transforming data into information so that the scientific question of interest can be answered.

Instrumental constraints and challenges as related to the goals of metabolomic studies will now be described. To set the stage, we provide a brief overview on chromatographic peak broadening theory and how it relates to the selection of the GC column length and inside diameter, GC oven temperature programming ramp rate, carrier gas flow rate with respect to MS flow rate compatibility, and MS duty cycle scanning rate with respect to the width of the peaks leaving the GC column and entering the MS. The general discussion is staged using the most commonly applied MS instruments for GC-based metabolomics, which are the qMS and TOFMS with electron impact (EI) ionization, with both providing unit mass resolution, with tremendous resources being available in the form of EI ionization-based mass spectral libraries of metabolites, and derivatized metabolites based on established derivatization chemistries [47]. However, the basic principles provided regarding the trade-offs, challenges, and benefits of using these two MS detectors can be readily extended to other MS detectors. It is noteworthy to mention that there is a growing interest in state-of-the-art high-resolution MS (primarily with GC–MS) to obtain the molecular formula from exact mass measurements, to aid in challenging analyte identification situations [44, 81].

Peak capacity of the separation is increased by decreasing the peak width to the extent the MS can scan fast enough for the data analysis software. To determine the lower limits for peak widths (and thus the upper limits for peak capacity production) for a column of given dimensions, it is necessary to further understand on-column band broadening. Excluding off-column sources of band broadening (injection, connections, detection, electronics, and so on), the on-column band broadening, H, for an analyte with a retention factor of k as derived by Golay is given by

$$H = \frac{2D_{G,o}jf}{u} + \frac{(1+6k+11k^2)}{96(1+k)^2} \frac{d_c^2 u f}{D_{G,o}j} + \frac{2k d_f^2 u}{3(1+k)^2 D_S} \quad (4.1)$$

where k is the retention factor of the analyte, u is the average carrier gas linear flow velocity (on column, at the oven temperature), d_c is the i.d. of the capillary, d_f is the thickness of the stationary phase film, $D_{G,o}$ is the diffusion coefficient of the analyte in the gas phase at the outlet of the column, j and f are gas compression correction factors, and D_S is the diffusion coefficient of the analyte in the stationary phase. The detected peak width at the base, w_b, in units of time is provided from a derivation from Eq. (4.1) and is given by Gross et al. [82], Reid and Synovec [83], and Wilson et al. [84]:

$$w_b = 4 \left[\frac{2D_{G,o}jf(1+k)^2 t_o^3}{L^2} + \frac{(1+6k+11k^2)}{96 D_{G,o}j} d_c^2 f t_o + \frac{2k d_f^2 t_o}{3 D_S} \right]^{1/2} \quad (4.2)$$

where L is the column length and t_o is the dead time of the unretained peak for the separation. From Eq. (4.2), we note that under isothermal temperature conditions, most of the on-column peak broadening is dominated by the $11k^2$ term to the $1/2$ power, so peak widths increase linearly with k. Therefore, a temperature program ramp (starting the separation at a low temperature and ending at a higher temperature) is needed to have each analyte peak elute from the column with a $k \sim 0$, the moment it leaves the column. This achieves narrow peak widths throughout the entire separation, that is, peak widths close to that of the unretained analyte. The flow rate of the carrier gas must be synchronized, so to speak, with the temperature programming rate in order to produce peaks with nearly the same narrow peak width as the unretained peak, which leaves the column first.

One key constraint for the MS is the GC carrier gas flow rate, as the maximum flow rate is limited by the MS vacuum system capabilities. The carrier gas (volumetric) flow rate that concerns the MS is at the GC column outlet, while the linear flow velocity, u as given in Eqs. (4.1) and (4.2), that concerns the GC separation performance is an average on-column value that is affected by the separation temperature, the column length and inside diameter, and the carrier gas due to the influence of gas phase viscosity. A typical qMS has the following constraints: a maximum flow rate of $\sim$1 ml min^{-1} that the vacuum system can handle and can provide $\sim$2–3 scans s^{-1} with a 40–600 m/z range per scan. Likewise, a typical TOFMS has a maximum flow rate of $\sim$1 ml min^{-1} (common) to 5 ml min^{-1} (top-of-the-line vacuum system), 500 scans s^{-1} with the same 40–600 m/z range per scan. As we shall see, the maximum flow rate constraint limits the ability to practice GC at, or above, the optimum linear flow velocity u_{opt} as per Eq. (4.1), where the optimum linear flow velocity is defined as the u where H is a minimum, referred to as H_{min}. The column length and inside diameter play key roles in determining the u_{opt} value where H is minimized. The primary dependence is that as the column diameter increases, the volumetric flow rate of, for example, 1 ml min^{-1} results in the GC system being operated at a u that is much slower than it ideally should be, producing relatively slow separations, and likely at a reduced peak capacity than could be achieved if the separation is performed at u_{opt}. In addition, the MS scan rate needs to be fast enough while covering the desired m/z range of $\sim$40–600, so that the number of scans collected across the chromatographic peak width is sufficient for the software to perform, ranging from 10 to 20 scans for the peaks entering the MS. The temperature programming rate for the GC instrument could also be a limiting factor in some cases, for example, if GC–TOFMS used in a high throughput mode is of interest.

We now describe the instrumental constraints for MS and GC, put into the context of peak broadening theory, by evaluation of a Golay plot (H vs u) following Eq. (4.1). In Figure 4.3 is shown a Golay plot for illustration purposes. Ideally, one should practice GC at the minimum, which is at H_{min} and u_{opt} indicated at point 1 or at least at a higher u (for higher throughput), for example, at point 2. At H_{min}, a maximum separation efficiency, N, is achieved because $N \alpha 1/H$, and hence a maximum chromatographic resolution R_s because $R_s \alpha N^{1/2}$. Poorer separation performance is provided to the left of H_{min}, and indeed at point 3, the R_s is identical to point 2; however, the separation run time is four times shorter at point 2 because the flow rate is four times faster. Basically, identical analytical

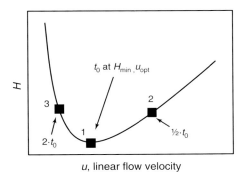

Figure 4.3 A Golay plot based on Eq. (4.1) for illustrative purposes, marked with points at the minimum H and u_{opt} (point 1), at twice the u_{opt} (point 2), and at half the u_{opt} (point 3).

information is provided four times faster at point 2 than at point 3. Ideally, if the run times are not too long, one should strive to operate the GC instrument at a flow rate that yields u_{opt} and H_{min} to yield the highest N, the highest R_s, thus the highest peak capacity. Unfortunately, as driven by constraints of MS flow rate capability and the scan rate to get enough scans across a GC peak width, many separations are performed to the left of point 1 and more likely to be performed in the vicinity of point 3 (well left of H_{min}, u_{opt}). This shortcoming is primarily a consequence of using a column i.d. that is not sufficiently narrow. By using narrower i.d. columns, a flow rate more compatible with the MS vacuum system will also result in a linear flow velocity that is nearer to u_{opt}.

The GC × GC–TOFMS with thermal modulation (which is the most common modulation system applied) is also constrained by flow rate selection, as the two columns are directly coupled [17–20]. It is challenging to combine two columns while still producing the most peak capacity production as possible from the thermally modulated 2D system. Although less used, a valve-based GC × GC–TOFMS allows the two columns to operate under separate flow controllers so that both the columns can be nearer to H_{min} and u_{opt}, hence a better 2D separation is produced [84–87]. However, with a valve-based modulator, not all of the analytes are sent to the second column so LODs are higher than with thermal modulation. This loss in LOD performance for valve-based GC × GC can be overcome to some extent by applying a smaller initial split on the first column when split injections are used. The peak capacity produced for the valve-based GC × GC can be optimized without sacrificing the peak capacity on the first column. Indeed, for all GC × GC separations, the analyst should strive to initially optimize the separation on the first column, so as not to reduce separation performance (so work at or above point 1 (H_{min}, u_{opt}) as shown in Figure 4.3), and then apply a modulator and second column that is also at or above H_{min}, u_{opt}. If this approach is taken then GC × GC can provide peak capacity production approaching 1000 peaks min^{-1}, while for 1D-GC done optimally, the peak capacity production is ∼100 peaks min^{-1}. This point is illustrated by referring back to the comparison between GC and GC × GC peak capacities produced in the example in Figure 4.2a,b, the peak capacities achieved are typical if the GC–MS is using a qMS, as the peaks need to be ~5 s wide to achieve ~15 mass spectra per peak width (at the base) at ~3 scan s^{-1}, as most alignment and retention time identification software relies on this scanning rate. Hence, GC × GC–TOFMS has about a 10- to 20-fold advantage in peak capacity over GC–qMS. However, if one applies GC–TOFMS then the peaks are narrower than the typical ~5 s, typically ~2 s at the base. This results in a GC–TOFMS peak capacity that is approaching the peak capacity of GC × GC–TOFMS. An additional advantage of GC × GC over 1D-GC is the additional chemical selectivity provided by the 2D separation, and the additional retention time obtained for each analyte, to aid in analyte identification.

4.4
Data Analysis Strategies and Software

Foremost, data analysis strategies are dictated by experimental design. Knowledge about the system and/or samples to be studied and the scientific questions to be addressed are essential to guide strategies for the sample preparation, instrumental parameters, and subsequent data analysis. In this section, considerations for experimental design and data analysis are discussed.

Although single sample analyses may be acceptable for exploratory purposes, statistical methods are usually employed in metabolomics, and thus multiple samples are typically run. When multiple samples should be run depend on what sources of variance need to be taken into account in the statistical analysis. For example, replicate samples of tissue or cell culture, can take into account biological variation; multiple extraction and derivatization of samples can give an idea of sample preparation error; and multiple injections can yield estimates of instrumental error. It might appear attractive at first to always run multiple samples in order to account for each source of error; however, the number of samples in metabolomics analysis already yields large data sets on GC–MS and especially GC × GC–TOFMS instrumentation, and multiplying the number of samples run at each step rapidly makes this approach impractical in most situations. Therefore, once a source of variation has been sufficiently characterized, it may no longer be necessary to always estimate that source of variation in each set of subsequent analyses. This is particularly the case for small consistent sources of variance that can be disregarded in the shadow of much larger sources of variance. For example, biological variance may be on the order of 20% RSD (relative standard deviation) or more across multiple samples that are nominally identical, while variance in volume of sample introduced to the GC is likely 5–10% when using an autosampler. In any case, variances are cumulative, that is, errors caused at an earlier step are further propagated (and possible confounded by) variances introduced later in sample handling or analysis. Ultimately, for purposes of data analysis, with the possible exception of some design-of-experiments and related analyses, only the cumulative variance is of importance. Thus, multiple runs of some sort are almost always performed for both targeted and nontarget analyses so that overall error can be estimated. Although it is impossible to eliminate all sources of variance, reasonable measures should be taken to reduce variance to levels that can yield the desired statistical confidence in the quantitative results. Specifically in relation to GC–MS and GC × GC–TOFMS analysis, this usually means reducing extraction and derivatization variance by performing these steps as reproducibly as possible, running samples as close together in time as possible, ensuring consistent instrument performance by proper maintenance and frequent mass spectral adjustment (tuning), and accounting for injection volume error by the use of one or more internal standard.

Generally speaking, metabolite experiments fall into one of the three categories in experimental design: (i) nontargeted analysis, in which the identities of chemicals of interest is not presumed known before the analysis (the identities are often part of the information being sought in this case), (ii) targeted analysis, in which the identities of the chemicals of interest are already known, and (iii) a mixture of categories (i) and (ii) in which the identities of some chemicals of interest are known, but information about other unknown chemicals may also be of interest. In each of these cases, the goal is often to obtain chemical information relating to changes in response to some experimental variable, for example, gene expression, growth conditions, environment, time course, and so on. Most data analysis methodologies are designed for comparative analysis of (at least) two samples or classes of samples, but some are specifically designed with a particular type of experiment in mind, for example, a time course study for which many pairs of samples would have to be analyzed independently to achieve what a more specialized method could achieve much more simply. Within the three categories of experimental design, there is also a choice that must be made as to how to translate the data into quantitative results (information). For the sake of discussion, we present two approaches to do so, one route pursues a peak-table-based approach, while another route pursues a pixel-based approach to analyzing the data. See Figure 4.4 for an outline of the parts of these two approaches. Further considerations specific to these two approaches are given in the appropriate sections below. Both of these approaches has challenges, for example, related to aligning one run to the next, properly identifying metabolites, and accurately quantifying the identified metabolites.

Although GC–MS and GC × GC–TOFMS produce data that is often considered well-behaved compared with data obtained using many other analytical techniques, there are still nonideal features of the data that should be addressed. Among these are baseline offsets and drift and misalignment between samples. Proper tuning of the mass spectrometer helps to reduce baseline offsets and drift (most instrument control software includes auto-tuning functions), and misalignment is much less of a problem for samples run close together temporally. Even with care in instrumental analysis, these two problems can cause significant trouble in data processing if not addressed. Baseline correction can be applied to deal with baseline offsets and drift within a run. There is often some correlation in

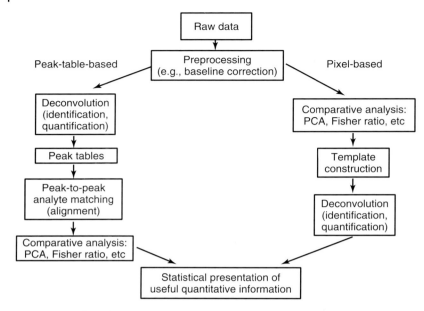

Figure 4.4 A flow diagram depicting analysis procedures using either peak-table-based or pixel-based approaches.

baseline between runs due to running with the same instrumental parameters, but baseline correction should be applied on each run to deal with the uncorrelated drift and often different offset. The goal of baseline correction is to zero-center the baseline throughout the run (without otherwise distorting or disturbing the chemical information in the data). Although it is trivial to accomplish this goal for blank samples in which there are no peaks present, the problem becomes more difficult as more peaks are present in the data and metabolite samples usually yield complex data with many peaks. It is usually not difficult for a skilled analyst to baseline correct data except in the most challenging cases, but it is undoubtedly time-consuming and may be somewhat subjective; so automated methods exist to accomplish this task and such functions are ubiquitously included in data analysis software. Typically, baseline correction is applied before any other kind of data processing. Without proper baseline correction, peak area and height calculations usually yield erroneous results with the magnitude of the error depending on the relative severity of the local baseline offset. Even with automated baseline correction, it behooves the analyst to check that the baseline offset and drift is correctly calculated and removed before subsequent analysis.

Misalignment of peaks from separate chromatograms poses a different type of challenge. Peaks from the same metabolite are expected to appear at the same retention time when run with the same method (on the same instrument, at least). Simply put, misalignment is the variation in retention time of the same metabolite peak between runs. This is caused by small deviations in run conditions (temperature and inlet pressure), degradation of the stationary phase inside the column, or sufficiently large changes in sample composition (e.g. matrix), among other things. With the improved parameter control of modern instrumentation, misalignment is less of a concern than it was many years ago when GC ovens had to be stabilized for hours to obtain reproducible runs. Presently running samples together in a batch is probably the best way to reduce misalignment from an instrumental standpoint; however, misalignment can still be a problem for many data processing methods, and many can be very sensitive to sample alignment (e.g., principal components analysis (PCA)). Retention time(s) are also very useful in combination with mass spectra in identifying metabolites, and so consistent and accurate retention times are desirable for purposes of analyte identification [47].

Correcting misalignment takes different forms depending on whether peak finding and deconvolution are performed before or after comparative analysis, that is, for a peak-table-based approach or a pixel-based approach (Figure 4.4). For methods performing deconvolution first, peak tables are produced for each sample. Alignment then consists of identifying the same peak (i.e., metabolite) across all the samples in which it appears and adjusting the retention time(s) so that they are uniform. For methods performing comparative analysis first, that is, the pixel-based approach, correcting misalignment is a matter of shifting, warping, or otherwise moving peaks such that the signal from the same metabolite peak in different runs has a retention time as reproducible as possible. In the pixel-based approach, the alignment is performed directly on the raw signal data and not peak table results. Again, in both the cases, it is possible for an analyst to do the alignment manually, but this is very time consuming, so automated methods exist for both peak table alignment and alignment of "raw" data [52, 88–91]. There are different benefits and challenges in each of these two approaches. With peak tables, peak distortion is not a concern, but peaks that were not properly deconvoluted (mathematically resolved) may either have poor retention time or mass spectra that make associating this peak with the same peak in the other samples difficult or impossible or cause the metabolite peak to not appear in the peak table for a particular sample at all.

For alignment methods operating on the raw signal "pixel" data, there is a significant difference in what is readily possible for GC–MS data and what is possible for GC × GC–TOFMS data. For GC–MS data, the situation is almost the reverse of that for baseline correction in that more peaks (especially more peaks in common) between sample runs engenders alignment to be more accurate. Distortion of peaks (and resulting changes in peak height, area and shape) is a potential concern, but the better automated alignment methods have addressed this challenge and so cause little distortion under common circumstances. Good results can be expected unless samples are very different in composition, which is not typically the case in metabolomic studies. For GC × GC–TOFMS data, the situation is very different with regard to aligning the raw signal data. Alignment of peaks from one second dimension run (i.e., modulation) to an adjacent one is possible (within a single sample run), but it is difficult, if not impossible, to align peaks from one sample run to the next in the first dimension under normal circumstances. This is because the misalignment is manifest not only in retention time but also in the magnitudes of each of the second dimension subpeaks, owing to a peak sampling phasing effect, such that in order to correct the misalignment not only do the position of the peaks need to be changed but the intensity of each peak on the second dimension also needs to be adjusted to deal with the phasing, and this cannot be done reliably without proper deconvolution of those peaks. Other methods of dealing with misalignment are necessary when it is sufficient to cause issues with GC × GC−TOFMS data.

Deconvolution or peak "mathematical resolution" steps are performed during most analyses, the point generally being to identify peaks and produce more accurate and precise peak attributes such as peak height, area, and retention time. A more accurate mass spectrum for each of these peaks can then be obtained from, for example, simple integration across a peak. If all the peaks in a chromatogram were sufficiently resolved at high enough signal-to-noise ratios (S/N), simple methods would suffice and deconvolution would not be necessary, but this is not typically the case, especially for complex metabolite samples. Many methods, with varying levels of automation, have been developed to perform deconvolution for GC−MS [49, 51, 92, 93] and GC × GC−TOFMS data [94].

Details of the different strategies taken to accomplish deconvolution are beyond the scope of this chapter, but generally speaking, they consist of approaches that rely on peak shape or apex identification, approaches that use multivariate resolution techniques, or hybrids of these two approaches. These various strategies each have different advantages and drawbacks, for example, some cannot easily deal with isotopically labeled metabolites, some require selective mass channels, some require multiple samples, some require higher S/N than others, some require more user interaction than others, and some take more time than others. Ignoring these differences, we can look at how deconvolution fits into the two different approaches to data analysis as shown in Figure 4.4, those being applying deconvolution on chromatographic data before performing comparative analysis or performing comparative analysis before deconvolution. When deconvolution is performed first, all the data must be submitted for processing. Ideally, the data complexity is then reduced to the number of peaks appearing in each sample. Subsequent comparative analysis can then work on a much reduced and filtered data set that takes advantage of the more accurate peak information provided by deconvolution. In practice, with deconvolution performed before comparative analysis, results depend greatly on the quality and capability of the deconvolution method employed and appropriate selection of associated parameters. If a peak is not properly deconvoluted, results of the comparative analysis will suffer; if a peak is missed in deconvolution, it is essentially lost and cannot be found in the subsequent comparative analysis. When deconvolution is performed first, fast deconvolution yielding accurate results is crucial. On the other hand, if comparative analysis is performed first on the data, much more data must be processed in the comparative analysis method. The risk of missing important peaks is much reduced because all the data is submitted for comparative analysis, but the results from the comparative analysis are more complicated. Results for overlapped peaks may be distorted, although there should be clear results on selective mass channels when those are present (which, incidentally, some deconvolution methods require to work properly anyhow). Deconvolution then only needs to be performed on the data locations indicated by the comparative analysis, although determining which locations are important may be more difficult than when deconvolution is performed first.

Under ideal conditions, results would be the same no matter in which order deconvolution is performed relative to comparative analysis, but in practice somewhat different results are obtained from the two general approaches. Another very important consideration in identification of a metabolite is not only the quality of the deconvoluted mass spectrum but also the quality and breadth of the mass spectral library to which it is compared. A great deal of effort has gone into collecting metabolite-specific mass spectral libraries with considerable success [47]. Almost always, for mass spectra collected on the same or a similar mass spectrometer, the better the quality of the matches between deconvoluted mass spectra and the library mass spectra, and the more mass spectra are available in a library (or libraries), the better the chances of identifying the metabolite(s). This is an especially important consideration for nontargeted analysis.

Comparative analysis can be simple or complex. Many useful chemometric and statistical methods have been developed for or applied to GC–MS and GC × GC–TOFMS data [55]. Details on these methods are beyond the scope of this chapter, but short descriptions with references are provided here. Many comparative methods have been developed that work on peak table data: subtractive and visual methods with t-tests [53], methods for targeted analysis across samples [54], methods using (functional) group type identification [48], and specialized methods for time-series data [43], for example. A number of comparative methods have also been developed for pixel-based data, such as a specialized method for time-course data [42], and especially chemometrics-based methods such as partial least-squares (PLS) [50] and PCA [26, 95]. Orthogonal partial least-squares (OPLS) and OPLS discriminant analysis (OPLS-DA) have been popular methods for GC–MS data analysis [40, 96–98]. GC–MS and GC × GC–TOFMS data can also be used in conjunction with data from other instrumentation in data synthesis methods using techniques such as hierarchical principal component analysis (HPCA) [31].

The last step in the overall analysis scheme for both peak-table-based and pixel-based analyses is the statistical presentation of the information obtained. Comparative analysis methods generally yield statistical values, and further analysis with statistical methods, for example, t-tests or Fisher ratios followed by analysis of variance (ANOVA), are often deemed appropriate. Even though the data may be much reduced and relevant information extracted, there still may be a large amount of useful information obtained because of the complexity of the metabolomic data. Therefore, some thought about how the information can best be presented (e.g., a heat map, a bar graph, or a table) would be the next step, although details of presentation considerations are beyond the scope of this chapter. See Chapter 11 for more on chemometric methods and data presentation considerations.

4.5
Illustrative Examples and Concluding Remarks

It is instructive to provide examples that illustrate the complete process from experimental design, sample preparation, instrumental setup, data collection, and data analysis. For the first example, GC × GC–TOFMS was applied to the comparative metabolic fingerprinting of a wild type versus a double mutant strain of *Escherichia coli* lacking the transhydrogenases UdhA and PntAB [89]. The aim of this study was the development and validation of data alignment and retention time correction software that uses peak tables generated by the commercial software as input and generates a data matrix as output for subsequent multivariate statistical analysis, including supervised classification for fingerprinting and PCA for data visualization. Hence, this study followed the peak-table-based data analysis path in Figure 4.4. The two sample class types, wild type versus double mutant, were handled identically in terms of growth conditions and metabolite derivatization chemistry. Isotopically labeled metabolite standards were also included for validation purposes. Appropriate columns were selected for the GC × GC, analogous to those in Figure 4.2. Each sample was prepared with sufficient replicates (six) to provide good statistical information from the GC × GC–TOFMS data. The heart of the project was the development of retention time correction software that utilized the retention time (from both columns) and mass spectral information provided by the peak tables to accurately align a given metabolite to itself, across all peak tables, so that any subsequent comparative analysis such as PCA could be optimally performed. The software performance was validated using metabolite standards and found to meet the goals of the study. This study demonstrates advances in developing data analysis software that otherwise may be a limiting factor in metabolomics investigations.

A second example of the complete process of applying GC × GC–TOFMS to metabolomics is the time-dependent profiling of metabolites in Snf1 mutant versus wildtype yeast cells [41]. The effect of sampling time (at four time points) was investigated, with two growth conditions (a high-glucose medium and in a low-glucose medium containing ethanol). The resulting dynamic metabolic systems were investigated to determine if useful kinetic information could be obtained and to determine if a single time point was sufficient to obtain broadly useful metabolic information. As there were two sample types (wild type and mutant) and two growth conditions, with four time points, this resulted in 16 sample classes from a chemometric data analysis perspective. This was a very challenging number of sample classes to confidently find upregulated or downregulated metabolites, requiring a novel software solution [42], which employed the pixel-based data analysis path in Figure 4.4. Sample preparation with derivatization was followed by GC × GC–TOFMS analysis of the 16 samples, in triplicate for a total of 48 runs. The pixel-based algorithm found locations in the 2D separation space in which the signal ratio, between any 2 of the 16 samples at a given m/z, was significantly larger than a user-specified threshold. Information at all m/z was ultimately taken into account to produce a list of hits. Locations of the hits in the 2D separation space were then subsequently analyzed using PARAFAC [94], which also worked on the data at the pixel level, and provided the purified mass spectrum for metabolite identification and the metabolite peak profiles from the first and second dimension separations for quantitative analysis. More than 50 kinetically changing metabolites were identified in this study.

These two examples illustrate some of the common steps and provide insight into ongoing challenges while using GC–MS and GC × GC–TOFMS for metabolomics, such as the need to carefully design the experiments, the need to properly prepare the samples for high-temperature gas phase separation and mass spectral detection, setting up and running the instrumentation to provide the desired analytical data, and improving the performance of the commercial software associated with the instrumentation or developing in-house software tools to overcome limitations of commercial software in order to glean useful information from the data. Indeed, there is considerable room for new advances in this area, such as better reagent chemistry for the derivatization step to minimize unwanted artifacts, improvements to the GC and the MS instrumentation to provide faster separations with lower detection limits and with higher mass resolution. For the faster separations, the analyst should strive to maintain or, better yet, increase the total peak capacity of the separations with shorter separation run times, thus providing higher sample throughput. Finally, there will always be a need to develop data analysis software that rapidly and accurately extracts as much information as possible from the data. Being able to conveniently extend the capabilities of the commercial software that is a part of the instrumentation platform by in-house modifications needs to be more readily possible for advances to be made more rapidly. Owing to wide applicability and outstanding dependability, it is apparent that GC–MS and GC × GC–TOFMS instrumental platforms will continue to be an extremely popular technology for metabolomics studies for many years to come.

References

1. Dunn, W.B. and Ellis, D.I. (2005) *Trends Anal. Chem.*, **24** (4), 285–294.
2. Sumner, L.W., Mendes, P., and Dixon, R.A. (2003) *Phytochemistry*, **62** (6), 817–836.
3. Guttman, A., Varoglu, M., and Khandurina, J. (2004) *Drug Discovery Today*, **9** (3), 136–144.
4. Winder, C.L., Dunn, W.B., and Goodacre, R. (2011) *Trends Microbiol.*, **19** (7), 315–322.
5. Chorell, E., Moritz, T., Branth, S., Antti, H., and Svensson, M.B. (2009) *J. Proteome Res.*, **8** (6), 2966–2977.
6. Oliver, S.G., Winson, M.K., Kell, D.B., and Baganz, F. (1998) *Trends Biotechnol.*, **16** (9), 373–378.
7. Bligny, R. and Douce, R. (2001) *Curr. Opin. Plant Biol.*, **4** (3), 191–196.
8. Bundy, J.G., Spurgeon, D.J., Svendsen, C., Hankard, P.K., Osborn, D., Lindon, J.C., and Nicholson, J.K. (2002) *FEBS Lett.*, **521** (1-3), 115–120.
9. von Roepenack-Lahaye, E., Degenkolb, T., Zerjeski, M., Franz, M., Roth, U., Wessjohann, L., Schmidt, J., Scheel, D., and Clemens, S. (2004) *Plant Physiol.*, **134** (2), 548–559.
10. Garcia, A., Barbas, C., Aguilar, R., and Castro, M. (1998) *Clin. Chem.*, **44** (9), 1905–1911.
11. Barsch, A., Patschkowski, T., and Niehaus, K. (2004) *Funct. Integr. Genomics*, **4** (4), 219–230.

12. Tetsuo, M., Zhang, C., Matsumoto, H., and Matsumoto, I. (1999) *J. Chromatogr. B Biomed. Sci. Appl.*, **731** (1), 111–120.
13. Lei, Z., Huhman, D.V., and Sumner, L.W. (2011) *J. Biol. Chem.*, **286** (29), 25435–25442.
14. Kind, T. and Fiehn, O. (2010) *Bioanal. Rev.*, **2** (1-4), 23–60.
15. Koek, M.M., Muilwijk, B., van der Werf, M.J., and Hankemeier, T. (2006) *Anal. Chem.*, **78** (4), 1272–1281.
16. Skogerson, K., Harrigan, G.G., Reynolds, T.L., Halls, S.C., Ruebelt, M., Iandolino, A., Pandravada, A., Glenn, K.C., and Fiehn, O. (2010) *J. Agric. Food. Chem.*, **58** (6), 3600–3610.
17. van Deursen, M., Beens, J., Reijenga, J., Lipman, P., Cramers, C., and Blomberg, J. (2000) *J. High. Resolut. Chromatogr.*, **23** (7-8), 507–510.
18. Shellie, R., Marriott, P., and Morrison, P. (2001) *Anal. Chem.*, **73** (6), 1336–1344.
19. Dalluge, J., van Rijn, M., Beens, J., Vreuls, R.J., and Brinkman, U.A.T. (2002) *J. Chromatogr. A*, **965** (1-2), 207–217.
20. Focant, J.F., Sjodin, A., and Patterson, D.G. Jr. (2003) *J. Chromatogr. A*, **1019** (1-2), 143–156.
21. Welthagen, W., Schnelle-Kreis, J., and Zimmermann, R. (2003) *J. Chromatogr. A*, **1019** (1-2), 233–249.
22. Welthagen, W., Shellie, R.A., Spranger, J., Ristow, M., Zimmermann, R., and Fiehn, O. (2005) *Metabolomics*, **1** (1), 65–73.
23. Hope, J.L., Prazen, B.J., Nilsson, E.J., Lidstrom, M.E., and Synovec, R.E. (2005) *Talanta*, **65** (2), 380–388.
24. Sinha, A.E., Hope, J.L., Prazen, B.J., Nilsson, E.J., Jack, R.M., and Synovec, R.E. (2004) *J. Chromatogr. A*, **1058** (1-2), 209–215.
25. Pierce, K.M., Hoggard, J.C., Hope, J.L., Rainey, P.M., Hoofnagle, A.N., Jack, R.M., Wright, B.W., and Synovec, R.E. (2006) *Anal. Chem.*, **78** (14), 5068–5075.
26. Mohler, R.E., Dombek, K.M., Hoggard, J.C., Young, E.T., and Synovec, R.E. (2006) *Anal. Chem.*, **78** (8), 2700–2709.
27. Beckstrom, A.C., Humston, E.M., Snyder, L.R., Synovec, R.E., and Juul, S.E. (2011) *J. Chromatogr. A*, **1218** (14), 1899–1906.
28. Kouremenos, K.A., Pitt, J., and Marriott, P.J. (2010) *J. Chromatogr. A*, **1217** (1), 104–111.
29. Li, X., Xu, Z., Lu, X., Yang, X., Yin, P., Kong, H., Yu, Y., and Xu, G. (2009) *Anal. Chim. Acta*, **633** (2), 257–262.
30. Snyder, L.R., Hoggard, J.C., Montine, T.J., and Synovec, R.E. (2010) *J. Chromatogr. A*, **1217** (27), 4639–4647.
31. Biais, B., Allwood, J.W., Deborde, C., Xu, Y., Maucourt, M., Beauvoit, B., Dunn, W.B., Jacob, D., Goodacre, R., Rolin, D., and Moing, A. (2009) *Anal. Chem.*, **81** (8), 2884–2894.
32. Buscher, J.M., Czernik, D., Ewald, J.C., Sauer, U., and Zamboni, N. (2009) *Anal. Chem.*, **81** (6), 2135–2143.
33. McGaw, E.A., Phinney, K.W., and Lowenthal, M.S. (2010) *J. Chromatogr. A*, **1217** (37), 5822–5831.
34. Skogerson, K., Runnebaum, R., Wohlgemuth, G., de Ropp, J., Heymann, H., and Fiehn, O. (2009) *J. Agric. Food. Chem.*, **57** (15), 6899–6907.
35. Yang, S., Sadilek, M., Synovec, R.E., and Lidstrom, M.E. (2009) *J. Chromatogr. A*, **1216** (15), 3280–3289.
36. Gu, Q., David, F., Lynen, F., Rumpel, K., Dugardeyn, J., Van Der Straeten, D., Xu, G., and Sandra, P. (2011) *J. Chromatogr. A*, **1218** (21), 3247–3254.
37. Antoniewicz, M.R., Kelleher, J.K., and Stephanopoulos, G. (2007) *Anal. Chem.*, **79** (19), 7554–7559.
38. Godin, J.P., Ross, A.B., Rezzi, S., Poussin, C., Martin, F.P., Fuerholz, A., Cleroux, M., Mermoud, A.F., Tornier, L., Arce Vera, F., Pouteau, E., Ramadan, Z., Kochhar, S., and Fay, L.B. (2010) *Anal. Chem.*, **82** (2), 646–653.
39. Huang, X.D. and Regnier, F.E. (2008) *Anal. Chem.*, **80** (1), 107–114.
40. Pohjanen, E., Thysell, E., Jonsson, P., Eklund, C., Silfver, A., Carlsson, I.B., Lundgren, K., Moritz, T., Svensson, M.B., and Antti, H. (2007) *J. Proteome Res.*, **6** (6), 2113–2120.
41. Humston, E.M., Dombek, K.M., Hoggard, J.C., Young, E.T., and Synovec, R.E. (2008) *Anal. Chem.*, **80** (21), 8002–8011.

42. Mohler, R.E., Tu, B.P., Dombek, K.M., Hoggard, J.C., Young, E.T., and Synovec, R.E. (2008) *J. Chromatogr. A*, **1186** (1-2), 401–411.
43. Peters, S., Janssen, H.G., and Vivo-Truyols, G. (2010) *Anal. Chim. Acta*, **663** (1), 98–104.
44. Fiehn, O., Kopka, J., Trethewey, R.N., and Willmitzer, L. (2000) *Anal. Chem.*, **72** (15), 3573–3580.
45. Shin, M.H., Lee do, Y., Liu, K.H., Fiehn, O., and Kim, K.H. (2010) *Anal. Chem.*, **82** (15), 6660–6666.
46. Vuckovic, D., Zhang, X., Cudjoe, E., and Pawliszyn, J. (2010) *J. Chromatogr. A*, **1217** (25), 4041–4060.
47. Kind, T., Wohlgemuth, G., Lee do, Y., Lu, Y., Palazoglu, M., Shahbaz, S., and Fiehn, O. (2009) *Anal. Chem.*, **81** (24), 10038–10048.
48. Castillo, S., Mattila, I., Miettinen, J., Oresic, M., and Hyotylainen, T. (2011) *Anal. Chem.*, **83** (8), 3058–3067.
49. Jiang, W., Qiu, Y., Ni, Y., Su, M., Jia, W., and Du, X. (2010) *J. Proteome Res.*, **9** (11), 5974–5981.
50. Jonsson, P., Gullberg, J., Nordstrom, A., Kusano, M., Kowalczyk, M., Sjostrom, M., and Moritz, T. (2004) *Anal. Chem.*, **76** (6), 1738–1745.
51. Jonsson, P., Johansson, A.I., Gullberg, J., Trygg, J., A, J., Grung, B., Marklund, S., Sjostrom, M., Antti, H., and Moritz, T. (2005) *Anal. Chem.*, **77** (17), 5635–5642.
52. Oh, C., Huang, X., Regnier, F.E., Buck, C., and Zhang, X. (2008) *J. Chromatogr. A*, **1179** (2), 205–215.
53. Shellie, R.A., Welthagen, W., Zrostlikova, J., Spranger, J., Ristow, M., Fiehn, O., and Zimmermann, R. (2005) *J. Chromatogr. A*, **1086** (1-2), 83–90.
54. Wojtowicz, P., Zrostlikova, J., Kovalcuk, T., Schurek, J., and Adam, T. (2010) *J. Chromatogr. A*, **1217** (51), 8054–8061.
55. Pierce, K.M., Hoggard, J.C., Mohler, R.E., and Synovec, R.E. (2008) *J. Chromatogr. A*, **1184** (1-2), 341–352.
56. Gröger, T. and Zimmermann, R. (2011) *Talanta*, **83** (4), 1289–1294.
57. Fiehn, O. (2008) *Trends Anal. Chem.*, **27** (3), 261–269.
58. Dietmair, S., Timmins, N.E., Gray, P.P., Nielsen, L.K., and Kromer, J.O. (2010) *Anal. Biochem.*, **404** (2), 155–164.
59. A, J., Trygg, J., Gullberg, J., Johansson, A.I., Jonsson, P., Antti, H., Marklund, S.L., and Moritz, T. (2005) *Anal. Chem.*, **77** (24), 8086–8094.
60. Farag, M.A., Ryu, C.M., Sumner, L.W., and Pare, P.W. (2006) *Phytochemistry*, **67** (20), 2262–2268.
61. Soini, H.A., Bruce, K.E., Klouckova, I., Brereton, R.G., Penn, D.J., and Novotny, M.V. (2006) *Anal. Chem.*, **78** (20), 7161–7168.
62. Humston, E.M., Zhang, Y., Brabeck, G.F., McShea, A., and Synovec, R.E. (2009) *J. Sep. Sci.*, **32** (13), 2289–2295.
63. Bolten, C.J., Kiefer, P., Letisse, F., Portais, J.C., and Wittmann, C. (2007) *Anal. Chem.*, **79** (10), 3843–3849.
64. Villas-Boas, S.G., Hojer-Pedersen, J., Akesson, M., Smedsgaard, J., and Nielsen, J. (2005) *Yeast*, **22** (14), 1155–1169.
65. Yang, S., Sadilek, M., and Lidstrom, M.E. (2010) *J. Chromatogr. A*, **1217** (47), 7401–7410.
66. Villas-Boas, S.G. and Bruheim, P. (2007) *Anal. Biochem.*, **370** (1), 87–97.
67. Schaub, J., Schiesling, C., Reuss, M., and Dauner, M. (2006) *Biotechnol. Progr.*, **22** (5), 1434–1442.
68. Spura, J., Reimer, L.C., Wieloch, P., Schreiber, K., Buchinger, S., and Schomburg, D. (2009) *Anal. Biochem.*, **394** (2), 192–201.
69. Yanes, O., Tautenhahn, R., Patti, G.J., and Siuzdak, G. (2011) *Anal. Chem.*, **83** (6), 2152–2161.
70. Canelas, A.B., ten Pierick, A., Ras, C., Seifar, R.M., van Dam, J.C., van Gulik, W.M., and Heijnen, J.J. (2009) *Anal. Chem.*, **81** (17), 7379–7389.
71. Halket, J.M., Waterman, D., Przyborowska, A.M., Patel, R.K., Fraser, P.D., and Bramley, P.M. (2005) *J. Exp. Bot.*, **56** (410), 219–243.
72. Kanani, H., Chrysanthopoulos, P.K., and Klapa, M.I. (2008) *J. Chromatogr. B, Anal. Technol. Biomed. Life Sci.*, **871** (2), 191–201.
73. Guo, X. and Lidstrom, M.E. (2008) *Biotechnol. Bioeng.*, **99** (4), 929–940.

74. Murphy, R.E., Schure, M.R., and Foley, J.P. (1998) *Anal. Chem.*, **70** (8), 1585–1594.
75. Khummueng, W., Harynuk, J., and Marriott, P.J. (2006) *Anal. Chem.*, **78** (13), 4578–4587.
76. Slonecker, P.J., Li, X., Ridgway, T.H., and Dorsey, J.G. (1996) *Anal. Chem.*, **68** (4), 682–689.
77. Synovec, R.E., Prazen, B.J., Johnson, K.J., Fraga, C.G., and Bruckner, C.A. (2003) *Adv. Chromatogr.*, **42**, 1–42.
78. Sharaf, M.A., Illman, D.L., and Kowalski, B.R. (1986) *Chemometrics*, John Wiley & Sons, Inc., New York.
79. Koek, M.M., Bakels, F., Engel, W., van der Maagdenberg, A., Ferrari, M.D., Coulier, L., and Hankemeier, T. (2010) *Anal. Chem.*, **82** (1), 156–162.
80. Koek, M.M., Muilwijk, B., van Stee, L.L., and Hankemeier, T. (2008) *J. Chromatogr. A*, **1186** (1-2), 420–429.
81. Peterson, A.C., McAlister, G.C., Quarmby, S.T., Griep-Raming, J., and Coon, J.J. (2010) *Anal. Chem.*, **82** (20), 8618–8628.
82. Gross, G.M., Prazen, B.J., Grate, J.W., and Synovec, R.E. (2004) *Anal. Chem.*, **76** (13), 3517–3524.
83. Reid, V.R. and Synovec, R.E. (2008) *Talanta*, **76** (4), 703–717.
84. Wilson, R.B., Siegler, W.C., Hoggard, J.C., Fitz, B.D., Nadeau, J.S., and Synovec, R.E. (2011) *J. Chromatogr. A*, **1218** (21), 3130–3139.
85. Bruckner, C.A., Prazen, B.J., and Synovec, R.E. (1998) *Anal. Chem.*, **70** (14), 2796–2804.
86. Sinha, A.E., Prazen, B.J., Fraga, C.G., and Synovec, R.E. (2003) *J. Chromatogr. A*, **1019** (1-2), 79–87.
87. Sinha, A.E., Johnson, K.J., Prazen, B.J., Lucas, S.V., Fraga, C.G., and Synovec, R.E. (2003) *J. Chromatogr. A*, **983** (1-2), 195–204.
88. Styczynski, M.P., Moxley, J.F., Tong, L.V., Walther, J.L., Jensen, K.L., and Stephanopoulos, G.N. (2007) *Anal. Chem.*, **79** (3), 966–973.
89. Almstetter, M.F., Appel, I.J., Gruber, M.A., Lottaz, C., Timischl, B., Spang, R., Dettmer, K., and Oefner, P.J. (2009) *Anal. Chem.*, **81** (14), 5731–5739.
90. Wang, B., Fang, A., Heim, J., Bogdanov, B., Pugh, S., Libardoni, M., and Zhang, X. (2010) *Anal. Chem.*, **82** (12), 5069–5081.
91. Nadeau, J.S., Wright, B.W., and Synovec, R.E. (2010) *Talanta*, **81** (1-2), 120–128.
92. Smith, C.A., Want, E.J., O'Maille, G., Abagyan, R., and Siuzdak, G. (2006) *Anal. Chem.*, **78** (3), 779–787.
93. Tautenhahn, R., Patti, G.J., Kalisiak, E., Miyamoto, T., Schmidt, M., Lo, F.Y., McBee, J., Baliga, N.S., and Siuzdak, G. (2011) *Anal. Chem.*, **83** (3), 696–700.
94. Hoggard, J.C. and Synovec, R.E. (2008) *Anal. Chem.*, **80** (17), 6677–6688.
95. Peters, S., Janssen, H.G., and Vivo-Truyols, G. (2011) *J. Chromatogr. A*, **1218** (21), 3337–3344.
96. Lee, S.H., Woo, H.M., Jung, B.H., Lee, J., Kwon, O.S., Pyo, H.S., Choi, M.H., and Chung, B.C. (2007) *Anal. Chem.*, **79** (16), 6102–6110.
97. Wiklund, S., Johansson, E., Sjostrom, L., Mellerowicz, E.J., Edlund, U., Shockcor, J.P., Gottfries, J., Moritz, T., and Trygg, J. (2008) *Anal. Chem.*, **80** (1), 115–122.
98. Bao, Y., Zhao, T., Wang, X., Qiu, Y., Su, M., and Jia, W. (2009) *J. Proteome Res.*, **8** (4), 1623–1630.

5
LC-MS-Based Nontargeted Metabolomics
Georgios A. Theodoridis, Helen G. Gika, and Ian D. Wilson

5.1
Introduction

Metabolomics as originally defined ultimately aims at the analysis of the whole small molecule (metabolite) complement of a sample in order to relate the determined metabolite levels with certain characteristics or properties of the sample. The related field of metabonomics is defined more in terms of seeking differences between the metabolic profiles of test and control groups. In practice, the terms are often used interchangeably [1, 2]. Metabolomics brings together high-level analytical chemistry with advanced statistical analysis, biochemistry, and other sciences such as medicine/life sciences, agricultural, or environmental sciences. To achieve these ends, metabolomics requires multidisciplinary research necessitating collaborative efforts from scientists from different fields. As described in this chapter, a useful "formula" to obtain good results is to apply meticulous preparations and sample handling to the preanalytical steps, holistic analysis using information-rich spectroscopic techniques, and finally excellence in computer handling and data mining. To reach the overall goal of elucidation of biochemical pathways and the generation of new knowledge on the various biochemical phenomena, careful data scrutiny is necessary to avoid biases and misinterpretation of the various findings. Metabolomics is basically a hypothesis-free approach; in fact, this type of research should lead to the generation of new hypotheses. In this aspect, linking the different markers into biochemical networks and fluxes is not an easy task and typically requires excellent knowledge of the biochemical pathways and often assistance of computer-based visualization tools. Such results may be combined with genomic or proteomic findings and provide a systems biology approach that would integrate the descriptions of the different systems to enhance the understanding of biochemistry in the studied system or organism.

Metabolomics and metabonomics have found use in biomarker discovery for life sciences, plant/food, and environmental sciences. The life science is probably the sector with the largest integration of *omics* technologies. Metabolomics methods have been used not only for mapping different specimens to search for early biomarkers of disease (diagnostic markers) or for disease progression biomarkers

Metabolomics in Practice: Successful Strategies to Generate and Analyze Metabolic Data, First Edition.
Edited by Michael Lämmerhofer and Wolfram Weckwerth.
© 2013 Wiley-VCH Verlag GmbH & Co. KGaA. Published 2013 by Wiley-VCH Verlag GmbH & Co. KGaA.

(prognostic markers) but also to find markers of drug efficacy or toxicity. In the plant sciences, metabolomics can assist in taxonomic studies or in the assessment of the origin, for example, of fruits or food products (e.g., wine, cheese, or olive oil) [3, 4].

Metabolomics provides good potential in these aspects because metabolite profiles offer a snapshot of the ongoing biochemical phenomena. Metabolites can be directly linked with physiology and phenotype because metabolites are further down the biochemical pathway compared with genes or proteins.

5.2
LC-MS-Based Untargeted Metabolomics

In untargeted metabolomics, a hypothesis-free study is performed to obtain and then analyze a sample set using an unbiased analytical approach. Multivariate statistical analysis is then employed to reduce data complexity and reveal underlying trends. The data is thus driving the analysis; when a marker is found, efforts are taken to identify and validate the potential biomarker [5, 6]. Figure 5.1 depicts a schematic of a workflow in untargeted metabolomics study based on LC-MS (mass spectrometry).

When a researcher considers nontargeted metabolomics analysis using LC-MS, there are some important factors that should be always kept in mind. (i) LC-MS is a hyphenated technique that means that despite the impressive advances of the last 20 years, there are still numerous analytical limitations and bottlenecks. LC-MS brings together two initially nonmatching techniques through a certain interface. In order to obtain a signal in MS, a molecule has to be ionized, and in the chosen mode, ionization performance is not the same even for similar compounds that share the same active moieties. It is often the case that, for example, certain organic acids are sufficiently ionized in certain conditions, while similar acids do not show proper signals. (ii) LC-MS offers a number of different combinations in either end: LC or MS. This provides many possibilities but at the same time may hinder standardization. (iii) Operating MS hardware and detection software may result in differences in the way that single scan MS data is interpreted.

Undoubtedly, LC-MS offers a series of advantages, and this is the reason for its widespread use. The most important and relevant to metabolomics are given in Section 5.2.2. Such considerations are discussed in more detail in the following sections. First, the chromatographic part is described, and then the various MS considerations are discussed.

5.2.1
LC Issues

The LC-MS for metabolomics should better be performed in the highest resolution power of both techniques, hence utilization of the smallest particle size (at present sub-2-μm termed ultra (high) performance liquid chromatography (U(H)PLC))

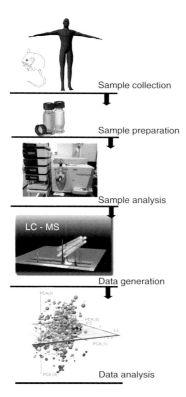

Figure 5.1 Simplified flow chart of the process of metabolomics/metabonomics research using LC-MS: painstaking preparation and sample collection phase, cautious "minimal" sample preparation, analysis in full scan MS, data extraction, and multivariate statistics to reach biomarker discovery.

and the highest temperature tolerated by the samples and the analytical system is generally suggested. Such measures offer the following potential advantages: (i) high peak capacity in LC, (ii) higher analytical sensitivity, and (iii) lower risk for matrix effect contribution (ion suppression/enhancement).

It should be pointed out that stand-alone MS using the direct infusion of the sample, even on the highest resolution MS instruments, has been used only in a handful of publications (see for example, Ref. [7]). Despite the significance of high mass accuracy in analyte identification and quantification in complex mixtures, direct introduction of complex biofluids (blood plasma/urine) or extracts (organs/feces) into the ion source means, in practice, that an unknown number of molecules of unknown properties and concentrations are subjected simultaneously to the ionization process. Hence in the very possible case of ion suppression, certain low abundance molecules, or analytes exhibiting a low ionization efficiency, may not be detected at all, no matter how sophisticated the mass analyzer is.

Chromatography is beneficial as it separates the different analytes (reducing the potential for poor ionization due to ion suppression), while it may also potentially be able to separate isobaric/isomeric metabolites thus enabling their detection. The emergence of new packing materials together with the proper advanced instrumentation has extended the limits of the technique by offering speed and higher peak capacity (number of peaks resolved per unit time in gradient separations). These parameters are of high importance in profiling/metabolomic studies because the superior resolution and sensitivity mean higher gain in information. In addition, higher robustness of the new instrumentation and software offers the possibility of using smaller sample volume and lower flow rates, and, along with better control of parameters applied and shorter runs, increases the advantages of applying U(H)PLC. Separations are currently typically performed using 2.1 mm i.d. columns (most used U(H)PLC column format). Nano LC [8, 9] and "chip"-based LC [10] are also now becoming of interest. With regard to the LC mode, the majority of the studies employs reversed-phase (RP) gradient chromatography [5, 6]. The reason that the RP-mode is the most widely used relates to the level of sophistication of the available RPLC stationary phases, the number of different chemistries available, the geometrical characteristics of particles and columns, and the ready compatibility of RPLC with aqueous samples. As a consequence, the penetration of the technology in the analytical market remains unchallenged by virtually any instrumental technique (either other LC modes or spectroscopic technique (e.g., NMR)). However, RPLC faces severe limitations in the analysis of polar and/or ionic species, which are typically poorly retained and elute in the solvent front. Such polar molecules include examples of practically all primary metabolites such as amines, organic acids, amino acids, and carbohydrates and metabolites involved in a multitude of biochemical pathways and fluxes. Alternative separation modes are urgently sought and as such hydrophilic interaction chromatography (HILIC) [11–15], ion-exchange LC, aqueous normal-phase chromatography [16], or "mixed mode" chromatography are increasingly being used coupled to MS [17].

The composition of the mobile phase is dependent on the selected LC mode (RP, HILIC, or other); however, researchers should apply basic chromatographic knowledge considering the molecular properties of the different solvents and are advised to experiment also on this side. In general, gradient elution with a run time from 10 to 30 min is applied. However, starting and ending conditions are not the same for all sample types (e.g., urine or blood plasma) and require optimization, that is, gradient programs for blood-derived samples or tissue extracts typically start from a higher organic content than samples such as urine and employ longer column washing steps to remove the various lipophilic substances present in these samples. Furthermore, removal of interferences from the system should not be taken for granted. Different solvents have different eluotropic properties and do not function equally in eluting strongly retained lipidic substances: methanol and other alcohols seem to work better on blood samples (compared to acetonitrile) in removing lipids much more efficiently possibly because of their better ability to disrupt hydrogen bonds. If the strongly retained lipids are allowed to build up on column, the stationary phase chemistry will soon be modified, thus resulting in analytical instability. In HILIC, the current LC mode of choice for the analysis of polar metabolites, one has to remember (i) not to exceed 50% of water in the mobile phase, (ii) to employ shallow gradients, and (iii) to allow sufficient re-equilibration time, otherwise analytical repeatability will be severely compromised.

5.2.2 Mass Spectrometry

MS has a major role in holistic metabolite profiling because of its high sensitivity and widespread availability. For these reasons, MS has become a leading analytical technology with the highest increase rate in both publication and citation numbers. This can be understood bearing in mind the higher abundance of LC-MS and GC-MS instruments and trained practitioners compared to the corresponding figures for high-field NMR instruments that have dominated global metabolic profiling until the rise of LC-MS. In addition, MS and LC-MS in particular offer a series of unparalleled advantages: high sensitivity, ever increasing resolution power (continuous technological advancements offer unprecedented mass accuracy and resolution), satisfactory sample throughput, suitability to the analysis of biological-plant-food extracts, flexibility, and versatility in method selection and development.

The choice of the ionization mode used in LC-MS has a significant effect on the metabolic profile that is obtained. As mentioned earlier, certain types of molecules are ionized more efficiently in one ionization mode or one polarity and some in a different mode. The most common ionization mode in LC-MS is undoubtedly positive electrospray ionization (ESI) as an effective means for the ionization of a wide range of molecules. Negative ionization is more effective in the analysis of organic acids and carbohydrates. Atmospheric pressure chemical ionization (APCI) is preferred for the analysis of more apolar analytes and has not been widely applied in metabolite profiling. In recent years, there has also been considerable interest in laser desorption ionization (LDI) techniques (for use on dry samples or tissues) such as MALDI (matrix-assisted laser desorption ionization) [18, 19], SELDI (surface-enhanced laser desorption ionization), or desorption/extractive electrospray ionization (EESI) [20, 21]. The application of these LDI techniques has mostly been on the so-called targeted metabolomics for the analysis of selected regions of a specimen to provide data on the localization of the detected metabolites; as such endeavors are out of the scope of this chapter, they are not further discussed.

The major tool for nontargeted metabolomics is the combination of U(H)PLC with time-of-flight mass spectrometry (TOF-MS). This combination provides the best available chromatographic resolution combined with the excellent sensitivity, fast data acquisition, and high mass accuracy (typically <5 ppm) that TOF-MS offers. This coupling offers the best conditions to exploit the potential of both technologies, as the TOF-MS is able to scan fast enough to cope with U(H)PLC type separations where chromatographic peaks with a base width of 3–5 s can be expected. Higher resolution MS machines (e.g., FT-ICR or Orbitrap-MS) typically require more time to achieve the highest resolution, and hence such apparatus can better be used mainly for biomarker identification. Lately, the Orbitrap-MS has found increasing use in metabolomics research. These instruments offer the following advantages: very high resolution and mass accuracy and MS^n capabilities because they are often combined with a preceding ion trap (lately also a quadrupole) where useful fragmentation data can be obtained. High mass accuracy is very useful in providing good atomic composition data thereby reducing the number of candidate identities of potential markers. The combination of high mass accuracy MS and MS/MS data with library searching and information from other experiments (e.g., LC retention time or NMR spectra) gives much higher level of confirmation ability.

Metabolomic analysis has also been performed with linear ion traps or the QTRAP [22]. Modern ion traps offer agreeable resolution along with MS/MS capabilities. The QTRAP also offers the potential of triple quadrupole MS functionalities in the same instrument.

The availability of such a wide variety of mass spectrometric solutions for metabolic profiling may appear very attractive, enabling the researcher to use a relatively low-resolution (and therefore inexpensive) instrument for profiling and a higher performance (and more expensive) spectrometer for more complex tasks; however, there are certain problems with this diversity that should be noted. (i) There are severe difficulties in correlating LC-MS data and comparing data obtained in different MS instruments [23]. (ii) Formation of adducts in LC-MS cannot be safely predicted or controlled. It is thus not surprising that when such systems operate in nontargeted mode they may provide different perspectives even when analyzing the same sample(s). As our group has recently shown, simultaneous LC-MS-based metabolite profiling on a QTOF and a QTRAP showed similar statistical group separation, but this was based on different markers [23]. (iii) Interlaboratory comparisons of data are difficult.

To conclude, the development of LC-MS methods for untargeted metabolomics should not be considered a one fit for all approach, otherwise there is a clear risk that expensive resources and time consumed on data mining are wasted. The method should be fit to the task, and to do this, researchers have a large variety of tools to select from. Table 5.1 provides a list of problems often addressed in LC-MS-based metabolite profiling together with possible causes.

5.3
Study Design

As metabolomics is often a quest into the unknown, it is strongly suggested that study design is taken very seriously. This should include different parameters: the number of samples necessary, the different conditions tested, the sampling process and handling, the analytical technologies used, the data treatment strategy, and most of all which are the answers and which are the questions asked in the first place. Hence an open dialogue from the involved parties should clarify issues such as training of the personnel, sample storage, materials to be used, duty assignment, and other often underestimated factors that may introduce biases. Meticulous preparations are a prerequisite especially in large [24] or multicenter

Table 5.1 List of common analytical problems and possible causes in the practice of LC-MS-based metabolomics.

Potential analytical problems	Possible causes
Background noise builtup	Unsuitable gradient, inadequate sample preparation, insufficient washing between injections, improper wash solvents in injector
Systematic signal drop	Source contamination
High background signal or series of mass peaks separated by 44 amu	Contamination, for example, from plasticizers such as PEG (vials, tubings, solvents, detergents, creams, and so on)
Inconsistent peaks	ESI parameters improper, LC-MS conditions not optimized
Mass imprecision	Spectrometer calibration needed
"Poor" chromatograms showing a small number of broad peaks	Low MS scanning speed, improper sample preparation
Shifting chromatographic peaks	Insufficient column conditioning, column failure, unstable LC system

studies. To ensure a high quality level of the generated data and thus hold promise for a valuable study, caution is needed to avoid mistakes or lack of skill at practicalities or sample collection and handling. Factors that should be considered in study design include age, gender, animal strain or plant species/cultivar type, diet, stress conditions, diurnal variation, and environmental factors, and so on. For studies involving humans, additional factors include body mass index (BMI) and habitual factors. Especially for diseased humans or those under treatment, matching with an appropriate control group is often a particular source of difficulty. Diurnal variation may provide a great source of variability in either humans or plants. Recently, the effect of circadian rhythm on the metabolite profile of rodent blood was investigated using LC-MS [25]. Hundreds of metabolites were found to be dependent on the time of sampling so that their concentration their concentrations could be used to calculate the internal time of mice.

For blood and tissue samples, the time until freezing is very important because ongoing biochemical reactions may modify the metabolic content. Minor issues should also be carefully considered (e.g., the brand of vials used for sampling). Vials or tubing may introduce contaminants such as surfactants and plasticizers (PEG or other such polymers) causing serious interference in MS detection.

To reach a satisfactory level of sample processing and handling, it may be necessary to explain the difference between holistic and targeted assays to the experimentalist in the field, the clinic, or the factory. For example, sampling of blood in hospitals, and so on, are typically performed by busy medical personnel. One may expect that the time left for clotting may not be identical to all samples. In current clinical analysis or drug metabolism assays, clotting times from 30 min to

1 h are typically used; however, in nontargeted omics assays, this 30 min difference can be expected to make a big difference as enzymatic reactions will keep going. Recent studies show that when using longer incubation times ongoing metabolism consumed nutrients such as glucose and aminoacids while at the same time, increasing metabolism products such as lactic and pyruvic acid [26, 27].

Sample stability on storage should also be investigated, although it may not be possible to identify and obtain a suitable profile point to compare results after some months of storage. In large studies where samples are collected and stored for long periods (i.e., more than a year), how can a researcher differentiate between markers of interest and markers of sample degradation? Blood plasma or plant crude extracts still contain a considerable amount of enzymes and even stored in the freezer, samples do change. Freeze-thaw cycles may also affect the sample integrity. The reader is directed to specific studies on this topic [28, 29], where stability of frozen (-20 or $-80\ °C$) urine is found acceptable for up to six months; however, the reader should also be advised that these studies report only the findings of the applied analytical technology and analysis via other technologies may have given a less positive result.

Overall, it is our experience that before commencing a metabonomics/metabolomics experiment, scientists from all disciplines should work together to clarify the smallest detail: this will help to ensure that appropriate measures are taken to avoid introducing unwanted biases. In this perspective, the experience gained in proteomics studies can be exploited [30]. It is also important that such matters are discussed in scientific meetings and that standard operating protocols are distributed and debated between metabolomics laboratories.

5.4
Sample Preparation

Sample preparation is a discrete and important part of any analytical process. In fact, it is the most error prone and the most time- and resources-consuming step of most chemical analyses. Despite major advances in automation and the utilization of robotic assays, a large part of sample preparation still relies on human work, often shared between different experimentalists. This fact increases the risks of mishandling and the possibilities for errors. In targeted analysis, the scope of sample preparation is often to purify the analyzed substances by selectively retaining these in the sample along the sample preparation procedure while at the same time removing interferences. In holistic analysis of the type undertaken in metabolomic work, a very different perspective should be adapted as sample preparation should avoid altering the sample composition. At the same time, sample preparation should promote the state of the sample from a raw sample or specimen to a sample ready for analysis. In accordance to the analytical method selected, different steps are taken, for example, derivatization in GC-MS. In LC-MS, important parameters that need to be considered are the following: (i) the need to preserve the performance of the LC system and column by removing particulate matter and macromolecules that may clog the system; (ii) the need to maintain the MS source clean and optimally functioning for the whole length of the run to avoid gradual decrease in the total signal; and (iii) matrix effects that may result in ion suppression/enhancement.

The effect of sample preparation on the obtained metabolic profile has been of increasing interest recently. It is understood than whatever method applied will edit the sample. It is recognized that it is of utmost importance to understand the effects of these manipulations in different types of samples either of human, mammalian, or plant origin [31, 32]. Otherwise there is the risk of misinterpreting biases introduced from sample preparation as real trends related to biology or physiology, and thereby be lead to a dead end or false hypothesis. A number of publications report the study of extraction methodologies before metabolite profiling.

The extent of sample preparation depends on the type of specimen analyzed: for urine very simple preparation schemes limited to centrifugation and dilution are often applied [22, 28]. In the case of protein-rich samples (such as blood), proteins have to be removed before injection. Blood being probably the most important biofluid in the life sciences has been the focus of numerous investigations. Although plasma and serum are considered practically equal with regard to the small molecule content, recent evidence shows that this is not true and there are certain differences [26, 33]. Plasma is the preferred specimen (Figure 5.2a) but two of the biggest metabolomics initiatives selected serum (Human metabolome project [34] and HUSERMET [35]). Protein removal from plasma/serum is typically done with the addition (3 : 1 v/v) of organic solvents (for which role methanol is preferred vs acetonitrile as shown in Figure 5.2b) resulting in protein precipitation and then centrifugation. However, solid phase extraction (SPE) [36] turbulent flow chromatography [37] and dried blood spots [38] can also be used.

Extracting small molecules from solid samples is not as straightforward as dealing with blood plasma/serum because the solvent should in this case solubilize the metabolites, removing them from a rather dense network of membranes and macromolecules where such molecules are often very strongly bound. To study such effects, different combinations of solvents have been tested for the extraction from solid tissue (liver, plant, fruit), as it has also been done with biofluids (blood, urine) [31, 32, 39]. Solid tissues such as organs are typically cryolysed before extraction with a combination of solvents. Tissue analysis may reveal new markers at the focal point of interest (e.g., in inflammation center or in a cancer tumor). For the profiling of solid tissues, it is important to remove as much blood as possible before storage (through rinsing of the tissue), or else there is the risk that the subsequent metabolic profile will be a combination of both blood and the tissue of interest [40, 41].

Overall, sample preparation before LC-MS-based metabolite profiling is recently recognized as an important topic. We can expect that research in this research line will continue to provide new data and useful information.

102 | *5 LC-MS-Based Nontargeted Metabolomics*

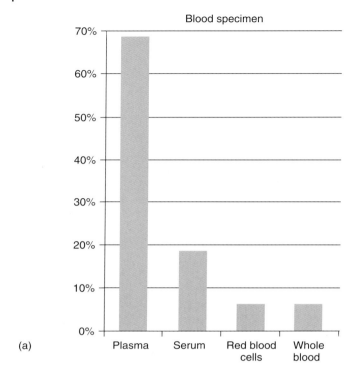

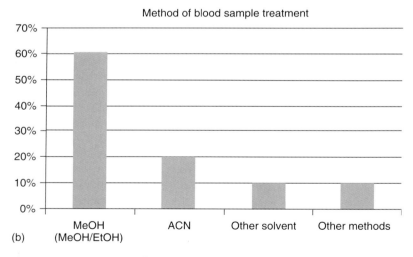

Figure 5.2 (a) Percentage of reports that utilize different types of blood-derived types in metabolomics analysis. (Data taken from bibliographic search in Scopus October 2011.) (b) Distribution of methods used for the sample preparation of blood prior to LC-MS metabolomics analysis. (Data taken from bibliographic search in Scopus October 2011.)

5.5
Analytical Strategies

LC-MS has become one of the most powerful tools in bioanalysis. LC-MS is used also routinely for monitoring pollutants, pharmaceuticals, or for the determination of key analytes in clinical chemistry. However, this does not mean that this is a job for robotics or untrained personnel. In contrast, it is widely accepted that in order to be compliant with the ever increasing demands with regard to sensitivity, specificity, and validation, the infrastructure and the personnel should function very effectively. The above-mentioned cases deal with targeted analysis where methods are developed in order to determine a small number of target analytes. In most cases, such assays involve LC-MS/MS analysis that, especially when used in multiple reaction monitoring mode (MRM) on triple quadrupole machines, provides high specificity and assurance of close to unambiguous quantitative determinations. Decades of years work in targeted analyses has lead to the development and thorough validation of methods and strategies to address virtually any potential pitfall: matrix effects can be evaluated, internal standards can be used to correct for analyte loss during sample preparation, and so on. However, in untargeted analysis, such measures cannot be applied, and hence alternative strategies should be sought. The addition of internal standard(s) for holistic metabolic profiling remains a field open for debate. Internal standards are used to correct for analyte losses or other unwanted effects during the analytical process. In targeted MS analysis, the ideal internal standard is a deuterated version of the analyte molecule, present at a concentration within the range of the expected target analyte concentrations. However, in holistic analysis, the addition of hundreds of internal standards is not an option as being impractical, bearing high cost in economics and sample throughput, while at the same time increasing the error risk. Addition of a number of internal standards representing selected important metabolite groups may provide a compromise but experience in the analysis of drugs using, for example, a compound with an extra methyl group as an internal standard for the target analyte suggest that this approach is not always valid.

The length of the analytical run is determined on the basis of (i) the stability of the column and (ii) the scale of contamination of the ion source on injection of a number of samples. It should be stressed that it is essential to maintain high retention time repeatability (to facilitate worthy peak alignment). As a result, the implementation of sufficient cleanup steps is essential, while, as also explained later in the text, the design of the experiment should seriously consider such limitations. It is wiser to preschedule shorter analytical batches employing a column control step or a "regeneration step" along with a fast ion source cleanup process, rather than risking days of experimental or data treatment work or having precious samples being analyzed on a noncompliant system.

The test samples should be analyzed in a random order in a way to help in recognizing and overcoming analysis time trends, which unfortunately are often observed. Quality control samples (QC) are now used by several groups in order to monitor the performance of the system. QC samples are often made by pooling from the biological test samples thus representing a bulk control sample [22]; however, commercially available samples have also been used for this purpose especially in the case of large studies. QC samples should be analyzed repeatedly throughout the sequence, that is, the set of injections in an analysis series, and the data obtained from them should be assessed against predefined criteria in order to facilitate compliance of the analytical batch. The QC samples are used to assess the analytical performance of the system, to correct possible drifts [35], and also to help fusing data from different analytical batches, for example, different time periods data. More details on the different ways for data scrutiny are given in Section 5.6.

A number of injections from the QC sample at the beginning of the analytical run can be employed in order to "condition" the system. Such conditioning injections can be used to assess system suitability before beginning the main run of test samples. Injection of a mixture of standards can also be used to provide data for the assessment of system suitability via examination of peak shape, retention time stability and required signal intensity, and so on.

5.6
Data Analysis

In holistic metabolite analysis, the mass spectrometer operates in full scan mode typically from about 70 to 1200 amu in order to cover the width of masses for small molecule metabolites. A typical raw data file contains all the information collected as a series of full scan mass spectra acquired in successive time points (a scan event typically last 2–20 ms). Hence an LC-MS data file can be considered as a cube constructed by joining a large number of planes: each plane represents a 2D full scan mass spectrum that is defined by mass (m/z) and signal intensity at the two axes (Figure 5.3). Such files contain a multitude of systematic noise along with the maximum possible information content provided by a single contemporary analytical platform. To be in the position to effectively investigate between hundreds of such bulky data files, a researcher needs to use special data treatment tools. For this reason, a number of software packages have been developed that involve a series of processes: noise filtering, baseline correction, centering, normalization, peak picking, peak integration, and alignment. The task for these tools is to detect all the peaks within a chromatographic run and produce a two-dimensional data matrix: a peak table where all samples and all peaks along with their signal intensities are reported. This is done by compressing the three dimensionality of LC-MS data to two dimensions, by combining the mass with the retention time information into a feature identity.

Some of the peak-picking software programs are open-source software, which are freely available on the internet. Examples of this type of software are found in programs such as MZmine [42], MetAlign [43], and XCMS [44]. These software programs analyze data from any mass spectrometer after "translation" to a universal format such as netCDF or mzXML. Other groups developed their own scripts (often in Matlab) to perform such tasks. With Matlab or, for example, XCMS (which operates in R), one can customize the peak-picking algorithm to fit the

5.6 Data Analysis | 105

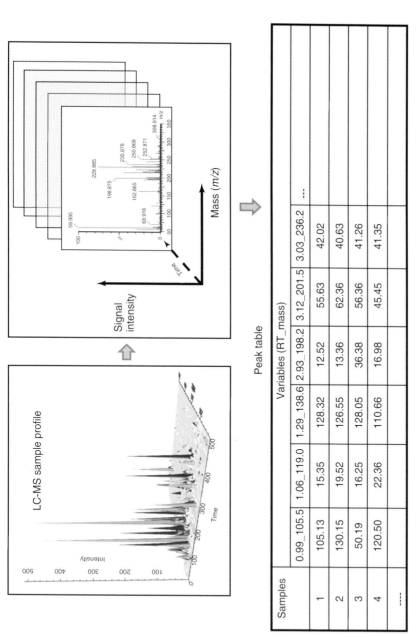

Figure 5.3 Simplified schematic of the process of peak-picking and alignment algorithm. A 3D chromatogram is "sliced" into planes by dividing retention time into spectral "events." Each spectrum is a 2D plane with m/z in one axis and signal intensity on the other. Signals found are collected in tables for each sample; in these tables ions are identified as features. Peak alignment fuses these tables into a large table, such as the one shown above where each row represents a sample and each column a feature. This table is subsequently subjected to multivariate analysis. Such peak tables can be exported to any spreadsheet program where they can be examined for, for example, number of zero values, accumulated total signal, and other factors.

specific needs of a certain analysis: for example, HILIC-MS profiling of small polar molecules eluting in rather broad peaks or GC-MS with narrow peaks of derivatized metabolites.

MS manufacturers have developed their own metabolomic software provided with their MS operating platform, which can process only data files generated in the same platform and in a mass spectrometer from the same manufacturer (e.g., MarkerLynx from Waters, analyses data from Waters instruments; MarkerView analyses data from AB Sciex instruments). Such software programs do not allow modifications or alterations in the way the algorithm operates, effectively limiting interventions from the researcher. Such software can also perform univariate and multivariate statistics such as principal component analysis (PCA) or even discriminant analysis, providing also advanced visualization tools (scores plots, trends plots, etc.) and marker identification functionalities (such as incorporated database search and formulae prediction).

Our experience is that parallel to examination by multivariate statistics, such peak tables should be inspected using different acceptance criteria, including univariate statistical analysis.

The number of features (variables) should always be scrutinized: large numbers of features do not mean that the analysis is of high quality but may rather mean failed peak alignment or that too small time window or inappropriate intensity threshold settings have been selected. The latter flaw may practically mean that the algorithm will assign noise as metabolite features. The number of features detected should not be taken as equivalent to the number of metabolites detected, as one metabolite may give rise to numerous features in the form of isotope peaks adducts, dimers, or fragments. To overcome such issues, one can use adduct-finding scripts and apply deisotoping of data. Such scripts can be developed in R, Matlab, or even in Excel (see, for example, CAMERA which works with XCMS). Most of the peak-picking software programs (including the commercially available ones) can perform deisotoping.

Possible error sources in data analysis include (i) varying mass accuracy often observed at the peak inflection points and/or in peaks of low intensity; (ii) failed peak alignment; and (iii) variations in chromatography. All these may also contribute in the detection of "ghost" features: for example, features detected as RT_m/z pair: 1.345_208.9045 and 1.332-208.9078 probably originate from the same molecule, but for one or more of the above-mentioned reasons, one feature can be detected in half of the samples and the second can be detected in the other half, or it can be that both features are detected in some samples.

Correcting or even controlling peak alignment can prove difficult. The selection of peak picking and alignment parameters (e.g., mass and retention time window, mass tolerance, and so forth) should better be made after the examination of a number of peaks that are spread in m/z and retention time axis. Such peaks should be examined in different analytical injections to obtain a trustworthy measure of the ideal parameters (mass or retention time windows and tolerances) to be applied. After peak extraction, thorough data scrutiny is mandatory. Different rules can be applied in order to check for data quality. (i) The number of zero values in a data set or a sample should not exceed a certain threshold, for example, 40%. (ii) The 80% rule dictates that features found in less than 20% of the analyzed samples are removed as being noise or irrelevant peaks [24]. (iii) Analysis of QC samples can provide a means for monitoring the quality of data and stability of the analytical system (Figure 5.4a) [22, 45]; in this case, the peak tables generated by the analyses of QC samples is subjected to statistical analysis versus predefined criteria: retention time and signal intensity should be repeatable, for example, by showing coefficient of variation less than 30% (Figure 5.4b) for the majority of the detected peaks (more than 70% of the peaks). (iv) Features should be considered as potential markers only if several different tests both multivariate (S-plots, loading plots, VIP value/plot) and univariate (ANOVA, T-test with $p < 0.05$) signify them as important. (v) Such markers should show signal stability in the QC samples.

The data should also be examined with conventional analysis as extracted ion chromatograms for a number of selected ions. Ions should be selected along the retention time and mass range to check for the performance of the analytical system along the run. In the case of significant retention time drift or time trends, one option is to develop special software tools to correct for such issues, as reported by Zelena *et al.* and Dunn *et al.* [35, 46] and is shown in Figure 5.4e.

For peaks highlighted by statistical analysis as potential markers, the stability of retention time, mass detected, and peak area in the QC samples should be evaluated with predefined acceptance criteria (e.g., retention time RSD $< 3\%$). Katajama and Oresic [47] have reviewed the data mining tools for LC-MS metabolomics. Figure 5.4 provides a nonexhaustive roadmap of strategies used for data scrutiny of LC-MS generated metabolomics data.

To conclude, data analysis is one of the most time-consuming steps of any metabolomics investigation. Typically, one can expect to spend a week on sample preparation and analysis (if methods are established) and then several weeks in data mining. Data analysis should better be performed by expert chemometricians/informaticians, but the contribution of analytical chemists/biochemists is also very important.

5.7
Metabolite Identification

With the current state of the art, biomarker identification represents after data analysis a major bottleneck in nontargeted LC-MS-based metabolomic studies. This is a result of a number of obstacles: (i) mass spectra may differ between experiments and instruments because of variation in adduct formation; (ii) the utility of spectral and retention time libraries in LC-MS is limited; and as a result, (iii) open access data repositories for LC-MS metabolomics are not at the level of genomic research. Consequently, LC-MS databases built by certain researchers are not interchangeable between laboratories. Hence the effort needed for the identification of markers is often disproportional as additional experiments (e.g., NMR or GC-MS analysis) may also be required.

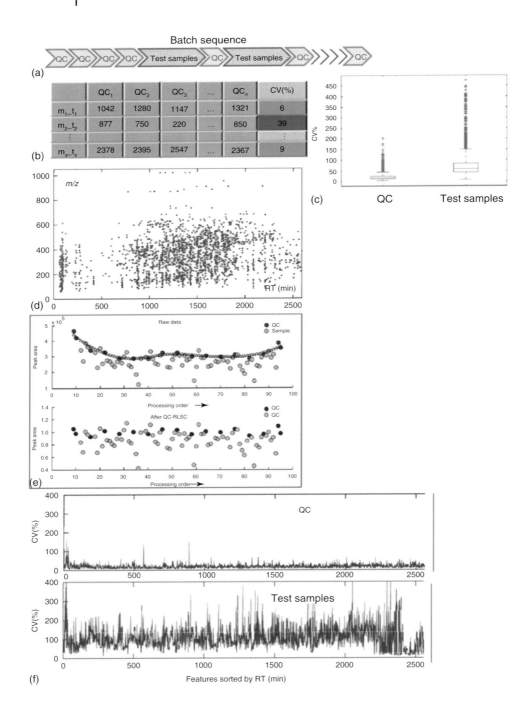

Fragmentation of work effort is a problem. We can expect that researchers from different institutes may spend weeks of work on the identification of the same metabolite in a certain type of samples. Web-based resources or data repositories for MS metabolomics are not as well established as for proteomic or genomic research [30] but developments in that direction are being made (see [48] for a recent review of the topic). Metabolite identification is often performed with the utilization of laboratory-built spectral databases. Tailor-made software scripts can automatically correlate high mass accuracy data with isotope ratio and MS/MS data. High mass accuracy data is necessary but the safest approach is to combine the latter with database searching in (i) general mass spectral databanks such as METLIN (http://metlin.scripps.edu/), (ii) specialized spectral databanks (e.g., lipidmaps) (http://www.lipidmaps.org/), and (iii) biochemical or other databases (HMDB (http://www.hmdb.ca/), Chemspider (http://www.chemspider.com/). With this approach, a shortlist of potential candidates is generated and this can be further reduced by applying the nitrogen rule or other calculations (e.g., Ring Double Bond Equivalent). In the end, however, the analysis of authentic standards is the best means to reach unambiguous identification: retention time, full scan mass, and product ion masses (in the same experimental conditions) should be compared between the unknown metabolite and the peak(s) obtained from the analysis of samples spiked with putatively identified metabolite(s). A comprehensive review covering marker identification based on MS data has been recently published [49].

5.8
Applications

LC-MS-based metabolomics research is increasingly used in order to find new biomarkers. Before reaching human studies, investigation of animal models of disease is often applied. There can be different experimental models for certain pathophysiological conditions, for example, cancer or arthritis, and one can expect that the metabolic profiles of such models differ substantially. However, animal

Figure 5.4 Schematic of data scrutiny strategy. (a) The batch sequence shows that QC samples are analyzed in the beginning and then sporadically in the batch. Test samples should be randomized. (b) A peak list from the QC samples should be examined for variability (CV < 30%). (c) Test samples should show much higher variability than the QC samples as shown in the corresponding box plots. (d) Variability of features (red dots represent features with CV > 30% in the QCs) should be inspected in the RT and m/z axis. (e) Use of QC injections to improve data quality: a low-order nonlinear locally estimated smoothing function (LOESS) is fitted to the QC data. A correction curve for the whole analytical run is interpolated (upper pane). The total data set is normalized (lower pane). Attenuation of peak responses over time can thus be minimized. (Source: Reproduced from Dunn et al. [46] with permission from Nature.) (f) The variability of features in the QCs can be plotted along the retention time. QCs show satisfactory stability while test samples show as indeed expected wider variation. Several other criteria can be applied to evaluate the goodness of the data in both QCs and test samples.

models offer the possibility to study organisms under controlled conditions thus eliminating a number of confound factors (e.g., genetic variation, food, or environmental parameters). Animal models of obesity [50], cancer hepatopathy [51], nephrotoxicity [52], and several other conditions have been studied using LC-MS. Before application in the clinic or the industry, we would expect that biomarkers are thoroughly validated and their concentrations are determined by targeted methods. In the following paragraphs, we try to illustrate the potential of the field highlighting the characteristic applications related to human health.

As a priority topic, cancer is a major field for biomarker discovery, and hence metabolomics approaches have been used to help in understanding cancer biochemistry. Among other investigations, the effects of renal cell carcinoma [53–55] and colorectal cancer [56] were studied by mapping the urinary metabolome. The composition of the serum metabolome has also been investigated in colorectal cancer [57]. As metabolomics is often a hypothesis-free study of the unknown, some rather "unusual" specimens have been the target of metabo-mapping. Hence metabolomics analysis of saliva samples from patients suffering from oral squamous cell carcinoma (OSCC) or oral leukoplakia (OLK) has been reported using RP gradient UPLC-MS [58]. Figure 5.5 provides an illustration of the utility of the technology in biomarker discovery depicting characteristic LC-MS traces from OSCC, OLK, and controls, along with OPLS-DA scores plots, S-plots used to find differentiating ions and box plots used to further verify the validity of two markers: lactic acid and valine. S-plots provide a plot of contribution of variables (covariance) versus variable confidence (correlation). Variables contributing highly to the differentiation are located at the ends of the plot. Receiver operating characteristic (ROC) curve evaluates the predictive power of discriminant metabolites. Metabolite signatures that have the largest area under the ROC curve are identified as those with the strongest predictive power (Figure 5.5d).

Fecal metabolome analysis aimed to compare the profiles from healthy individuals and patients with liver cirrhosis or hepatocellular carcinoma [59]. A human breast cancer cell line (MCF-7) was studied with the aim of developing efficient methods for sample extraction to facilitate the profiling of intracellular metabolites [60]. Investigation of tissues urine and plasma from prostate cancer patients has recently shown increased levels of sarcosine during prostate cancer progression to metastasis. This study offered high potential because of the fact that the marker could also be detected noninvasively in urine [61], although both claims have recently been disputed [62, 63].

Infection represents a major health threat especially against "weak" or "weakened" human patients. A pilot study used UPLC-MS to profile plasma and urine from 11 children and community controls with severe pneumonia [64] in an attempt to find disease markers.

Alzheimer's disease (AD) becomes an increasing problem especially for the Western world. The onset of the disease typically precedes clinical symptoms (dementia) by decades, and hence diagnostic biomarkers are sought. In a recent study, plasma of AD patients was UPLC-MS analyzed for potential biomarkers and was compared to control group of matched healthy individuals [65].

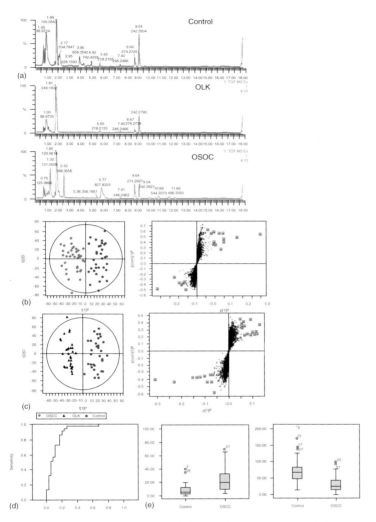

Figure 5.5 Utility of UPLC-MS metabolic profiling in the discovery of disease biomarkers. (a) UPLS-QTOF-MS positive ionization chromatograms of saliva samples. From top to the bottom: oral squamous cell carcinoma (OSCC), leukoplakia (OLK), and healthy control. (b) OPLS-DA scores plots (left) and S-plots (right) of OSCC group (red diamonds) and control group (blue dots). S-plot highlights discriminating metabolites (highlighted in red boxes). (c) OPSL-DA scores plots (left) and S-plots (right) of OLK group (black triangles) and control group (blue dots). S-plot highlights discriminating metabolites (in red boxes). (d) Receiver operating characteristic (ROC) curve analysis for the predictive power of combined salivary biomarkers for distinguishing OSCC from healthy control. The final model included two salivary biomarkers, lactic acid and valine. Using a cutoff probability of 50%, sensitivity of 86.5% and specificity of 82.4% was obtained. The calculated area under the ROC curve was 0.89 (95% confidence intervals, 0.813–0.972). (e) Box plots of the two characteristic markers that distinguish OSCC patients from healthy controls.

Gynecology and obstetrics represent a major market for biomarker discovery. In fact one of the first commercialized applications of metabolomics was the NIR-spectroscopy-based kit developed by Molecular Biometrics for the increase of success rate for *in-vitro* fertilization [66]. Metabolomics has also been used in order to study pregnancy complications such as pre-eclampsia (PE) [67, 68].

Metabolic disorders such as diabetes and obesity can be considered a clinical topic where metabonomics offers high potential. Indeed UPLC-Q-TOF-MS was used to [69] compare the metabolic profiles of overweight/obese men and age-matched normal-weight men. Overweight/obese men were found with higher levels of Homeostasis Model Assessment Insulin Resistance (HOMA-IR) triglycerides, total cholesterol, and LDL-cholesterol, and lower concentrations of HDL-cholesterol and adiponectin compared to control subjects. Type II diabetes is becoming an increasing problem for both the western societies and the developing world. As a result, numerous studies of the metabolic consequences of diabetes or therapeutic schemes have been performed in human individuals mostly using UPLC-MS [70, 71].

5.9
Synopsis

In the past years, MS-based metabolomics has progressed greatly from a small topic in between analytical chemistry and plant biotechnology or life sciences to a major intersectorial research theme with growing application and penetration in life science, food, environmental, and other sciences [72]. LC-MS in particular is the analytical platform with the highest increase in applications in metabolomics. Large investments are directed by instrument/software developers and research institutes in the development of analytical equipment, software tools, and methodologies in order to enhance the level of sophistication of analytical science applied to metabolomics. Mass resolution is continually increased, while the development of more advanced liquid phase separation platforms (U(H)PLC or nanoLC type) and separation media (e.g., HILIC) provide new alternative and powerful tools that allow profiling of samples and/or metabolites, which till recently could not be reached. For example, new HILIC-MS(/MS) methodologies allow the determination of large numbers of polar metabolites in a single run; recent high-resolution MS (either Orbitrap or TOF-MS) provides sub-2-ppm mass accuracy and facilitates the identification of unknown molecules combining with different MS/MS functionalities. Detection sensitivity is increasing revealing markers that had been under the detection limit of previous technologies or equipment. Data treatment and visualization tools have been developed that improve the data mining process and allow statistical analysis of complex data and extraction/perception of information by nonexperts.

Validation of findings of metabolomics research using different approaches is necessary. It is strongly recommended that LC-MS analyses are performed again on different methods/instruments to verify the dependability of found markers. Finally, clear quantification should be performed and concentrations should be compared. Repeating the experimental clinical/animal study is also strongly recommended. Remaining challenges concern protocol and data reporting standardization, system robustness, and lack of reproducibility, and these matters can cause difficulties as profiling results are not interchangeable between researchers. Furthermore, tracking back to previous analyses shows little promise for the unambiguous identification of key markers. The field is growing fast, and we can expect more coordinated large-scale efforts in this direction in the coming years. Such initiatives are necessary to overcome fragmentation and to promote the field.

References

1. Nicholson, J.K., Connelly, J., Lindon, J.C., and Holmes, E. (2002) *Nat. Rev. Drug Discov.*, **1**, 153–161.
2. Nicholson, J.K. and Lindon, J.C. (2008) *Nature*, **455**, 1054–1056.
3. Last, R.L., Jones, A.D., and Shachar-Hill, Y. (2007) *Nat. Rev. Mol. Cell Biol.*, **8**, 167–174.
4. Cavaliere, B., De Nino, A., Hayet, F., Lazez, A., Macchione, B., Moncef, C., Perri, E., Sindona, G., and Tagarelli, A. (2007) *J. Agric. Food Chem.*, **55**, 1454–1462.
5. Theodoridis, G., Gika, H.G., and Wilson, I.D. (2008) *Trends Anal. Chem.*, **27**, 251–260.
6. Theodoridis, G.A., Gika, H., and Wilson, I.D. (2011) *Mass Spectrometry Reviews*, in press. **30**, 884–906.
7. Overy, D.P., Enot, D.P., Tailliart, K., Jenkins, H., Parker, D., Beckmann, M., and Draper, J. (2008) *Nat. Protoc.*, **3**, 471–485.
8. Myint, K.T., Uehara, T., Aoshima, K., and Oda, Y. (2009) *Anal. Chem.*, **81**, 7766–7772.
9. Uehara, T., Yokoi, A., Aoshima, K., Tanaka, S., Kadowaki, T., Tanaka, M., and Oda, Y. (2009) *Anal. Chem.*, **81**, 3836–3842.
10. Boernsen, K.O., Gatzek, S., and Imbert, G. (2005) *Anal. Chem.*, **77**, 7255–7264.
11. Yin, P., Wan, D., Zhao, C., Chen, J., Zhao, X., Wang, W., Lu, X., Yang, S., Gu, J., and Xu, G. (2009) *Mol. BioSyst.*, **5**, 868.
12. Spagou, K., Tsoukali, H., Raikos, N., Gika, H., Wilson, I.D., and Theodoridis, G. (2010) *J. Sep. Sci.*, **33**, 716–727.
13. Spagou, K., Wilson, I.D., Masson, P., Theodoridis, G., Raikos, N., Coen, M., Holmes, E., Lindon, J.C., Plumb, R.S., Nicholson, J.K., and Want, E.J. (2011) *Anal. Chem.*, **83**, 382–390.
14. Gika, H.G., Theodoridis, G.A., and Wilson, I.D. (2008) *J. Sep. Sci.*, **31**, 1598–1608.
15. Idborg, H., Zamani, L., Edlund, P.-O., Schuppe-Koistinen, I., and Jacobsson, S.P. (2005) *J. Chromatogr. B*, **828**, 9–13.
16. Matyska, M.T., Pesek, J.J., Duley, J., Zamzami, M., and Fischer, S.M. (2010) *J. Sep. Sci.*, **33**, 930–938.
17. Hinterwirth, H., Lämmerhofer, M., Preinerstorfer, B., Gargano, A., Reischl, R., Bicker, W., Trapp, O., Brecker, L., and Lindner, W. (2010) *J. Sep. Sci.*, **33**, 3273–3282.
18. Nordström, A., Want, E., Northen, T., Lehtiö, J., and Siuzdak, G. (2008) *Anal. Chem.*, **80**, 421–429.
19. Li, Y., Shrestha, B., and Vertes, A. (2008) *Anal. Chem.*, **80**, 407–420.
20. Li, X., Hu, B., Ding, J., and Chen, H. (2011) *Nat. Protoc.*, **6**, 1010–1025.
21. Jackson, A.U., Werner, S.R., Talaty, N., Song, Y., Campbell, K., Cooks, R.G., and Morgan, J.A. (2008) *Anal. Biochem.*, **375**, 272–281.
22. Gika, H.G., Theodoridis, G.A., Wingate, J.E., and Wilson, I.D. (2011) *J. Proteome Res.*, **6**, 3291–3303.

23. Gika, H.G., Theodoridis, G.A., Earll, M., Snyder, R.W., Sumner, S.J., and Wilson, I.D. (2010) *Anal. Chem.*, **82**, 8226–8234.
24. Bijlsma, S., Bobeldijk, I., Verheij, E.R., Ramaker, R., Kochhar, S., Macdonald, I.A., van Ommen, B., and Smilde, A.K. (2006) *Anal. Chem.*, **78**, 567–574.
25. Minami, Y., Kasukawa, T., Kakazu, Y., Iigo, M., Sugimoto, M., Ikeda, S., Yasui, A., van der Horst, G.T.J., Soga, T., and Ueda, H.R. (2009) *Proc. Natl. Acad. Sci.*, **106**, 9890–9895.
26. Dettmer, K., Almstetter, M.F., Appel, I.J., Nürnberger, N., Schlamberger, G., Gronwald, W., Meyer, H.H.D., and Oefner, P.J. (2010) *Electrophoresis*, **31**, 2365–2373.
27. Liu, L., Aa, J., Wang, G., Yan, B., Zhang, Y., Wang, X., Zhao, C., Cao, B., Shi, J., Li, M., Zheng, T., Zheng, Y., Hao, G., Zhou, F., Sun, J., and Wu, Z. (2010) *Anal. Biochem.*, **406**, 105–112.
28. Gika, H.G., Theodoridis, G.A., and Wilson, I.D. (2008) *J. Chromatogr. A*, **1189**, 314–322.
29. Maher, A.D., Zirah, S.F.M., Holmes, E., and Nicholson, J.K. (2007) *Anal. Chem.*, **79**, 5204–5211.
30. Arita, M. (2009) *Curr. Opin. Biotechnol.*, **20**, 610–615.
31. Bruce, S.J., Tavazzi, I., Parisod, V., Rezzi, S., Kochhar, S., and Guy, P.A. (2009) *Anal. Chem.*, **81**, 3285–3296.
32. Theodoridis, G., Gika, H., Franceschi, P., Caputi, L., Arapitsas, P., Scholz, M., et al. (2012) *Metabolomics*, **8**, 175–185.
33. Denery, J.R., Nunes, A.A.K., and Dickerson, T.J. (2011) *Anal. Chem.*, **83**, 1040–1047.
34. Psychogios, N., Hau, D.D., Peng, J., Guo, A.C., Mandal, R., Bouatra, S., Sinelnikov, I., Krishnamurthy, R., Eisner, R., Gautam, B., Young, N., Xia, J., Knox, C., Dong, E., Huang, P., Hollander, Z., Pedersen, T.L., Smith, S.R., Bamforth, F., Greiner, R., McManus, B., Newman, J.W., Goodfriend, T., and Wishart, D.S. (2011) *PLoS ONE*, **6**, e16957.
35. Zelena, E., Dunn, W.B., Broadhurst, D., Francis-McIntyre, S., Carroll, K.M., Begley, P., O'Hagan, S., Knowles, J.D., Halsall, A., Wilson, I.D., and Kell, D.B. (2009) *Anal. Chem.*, **81**, 1357–1364.
36. Michopoulos, F., Lai, L., Gika, H., Theodoridis, G., and Wilson, I. (2009) *J. Proteome Res.*, **8**, 2114–2121.
37. Michopoulos, F., Edge, A.M., Theodoridis, G., and Wilson, I.D. (2010) *J. Sep. Sci.*, **33**, 1472–1479.
38. Michopoulos, F., Theodoridis, G., Smith, C.J., and Wilson, I.D. (2010) *J. Proteome Res.*, **9**, 3328–3334.
39. Bruce, S., Jonsson, P., Antti, H., Cloarec, O., Trygg, J., Marklund, S., and Moritz, T. (2008) *Anal. Biochem.*, **372**, 237–249.
40. Masson, P., Alves, A.C., Ebbels, T.M.D., Nicholson, J.K., and Want, E.J. (2010) *Anal. Chem.*, **82**, 7779–7786.
41. Masson, P., Spagou, K., Nicholson, J.K., and Want, E.J. (2011) *Anal. Chem.*, **83**, 1116–1123.
42. Katajamaa, M., Miettinen, J., and Orešiè, M. (2006) *Bioinformatics*, **22**, 634–636.
43. Lommen, A. (2011) *Anal. Chem.*, **81**, 3079–3086.
44. Smith, C.A., Want, E.J., O'Maille, G., Abagyan, R., and Siuzdak, G. (2006) *Anal. Chem.*, **78**, 779–787.
45. Lai, L., Michopoulos, F., Gika, H., Theodoridis, G., Wilkinson, R.W., Odedra, R., Wingate, J., Bonner, R., Tate, S., and Wilson, I.D. (2010) *Mol. Biosyst.*, **6**, 108–120.
46. Dunn, W.B., Broadhurst, D., Begley, P., Zelena, E., Francis-McIntyre, S., Anderson, N., Brown, M., Knowles, J.D., Halsall, A., Haselden, J.N., Nicholls, A.W., Wilson, I.D., Kell, D.B., and Goodacre, R. (2011) *Nat. Protoc.*, **6**, 1060–1083.
47. Katajamaa, M. and Oresic, M. (2007) *J. Chromatogr. A*, **1158**, 318–328.
48. Tohge, T. and Fernie, A.R. (2009) *Phytochemistry*, **70**, 450–456.
49. Werner, E., Heilier, J.-F., Ducruix, C., Ezan, E., Junot, C., and Tabet, J.-C. (2008) *J. Chromatogr. B*, **871**, 143–163.
50. Loftus, N., Miseki, K., Iida, J., Gika, H.G., Theodoridis, G., and Wilson, I.D. (2008) *Rapid Commun. Mass Spectrom.*, **22**, 2547–2554.
51. Whitfield, P.D., Noble, P.-J.M., Major, H., Beynon, R.J., Burrow, R., Freeman, A.I., and German, A.J. (2005) *Metabolomics*, **1**, 215–225.

52. Chen, M., Su, M., Zhao, L., Jiang, J., Liu, P., Cheng, J., Lai, Y., Liu, Y., and Jia, W. (2006) *J. Proteome Res.*, **5**, 995–1002.
53. Ganti, S., Taylor, S.L., Kim, K., Hoppel, C.L., Guo, L., Yang, J., Evans, C., and Weiss, R.H. (2012) *Int. J. Cancer*, **130**, 2791–800.
54. Kim, K., Aronov, P., Zakharkin, S.O., Anderson, D., Perroud, B., Thompson, I.M., and Weiss, R.H. (2009) *Mol. Cell Proteomics*, **8**, 558–570.
55. Kind, T., Tolstikov, V., Fiehn, O., and Weiss, R.H. (2007) *Anal. Biochem.*, **363**, 185–195.
56. Wang, W., Feng, B., Li, X., Yin, P., Gao, P., Zhao, X., Lu, X., Zheng, M., and Xu, G. (2010) *Mol. BioSyst.*, **6**, 1947.
57. Qiu, Y., Cai, G., Su, M., Chen, T., Zheng, X., Xu, Y., Ni, Y., Zhao, A., Xu, L.X., Cai, S., and Jia, W. (2009) *J. Proteome Res.*, **8**, 4844–4850.
58. Wei, J., Xie, G., Zhou, Z., Shi, P., Qiu, Y., Zheng, X., Chen, T., Su, M., Zhao, A., and Jia, W. (2011) *Int. J. Cancer*, **129**, 2207–2217.
59. Cao, H., Huang, H., Xu, W., Chen, D., Yu, J., Li, J., and Li, L. (2011) *Anal. Chim. Acta*, **691**, 68–75.
60. Sheikh, K.D., Khanna, S., Byers, S.W., Fornace, A.J., and Cheema, A.K. (2011) *J. Biomol. Tech.*, **22**, 1–4.
61. Sreekumar, A., Poisson, L.M., Rajendiran, T.M., Khan, A.P., Cao, Q., Yu, J., Laxman, B., Mehra, R., Lonigro, R.J., Li, Y., Nyati, M.K., Ahsan, A., Kalyana-Sundaram, S., Han, B., Cao, X., Byun, J., Omenn, G.S., Ghosh, D., Pennathur, S., Alexander, D.C., Berger, A., Shuster, J.R., Wei, J.T., Varambally, S., Beecher, C., and Chinnaiyan, A.M. (2009) *Nature*, **457**, 910–914.
62. Struys, E.A., Heijboer, A.C., van Moorselaar, J., Jakobs, C., and Blankenstein, M.A. (2010) *Ann. Clin. Biochem.*, **47**, 282.
63. Jentzmik, F., Stephan, C., Miller, K., Schrader, M., Erbersdobler, A., Kristiansen, G., Lein, M., and Jung, K. (2010) *Eur. Urol.*, **58**, 12–18; discussion 20–21.
64. Laiakis, E.C., Morris, G.A.J., Fornace, A.J., and Howie, S.R.C. (2010) *PLoS ONE*, **e12655**, 5.
65. Li, N.-J., Liu, W.-T., Li, W., Li, S.-Q., Chen, X.-H., Bi, K.-S., and He, P. (2010) *Clin. Biochem.*, **43**, 992–997.
66. Seli, E., Sakkas, D., Scott, R., Kwok, S., Rosendahl, S., and Burns, D. (2007) *Fertil. Steril.*, **88**, 1350–1357.
67. Dunn, W.B., Brown, M., Worton, S.A., Crocker, I.P., Broadhurst, D., Horgan, R., Kenny, L.C., Baker, P.N., Kell, D.B., and Heazell, A.E.P. (2009) *Placenta*, **30**, 974–980.
68. Dunn, W.B., Broadhurst, D., Brown, M., Baker, P.N., Redman, C.W.G., Kenny, L.C., and Kell, D.B. (2008) *J. Chromatogr. B*, **871**, 288–298.
69. Kim, J.Y., Park, J.Y., Kim, O.Y., Ham, B.M., Kim, H.-J., Kwon, D.Y., Jang, Y., and Lee, J.H. (2010) *J. Proteome Res.*, **9**, 4368–4375.
70. Zhang, J., Yan, L., Chen, W., Lin, L., Song, X., Yan, X., Hang, W., and Huang, B. (2009) *Anal. Chim. Acta*, **650**, 16–22.
71. Wang, T.J., Larson, M.G., Vasan, R.S., Cheng, S., Rhee, E.P., McCabe, E., Lewis, G.D., Fox, C.S., Jacques, P.F., Fernandez, C., O'Donnell, C.J., Carr, S.A., Mootha, V.K., Florez, J.C., Souza, A., Melander, O., Clish, C.B., and Gerszten, R.E. (2011) *Nat. Med.*, **17**, 448–453.
72. Wishart, D.S., Knox, C., Guo, A.C., Eisner, R., Young, N., Gautam, B., Hau, D.D., Psychogios, N., Dong, E., Bouatra, S., Mandal, R., Sinelnikov, I., Xia, J., Jia, L., Cruz, J.A., Lim, E., Sobsey, C.A., Shrivastava, S., Huang, P., Liu, P., Fang, L., Peng, J., Fradette, R., Cheng, D., Tzur, D., Clements, M., Lewis, A., De Souza, A., Zuniga, A., Dawe, M., Xiong, Y., Clive, D., Greiner, R., Nazyrova, A., Shaykhutdinov, R., Li, L., Vogel, H.J., and Forsythe, I. (2009) *Nucleic Acids Res.*, **37**, D603–D610.

6
The Potential of Ultrahigh Resolution MS (FTICR-MS) in Metabolomics

Franco Moritz, Sara Forcisi, Mourad Harir, Basem Kanawati, Marianna Lucio, Dimitrios Tziotis, and Philippe Schmitt-Kopplin

6.1
Introduction

Metabolomics is an analytical discipline for the top-down analysis of metabolic diversity in biological samples such as biofluids, cells, and biopsies. It enables tracking of metabolic changes in time and space in biochemical mechanisms and disease progression and helps follow environmental effects. The "metabolome" is the set of metabolites synthesized by a biological system regulated by the transcriptome and the proteome [1]. Bioinformatics approaches such as pattern recognition and network analysis allow us to identify such metabolic changes in order to enrich our understanding of the metabolism.

In recent years, metabolomics analysis has been applied on different matrices and several research fields such as human and animal nutrition [2–4], cancer diagnosis and therapy [5, 6], biomarker discovery [7, 8], toxicology [9, 10], obesity studies [11], enzyme discovery [12, 13], drug discovery [14], transplantation [15], agriculture [16, 17], and bioremediation [18]. While the total number of different metabolites is still unknown, some estimates range between a few thousands and almost a million [19, 20]. Such estimates, however, may be conservative, and if we take into account plant and bacterial secondary metabolites, this number is probably significantly larger [19]. The probable number of metabolites is also considerably larger than the number of corresponding genes [21], so it seems that the currently available databases cover at best 2% of the total chemical diversity of existing metabolites [21].

In order to have a holistic view of metabolites in a biological system (metabolite cartography; Figure 6.1), metabolomics can be viewed as the complementary part of the other omics sciences such as genomics, transcriptomics, or proteomics. Owing to the chemical structure and function flexibility it is filling the possible gap between phenotype and genotype [22, 23]. There is an estimation that 30% of the identified genetic disorder involves diseases of small molecules metabolism [24]. The relevance of metabolomics has been related mainly to the fact that it measures the metabolite flux, thereby monitoring the enzymatic kinetics and identifying the

Metabolomics in Practice: Successful Strategies to Generate and Analyze Metabolic Data, First Edition.
Edited by Michael Lämmerhofer and Wolfram Weckwerth.
© 2013 Wiley-VCH Verlag GmbH & Co. KGaA. Published 2013 by Wiley-VCH Verlag GmbH & Co. KGaA.

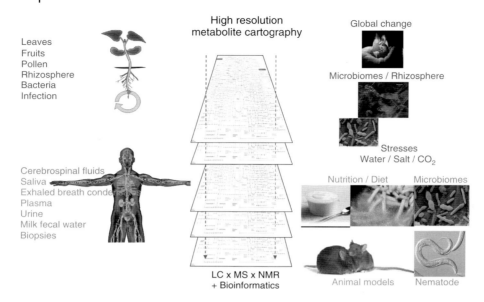

Figure 6.1 Metabolite cartography in metabolomics.

phenotypes. Consequently, the metabolome is most predictive of phenotype [1, 25].

6.2
Metabolomics Technologies

Several approaches have been proposed for metabolomics investigations, all having the same goal of describing the chemical diversity of the samples in various dimensionality and resolutions (Figure 6.2). We can recapitulate these approaches into two prevalent strands: (i) separation-mass spectrometry "for example, Liquid chromatography-mass spectrometry (LC-MS), Capillary electrophoresis-mass spectrometry (CE-MS), and Gas chromatography-mass spectrometry (GC-MS)" and (ii) Nuclear magnetic resonance (NMR). Accordingly, LC-MS, CE-MS, and GC-MS are extensively used in *target analysis* for quantitative metabolite profiling. Target analysis is restricted to the substrate and/or the direct product of a specific metabolic step [26]. GC-MS techniques allow the identification and robust quantification of a few hundred metabolites within a single extract [27, 28]. Compared to GC-MS, LC-MS offers several distinct advantages because it is adapted to a wider array of molecules including a range of secondary metabolites in phytochemistry, for example, alkaloids, flavonoids, isoprenes, glucosinolates, oxylipins, phenylpropanoids, pigments, and saponins [29, 30].

A further approach is *metabolic profiling*, where the analysis is restricted to the identification and the quantification of a selected number of predefined

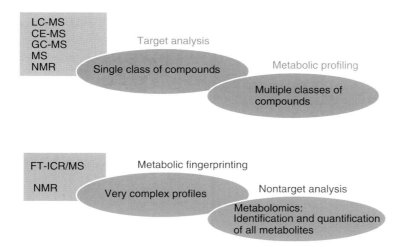

Figure 6.2 Types of instrumentations used for target and/or nontarget analysis.

metabolites in biological matrices [26]. In this case, the targeted components are larger in number and focused on classes of metabolites (i.e., lipids, sugars, peptides, proteins, etc.), while the information is of a structural basis with the possibility of further target quantification [31]. NMR is a consolidate technique in the metabolomics field for global *metabolic fingerprinting* and biomarker identification. Despite its lower sensitivity and resolution compared to mass spectrometry techniques, NMR is a fundamental technique in metabolite profiling. Furthermore, NMR has the advantage of being noninvasive and can be used *in vivo* under certain conditions.

The two-dimensional projection of the analytical volumetric pixel space covering NMR spectroscopy, mass spectrometry, and separation techniques is shown in Figure 6.3 and describes our current potential to represent variance in complex matrices through molecular resolution. The currently accessible discrete volumetric pixel (voxel) space for the characterization of complex matrices is in the range of 10^{8-14} voxels and its extension is defined by the significant resolution of the complementary techniques of NMR (depicting the order of molecules; inset cube in Figure 6.1), ultrahigh resolution Fourier Transform ion cyclotron resonance (FTICR) mass spectrometry (depicting molecular masses and formulae of gas-phase ions; inset cube in Figure 6.3), and high performance separation (depicting both ions and molecules; inset cube in Figure 6.3, this provides a way to validate NMR against MS data). The various projections of this voxel space, such as separation/MS, separation/NMR, and NMR/MS can be realized in the form of direct hyphenation via mathematical analysis [33–40]. In general, the level of importance of the produced data using any analytical technique will depend on both the intrinsic resolution of the respective method and the characteristics of the matrices used. Any insufficient correlation between the resolving power of the technique and the intrinsic analyte properties will be inefficient.

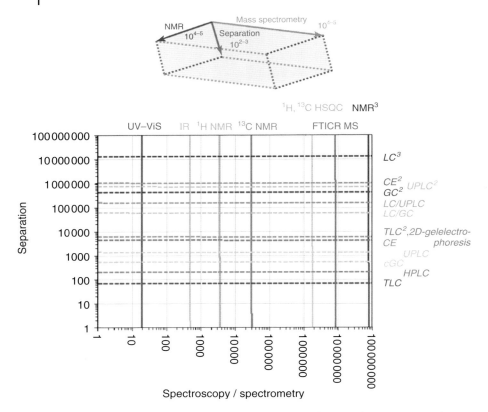

Figure 6.3 Characteristic resolutions (peak capacity: total range/half width [32]) of different separation technologies and organic structural spectroscopic methods. Inset cube at the top shows the significant resolution of the complementary techniques, for example, NMR, FTICR/MS, and other separation techniques; further dimensions may be the various ionization sources in MS and ion mobility MS. (Source: Reprinted with permission from Hertkorn, N., N. Hertkorn, C. Ruecker, M. Meringer, R. Gugisch, M. Frommberger, E. M. Perdue, M. Witt and P. Schmitt-Kopplin. High-precision frequency measurements: indispensable tools at the core of the molecular-level analysis of complex systems. Analytical and Bioanalytical Chemistry. (2007) 389:1311–1327. Copyright 2007 Springer-Verlag.)

Therefore, any organic structural spectroscopy with a limited peak capacity will inevitably lead to a summary bulk-type description of complex matrices and considerable averaging rather than a meaningful molecular-level resolution analysis (Figure 6.3).

To date (FTICR-MS), mass spectra afford the most credible direct experimental proof for the extraordinary molecular diversity of the most complex matrices. The complexity of the molecular level of the most studied matrices is effectively transformed into very highly resolved and extremely information-rich features. Additionally, FTICR-MS is presented as the latest technique among the nontargeted approaches. FTICR-MS is regarded as a rapidly emerging alternative to other types of mass spectrometers capable of nontargeted metabolic analysis and suitable for rapid screening of similarities and dissimilarities in large collections of biological matrices [41].

6.3
Principles of FTICR-MS

6.3.1
Natural Ion Movement Inside an ICR Cell Subjected to Magnetic and Electric Fields

The concept and application of nontargeted analysis in a system-wide hypothesis-driven approach has transformed the methodological strategies in different areas of life sciences, with sophisticated physical techniques. The attention will be focused on ion cyclotron resonance mass spectrometry that allows identification and quantification of metabolites based on their accurate mass determination. This technique of ion cyclotron resonance mass spectrometry refers to the measurement of the cyclotron frequency of ions trapped inside a confined cylindrical geometry located inside a magnet [42]. If a moving charged particle is subjected to a magnetic field, a Lorentz force is generated and forces the ion to move in rotational tracks in a cyclotron motion in the XY plane perpendicular to the magnetic field lines oriented along the Z axis (Figure 6.4). While ions can be confined in circular cyclotron motion in the XY plane, the magnetic field alone is not sufficient to prevent ions from escaping the ion cyclotron resonance (ICR) cell along the Z axis where the magnetic field lines are oriented. Therefore, an electric field is needed on both end cap electrodes. This is feasible by applying a DC electric voltage of $+1$ V on both end caps for trapping positive ions, while keeping the central ring electrode, which is segmented into four slices, grounded (0 V).

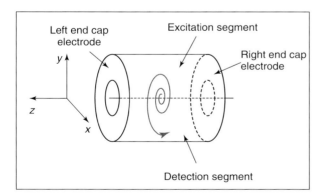

Figure 6.4 A cylindrical geometry of an ICR cell with two bored end cap electrodes. The blue spiral indicates a trajectory of an ion during radial (XY) ion excitation before detection. The central electrode is four times segmented. Both counterpart segments represent either detector pair or transmitter pair.

The formed electric potential well along the Z axis represents a stability region for trapping positive ions inside the central region of the ICR cell. If −1 V is applied on both end caps, an electric potential hill along the Z axis will be produced and this will allow trapping negative ions in the center of the cell.

Once trapped inside the cell center, ions can then be detected if they induce image charge on two detector segments located in the central ring electrode. This image charge induction cannot take place if the trapped ions remain in the center of the cylinder $(X, Y) = (0,0)$ because they are too far from the detector segments. Therefore, ion excitation in the radial (XY) plane is necessary in order to let all trapped ions increase their cyclotron radius so that they can fly in proximity to the detector segments. After the end of the excitation pulse, the generated sinusoidal curve which represents the image charge behavior can then be plotted as a function of time, hence a time domain spectrum is produced. A mathematical Fourier transform is necessary to move forward from the time domain spectrum to produce a frequency domain spectrum represented by a histogram showing the revealed cyclotron frequencies and the ion intensities for each measured frequency. This frequency domain spectrum can be simply converted to a high-resolution mass spectrum by the use of Eq. (6.1).

$$\omega_c = \frac{q \times B}{m} \tag{6.1}$$

The unperturbed cyclotron frequency ω_c is given by Eq. (6.1) where m is the ion mass in amu, q is the charge state, and B is the magnetic field strength in Tesla. It is called *unperturbed* to point out that it is the pure cyclotron frequency caused by rotation of a charged particle inside a magnetic field with no electric fields imposed (as when +1 V is applied on both end caps to trap positive ions inside the ICR cell).

The application of +1 V on both end caps to confine ions in the 3D space causes a new type of motion "axial ion vibration" along the Z axis, which alters the measured cyclotron frequency. The trapping frequency ω_z, which characterizes this new type of motion is independent of the magnetic field strength and is given in Eq. (6.2), where α is a constant which depends on the cell shape, a is a constant which depends on the cell size, V_{trap} is the trapping DC voltage applied on both end caps, and q is the charge state of the ion.

$$\omega_z = \sqrt{\frac{2qV_{trap}\alpha}{ma^2}} \tag{6.2}$$

Although ion detection can take place through the measurement of the trapping frequency, it has been decided to measure the "radial" cyclotron frequency on the XY plane and not the "axial" vibrational frequency along the Z axis (where the magnetic field lines are oriented) because the axial vibrational frequency is much lower than the cyclotron frequency. The higher the measured frequency is, the higher will be the achieved accuracy. This improves the resolution in the frequency measurement. One example can explain this: While an ion with $m/z = 200$ has an unperturbed cyclotron frequency of 921.361 KHz at a magnetic field strength of 12 T, the axial "trapping" frequency is only 4.518 KHz.

The trapping frequency is important because it influences the pure cyclotron frequency and causes a perturbation as shown in Eq. (6.3), where ω_c is the unperturbed cyclotron frequency, ω_z is the trapping frequency along the Z axis, ω_+ is the perturbed "modified" cyclotron frequency, and ω_- is called a *magnetron frequency*.

$$\omega_+ = \frac{\omega_c}{2} + \sqrt{\left(\frac{\omega_c}{2}\right)^2 - \frac{\omega_z^2}{2}} \tag{6.3a}$$

$$\omega_- = \frac{\omega_c}{2} - \sqrt{\left(\frac{\omega_c}{2}\right)^2 - \frac{\omega_z^2}{2}} \tag{6.3b}$$

Equations (6.3a,b) summarize two different types of perturbation: A reduction in the measured pure cyclotron frequency which depends on the trapping "axial" Z vibrational frequency. The higher the trapping frequency ω_z, the lower is the perturbed cyclotron frequency ω_+. ω_z can be increased by applying higher electric voltages on both end cap electrodes of the ICR cell. The second type of perturbation caused by an existing electric field implied a drift of the center of the cyclotron "rotational motion" on the XY plane from the cell center outwards. This causes ions not only to draw cyclotron motion in the XY plane but also to allow the center of the cyclotron circle to shift and also move circularly in the XY plane in what so-called a magnetron motion, whose frequency is denoted by ω_- and is given in Eq. (6.3b). The drift of the cyclotron center from the cell center outwards in the XY plane defines the magnetron radius.

Trapped ions inside the ICR cell can be detected through two detector segments located in the central ring electrode of the ICR cell when ions expand their cyclotron radius so that they fly in proximity to these two detector metal arcs. This requires that the thermal trapped ions in the X,Y (0,0) plane be radially excited (in the XY plane) to higher cyclotron radius by letting ions absorb energy from an external RF waveform generator. In order for the trapped ions to absorb energy and expand their cyclotron radius, the triggered RF should match the perturbed cyclotron frequency of the trapped ions inside the cell. This is defined as resonance; hence, this technique is called *ion cyclotron resonance*. The produced transient can be then Fourier transformed to change from the time domain to the frequency domain and the resolved cyclotron frequencies of all trapped ions can be easily converted to corresponding m/z ratios as indicated in the previous equations.

6.3.2
Applied Physical Techniques in FTICR-MS

Trapped ions inside an ICR cell can be radially excited in the XY plane for one of the following purposes: (i) Increase the cyclotron radius of all ions to 70% of the cell radius for ion detection to occur [43]. (ii) Eject ions out of the cell by applying high power radial ion acceleration pulse so that unwanted ions collide with the inner metallic walls of the central ring electrode, whereas only ions of interest remain in the cell for further investigations. (iii) Isolate interesting ion by technique B and then accelerate this specific ion radially and open an argon pulse gas valve simultaneously for inducing collisions between the accelerated ion and argon atoms for fragmentation studies delivering crucial structural information of the precursor ion. (iv) Measure the same ions several times inside the cell by applying coaxialization pulse on the ions so that they reduce their cyclotron radius to zero and then can be excited again in the XY plane for remeasurements [44].

It should be mentioned that the ICR cell is installed inside a vacuum chamber with efficient turbo molecular pumps for maintaining a base pressure in ultrahigh vacuum region (below 1×10^{-9} mbar). The lower the pressure, the higher is the achieved resolution. In broadband excitation, we can achieve a resolution of 400 000 at mass m/z 400 with a time domain transient length of 1.6 s and 4 M data points. However, ICR cells were also used as an electronic chemical reactor for running gas-phase ion–molecule reactions in the pressure region (1×10^{-7} to 1×10^{-8} mbar) [45]. The new ICR cells are capable of accommodating incoming ions which are generated by external ionization sources which work in different pressure regions. By implementation of a couple of stacked ion beam guides as well as quadrupole and hexapole electric devices, ions produced from atmospheric pressure ionization sources such as electrospray and photoionization sources can be delivered to the ICR cell in a sophisticated differential pumping vacuum system. Ions can also be enriched inside the quadrupole device and selected before running possible collision-induced dissociation (CID) experiments through linear ion acceleration and collision with argon atoms inside a hexapole device.

6.3.3
Practical Advantages of FTICR-MS

The capability of ICR cells to trap ions for a couple of seconds or even minutes can be invested not only for performing gas-phase ion–molecule reactions but also to study ion–electron interactions which cause fragmentation of multiple charged positive ions such as peptides [46]. This technique is known as electron capture dissociation (ECD) and can be implemented inside the ICR cell with no need to increase the pressure 2 orders of magnitude as is the case for performing ion–molecule reactions. No ECD experiments can be performed inside Q-time-of-flight (TOF) mass spectrometers because ions need to be trapped for a sufficient time before a strong interaction between ions and electrons can take place. The ultrahigh resolution of the ICR technique is unique when compared to other mass analyzers such as Orbitrap or Q-TOF systems. Figure 6.5 shows the power of long transients in achieving ultrahigh resolution for nominal masses under study.

On the other hand, Fourier Transform mass spectrometry (FTMS) instruments are not fast enough to keep the ultrahigh resolution when coupled to liquid chromatographic systems such as high performance liquid chromatography (HPLC) or ultra performance liquid chromatography (UPLC), as high resolution is always associated with longer time domain transients (as shown in Figure 6.5). LC-MS

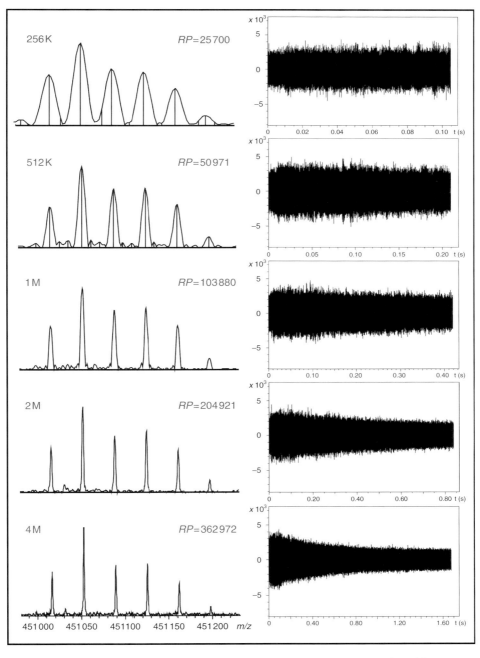

Figure 6.5 Acquisition of standard sample (from IHSS: Suwannee River Fulvic Acid) at different time domain transient lengths ranging from 0.2 to 1.6 s. The numbers of digitized points are given on the left from 256 K data points to 4 M data points.

coupling can be achieved if a moderate resolution is accepted (512 K with transient length of 0.2 s achieving a resolution of around 50 000). This resolution is the maximum that can be achieved in TOF mass analyzers to date and is sufficient for sum formulae determination. It should be mentioned that Orbitrap mass analyzer can achieve a mass resolving power up to 150 000.

6.4
Proceeding in Metabolomics

6.4.1
Network Analysis and NetCalc Composition Assignment

Exact masses combined with ultrahigh resolution are the key issue in enabling an attribution of all elementary compositions out of a mass spectra and further interpretation in chemical structure; even a ppb-level mass accuracy is not sufficient if the resolution is not enough and if the signal differentiation in a complex mixture is not assured [47, 48]. Various approaches are offered in commercial available softwares, all with the limitation of multiple composition annotations, especially in higher masses. We recently proposed calculation based on mass differences and involving a network approach [49].

Graph theory is widely used in bioinformatics and chemometrics because of its ability to provide efficient means of modeling and visualizing real-world scenarios. Many pragmatic situations can be represented in the form of a diagram consisting of a set of points (nodes) and a set of lines (edges) connecting parts of these points; a mathematical abstraction which yields the concept of a graph [50]. Such an abstraction can be represented graphically, and through this graphical representation we are able to study some of its properties. A graph (also called a network) is, in addition, associated with a specialized matrix which allows us to store it inside computers and apply mathematical methods to analyze our data more thoroughly; a procedure known as *network analysis*.

Network analysis can be applied on almost every scenario of FTICR-MS spectra in a number of ways. An important approach involves the *mass–mass difference networks*, in which each node represents an exact experimental mass and each edge represents a selected mass difference either taken from a predefined list of potential transformations, or detected on the fly through mass–mass difference clustering and correlation analysis [51]. Such a network model can be divided into *structural* and *functional* networks [49]. In the case of structural networks, a list of selected theoretical mass differences is used in order to determine the adjacency relation between the nodes, that is, detected transformations between the experimental masses. The resulting network can be described as a simulation of the real biochemical system that can reflect the structural information expressed in an FTICR-MS dataset.

In the context of metabolomics, we have the additional possibility of modeling FTICR-MS datasets in the form of correlation networks. Such a task is achieved by treating mass spectra either as row or column vectors out of which a correlation matrix is extracted (usually using Pearson correlation). By setting a threshold value on the correlation coefficient, the correlation matrix can be converted into a binary adjacency matrix which represents a network. In the case of row vector correlation, the resulting network is a metabolic correlation network which, through various methods of node quantification, hierarchization, and clustering, has the potential of contributing in biomarker identification. Combined with the mass–mass difference network approach, this method has a great potential of nontargeted data reduction and comparison of the clusters to known KEGG pathways. In the case of column vector correlation, the mass spectra of the various samples can be modeled into a sample correlation network, which may be used for clustering samples into biologically significant groups of individuals. Such a network analysis approach is flexible enough to be used in both supervised and unsupervised ways.

6.4.2
Statistics on FTICR-MS Datasets

The application of statistics on ultrahigh resolution mass spectrometric data should start with the consideration that the complexity of the dataset is due to the presence of structure of correlation. Considering this, the inspection and the evaluation should start with multivariate analysis. These techniques can cope with challenges regarding the data amount, the presence of notable noise and collinearity. After the basic data processing is achieved (peak picking and alignment, baseline correction, smoothing), the next obligatory steps are the data cleaning and preprocessing, which are dependent on the type of instruments used to generate the data. Different normalization and scaling techniques should be evaluated a priori depending on the type of data (GC-MS, LC-MS, CE, NMR, and FTICR-MS). This quality control stabilizes the data and prepares it for statistical analysis. Multidimensional and multivariate statistical analysis and pattern-recognition programs have been developed to distill the large amount of data in an effort to interpret the complex metabolic pathway information [52–54]. The multivariate analysis up to date is one of the most powerful tools able to interpret biological phenomena. This is divided into unsupervised analysis and supervised analysis. The resulting classification depends on whether or not there is information available when the investigation starts. For data overview many tools can be used in the metabolomics field, commonly dendrograms from clustering results [27], principal components analysis (PCA) [27, 52, 55] and heat maps [56]. For classification challenges the most common utilities are: partial least square-discriminant analysis (PLS-DA), orthogonal partial least square (OPLS), random forest, or support vector machine (SVM). The validation process should be carried on with internal (cross-validation and permutation test) and, if possible, external validation (training and external set) to test the predictability of the model and possible presence of overfitting. The crucial point is the identification of metabolites in order to give a biological interpretation and pathway mapping. The use of different databases made it possible to obtain information about the small molecule metabolites.

The comparison of the results obtained from different instruments and the integration into other "omics" data can contribute to biological findings. The complementarity of those techniques and information is still a turning point in metabolomics studies.

6.5
Application Example in Metabolomics Using FTICR-MS Exhaled Breath Condensate

Biofluids are important matrices in medical diagnosis because their sampling process is simple and minimally invasive. Still, plasma sampling requires the medic to pierce through tissue which may compromise the patient's well being. Exhaled breath condensate (EBC) in turn is being sampled by simply channeling the patient's breath through a cooled trap, which condenses H_2O vapor and traps aerosols and particulate matter derived from the airway lining fluid (ALF). EBC has been screened for several pulmonary illnesses such as cystic fibrosis, chronic obstructive pulmonary disease, asthma, and gastric acid reflux syndrome. Clinical markers such as pH, H_2O_2 concentration, nitric oxide levels, chemokines, and arachidonic acid derived oxidative stress markers have been proven to aid differentiation of healthy and ill phenotypes [57]. A small amount of research has been undertaken to reveal metabolite fluxes between ALF and plasma. The potential metabolic setup of EBC is therefore not yet known.

It is known that the quantitative analysis of EBC metabolites becomes complicated by strong variation in total metabolite concentration. As mentioned by R.M. Effros [58], the total concentration of non-volatile small metabolites – which are the subject of metabolomics – may vary more than fivefold as deduced by the amount of ALF deriving aerosols relative to the amount of condensed H_2O vapor. We hypothesized that direct flow injection electrospray ionization (ESI)-FTICR/MS provides a solution to this problem. Owing to its high resolution, accuracy, and sensitivity over a mass range of 150 to 1500 Da, metabolites can directly be annotated by virtue of their exact mass. Multivariate statistics and informatics help identifying general trends discriminative to the experimental task. In order to yield the most benefit from FTICR-MS measurements, samples need to be pre-processed in a way that the concentration of salts and proteins is minimized while the conservation of the intrinsic metabolic information is maximized. We have found that any manipulation of EBC using solid phase extraction methods does not provide any advantage over simple dilution in methanol.

6.5.1
The Experiment

In this application, we were asking two questions: "To which extent can we differentiate a lower airway EBC (alveolar or "AL") from an upper airway EBC (bronchial or "BR")?" and "Is a differentiation between two phenotypic groups such as smokers and non-smokers stable throughout lower airway condensate and upper airway condensate?" To answer these questions, 13 smokers and 13 non-smokers were sampled using the EcoScreen 2 EBC sampler (Jäger, Germany). This sampler divides the breath air into an upper airway condensate and a lower airway condensate yielding a final number of 52 samples in this study. The volume yielded in the upper airway section is usually one-tenth of the lower airway volume.

6.5.2
FT-ICR/MS Measurement

EBC samples were diluted in methanol (1 : 1, v:v). As the extent of peak suppression is a function of sample dilution we used the chip based nanoESI system Tiversa NanoMate from ADVION for sample introduction and ionization. Nano electrospray ionization effectively reduces adverse effects introduced by varying dilution and the presence of suppressive agents such as fatty acids. In order to fully exploit the advantages of FTICR-MS, we routinely control the instrument's performance by means of internal calibration on arginine clusters before any analysis. Relative m/z errors are usually smaller than 0.1 ppm over a range of $150 < m/z < 1500$. To depict the extent of mass accuracy for readers not familiar with MS; the error range is as much as 1/30 to 1/5 of an electron mass. FTICR mass spectra are superimpositions of several single scans. Consequently, noise amplitudes stay small and real m/z peaks are being enhanced as a function of scan number, which increases the sensitivity of this technique. EBC spectra in our example were measured in positive ionization mode and 1000 scans were accumulated.

6.5.3
Data Preprocessing

Each above accumulated spectrum was internally calibrated on m/z values of ubiquitous and known solvent impurities. After internal calibration and signal to noise filtering at $S/N = 3$, we yielded mass spectra of average 4000 m/z peaks and an error standard deviation of <0.1 ppm.

The above created spectra were stored as ASCI files in the form of an $n*2$ matrix where the centroid m/z values and their peak intensities are stored in the first column and the second column, respectively. The alignment of the spectra was performed using an algorithm called *Matrix Generator*, an in-house program written in Python [59]. The alignment is performed by "clustering" of masses identical within an error margin of 1 ppm and concatenating their intensities. Until this point, our application yielded an intensity matrix of 28 500 m/z values and 52 samples (4×13 samples). The alignment of mass spectra leads to an enormous amount of zero intensity values for masses, which were not detected in all mass spectra. In order to stabilize the variance, we removed isotopic mass peaks and deleted variables occurring at a frequency of less than 10% throughout all the 52 samples. At the end, the reduced matrix counted 7425 masses.

6.5.4
C–H–N–O–S–P Formula Annotation

Independent of statistical analysis, we uploaded the initial 28 500 experimental masses to MassTRIX [60], a web interface enabling the annotation of the exact masses to possible metabolites in pathways using the organism specific databases of KEGG, LipidMaps, HMDB. The annotation of these positive ionization mode data was performed at 1 ppm error tolerance. Aside from H^+ adducts we chose Na^+ adducts to be considered. This step yielded 1639 monoisotopic CHNOSP mass annotations which equals to about 8% annotation efficiency relative to the amount of masses uploaded to MassTRIX. A holistic representation of involved pathways is problematic at an annotation rate of only 8%. In order to increase the annotation rate, we compared the EBC masses to a metabolite list downloaded from HMDB [61]. Here, we could additionally consider the possible presence of $[M + K^+]^+$ and $[M + [H^+] - H_2O]^+$ ions to occur [62]. This comparison yielded 2370 monoisotopic CHNOSP annotations, that is, 11.4% coverage. This coverage still renders 88.6% of all peaks to be empty numbers. Another option, one of our group's recent developments – the NetCalc algorithm – allows for a much greater coverage. As described above, NetCalc uses the hypothesis that metabolites are involved in biochemical reactions which in turn are attributed to exact mass differences. Applying NetCalc to our data set finally yielded as much as 13 336 putatively annotated monoisotopic formulae exclusively based on CHONSP. This equals to an annotation rate of 64.4% in the whole data set.

6.5.5
Statistical Analysis

Data naturally contains biologically relevant variance as well as variance because of random and systemic errors. Therefore, the next stage of data analysis is statistical processing [59].

6.5.5.1 Statistical Preprocessing
Initial PCA plotting – an unsupervised method of multivariate statistics – was not able to show latent trends in the data. Consequently, we performed a supervised classification starting with orthogonal signal correction (OSC). This way we were able to remove information orthogonal (independent) to the predefined metabotypes (see the statistics section). In addition to removal of nonimportant variance this step led to the discovery of mass peaks with too strong leverage. These signals were removed as well. Consequently, we could construct a PLS model and OPLS/O2PLS-DA models for the differentiation between the lower and upper

airways. As shown in Figure 6.6a, we managed to define a model classifying the samples into the desired classes. To further test the model we "contaminated" the data set with a duplicate spectrum of the AL group but labeled it as being a BR EBC spectrum. In the (OSC)PLS model, this duplicate was drawn into the direction of its original group – AL – and was therefore recognized as an outlier. Interestingly, the created model led to the – at that point unintended – separation between smokers and non-smokers showing the natural trend of the respective data. In order to examine the validity of these second trends we divided the data into an AL section and a BR section. We observed that OPLS/O2PLS-DA models were only found to be valid in the AL breath partition. This very fact indicates that the AL section carries the majority of biological information. Despite this, we found that some compounds exhibited the tendency to be a smoker classifier in AL but a non-smoker classifier in BR. All analyses were performed using SIMCA-P 11.5 (Umetrics, Umea, Sweden). For further analysis, features were attributed to their respective S-plot parameters, stored as tab separated text file and forwarded to network analysis.

6.5.6
Synthesis of Biochemical Mass Difference Networking and Statistical Results

Metabolomics especially aims at the formation of new hypotheses. In order to do so, the advantages of the techniques applied so far have to be combined. Similar to the combination of separation techniques with high accuracy mass spectrometry (Section 6.1) we move toward a space of higher "analytical resolution." On the basis of the S-plot of the AL breath partition model we sorted all variables according to the highest magnitude in covariation and correlation to the experimental context. Owing to this sorting we could ascribe the uppermost 300 and the lowermost 300 masses out of 7425 as being characteristic for non-smokers and smokers, respectively. At this point, we were rerunning the NetCalc algorithm with a slight modification. We were not allowing any reactions among masses of the statistically nonsignificant variables and by this means eliminated reactions of low importance for the experiment. *Reactions* were defined as being valid if either a smokers marker or a non-smokers marker was involved. Invariant to any cropping the degree distributions of the summarized chemical scenario were all scale free. Scale freeness indicates robustness of proportions against scaling, removal of connections or false positives. In such networks usually the probability of finding a highly connected node follows the relation $P(\text{connectivity}) \approx k - \gamma$, where k is the connectivity and γ the scaling factor. When $\gamma \leq 3$ the "meaning" of a network is conserved, nonhierarchical, and nonrandom. The EBC networks exhibited $\gamma = 1.7$ in case of the overall network and $\gamma = 2.2$ in case of the cropped network [63]. This workflow finally yielded a network of 5129 edges (reactions, proteins, or genes). Nodes of the same statistical/phenotypic group which got appointed a valid connection are well suited for classification of data, that is, for diagnostics. In fact, we yield a higher dimensional classifier space which is not of random nature but directly projectable to other omics disciplines. In

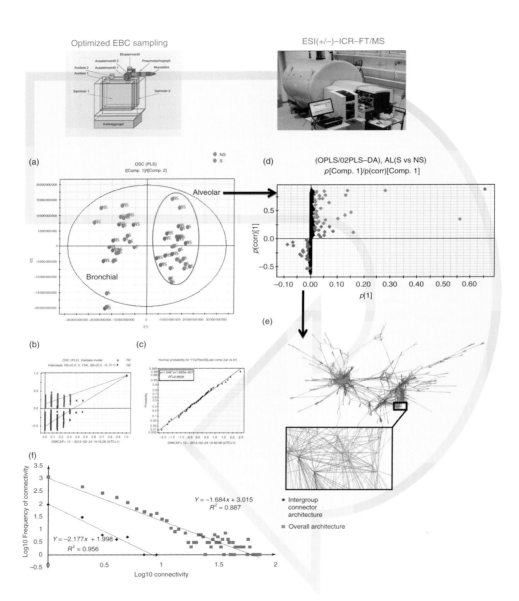

6.5 Application Example in Metabolomics Using FTICR-MS Exhaled Breath Condensate

genetics, the accumulative effect of even weak classifiers on overall classification performance is a well-established concept [64]. Within-group reactions are fairly interesting; however, reaction pairs of probably highest importance for diagnostic and pharmaceutical target search may be reaction pairs of which one entity exhibits a pro-smoker trend, whereas the other entity does the opposite. Reactions like these are naturally rare. But as can be seen in Figure 6.6, there are regions within the network, where such reactions accumulate. Also depending on these structures' interconnectivity these pathways and compounds may be hallmarks for future investigations.

The fact that smokers and non-smokers can be differentiated via EBC analysis is not surprising. The striking result of this small example is the actual richness in molecular species found in this matrix. Evidently, EBC has much more to offer than the standard clinical markers. This richness may potentially enable scientists to profile and clearly distinguish specific pulmonary and even systemic diseases. Just as EBC glucose levels may be indicative of whole body glucose homeostasis [65], all other compounds found are biomarker candidates. Additionally, we can state that the AL EBC partition holds the major partition of metabolic information, clearing wide spread concerns on possible saliva contaminations [66]. Several models calculated in our group show the same trend: weak test statistics for BR breath condensate and small impact on networking approaches. Additionally, the BR EBC partition usually makes up only 10% of the whole breath volume, which strongly dilutes the upper airways impact. In order to cope with the strong variation in dilution future research will center around integrated and multivariate characterizations of the ESI process. Manuscripts explaining the matters discussed in this section are in progress.

Figure 6.6 (a) PLS Score Scatter plot (Q^2(cum) = 0.51 and $R^2(Y)$ = 0.75). The p-value indicates the probability level where a model with this F-value may be the result of just chance (cross-validation ANOVA: $p = 1.24138E^{-20}$), moreover the permutation test with 200 permutation shows no overfitting on the data. Plot (b) shows the normal probability plot of the Y-residuals. Conceptually the residuals have been seen as a form of error and they should be normal distributed. When they are coming from a normal distribution the plot (c) should produce an approximately straight line. Moreover no outliers have been detected. (d) S-Plot of OPLS/O2PLS-DA ((Q^2(cum) = 0.72 and $R^2(Y)$ = 0.86)). The p-value indicates the probability level where a model with this F-value may be the result of just chance (cross-validation ANOVA: $p = 0.000956986$). (e) Biochemical mass difference network showing the interaction of smoker related and nonsmoker related compounds. Both groups clearly populate different network sections. Intergroup connectors could be considered to pronounce reactions/enzymes/genes of importance for the experimental context and therefore provide an important link for target search in other omics departments. (f) Network architectures after cropping. The network is scale free, indicating a realistic representation of a real-world scenario.

6.6
Conclusion and Remarks

Metabolomics becomes a progressively influential field in biochemistry and environmental science, as analysis of the metabolome can reveal all upstream regulatory events from the external environment and provide insights into metabolic dynamics. This chapter aims to put "Ultrahigh resolution mass spectrometry (FTICR-MS)" into perspective as an upcoming technology in metabolomics. FTICR-MS provides the highest accuracy and resolution known in mass spectrometry. Its capability to trap ions for long period of time enables experimental insights which are not accessible with other types of mass spectrometers. The accumulation of consequent scans is gradually increasing the sensitivity of FTICR-MS experiments. These properties make this instrument well fit to metabolomics. Using this technique, poorly characterized sample matrices such as EBC can be attributed to thousands of putatively annotated analytes. Computational approaches and statistics furthermore help to appoint biochemical meaning to this ensemble of analytical features. With its ultrahigh resolution, FTICR-MS allows for annotation of elementary compositions but no molecular structures can be addressed. Hyphenated analytical strategies are a mainstay in metabolomics research (Chapters 6, 10, and 11). In conclusion, future developments regarding the online and offline combination of CE, LC, and NMR with FTICR-MS would greatly assist researchers in unambiguously identifying and quantifying metabolites on a metabolome-wide scale.

References

1. Fiehn, O. (2002) *Plant Mol. Biol.*, **48**, 155–171.
2. Whitfield, P.D., German, A.J., and Nobel, P.J.M. (2004) *Br. J. Nutr.*, **92**, 549–555.
3. Gibney, M.J., Walsch, M., and Brennan, L. (2005) *Am. J. Clin. Nutr.*, **82**, 497–503.
4. Rist, M.J., Wenzel, U., and Daniel, H. (2006) *Trends Biotechnol.*, **24**, 172–178.
5. Hartmann, M., Baumbach, J., and Nolte, J. (2006) *Ann. Oncol.*, **17**, 36.
6. Malhi, H. and Gores, G. (2006) *Cancer Biol. Ther.*, **5**, 986–987.
7. Goodacre, R. (2005) *J. Med. Gen.*, **42**, 16.
8. Schlotterbeck, G., Ross, A., and Dieterle, F. (2006) *Pharmacogenomics*, **7**, 1055–1075.
9. Robertson, D. (2005) *Toxicol. Sci.*, **85**, 809–822.
10. Gerner, C., Teufelhofer, O., and Parzefall, W. (2006) *Toxicol. Lett.*, **164**, 3–4.
11. Hochberg, Z. (2006) *Int. J. Obesity*, **30**, 4.
12. Saito, N., Robert, M., and Kitamura, S. (2006) *J. Proteome Res.*, **5**, 1979–1987.
13. Villas-Bôas, S., Noel, S., and Lane, G. (2006) *Anal. Biochem.*, **349**, 297–305.
14. Harrigan, G. (2006) *NC IDrugs*, **9**, 28–31.
15. Wishart, D. (2005) *Am. J. Transplant.*, **5**, 2814–2820.
16. Bender, D. (2005) *J. Sci. Food Agric.*, **85**, 7–9.
17. Dixon, R.A., Gang, D.R., and Charlton, A.J. (2006) *J. Agric. Food Chem.*, **54**, 8984–8994.
18. Singh, O. (2006) *Proteomics*, **6**, 5481–5492.
19. Ott, M. and Vriend, G. (2006) *BMC Bioinformatics*, **7**, 517.

20. Wink, M. (1988) *Theor. Appl. Genet.*, **75**, 225–233.
21. Green, J.L., Bohannan, B., and Whitaker, R.J. (2008) *Science*, **320**, 1039–1043.
22. Holmes, E., Loo, R.L., Stamler, J., Bictash, M., Yap, I.K.S., Chan, Q., Ebbels, T., De Iorio, M., Brown, I.J., Veselkov, K.A., Daviglus, M.L., Kesteloot, H., Ueshima, H., Zhao, L.C., Nicholson, J.K., and Elliott, P. (2008) *Nature*, **453**, 396–400.
23. Keurentjes, J.J.B. (2009) *Curr. Opain. Plant Biol.*, **12**, 223–230.
24. Goldsmith, P., Fenton, H., Morris-Stiff, G., Ahmad, N., Fisher, J., and Prasad, K.R. (2010) *J. Surg. Res.*, **160**, 122–132.
25. Weckwerth, W. (2003) *Annu. Rev. Plant Biol.*, **54**, 669–689.
26. Bhalla, R., Narasimhan, K., and Swarup, S. (2005) *Plant Cell Rep.*, **24**, 562–571.
27. Roessner, U., Willmitzer, L., and Fernie, A.R. (2001) *Plant Physiol.*, **127**, 749–764.
28. Halket, J.M., Przyborowska, A., Stein, S.E., Mallard, W.G., Down, S., and Chalmers, R.A. (2003) *Rapid Commun. Mass Spectrom.*, **13**, 279–284.
29. Aharoni, A., Giri, A.P., Deuerlein, S., Griepink, F., de Kogel, W.J., Verstappen, F.W., Verhoeven, H.A., Jongsma, M.A., Schwab, W., and Bouwmeester, H.J. (2003) *Plant Cell*, **15**, 2866–2884.
30. Matuszewski, B.K., Constanzer, M.L., and Chavez Eng, C. (2003) *Anal. Chem.*, **75**, 3019–3030.
31. Schmitt-Kopplin, P., Hertkorn, N., Frommberegr, M., Lucio, M., Englmann, M., Fekete, A., and Gebefugi, I. (2007) *Adv. Tech. Soil Biol.*, **11**, 281–293.
32. Shen, Y.F. and Lee, M.L. (1998) *Anal. Chem.*, **70**, 3853–3856.
33. Zimmer, J.S.D., Monroe, M.E., Qian, W.J., and Smith, R.D. (2006) *Mass Spectrom. Rev.*, **25**, 450–482.
34. Lindon, J.C., Nicholson, J.K., and Wilson, I.D. (2000) *J. Chromatogr. B*, **748**, 233–258.
35. Spraul, M., Freund, A.S., Nast, R.E., Withers, R.S., Maas, W.E., and Corcoran, O. (2003) *Anal. Chem.*, **75**, 1536–1541.
36. Chalmers, M.J., Mackay, C.L., Hendrickson, C.L., Wittke, S., Walden, M., Mischak, H., Fliser, D., Just, I., and Marshall, A.G. (2005) *Anal. Chem.*, **77**, 7163–7171.
37. Smith, C.A., Want, E.J., O'Maille, G., Abagyan, R., and Siuzdak, G. (2006) *Anal. Chem.*, **78**, 779–787.
38. Want, E.J., O'Maille, G., Smith, C.A., Brandon, T.R., Uritboonthai, W., Qin, C., Trauger, S.A., and Siuzdak, G. (2006) *Anal. Chem.*, **78**, 743–752.
39. Cloarec, O., Campbell, A., Tseng, L.H., Braumann, U., Spraul, M., Scarfe, G., Weaver, R., and Nicholson, J.K. (2007) *Anal. Chem.*, **79**, 3304–3311.
40. Cloarec, O., Dumas, M.E., Trygg, J., Craig, A., Barton, R.H., Lindon, J.C., Nicholson, J.K., and Holmes, E. (2005) *Anal. Chem.*, **77**, 517–526.
41. Aharoni, A., Ric de Vos, C.H., Harrie, A.V., Maliepaard, C.A., Kruppa, G., Bino, R., and Goodenowe, D.B. (2002) *J. Integr. Biol.*, **6**, 217–234.
42. Marshall, A.G., Hendrickson, C.L., and Jackson, G.S. (1998) *Mass Spectrom. Rev.*, **17**, 1–35.
43. Gorshkov, M.V. and Nikolaev, E.N. (1993) *J. Mass Spectrom.*, **125**, 1–8.
44. Speir, J.P., Gorman, G.S., Petsenberger, C.C., Turner, C.A., Wang, P.P., and Amster, I.J. (1993) *Anal. Chem.*, **65**, 1746–1752.
45. Kanawati, B. and Wanczek, K.P. (2007) *Int. J. Mass Spectrom.*, **264**, 164–174.
46. Zubarev, R.A. (2003) *Mass Spectrom. Rev.*, **22**, 57–77.
47. Schmitt-Kopplin, Ph., Kiss, G., Dabek-Zlotorzynska, E., Gelencsér, A., Hertkorn, N., Harir, M., Hong, Y., and Gebefügi, I. (2010) *Anal. Chem.*, **82**, 8017–8026.
48. Schmitt-Kopplin, Ph., Gabelica, Z., Gougeon, R.D., Fekete, A., Kanawati, B., Harir, M., Gebefuegi, I., Eckel, G., and Hertkorn, N. (2010) *Proc. Natl. Acad. Sci.*, **107**, 7.
49. Tziotis, D., Hertkorn, N., and Schmitt-Kopplin, Ph. (2011) *Eur. J. Mass Spectrom.*, **17**, 415–421.
50. Bondy, J.A. and Murty, U.S.R. (2007) *Graph Theory*, Graduate Texts in Mathematics, Springer-VerlagISBN: 978-1-84628-969-9.

51. Breitling, R., Ritchie, S., Goodenowe, D., Stewart, M.L., and Barrett, M.P. (2006) *Metabolomics*, **2**, 155–164.
52. Nicholson, J.K., Lindon, J.C., and Holmes, E. (1999) *Xenobiotica*, **29**, 1181–1189.
53. Boutilier, K., Ross, M., Podtelejnikov, A.V., Orsi, C., Taylor, R., Taylor, P., and Figeys, D. (2005) *Anal. Chim. Acta*, **534**, 11–20.
54. Smith, C.A., Want, E.J., O'Maille, G., Abagyan, R., and Siuzdak, G. (2006) *Anal. Chem.*, **78**, 778–787.
55. Fiehn, O., Kopka, J., and Dormann, P. (2000) *Nat. Biotechnol.*, **18**, 1157–1161.
56. Jansson, J., Willing, B., Lucio, M., Fekete, A., Dicksved, J., Halfvarson, J., Tysk, C., and Schmitt-Kopplin, P. (2009) *PLoS ONE*, **4** (7), e6386.
57. Malerba, M. and Montuschi, P. (2012) *Curr. Med. Chem.*, **10**, 187–196.
58. Effros, R.M. (2010) *Chest*, **138**, 471–472.
59. Lucio, M., Fekete, A., Weigert, C., Wagele, B., Zhao, X.J., Chen, J., Fritsche, A., Haring, H.U., Schleicher, E.D., Xu, G.W., Schmitt-Kopplin, P., and Lehmann, R. (2010) *PLoS ONE*, **5** (10), e13317. doi: 10.1371/journal.pone.0013317
60. Suhre, K. and Schmitt-Kopplin, P. (2008) *Nucleic Acids Res.*, **36**, 481–484.
61. Wishart, D.S., Tzur, D., Knox, C., Eisner, R., Guo, A.C., Young, N., Cheng, D., Jewell, K., Arndt, D., Sawhney, S., Fung, C., Nikolai, L., Lewis, M., Coutouly, M.A., Forsythe, I., Tang, P., Shrivastava, S., Jeroncic, K., Stothard, P., Amegbey, G., Block, D., Hau, D.D., Wagner, J., Miniaci, J., Clements, M., Gebremedhin, M., Guo, N., Zhang, Y., Duggan, G.E., MacInnis, G.D., Weljie, A.M., Dowlatabadi, R., Bamforth, F., Clive, D., Greiner, R., Li, L., Marrie, T., Sykes, B.D., Vogel, H.J., and Querengesser, L. (2007) *Nucleic Acids Res.*, **36**, 521–526.
62. Camera, E., Ludovici, M., Galante, M., Sinagra, J.L., and Picardo, M. (2010) *J. Lipid Res.*, **51**, 3377–3388.
63. Barabàsi, A.-L. and Oltvai, Z.N. (2004) *Nat. Rev. Genet.*, **5**, 101–113.
64. Omri, T. (2012) *J. Theor. Biol.*, **293**, 206–218.
65. Baker, E.H., Clarck, N., Brennan, A.L., and Fisher, D.A. (2007) *J. Appl. Physiol.*, **102**, 1969–1975.
66. Ichikawa, T., Matsunaga, K., and Minakata, Y. (2007) *Anal. Chem. Insights*, **2**, 85–92.

7
The Art and Practice of Lipidomics

Koen Sandra, Ruben t'Kindt, Lucie Jorge, and Pat Sandra

Abbreviations

A	*Alpha*-hydroxy fatty acid
AD	Alzheimer's disease
amu	Atomic mass unit
ALS	Automatic liquid sampler
AMDIS	Automatic mass spectral deconvolution and identification system
AMRT	Accurate mass retention time
APCI	Atmospheric pressure chemical ionization
APPI	Atmospheric pressure photoionization
BAME	Bacterial fatty acid methyl ester
CER	Ceramide
CE	Cholesterol ester
CI	Chemical ionization
CID	Collision-induced dissociation
CL	Cardiolipin
CLA	Conjugated linoleic acids
Da	Dalton
DDA	Data-dependent acquisition
DESI	Desorption electrospray ionization
DG	Diacylglycerol
DIE	Diethyl ether
DM-2	Diabetes mellitus type 2
DS	Dihydrosphingosine
ECC	Extracted compound chromatogram
EI	Electron ionization
EIC	Extracted ion chromatogram
EO	Esterified *omega*-hydroxy fatty acid
ESI	Electrospray ionization
FA	Fatty acid
FAME	Fatty acid methyl ester
FT-ICR	Fourier-transform ion cyclotron resonance
GC	Gas chromatography
GL	Glycerolipids
GP	Glycerophospholipids

Metabolomics in Practice: Successful Strategies to Generate and Analyze Metabolic Data, First Edition.
Edited by Michael Lämmerhofer and Wolfram Weckwerth.
© 2013 Wiley-VCH Verlag GmbH & Co. KGaA. Published 2013 by Wiley-VCH Verlag GmbH & Co. KGaA.

H	6-Hydroxy-sphingosine
HEX	Hexane
HILIC	Hydrophilic interaction chromatography
HMDB	Human metabolome database
ILCNC	International Lipid Classification and Nomenclature Committee
IPA	Isopropanol
L	Lyso
LC	Liquid chromatography
LMSD	LIPID MAPS Structure Database
LPC	lyso-glycerophosphocholine
LPE	lyso-glycerophosphoethanolamine
MALDI	Matrix-assisted laser desorption ionization
MFE	Molecular feature extraction
MG	Monoacylglycerol
MRM	Multiple reaction monitoring
MS	Mass spectrometry
MTBE	Methyl-*tert*-butylether
MW	Molecular weight
m/z	Mass-to-charge ratio
N	Nonhydroxy fatty acid
NL	Neutral loss
NP	Normal phase
O	*Omega*-hydroxy fatty acid
OzID	Ozone-induced dissociation
P	Phytosphingosine
PA	Glycerophosphatidic acid
PBS	Phosphate-buffered saline
PC	Glycerophosphocholine
PE	Glycerophosphoethanolamine
PGA2	Prostaglandin A2
PG	Glycerophosphoglycerol
PI	Glycerophosphoinositol
PK	Polyketides
ppm	Parts per million
PR	Prenol lipids
PS	Glycerophosphoserine
pSFC	Packed column supercritical fluid chromatography
PTV	Programmed temperature vaporizing
Q	Quadrupole
RP	Reversed phase
RSD	Relative standard deviation
RTL	Retention time locking
S	Sphingosine
SB	Sphingoid base
SC	Stratum corneum
SFC	Supercritical fluid chromatography
SI	Secondary ion
SL	Saccharolipids
SM	Sphingomyelin
SP	Sphingolipids
SPE	Solid-phase extraction

SRM	Selected reaction monitoring
ST	Sterol lipids
TFA	Trifluoroacetic acid
TG	Triacylglycerol
TOF	Time-of-flight
UHPLC	Ultrahigh pressure liquid chromatography
WAX	Weak anion exchange

7.1 Introduction

In recent years, the global analysis of different types of biomolecules in a variety of sample sources gained momentum. Genomics has indispensably initiated this so-called omics cascade and the concept of unbiased analysis subsequently attracted a vast amount of researchers to the field. The potential of these relatively new disciplines is indeed enormous as they might impact on biomarker discovery, drug discovery/development, and system knowledge [1–4], among others. The realization that lipids not only serve as building materials of membranes and energy providers but are also involved in biological processes such as signaling, cell–cell interactions and, moreover, are linked to diseases such as diabetes, obesity, atherosclerosis, Alzheimer, has led to the emergence of the discipline of lipidomics [4–9]. Lipidomics aims at the comprehensive measurement of the lipids present in a biological matrix and the concomitant detection of the individual lipid responses to various stimuli, for example, disease and pharmaceutical treatment. This holistic approach, simultaneously measuring hundreds of compounds, is revolutionary because it allows to reveal differences between conditions without a priori knowledge. From a historical perspective, lipids have been analyzed in a targeted manner to test certain hypotheses. This has evidently resulted in an enormous knowledge, but the behavior of the thousands of individual lipid species in health and disease is currently underevaluated [10]. Lipidomics, as a hypothesis generating tool, is expected to open a critical door that will greatly increase our understanding regarding the roles of individual lipid species and/or lipid classes. Impressive results have already been obtained. Profiling of plasma phospholipids of healthy and type 2 diabetes mellitus (DM-2) individuals identified four lipids, two phosphoethanolamines and two lysophosphocholines, as potential biomarkers for classifying DM-2 patients from the control population [11]. Han *et al.* applied shotgun lipidomics onto mouse myocardial samples at different time intervals after induction of the diabetes state. A dramatic loss of abundant cardiolipin (CL) species could be detected at the very earliest stages of diabetes, in two independent models of diabetes. As these CL alterations precede the lipotoxic hallmarks of diabetic cardiomyopathy (such as triglyceride accumulation), they will bring new insights for identifying treatment efficacy and might provide novel biomarkers of the diabetes state [12]. A recent study showed an altered plasma sphingolipidome in early Alzheimer's disease (AD). In about 800 plasma lipid species covered, the authors demonstrated a significant reduction of sphingomyelin (SM) mass and significant increase in ceramide (CER) content in plasma of AD patients. Moreover, ratios of SM and CER species with identical fatty acid (FA) chains resulted in a more robust discrimination of AD patients and healthy controls than either metabolite alone [13]. Naturally, not just research on human diseases benefits from global lipidomics. Comparative shotgun lipidomics covered about 95% of the yeast lipidome, quantifying 250 lipid species of 21 major lipid classes [14]. The authors demonstrated that differences in growth temperature and defects in the lipid biosynthesis machinery produced alterations throughout the yeast lipidome. Surprisingly, deletion of genes showed seemingly unrelated compensatory changes within the lipidome of the investigated phenotypes. The latter results emphasize the value of the simultaneous measurement of lipids in biological systems. Despite this, enthusiasm needs to be tempered because the explosion of omics activities also resulted in an increase of ambiguous data in literature. In addition, omics disciplines regularly fail in validating potential markers or data validated by one group cannot be confirmed at another site. By way of example, the amino acid sarcosine has recently been described to be a highly specific marker for prostate cancer progression [3]. This has been disputed in several independent essays [15–17].

This chapter sheds some light on the art and practice of lipidomics. After introducing the target compounds and the complexity one is confronted with, an overview of state-of-the-art methods to unravel this complexity will be provided. The subsequent sections are devoted to a more elaborate description of both a state-of-the-art liquid chromatography (LC) – and gas chromatography (GC) – mass spectrometry (MS)-based lipidomics methodology developed and applied on a routine basis at the authors' laboratory [18–20].

7.2
Lipid Diversity

In lipidomics, the researcher is confronted with a substantial complexity. The size of the lipidome is gargantuan, but unfortunately still speculative. A rather loose approximation of the total number of lipid species is believed to be on the order of 200 000 [21]. To date, the LIPID MAPS Structure Database (LMSD) is populated with over 30 000 structures [22, 23]. The lipidome, considered as a subfraction of the metabolome, in fact occupies circa 70% of the Human Metabolome Database (HMDB) [24]. This immediately explains the necessity of a separate discipline besides metabolomics. The majority of biologically occurring lipids are linear combinations of aliphatic chains and polar head groups attached to glycerol or sphingoid backbones as well as aliphatic chains covalently attached to cholesterol [25]. Knowing this, the estimated, and already observed, size of the lipidome arises through alterations in the nature of the head group (defining the lipid classes); the chain length of the individual aliphatic chains; the number, position, and stereochemistry of double bonds; hydroxyl groups; and other functionalities in the individual aliphatic chains, the nature of the covalent bond (ester, ether, and vinyl ether) to the head group, this to name just a few. Besides describing the lipidome in terms of numbers, it is important to realize that the different lipid species possess diverse physicochemical properties, an attractive feature for the analytical chemist. Lipids span a substantial polarity and molecular weight (MW) range with the cholesterol esters (CEs) and triacylglycerols (TGs) being very apolar and neutral, the glycerophospholipids (GPs) being more polar and charged and the individual fatty acids being of lower MW. On top of the above-described complexities, one is inevitably confronted with a substantial dynamic range within biological matrices. As an example, a recent comprehensive human blood plasma lipidomics study showed that 1 ml of human plasma contains lipids at low femtomole level (eicosanoids) up to micromole level (cholesterol), representing a dynamic concentration range of 8 orders of magnitude [10].

The International Lipid Classification and Nomenclature Committee (ILCNC) developed a "Comprehensive Classification System for Lipids" [22, 23, 26]. The lipidome has been divided in eight primary categories based on their chemical structure and biosynthesis, comprising fatty acyls (FA), glycerolipids (GLs), GPs, sphingolipids (SPs), sterol lipids (STs), prenol lipids (PRs), saccharolipids (SLs), and polyketides (PKs). The first six classes are found in all organisms, whereas the last two are found only in plants and bacteria. Figure 7.1 gives an overview of all primary categories and their lipid classes. More structural details on representative lipid species typically covered by current lipidomics approaches are provided in Figure 7.2.

7.3
Tackling the Lipidome: State-of-the-Art

Lipidomics involves a multidisciplinary approach requiring input from physicians, biologists, analytical chemists, statisticians, (bio)informaticians, and so on. As elaborated on in other chapters (e.g., Figure 14.1), it all starts with defining a biological question and designing experiments in order to answer that question in an unambiguous manner. Next in line are sampling and sample preparation followed by data acquisition, processing, analysis, and interpretation. It is not our intention to provide a detailed discussion on all of these aspects because this would result in a substantial overlap with other chapters. Instead, in providing the reader with an overview of the state-of-the-art lipidomics, we would like to focus on the heart of the pipeline, the data acquisition part, requiring the expertise of the analytical chemists.

MS, particularly using electrospray ionization (ESI) and to a lesser extend atmospheric pressure chemical ionization (APCI), is the principal enabling technology to tackle the lipidome [9, 25, 27–30]. The complexity of the sample under investigation evidently benefits from the use of high-resolution, accurate mass, and tandem mass spectrometric equipment. Therefore, Fourier-transform ion cyclotron resonance (FT-ICR), orbitrap, and (quadrupole) time-of-flight ((Q)-TOF) MS systems are dominating the lipidomics literature [14, 18, 19, 31–37]. Triple quadrupole instruments [10, 38–43] have as well proved their value for the class-specific detection

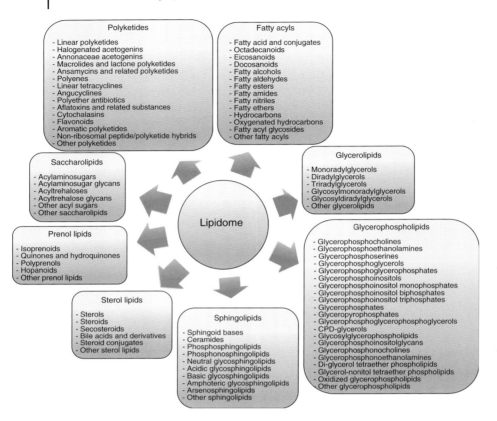

Figure 7.1 Overview of all primary lipid categories and their subclasses present in LIPID MAPS.

through precursor ion and neutral loss (NL) scanning and for the detection of minor species via selected or multiple reaction monitoring (S/MRM). A number of researchers solely rely on MS to measure the lipidome, and in the so-called shotgun lipidomics approach, intrasource separation of the lipids is exploited to widen the lipidome coverage [14, 25, 35, 39, 44]. This intrasource separation is based on the fact that different lipid classes exhibit different charge properties, largely depending on the nature of the head group. As such, lipids can be classified into three categories: anionic (e.g., PA, PG, PI, PS, and CL), weakly anionic (CER and PE), and "neutral" lipids (MG, DG, TG, CE, PC, and SM). The former category is readily measured from a crude lipid extract in negative ionization mode. The latter two categories are measured, respectively, in negative and positive ionization mode, following the generation of alkaline conditions (LiOH). Shotgun lipidomics is currently evolving toward a highly multidimensional MS-based approach with the addition of an endless number of variables including multiplexed lipid extractions, various tandem MS scan modes, and various fragments. As such

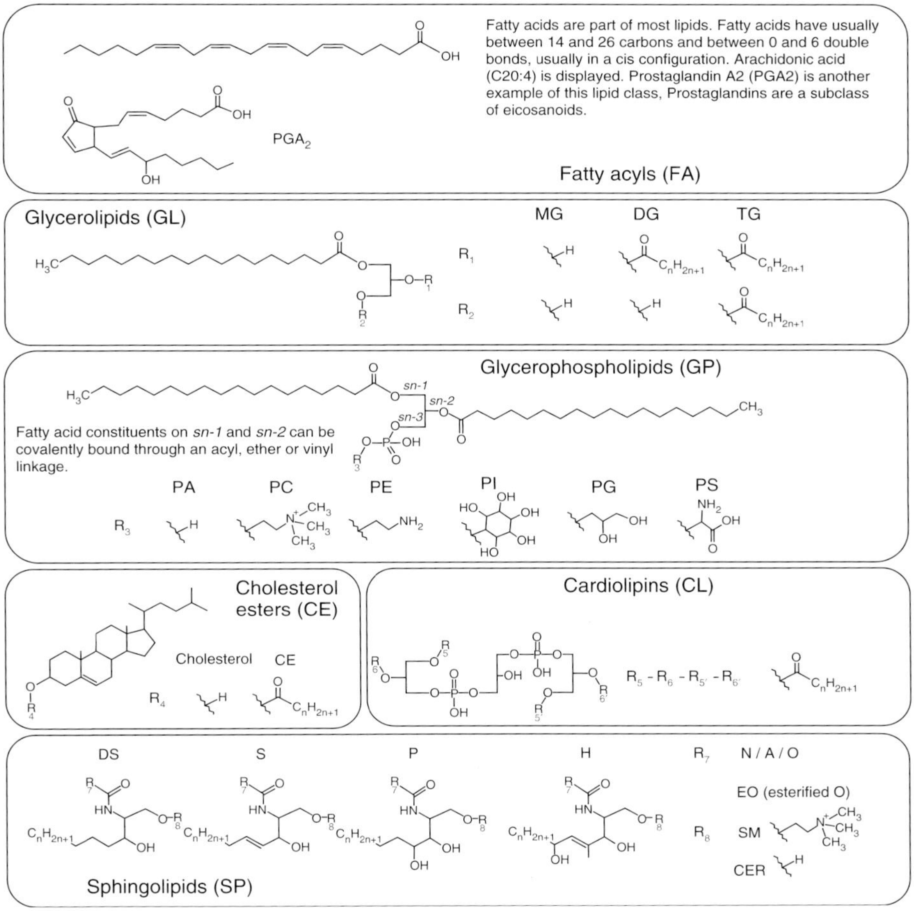

Figure 7.2 Structures of lipid species typically covered by current lipidomics approaches. FA chains of various lipid classes are annotated as C_nH_{2n+1} (saturated FA), but these can be unsaturated as well. (PGA2 – Prostaglandin A2, MG – monoacylglycerol, DG – diacylglycerol, TG – triacylglycerol, PA – glycerophosphatidic acid, PC – glycerophosphocholine, PE – glycerophosphoethanolamine, PI – glycerophosphoinositol, PG – glycerophosphoglycerol, PS – glycerophosphoserine, CE – cholesterol ester, DS – dihydrosphingosine, S – sphingosine, P – phytosphingosine, H – 6-hydroxysphingosine, N – non-hydroxy fatty acid, A – *alpha*-hydroxy fatty acid, O – *omega*-hydroxy fatty acid, EO – esterified *omega*-hydroxy fatty acid, SM – sphingomyelin, and CER – ceramide). Individual lipids are assigned with their carbon number and degree of saturation following the specific lipid class: for example, PC(36 : 0) or more specific PC(18 : 0/18 : 0). Acyl linkages are not specified but plasmenyl or plasmanyl linkages are reported as PC(18 : 0/P18 : 0) or PC(18 : 0/O18 : 0), respectively. GPs with only one fatty acid attached to the glycerol backbone are termed lysoGP and are abbreviated as LPC, LPE, and so on. Ceramides are reported in the format CER[N(26)DS(16)]. The fatty acid esterified to the omega fatty acid in EO species is typically linoleic acid (C18 : 2).

an informative multidimensional array is generated this with an inherent high coverage. This has recently been compiled in a noteworthy review by the pioneers of shotgun lipidomics [25]. Owing to its speed, sensitivity, and ease of automation, shotgun lipidomics is nowadays practiced by various research groups at different levels of sophistication. Although impressive results have already been reported, one is inevitably confronted with the fact that the MS can only tolerate a certain complexity, encompasses a limited in-spectrum dynamic range, and is sensitive toward ion suppression. These constraints, combined with the knowledge that lipid structures often come in a variety of isomers that are difficult, if not impossible, to distinguish solely relying on MS, justify the combined use of chromatography and MS. Various reports describe the combination of reversed-phase (RP), normal-phase (NP), or hydrophilic interaction liquid chromatography (HILIC) with MS [31–34, 45–49]. NP-LC and HILIC typically distinguish lipid species according to their hydrophilic functionalities; as such, it resolves lipids in their representative classes. Under certain conditions, that is, silver ion NP-LC, regioisomeric species, differing in the positioning of FAs on to the glycerol backbone (sn-1, sn-2, and sn-3), can be highlighted [50]. In RP-LC, the separation is mainly based on the hydrophobic properties of lipids, in essence based on the number of carbons and the degree of saturation. In theory, the use of RP-LC over NP-LC or HILIC substantially increases the separation efficiency of lipids, as the intrinsic hydrophilic content of the lipidome is poor compared to the hydrophobic content present. This assumption has been confirmed by Taguchi and coworkers with the identification of two times more phospholipids using RP-LC ESI-MS compared to using NP-LC ESI-MS [46]. The fact that chromatographic techniques exist that separate lipids according to two independent molecular properties offers great opportunities toward multidimensional chromatography allowing a further unscrambling of the lipidome complexity. Indeed, various researchers start to exploit this in either an on-line or off-line manner [45–47]. Off-line multidimensional chromatography is the preferred mode of operation because of its flexibility and the absence of time constraints. On-line multidimensional chromatography requires fast second-dimension separations at high flow rates in order to obtain sufficient samplings. This is, evidently, detrimental toward resolution and highly sensitive mass spectrometric detection. We refer readers interested in multidimensional chromatographic techniques for the analysis of fractions of the lipidome to a recently published book [51].

Packed column supercritical fluid chromatography (pSFC) has been used as alternative to NP-LC for analysis of lipids. Bamba *et al.* [52] used pSFC on a cyanopropyl silica column for the separation of phospholipids, glycolipids, neutral lipids, and sphingolipids. In the same publication, an octadecyl silica column was used for analysis of unsaturation and isomerism in the fatty acid chains. However, the RP mechanism in pSFC shows less performance compared to RP-LC. Online pSFC-RP-LC and off-line pSFCxRP-LC using a silver-ion-loaded first-dimension column were described by François *et al.* [53] for in-depth analysis of triglycerides in fish oil.

GC in combination with either electron ionization (EI) or chemical ionization (CI) MS has proved great value in studying the lipidome [20, 54, 55]. While less suited for intact lipid analysis, it is an ideal tool for fatty acid characterization following hydrolysis and methylation. Fatty acid methyl esters (FAMEs) can be separated on nonpolar and polar stationary phases that enable separation according to carbon number and saturation [20, 55]. Positional isomers, that is, $\omega-3$, $\omega-6$, $\omega-9$, and so on, can be resolved, a feature that is not readily offered by HPLC. Cis–trans isomers are commonly separated on highly polar biscyanopropyl polysiloxane stationary phases [20, 55]. The recently introduced ionic liquid SLB-IL 111 is complementary to the latter phase [56]. Owing to the complexity of the FAs originating from the different lipids, multidimensional GC, and especially GCxGC, has been explored making use of different column selectivities to further mine the lipidome. GCxGC for FAME analysis was pioneered by the groups of Brinkman and Mondello [57, 58]. Our group described the thermal modulated GCxGC separation of bacterial fatty acid methyl esters (BAMEs) based on carbon number and unsaturation in the first dimension and OH-functionalities in the second dimension for the chemotaxonomic characterization of bacteria [59]. The fatty acid composition in marine biota was studied with flow modulated GCxGC using an apolar column in the first dimension and a cyanopropyl silicone or ionic liquid phase in the second dimension [60]. The chromatographic efficiency provided by capillary GC seems to obviate the immediate need for high-resolution MS. Most reports deal with quadrupole instrumentation. TOF systems are often used as back-end detector in multidimensional chromatography because of their acquisition speed and compatibility with the fast second-dimension separations [61]. Nevertheless, the recent introduction in GC of Q-TOF-MS systems will further deepen our knowledge on FA compositions.

Ion mobility [62, 63] and imaging MS [64–66] represent emerging mass spectrometric tools in lipidomics. In the former technology, ions can be resolved based on size and shape in a gas-filled mobility drift cell region before mass-to-charge ratio (m/z) analysis. The major impact of this technology on lipidomics lies in its ability to increase peak capacity and separate isomers and conformers in a very short timescale. The technology of imaging MS possesses the capability to visualize the spatial distribution of lipids, for example, in tissues. Matrix-assisted laser desorption ionization (MALDI), secondary ion (SI), and desorption electrospray ionization (DESI) are typical ionization techniques. Currently, the number of lipids detected is only modest compared to other methods used in lipidomics. As is the case in the field of proteomics, ion mobility and imaging MS are expected to become widely employed.

Despite all technological advancements made, the enormous complexity inherent to the lipidome provides a huge and almost insurmountable challenge toward comprehensive analysis when using a single analytical methodology. Consequently, current interest is shifting to a combination of approaches to comprehend the full lipidome, where different analytical platforms do target one or several lipid classes. In this perspective, Quehenberger *et al.* [10] recently unraveled the enormous diversity of the human plasma lipidome using a spectrum of several GC-MS- and

LC-MS-based approaches. In their text book example, 73 GLs, 160 GPs, 204 SPs, 76 eicosanoids, 31 fatty acids, 36 sterols, and 8 prenols were quantitatively determined down to low femtomoles per milliliter levels.

7.4
LC-MS-Based Lipidomics

The tools developed by lipid researchers in the past, complemented with the enormous advancements in analytical instrumentation and the implementation of sophisticated (bio)-informatics tools (databases, algorithms, statistics, data management and analysis, pathway interrogation, etc.), form the basis of current lipidomics technologies.

Our lipidomics platform combines established lipid sample preparation strategies with both state-of-the-art LC- and GC-MS-based analyses together with commercial and in-house build informatics tools [18–20]. The platform is currently applied on a routine basis on different biological samples including blood plasma/serum and other biological fluids and mammalian cells and tissues to address a variety of biological questions. The LC-MS-based branch of the platform combines high-resolution RP-LC with accurate mass and high-resolution Q-TOF-MS and covers a broad range of lipid species with high precision. A general scheme is shown in Figure 7.3. The following paragraphs are devoted to a description of the different steps and features of the methodology and as such provide the reader with an insight in state-of-the-art LC-MS-based lipidomics. In addition, critical comments and some alternatives/deviations to the methodology based on highly relevant scientific publications have been added.

7.4.1
Lipid Extraction

The first and very important step in lipidomics involves the extraction of lipids from the biological material. Repeatability, recovery, and dilution are key parameters to take into consideration. The standard method of lipid extraction was introduced by Folch *et al.* [67] and later modified by Bligh and Dyer [68]. This method is based on the use of chloroform/methanol/water as ternary solvent mixture that separates into two layers: an upper aqueous layer generally containing non-lipid compounds and a lower chloroform layer containing lipid compounds for the greater part. In our laboratory, this methodology is applied to extract lipids from biological fluids and cells.

7.4.1.1 Biological Fluids and Cellular Material
Biological fluids (e.g., plasma) are generally extracted by adding 200 µl of ice-cold ($-20\,°C$) chloroform/methanol 1/2 (v/v) to 10 µl of biological fluid (volumes can be adapted accordingly). After vortex mixing, 200 µl of water is added followed by a second round of vortex mixing after which the mixture is centrifuged at 10 000 rpm

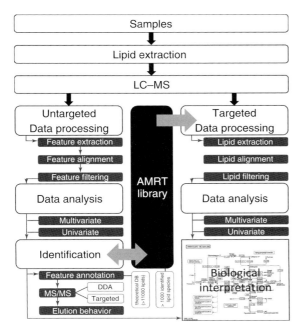

Figure 7.3 Flowchart representing the successive steps in a lipidomics experiment. Lipids are extracted from the biological samples and analyzed via high-resolution LC–MS. Acquired data is subsequently processed in either an untargeted or targeted manner. The former involves the extraction, alignment, and filtering of features with the generation of a data matrix accessible to univariate or multivariate analysis. Differential features are subsequently subjected to an identification strategy fully exploiting the capabilities of a Q-TOF-MS system. Differential features can be searched in an accurate mass retention time (AMRT) library or identified *de novo*. Targeted data processing involves the use of the AMRT library to extract lipids out of the LC–MS data set. Identified lipids in the different samples can subsequently be aligned, filtered, and subjected to statistical analysis. The final step involves the biological interpretation. Note that a set of internal standards can be added before lipid extraction.

for 10 min, thereby generating a lower lipophilic and upper hydrophilic phase separated by a protein layer. Fifty microliters of the lower phase is removed, and an equal amount of isopropanol is added before injection. Some groups perform a double extraction to maximize the recovery. To extract lipids from cells, cell pellets are typically washed several times with ice-cold phosphate-buffered saline (PBS) before adding chloroform/methanol. The Folch and Bligh and Dyer recipe simultaneously recovers all major lipid classes in the chloroform phase, which is relevant when aiming at a comprehensive lipidome analysis. Deviations from this recipe exist for a more class-directed recovery of lipids with tuning of pH or solvent polarity [25]. In our hands, the lipid extraction methodology results in a more than acceptable precision with relative standard deviation (RSD) values typically well below 10% ($n = 3$) both for extractions from blood plasma and liver cells. Evidently, the higher the precision, the smaller the biological variability that can be highlighted. Note that solvents (chloroform) used are not compatible with all plastic tubes and tips typically used in biology laboratories. Moreover, pipetting of organic solvents (chloroform and methanol) using plastic air displacement micropipettes is typically prone to volumetric errors due to leakage. Alternatives exist and the use of robotic systems using a glass syringe tends to be more precise for liquid handling [69]. Because of the higher density of chloroform compared to a water/methanol mixture, it is recovered from the lower phase of the two-phase partitioning system. While collecting the chloroform fraction, a voluminous layer of nonextractable insoluble matrix (e.g., proteins), usually residing at the interface of water/methanol and chloroform phases, needs to be pierced. Lipid extraction using methyl-*tert*-butylether (MTBE) instead of chloroform has been described as an alternative to simplify collection and minimize dripping losses [70]. Owing to the lower density of MTBE compared to methanol, lipids can be recovered from the upper layer. The insoluble matter resides at the bottom. It has been demonstrated that lipid recovery is very similar compared to the golden standard Folch and Bligh and Dyer recipes.

7.4.1.2 Skin (Stratum Corneum)

A more specific protocol is used in our laboratory for the extraction of lipids from the outermost layer of the skin epidermis, that is, stratum corneum (SC) [19]. SC is usually sampled on the inner forearm of individuals. The skin of the subjects is stripped with a patch coated with an adhesive layer, named Corneofix® (Courage + Khazaka electronic GmbH, Köln, Germany). Generally, three patches are placed on 1 cm distance on the lower arm of every individual, pushed down, and removed (three patches equal one sample). The three skin patches are immersed in 5 ml of methanol, vortexed, and sonicated for 10 min. After sonication, the patches are removed and the remaining methanol fractions are combined in one tube. The combined methanol fraction is dried at 37 °C under a nitrogen stream.

7.4.1.3 Solid-Phase Extraction (SPE)

The sample complexity in lipidomics may require additional prefractionation steps to isolate one or more lipid classes from biological matrices. NP stationary phases are able to separate lipids according to their hydrophilic functionalities. These are typically used to fractionate the lipidome in individual lipid classes. Both NP-SPE (solid-phase extraction) and NP-LC are mostly used to obtain a prefractionation of lipids before RP-LC separation [45–47, 71]. The parallel nature of the former allows high-throughput sample preparation. By way of example, the following methodology, based on aminopropyl silica SPE, is used to separate and/or isolate different lipid classes from skin and to remove contaminants originating from the skin patch (Figure 7.4). Dried skin extract (see above) is reconstituted in 300 µl of 11 : 1 hexane/isopropanol (v/v). After the sample load, the cartridge is washed with hexane, which elutes the most apolar lipids such as CEs and TGs from the sample. Only a minor fraction of DG ($\pm 20\%$) is retained after eluting the cartridge with

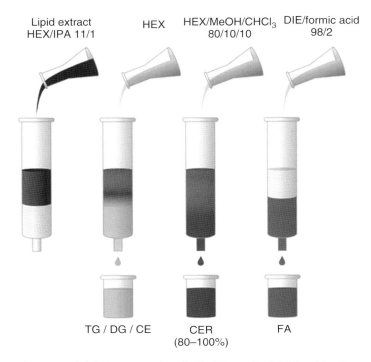

Figure 7.4 Solid-phase extraction of skin SC samples (CHCl$_3$ – chloroform, IPA – isopropanol, HEX – hexane, MeOH – methanol, and DIE – diethyl ether).

hexane. CERs are eluted using hexane/methanol/chloroform 80 : 10 : 10 (v/v). The more polar CERs such as CER[NP] and CER[AS] have a higher extraction yield than the more apolar CERs such as CER[NS] and CER[NDS], which show a small loss in the washing step with hexane. For subsequent LC–MS analysis, the eluted fraction is dried under nitrogen and dissolved in isopropanol/chloroform 50 : 50 (v/v). FAs stay on the SPE cartridge, even after elution with methanol. These molecules can be eluted by diethyl ether/formic acid 98 : 2 (v/v). Similar strategies exist for isolating other lipid classes such as GPs. A weak anion exchange (WAX) SPE approach can be used to fractionate GPs one by one in, respectively, PCs and lysoPCs, PEs, and the acidic GPs (i.e., PA, PS, PG, PI, and CL) [72]. Low-abundant GPs are as such separated from high-abundant lipid classes, enabling a better detection of the former lipids. Another example of a systematic lipid fractionation was published by Kaluzny et al. [73]. They described a rather complex SPE-based methodology to separate cholesterol, CE, TG, DG, MG, FA, and GP based on a SPE setup that combined three NH$_2$ columns and eight different eluents. Next to the fractionation of complex lipid samples into individual lipid classes, SPE is also employed for the enrichment of low-abundant lipid species, such as eicosanoids [74].

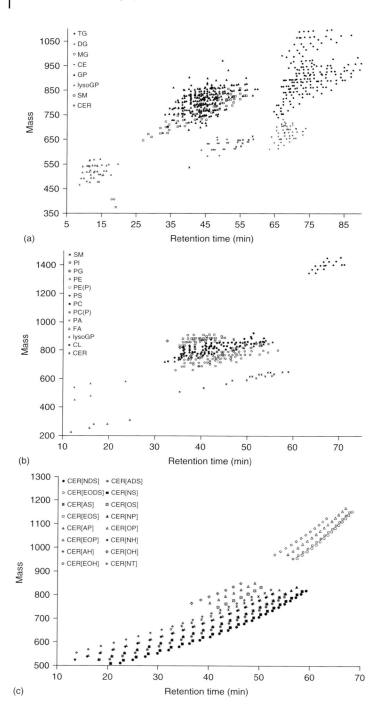

7.4.2
LC–MS(/MS)

7.4.2.1 Retention Time Characteristics

Following extraction, the lipids are subjected to an LC–MS methodology. The complexity of the sample under investigation evidently benefits from the use of high-resolution chromatography combined with high-end mass spectrometric equipment with inherent high-resolution, accurate mass, and tandem MS capabilities. In our case, a 1200 HPLC system, equipped with binary pump, is hyphened to a 6530 Q-TOF system (Agilent Technologies, Waldbronn, Germany). To maximize inter- and intraspecies/class resolution, an RP-LC methodology using sub 2-µm particles, elevated temperature (80 °C) and carefully designed elution conditions, has been developed. The solvent system, composed of 20 mM ammonium formate pH 5 (solvent A) and methanol (solvent B), is delivered at a rate of 0.5 ml min^{-1} onto a 100 mm l × 2.1 mm ID × 1.7 µm dp XBridge C18 Shield column (Waters, Milford, MA). The setup allows working at pressures <600 bar, thereby eliminating the need of a ultrahigh pressure liquid chromatography (UHPLC) pump. The simultaneous detection of FA, (lyso)GP, SP, CE, MG, DG, and TG species in, for example, a total lipid extract of blood plasma, requires a multistep gradient to deconvolute the lipids: 50–70% B in 14 min followed by a slow gradient of 70–90% B in 50 min and an isocratic separation at 90% B during 15 min. The latter step can be omitted if apolar lipids (TG or CE) are not expected in the sample. A dedicated method for skin CER (SPE fraction) profiling deviates from the above-reported method and uses a linear gradient between 70 and 100% methanol.

The interspecies resolution in the lipid profiling method is shown in Figure 7.5. Figure 7.5a–c displays the positioning of identified blood plasma, liver cell, and skin lipid species in the LC–MS plane. A separation into different lipid classes is visualized. LysoGPs and MGs are clearly separated from the GPs, SMs, CERs, and DGs, which on their turn are cut off from the elution region of the more apolar lipids covering CEs and TGs. CEs and TGs are very retentive on the RP column. CEs only bear one FA moiety but the cholesterol headgroup substantially contributes to the retention. The elution order of GLs is MG < DG < TG. The latter lipid class has three FAs that contribute to the interaction with the stationary phase, in contrast to two and one FA group(s) for DGs and MGs, respectively. Similarly, lyso- or monoacyl-GPs with one FA attached to the glycerol backbone elute earlier than diacyl-GPs because of the contribution of only one FA to the

Figure 7.5 2D LC–MS plot of identified lipids in (a) blood plasma, (b) liver cells, and (c) skin (stratum corneum). Graphs (a) and (b) originate from the LC–MS acquisition of chloroform extracted blood plasma lipids and cellular lipids using the multistep gradient in, respectively, positive and negative ionization modes. Graph (c) results from the LC–MS analysis of SPE enriched skin CER in negative ionization mode using the adapted gradient program (linear between 70 and 100% methanol). Lipids reported have been detected with 100% frequency in a triplicate experiment.

interaction with the stationary phase. As shown in Figure 7.5b, CLs, bearing four FA constituents and two phosphate moieties, elute later than GPs containing two FAs and one phosphate. Despite the different head group, that is, choline versus ethanolamine, PCs and PEs with an identical carbon number nearly elute at the same retention time. However, these species are easily resolved by their masses. Note that adapting mobile phase B to a mixture of methanol/acetonitrile (1 : 1 v/v) governs the separation of these two classes. Compared to PCs and PEs, other GPs with identical carbon number elute earlier. The following elution order is observed: $PG < PI \leq PS \ll PE = PC$. Nevertheless, considering all GPs a substantial overlap exists between the different classes because of the broad range of hydrophobicity created by the FA constituents. PCs and PEs differing by three carbons are true isomeric species showing identical molecular formulas, but these isomers are separated in the chromatographic step.

In the adapted skin CER method (Figure 7.5c), a similar elution pattern is observed. The more polar CERs (e.g., CER[NP]) elute earlier than their apolar counterparts with identical carbon atom numbers (e.g., CER[NDS]). Because of the additional FA chain, esterified CERs (EO) are more retentive than their nonesterified counterparts. An earlier study reported that RP-LC in combination with methanol has difficulties in eluting the more hydrophobic esterified ω-hydroxy FA containing species (CER[EO]) [75]. Nevertheless, the current protocol allows eluting CER[EO] using methanol due to the use of the elevated column temperature of 80 °C, without any carryover between successive analyses.

Intraspecies resolution is readily obtained on RP-LC, which is absent in NP-LC that separates lipids based on their head group. This is demonstrated in Figures 7.6 and 7.7. The more carbon atoms, the more retention is attained on RP-LC [76]. Figures 7.6a and 7.7 illustrate the increased retention for lipid species with longer FA chains for, respectively, the CER class CER[NS] and PE and LPG species. The resulting unique elution pattern originates from the differences in the number of CH_2 groups. Lipid species containing an equal carbon atom number but a different number of double bonds in the FA chain are also resolved, as is displayed in Figures 7.6b and 7.7. The more double bonds a lipid contains, the less retention it displays on RP-LC. Furthermore, isomeric species are observed that vary in their FA content on the *sn*-1 or *sn*-2 position (e.g., PC(18 : 1/18 : 2) and PC(16 : 0/20 : 3) in Figure 7.6b and PE(16 : 1/20 : 4) and PE(16 : 0/20 : 5) in Figure 7.7a; see further for identification of these isomers). Remarkably, isomeric species containing one saturated FA always elute later than species solely containing unsaturated FAs. Isomeric species resulting from the differential positioning of identical FAs onto the glycerol backbone (*sn*-1 or *sn*-2) are also resolved as is clearly demonstrated in Figure 7.7b for LGPs. On surveying the 2D LC–MS maps of PE and LGP lipids, it can be observed that a clear correlation exists between retention time and carbon chain length and degree of saturation. This facilitates the lipid identification process. From the 2D plot of the LPG lipids, it becomes apparent that the nature of the covalent bond (ester, ether, and vinyl ether) of the FA to the head group also has its influence on retention. Plasmanyl lipids, with the aliphatic chain attached to the glycerol backbone via an alkyl ether linkage (e.g., LPC(O-18 : 0)),

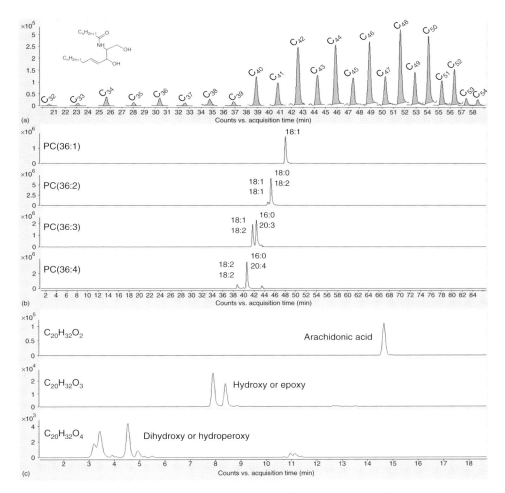

Figure 7.6 Intraspecies resolution of the chromatographic setup: (a) extracted compound chromatograms (ECCs; negative ESI mode) of skin derived CER[NS]C32 up to CER[NS]C54, clearly showing the unique elution pattern because of the difference in the number of carbon atoms (each CER differs one CH_2 group with the following CER). (b) Extracted ion chromatograms (EICs; positive ESI mode) of a number of PCs detected in blood plasma differing in the number of double bonds. Ions were extracted at 5 ppm mass accuracy (from the exact lipid mass), which is sufficient to eliminate the isotopes associated with other PC species. The FA composition, as revealed by performing MS/MS, is reported on top of the peaks. (c) EIC (negative ESI mode) of arachidonic acid and its derivatives detected in lung epithelial lining fluid. These species are separated in the first part of the multistep gradient (50–70% B).

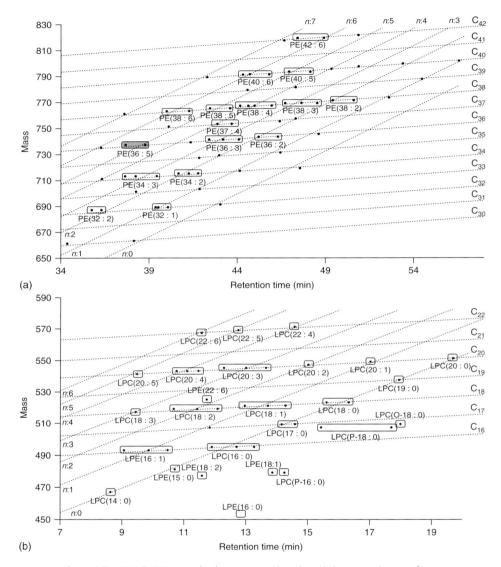

Figure 7.7 2D LC–MS map displaying (a) the glycerophosphoethanolamines (PE) detected in liver cells and (b) the lyso-glycerophosphocholines (LPC) and ethanolamines (LPE) in human plasma (Figure 7.5). The total carbon atom number and the degree of unsaturation are indicated. PE and lysoPC with the same carbon chain length and the same degree of unsaturation are connected, illustrating the correlation between elution and number of carbons and double bond. The FA composition of the two isomers of PE(36 : 5) (gray colored) is revealed in the MS/MS spectra presented in Figure 7.8.

are more retentive than their plasmalogen (plasmenyl phospholipid) counterparts (e.g., LPC(P-18 : 0)), which have the aliphatic chain attached through a vinyl ether linkage. As is the case for GPs, LGPs, SMs, CERs, and so on, intraspecies resolution of TG is readily obtained. In contrast to the former, it is achieved in the isocratic part at 90% methanol. Applying a gradient in that specific region would result in a substantial coelution of TG species. Hydroxylated derivatives of FAs, playing essential biological functions, elute earlier than their nonhydroxylated counterparts. This is demonstrated in Figure 7.6c for FAs taking part in the arachidonic acid metabolism. These species are typically less accessible in, for example, blood plasma without enrichment (via SPE) and targeted measurements (via MRM), but their measurement at specific locations, for example, lung epithelial lining fluid, using an untargeted approach is feasible. Mediators of the arachidonic acid metabolism have been shown to be involved in asthma, in which airways get restricted [77].

It is obvious that the LC–MS methodology allows a thorough infiltration into different lipidomes. In real-life experiments, processing a substantial number of samples, retention time repeatability is important. Run-to-run retention time repeatability of our lipidomics methodology is typically well below RSD 0.1%. This greatly facilitates subsequent data processing because alignment of identical lipids in different samples/runs becomes straightforward. Throughput of the described chromatography is only moderate. Some studies might benefit from a higher throughput. Note that the use of sub-2 μm particles, allows the use of higher flow rates and reduced run times without substantially affecting separation. On the other hand, it might impact on mass spectrometric sensitivity. In any case, one has to find a balance between speed and content.

7.4.2.2 Ionization Characteristics

In the lipidomics platform, a Jetstream ESI source, operated in positive and negative ESI mode, is applied for ionization of the various charged and uncharged lipid classes. This specialized ESI source, equipped with thermal gradient focusing technology, uses superheated nitrogen to improve ion generation and desolvation. The intensity for all lipid classes thereby increases up to a factor 4, compared to a classical ESI source, without displaying increased in-source fragmentation [18].

The choice of the modifier in the solvent system can substantially affect the ionization sensitivity and selectivity as well as the ion profiles of the lipid species [25]. The addition of a modifier ammonium formate, acidified to pH 5, allows us to cover lipid classes in both positive and negative ESI mode. The inclusion of ammonia hereby greatly reduces sodium adduct formation [78]. CEs, MPs, DGs, and TGs are only detected in positive ESI mode as $[M + NH_4]^+$ ions, with traces of $[M + Na]^+$ and $[M + K]^+$. GPs and SPs are detected as protonated ions, again with traces of $[M + Na]^+$ and $[M + K]^+$. Phosphoinositol (PI) lipids are the exception here, they only appear as ammonium adducts in positive ionization mode. More acidic lipids (e.g., PE, PS, PG, PA) show substantial lower signal intensities in positive ionization mode compared to, for example, PC or SM. For the same concentration injected, the detectability of the PE species in positive ESI mode is typically less than 10% of the detectability of the corresponding PC species. In negative ESI mode, the ionization efficiency increases to 65% of that of the PC species, which justifies the combination of two ionization modes in lipidomics. In negative ionization mode, (acidic) GPs are mostly detected as $[M-H]^-$, except for PCs and SMs that appear as $[M+HCOO]^-$. Owing to the presence of two phosphate moieties, CLs are detected as both singly and doubly charged species. FAs can be easily detected as deprotonated ions. While CERs predominantly appear as protonated species in positive ESI mode, a CER spectrum in negative ESI mode displays ions with an increasing signal intensity from $[M-H]^- < [M+HCOO]^- \approx [M+CF_3COO]^- < [M+Cl]^-$. This adduct formation occurs independently of the sample, instrument, and ionization source (APCI, ESI, or JetStream ESI). Although adduct formation is generally an unwanted phenomenon in MS, it can also aid in the lipid identification process. As an example, measurements in negative ESI mode discriminate between PCs and PEs because the former are detected as formate adducts, while the latter appear mainly as deprotonated species.

For more polar and charged lipids such as CER, SM, GP, LGP, and CL, ESI is the ionization method of choice. For neutral or nonpolar lipids such as MGs, DGs, TGs, and CEs, APCI and atmospheric pressure photoionization (APPI) have been demonstrated as valuable alternatives to ESI. Neutral lipids are typically detected as protonated species under these conditions. It has been demonstrated that, for these compounds, APPI in combination with NP-LC can offer lower detection limits and an extended linear dynamic range over both ESI and APCI [79, 80]. It might be interesting to determine the influence of APPI ionization in the case of RP-LC conditions, especially with methanol as a mobile phase solvent [81]. However, the universality of the lipidomics method is not maintained because other lipid classes are expected to be ionized only marginally using both APCI and APPI.

7.4.2.3 Identification of Lipids

The annotated lipid species presented in Figure 7.5 result from an identification strategy that fully exploits the features of the Q-TOF-MS system. Molecular formulas, based on accurate mass, isotopic abundance, and spacing both in positive and negative ionization modes, are complemented with accurate mass database searching in an in-house build database (populated with LIPID MAPS and HMDB entries and theoretical lipid structures) and MS/MS measurement in both modes. Other parameters such as lipid elution behavior and adduct formation interpretation (e.g., difference in adduct formation between isomeric PCs and PEs in negative ionization mode) further assist in the identification.

Accurate mass measurement is of great importance toward identifying lipids. By way of example, the mass difference not only between PE(36 : 2) and PC(18 : 1/P16 : 0) but also between CER[NH] and CER[NDS] with one additional CH_2 group is only 0.0364 Da. On the other hand, lipids come in a variety of isomers and multiple annotations can, hence, occur for a particular molecular formula.

For example, $C_{44}H_{86}NO_8P$ annotated as PC(36 : 0) will also have a PE(39 : 0) annotation in positive ionization mode, a formate adduct of PC(36 : 1) will also be annotated as PS(39 : 0), PC(36 : 2) can be the result of a variety of fatty acid combinations, CER[NH] is an isomer of CER[AS], and so on. The possibility of a variety of structures per *m/z* illustrates the limitation of using only accurate mass for lipid identification. It underlines the importance of performing MS/MS to discriminate between isomeric lipid species. Wide scale lipid identification typically proceeds through MS/MS experiments in the data-dependent acquisition (DDA) mode. In our case, one survey MS measurement is complemented with three data-dependent MS/MS measurements resulting in a cycle time of 4 s. After being fragmented twice, a particular *m/z* value is excluded for 30 s. Selecting the same *m/z* value twice increases the chance of measuring a particular precursor at its maximum intensity, while an exclusion time of 30 s allows to obtain MS/MS information on chromatographically resolved isomers. Remaining unidentified species are subsequently identified in either targeted MS/MS or further DDA experiments, thereby adding previously fragmented precursors in an exclusion list.

Lipid fragmentation mechanisms/spectra in both positive and negative ionization modes have extensively been reported in literature [33, 38–40, 49, 82–86]. Table 7.1 displays the qualitative fragment ions in positive and negative ESI modes for most covered lipid classes in LC-MS-based lipidomics. Note that the information captured in Table 7.1 is often used for the class-specific detection through NL and precursor ion scanning [38]. Collision-induced dissociation of protonated PCs typically gives rise to an intense phosphorylcholine ion at 184 amu [83, 84]. SM species follow the closely related choline-containing species and show the same abundant fragment in positive ESI mode. PEs show an NL of 141 amu when fragmented in positive ESI mode, corresponding to the PE polar head group phosphoethanolamine. As such, isomeric PC and PE species can be distinguished in positive ESI mode fragmentation. Formate adducts of PCs and SMs show a typical loss of a methyl group and the formate ion when fragmented in negative ESI mode. Glycerophosphoserines (PSs) are recognized by the NL of serine. This allows the discrimination of isomeric PC and PS species in negative ionization mode. Several ions characterize PI species after fragmentation in negative ESI, with 241 and 223 amu as being typical for the PI class (corresponding to inositol phosphate ions) [84]. Glycerophosphoglycerols (PGs), glycerophosphatidic acids (PAs), and CLs show a typical fragment ion of 153 amu in negative ESI mode, which corresponds to glycerol phosphate minus water. Information on the FA composition of GPs is obtained in the negative ionization mode through the formation of FA-derived carboxylate anions. Isomers that are associated with the same head group but vary in the FA content or position on the glycerol backbone, that is, *sn-1* or *-2*, are resolved by the RP-LC method (e.g., PE(36 : 5) in Figure 7.7). To obtain their individual FA content, MS/MS has to be performed in the negative ESI mode. Figure 7.8 displays the MS/MS spectra in negative ESI mode of two isomers of PE(36 : 5) that could be identified as PE(16 : 1/20 : 4) (eluting first) and PE(16 : 0/20 : 5) (eluting second) (see Figure 7.7 for their retention time difference).

Table 7.1 Different lipid classes and their qualitative fragment ions, in positive and/or negative ESI, when MS/MS fragmentation is performed.

Lipid class	Precursor ion	MS/MS mode	Fragment
PA	$[M - H]^-$	152.9953 amu	Glycerol phosphate $-H_2O$ $[C_3H_6O_5P]^-$
PC/lysoPC	$[M + H]^+$	Species dependent 184.0739 amu	FA carboxylate anions Phosphocholine $[C_5H_{15}NO_4P]^+$
	$[M + HCOO]^-$	60.0211 amu	Neutral loss of methyl group and formate ion $[C_2H_6O_2]$
PE/lysoPE	$[M - H]^-$	Species dependent 196.0375 amu	FA carboxylate anions Glycerol phosphoethanolamine $-H_2O$ $[C_5H_{11}NO_5P]^-$
	$[M + H]^+$	Species dependent 141.0191 amu	FA carboxylate anions Neutral loss of phosphoethanolamine $[C_2H_8NO_4P]$
PG	$[M - H]^-$	152.9953 amu	Glycerol phosphate $- H_2O$ $[C_3H_6O_5P]^-$
		227.0321 amu	Glycerol phosphoglycerol $- H_2O$ $[C_6H_{12}O_7P]^-$
PI	$[M - H]^-$	Species dependent 223.0008 amu	FA carboxylate anions Cyclic inositol phosphate $- H_2O$ $[C_6H_8O_7P]^-$
		241.0113 amu	Cyclic inositol phosphate $[C_6H_{10}O_8P]^-$
	$[M + NH_4]^+$	Species dependent 277.0563 amu	FA carboxylate anions Neutral loss of NH_3 and phosphoinositol $[C_6H_{16}NO_9P]$
PS	$[M+H]^+$	185.0089 amu	Phosphorylserine $[C_3H_8NO_6P]^+$
	$[M - H]^-$	87.0320 amu	Neutral loss of serine $- H_2O$ $[C_3H_5NO_2]$
SM	$[M + H]^+$	Species dependent 184.0739 amu	FA carboxylate anions Phosphocholine $[C_5H_{15}NO_4P]^+$
	$[M + HCOO]^-$	60.0211 amu	Neutral loss of methyl group and formate ion $[C_2H_6O_2]$
TG, DG, and MG	$[M + NH_4]^+$	Species dependent Species dependent	FA carboxylate anions Neutral loss of NH_3 and FA
Cholesterol, CE	$[M + NH_4]^+$	369.3521 amu	Neutral loss of NH_3 and $-OH$ or FA, resulting in $[C_{27}H_{45}]^+$

It has been described that the actual FA group position on the glycerol backbone can be determined by MS/MS on consulting the carboxylate anion ratio [84]. This is rather difficult in real-life situations. Fragmentation of SM in negative ionization mode does not yield valuable FA carboxylate anions. Aliphatic chains connected via ether or vinyl linkages in plasmanyl and plasmenyl phospholipids cannot be revealed following MS/MS. Fatty acyls present in these species, on the other hand, are revealed in negative ESI mode.

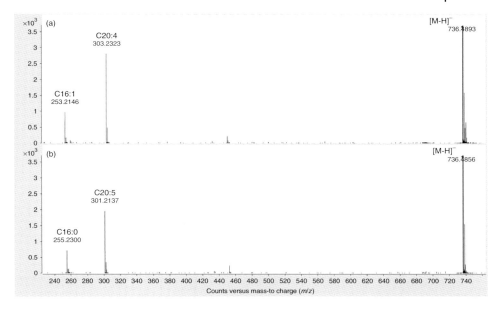

Figure 7.8 MS/MS measurements performed on two isomers of PE(36 : 5) in negative ESI mode (Figure 7.7), revealing the FA composition of the two isomers. PE(16 : 1/20 : 4) is the first eluting lipid of the two isomers.

Lipid species such as MGs, DGs, TGs, and CEs give rise to an informative NL of the FA chains in positive ESI-MS/MS [49]. This ends up in a typical fragment ion for CEs displaying a mass of 369 amu [87]. Figure 7.9 shows the MS/MS spectra of several chromatographically resolved TG isomers.

The interpretation of CER fragmentation spectra is somewhat more complicated compared to other lipids, because of the presence of numerous isomeric species (class isomerism) [75, 88–93]. The sphingoid base (SB) part can usually be built up from the fragmentation spectra in positive ESI mode, for example, the number of hydroxyl groups is easily derived from the loss of water molecules in the fragmentation spectrum, both from the molecular ion and from the SB fragment (Figure 7.10). Negative ESI spectra of $[M-H]^-$ ions are easier in deducing the FA part, because of the presence of carboxylate ions. The adduct ions (trifluoroacetic acid (TFA), chloride, and formate) do not always give rise to relevant fragment ions. Figure 7.10 displays the MS/MS spectra of standard CER[N(24)DS(18)] and of CER[NDS]C_{42} present in skin, which is actually composed out of 10 different skeletal isomers varying in carbon length of SB and FA building blocks, being CER[N(26)DS(16)], CER[N(25)DS(17)], CER[N(24)DS(18)], CER[N(23)DS(19)], CER[N(22)DS(20)], CER[N(21)DS(21)], CER[N(20)DS(22)], CER[N(19)DS(23)], CER[N(18)DS(24)], CER[N(17)DS(25)], and CER[N(16)DS(26)]. Fragmentation spectra of CER[EO] in negative ESI mode show a very intense signal of linoleic

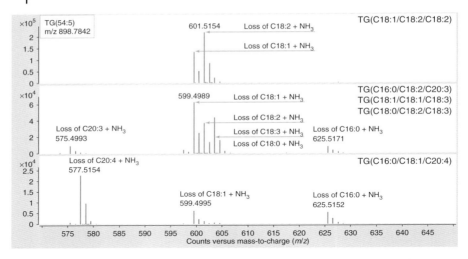

Figure 7.9 MS/MS spectra associated with blood plasma derived chromatographically resolved TG(54 : 5) isomers acquired in the positive ionization mode on the $[M + NH_4]^+$ precursors. Proposed structures are incorporated as well. Only the region of interest is shown to allow a clear distinction between the different ions that result from the neutral loss of the fatty acids from the precursor. The middle spectrum clearly reveals several compounds.

acid (N18 : 2) that originates from cleavage of the ester-linked FA of the CER[EO] species.

ESI-MS/MS-based lipidomics can easily resolve the lipid class, carbon chain length, and degree of unsaturation of FA components of lipid species but is unable to the ready identification of sites of unsaturation (i.e., double-bond position) of the latter FA components. Specialized approaches are requisite in order to determine the exact location of the double bonds. One technique to identify the location of the double bond(s) in unsaturated lipids is ozone-induced dissociation (OzID) [94, 95]. Hereby, mass-selected lipid ions are exposed to ozone vapor within an ion-trap mass spectrometer. Although this technique has several limitations, that is, requirement of specialized equipment, instrument modification, and strict safety considerations in using highly reactive ozone, it is capable of identifying the lipid double-bond positions in even complex biological mixtures [96]. More recently, it has been demonstrated that straightforward MS/MS is able to sort out the location of double bonds in FA and FA chains of complex lipids. Here, unsaturated FA isomers can be discriminated based on the characteristic intensity distribution of the specific fragment ions, resulting from the loss of H_2O or CO_2, from FA anions with varying Collision-induced dissociation (CID) conditions (i.e., either collision energy or collision gas pressure) [97].

In the absence of MS/MS information or in case of ambiguity, elution behavior furthermore aids in identifying lipids. As shown in Figures 7.6 and 7.7, a clear correlation exists between retention time and number of carbons and degree of saturation. This correlation, for example, aids in the identifying LPC (17:1), a minor

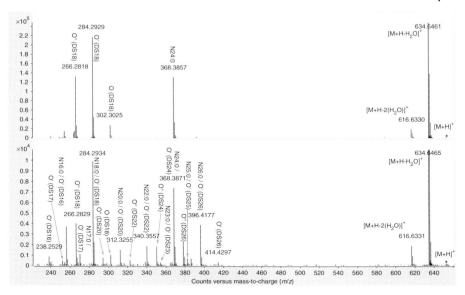

Figure 7.10 MS/MS fragmentation pattern of (a) a standard CER[N(24)DS(18)] and (b) CER[NDS]C42 in skin (SC) at a collision energy of 35 eV in positive ESI mode (O = loss of fatty acyl group, O' = loss of fatty acyl group and 1 mol of water, and O'' = loss of fatty acyl group and 2 mol of water). Owing to skeletal isomerism, CER[NDS]C42 can be defined as the result of various CER species varying in carbon length of SB and FA building blocks, that is, CER[N(26)DS(16)], CER[N(25)DS(17)], CER[N(24)DS(18)], CER[N(23)DS(19)], CER[N(22)DS(20)], CER[N(21)DS(21)], CER[N(20)DS(22)], CER[N(19)DS(23)], CER[N(18)DS(24)], CER[N(17)DS(25)], and CER[N(16)DS(26)].

species, with molecular formula $C_{25}H_{50}NO_7P$. It can be discriminated from its isomer LPE(20:1) by its position in the plot. These two species can as well be discriminated in negative ionization mode because LPC species appear as formate adducts and LPE species as deprotonated ions.

7.4.3
Data Processing and Analysis

A general scheme for our data handling workflow is provided in Figure 7.3. As lipidomics involves the global analysis of lipids, all individual lipid compounds need to be extracted out of the data set. This is generally accomplished by the so-called feature extraction algorithm that localizes and combines related covarying ions (isotopes, adducts, and charge states) with the generation of a single mass (median), retention time (peak apex), and an abundance (sum of ions in clusters). Several feature extraction algorithms have been described in the literature, for example, XCMS, MZmine, MetAlign, and mzMatch [98–101]. We typically opt for the molecular feature extraction (MFE) algorithm incorporated in the MS vendor software (MassHunter Qualitative Analysis software, Agilent Technologies). Feature lists of all samples are subsequently converted to a data matrix this by aligning features based on retention time and mass, typically followed by filtering (e.g., frequency filter, RSD filter) and/or normalization (e.g., internal standards, median intensity), and so on. These steps can be performed in open-source (XCMS, MZmine, MetAlign, and mzMatch) or vendor-specific software programs (in our case MassProfiler Professional, Agilent Technologies). The generated data matrix can subsequently be subjected to statistical analysis to reveal lipid features that differentiate the groups under investigation. A more elaborate description of the tools that can be used is provided in Chapter 14. Identification of differential features is typically the step that precedes biological interpretation. The quality of this so-called untargeted data processing is more than satisfactory as is demonstrated in Figure 7.11 and Table 7.2, showing the precision of feature extraction and subsequent alignment from the triplicate analyses of a blood plasma, a liver cell, and a skin extract. Of the 890 aligned features obtained in negative ionization mode from the liver cell extract, 675 appeared to be present with a frequency of 100%. Of these, 84.1% (568 features) had an RSD% <15%, confirming the precision. For a blood plasma and a skin extract, measured in positive and negative ionization modes, respectively, these values were 81.9% (1194 features) and 62.1% (728 features). The more precise the data processing is, the smaller are the differences in lipid concentration that can be highlighted between samples. To illustrate this, two phospholipids (PE(18 : 1/18 : 1) and PC(18 : 1/18 : 1)) were spiked in the same blood plasma at different concentrations and subjected to the lipidomics strategy in both positive and negative ionization modes. Every sample was treated in triplicate and samples were randomized both at sample preparation and LC–MS level. As is shown in Figure 7.6b, PC(18 : 1/18 : 1) is a minor isomer of the closely eluting PC(18 : 0/18 : 2). After the generation of a data matrix, a t-test was performed to extract these spiked compounds out of the matrix. Differences in phospholipid concentration as small as 25% could readily be discriminated with p-values well below 0.01 giving confidence in the untargeted approach to detect subtle biological differences between samples.

Table 7.2 Features detected in positive ionization mode for plasma and in negative ionization mode for liver cells and skin (stratum corneum) for extraction replicates using both untargeted and targeted data processing.

	All features	Features in 3/3	RSD < 15%	Identified lipids	RSD < 15%
Plasma	2334	1458	1194	528	472
Liver cells	890	675	568	346	308
Skin	1947	1172	728	650	474
	Untargeted data processing			Targeted data processing	

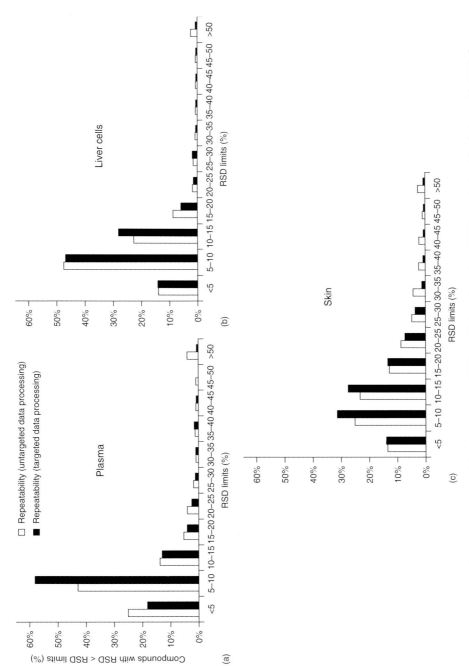

Figure 7.11 (a–c) Technical repeatability of the lipid analysis of different biological matrices: RSD% is calculated for features with 100% frequency after untargeted or targeted data processing.

The repeated analysis of dozens of samples from different biological sources has allowed us to identify a substantial number of features. All thus far identified lipids have been listed in two separate accurate mass and retention time (AMRT) libraries, namely, a lipid and skin CER AMRT library compatible with the general lipid and skin CER LC–MS methodology. An AMRT library can be built easily in Excel (in comma separated value (.csv) format) and consists of the molecular formula, exact mass, and retention time of all previously identified lipids. Furthermore, an AMRT library can be extended after each new lipidomics experiment by adding additional identified lipid species. Lipids included in the library (by now over 1000 species) can be readily identified in samples in a so-called targeted data processing workflow. Targeted data processing in an untargeted data set (Q-TOF-MS) searches for specific lipid ions in the raw data files in an automated manner. We use the Find By Formula application in the MassHunter Qualitative Analysis software to localize lipids from the library in two dimensions, retention time and accurate mass, both within a user-defined window (±25 ppm mass accuracy and ±15 s retention time accuracy, depending on the quality of the data). Open-source tools are also accessible for these purposes [33]. The two-dimensional plots presented in Figure 7.5 are the result of matching LC–MS runs of blood plasma, liver cells, and skin onto the lipid and CER AMRT libraries. Note that these plots originate from triplicate experiments. Annotated lipids from every run were aligned and filtered based on 100% frequency. In this way, only data that matters are presented. As is the case with features, lipids, extracted via targeted data processing, are associated with an intensity. Figure 7.12 shows the contribution of the different lipid classes to the overall lipid intensity within blood plasma, liver cells, and skin lipid extracts. Note that differences in ionization efficiency have not been taken into account.

Targeted data processing equals a faster data processing step and an increased sensitivity and quality of the data. The enhanced precision of targeted versus untargeted processing is presented in Figure 7.11. While it is less pronounced for the liver cell extracts, targeted processing of blood plasma lipids and skin CERs

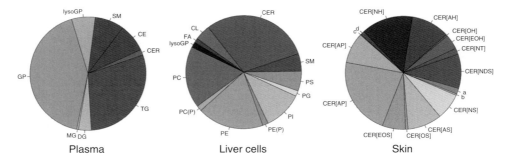

Figure 7.12 Pie chart representing the distribution of identified lipid classes in liver cells, plasma, and skin (potential differences in ionization efficiency have not been taken into account). Some classes are appointed as letters: a – CER[ADS], b – CER[EODS], c – CER[OP], and d – CER[EOP].

clearly shows an enhanced precision. The targeted data processing of the liver cell sample set extracted 346 lipids with 100% frequency out of the extraction replicates, of which 89.0% showed an RSD < 15%, an increase of 5% compared to the untargeted approach. For a blood plasma and a skin extract, these values were 89.4% (528 identified lipids) and 72.9% (650 identified lipids – note that, in this case, adducts are considered as separate species overestimating the real number of CERs), which provided an increase in precision of 7.5 and 10%, respectively (Table 7.2). Surely, new lipid species will be missed if a targeted data processing approach is used. The combined use of both targeted and untargeted data processing is hereby justified.

7.5
GC-MS-Based Lipidomics

The GC-MS-based branch of our lipidomics platform combines high-resolution capillary GC with quadrupole MS. It is specifically directed to the analysis of the total fatty acids (free or lipid-associated) as their methyl esters (FAME). Fatty acids containing hydroxyl functions are not targeted with this methodology. The different steps of the methodology are presented in the following paragraph. Note that targeted fatty acid analysis such as tuberculostearic acid in sputum samples requires dedicated analytical schemes [102]. We refer the readers to the literature as this is out of the scope of this contribution.

7.5.1
Sample Preparation

Fatty acids can be analyzed directly by GC; but for routine analysis, we prefer deactivation of the acidic functionality. This makes the method much more robust, and, at the same time, better resolution according to the degree of unsaturation and cis–trans configuration is obtained. Numerous derivatization procedures have been described, but the most common one is methyl esterification. Lipid hydrolysis and esterification can typically be performed in a single step. In our hands, the use of a methanolic solution of acetyl chloride provides a high yield of FAMES from CEs, GPs, and TGs [20]. We classically add 500 µl of acetyl chloride (5%) in methanol to 10–50 µl of biological sample (e.g., whole blood, plasma/serum) in a glass reaction vial. The mixture is incubated for 30 min at 90 °C allowing lipid hydrolysis and methylation. One milliliter of hexane is subsequently added giving rise to a biphasic system. The hexane phase is transferred to an injection vial. Care should be taken at all times to avoid fatty acid contamination (typically palmitic and stearic acid). We have recently automated this sample preparation approach on an Automatic Liquid Sampler (ALS, Agilent Technologies) placed on top of a GC–MS system [103]. Hydrolysis and methylation, followed by micro liquid–liquid extraction, is performed using the back injection tower with the tray enabling vortex mixing and sample heating. Injection from the upper extract layer was performed using the front injector. This prevents all user intervention, thereby streamlining sample preparation and GC–MS analysis. The precision of this approach gives typically average area RSD values of 2% for the replicate analysis ($n = 6$) of 50 µl of blood plasma.

SPs containing amide bonds instead of ester bonds are less susceptible to hydrolysis under these conditions. Acid hydrolysis of CERs has been reported using 0.5 M HCl in 90/10 acetonitrile/water (v/v) and incubation at 100 °C for 1 h [104]. Following solvent evaporation, the hydrolysate can be subjected to methylation as described above. However, this hydrolysis procedure results in a limited hydrolysis yield when applied onto SPE isolated skin CERs (10–60% depending upon the CER class). Nevertheless, interesting qualitative information can be obtained [19].

Note that free fatty acid concentrations in plasma and tissues can be a useful measure of the metabolic status. Aminopropyl SPE cartridges are especially useful for the isolation of the free fatty acid fraction from lipid extracts as described earlier [105]. Care should be taken not to generate free acids by faulty storage, during extraction or by contamination.

7.5.2
GC–MS

GC is a widely adopted tool for the separation of complex fatty acid mixtures dating back to the 1950s [106, 107]. For the separation of complex FAME mixtures, a variety of stationary phases can be used [20, 54–56, 59, 60]. Our FAME GC–MS method, performed on an Agilent 6890 GC and 5975 MSD, uses a highly polar cyanopropyl silicone column (HP-88 100 ml × 0.25 mmID × 0.2 µm d_f, Agilent Technologies) providing separation according to carbon number and number, position, and configuration (cis and trans) of double bonds. The use of a 100 m column provides excellent resolution as is demonstrated in Figure 7.13 showing a typical GC–MS chromatogram of a human blood plasma sample (50 µl). To maximize the detection of the minor species, a programmed temperature vaporizing (PTV) inlet with cryo cooling is used allowing the injection of 5 µl of sample. Injection is performed in the solvent vent mode at 50 °C for 15 s and injection temperature is subsequently ramped to 250 °C at $12 \, \text{s} \, °\text{C} \, \text{s}^{-1}$, while the sample is transferred in the splitless mode. Analyses are performed in constant pressure mode at a nominal velocity of $28 \, \text{cm} \, \text{s}^{-1}$ using helium as a carrier gas. The oven temperature program is as follows: 50 °C for 1 min, 15 °C min^{-1} to 175 °C, and 1 °C min^{-1} to 225 °C. Electron ionization (EI) analysis is performed in quadrupole scan mode. The entire method is retention time locked (RTL). RTL uses electronic pressure control to calibrate column head pressure versus retention time and to fix the retention time of a chosen locking compound, in our case methyl stearate. It is based on defined columns and defined chromatographic conditions and leads to excellent retention time stability, a highly appreciated feature in lipidomics. Table 7.3 displays the more abundant compounds detected in 50 µl of human blood plasma with their locked retention times. As described in an earlier essay by our group [20],

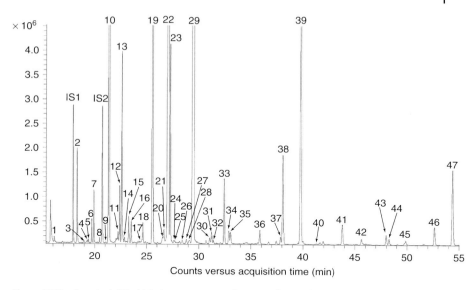

Figure 7.13 A typical GC–MS chromatogram of FAMES from a human blood plasma sample (50 µl). Identity of labeled peaks is listed in Table 7.3 with their locked retention times.

Table 7.3 The most abundant FAMES detected in 50 µl of blood plasma with their locked retention times.

Label	Compound	tR (min)	Label	Compound	tR (min)	Label	Compound	tR (min)
IS1	Myristic acid d27	17.919	16	C18:0–DMA	23.475	33	α C18:3 n3c	32.400
IS2	Palmitic acid d31	20.700	17	C18:0 iso	24.290	34	C20:1 n9c	32.848
1	C12:0	16.134	18	C18:1–DMA	24.598	35	c9, t11–CLA	33.015
2	C14:0	18.307	19	C18:0	25.517	36	C20:2 n6c	35.810
3	C15:0 iso	18.981	20	C18:1 n8t	26.543	37	C22:0	37.785
4	C15:0 anteiso	19.279	21	C18:1 n7t	26.730	38	C20:3 n6	38.009
5	C14:1	19.410	22	C18:1 n9c	27.005	39	C20:4 n6	39.722
6	C15:0	19.717	23	C18:1 n7c	27.188	40	Unknown – DMA	41.297
7	C16:0–DMA	19.903	24	C18:1 n5c	27.566	41	C20:5 n3c	43.723
8	C16:0 iso	20.460	25	C18:1 n4c	27.665	42	C24:0	45.539
9	C16:1–DMA	21.035	26	C19:0	28.073	43	C24:1 n9c	47.901
10	C16:0	21.359	27	C18:2 n6ct	28.987	44	C22:4 n6c	48.188
11	C17:0 iso	22.210	28	C18:2 n6tc	29.091	45	C22:5 n6c	49.802
12	C16:1	22.375	29	C18:2 n6cc	29.432	46	C22:5 n3c	52.554
13	C16:1	22.569	30	C20:0	30.998	47	C22:6 n3c	54.284
14	C16:1	22.843	31	γ C18:3 n6c	31.191	—	—	—
15	C17:0	23.256	32	C18:3	31.359	—	—	—

more than 100 FAMEs have thus far been identified in blood plasma using the described methodology. The identification strategy consisted of the analysis of FAME reference mixtures complemented with spectral searches in the NIST05 and WILEY275 libraries and manual interpretation of EI and positive CI spectra. The latter is a definite requirement because mass spectra of FAMEs are very similar and show strong fragmentation making it often hard to reveal the molecular ion. Literature information describing the elution of isomeric species, among others, further completed the identification cycle [55]. The strength of using complementary ionization techniques in the identification process is illustrated in Figure 7.14 showing the EI and positive CI spectra of the FAME detected at 52.55 min. The base peak ion at m/z 79 in the EI spectrum is indicative for a polyunsaturated fatty acid. Note that saturated species give rise to the characteristic base peak ion at m/z 74 corresponding to the McLafferty rearrangement ion, that monounsaturated species give rise to the characteristic base peak ion at m/z 55 and that dimethylacetal species give rise to a characteristic ion at m/z 75 corresponding to the dimethoxy methyl radical ion [55]. Performing an EI spectral library search results in a variety of possibilities including C20 : 4, C20 : 5, C22 : 5, and C22 : 6. From the EI spectrum, the molecular ion is barely visible which makes its identification difficult. The positive CI spectra using ammonia and methane, on the other hand, allow the unambiguous identification of the fatty acid as C22 : 5 due to the detection of the $[M + 1]^+$ ion at m/z 345 in case methane is used as reagent gas and the $[M + 18]^+$ ion at m/z 362 in case ammonia is used indicative for an MW of 344.

Saturated fatty acids ranging from 6 to 26 carbon atoms and unsaturated species with up to six double bonds are covered by the methodology. The more carbons and the more double bonds, the more retention is observed. Cis–trans isomers can readily be discriminated, with the former showing longer retention. Positional isomers (e.g., C18 : 1 n9c, C18 : 1 n7c, etc.) are also differentiated with the current methodology. The closer the double bond to the omega position, the more retention is observed. Conjugated linoleic acids (CLAs) elute later than their nonconjugated counterparts. Plasmenyl (vinyl ether) derivatives are detected as dimethylacetals and elute earlier than their methyl ester counterparts. The detection of these products results from the formation of a free aldehyde under the transesterification conditions that is rapidly converted to dimethylacetal. Branched chain fatty acids (iso and anteiso) can be observed as well typically eluting earlier than their linear counterparts.

Several biological samples contain cholesterol, either free or esterified. Following hydrolysis and methylesterification, cholesterol is coextracted in hexane and is subsequently injected into the GC–MS system. Cholesterol has a higher boiling point than FAMES and, under the given conditions, is not eluted. Nevertheless, cholesterol starts to elute as a broad peak in subsequent runs masking the detection of various fatty acids. Several options are available to prevent this event. Cholesterol and its esters can be removed before GC-MS analysis, but this is a time-consuming procedure. Of more interest is to increase the oven temperature to elute cholesterol within the same run. However, cyanopropyl silicone stationary

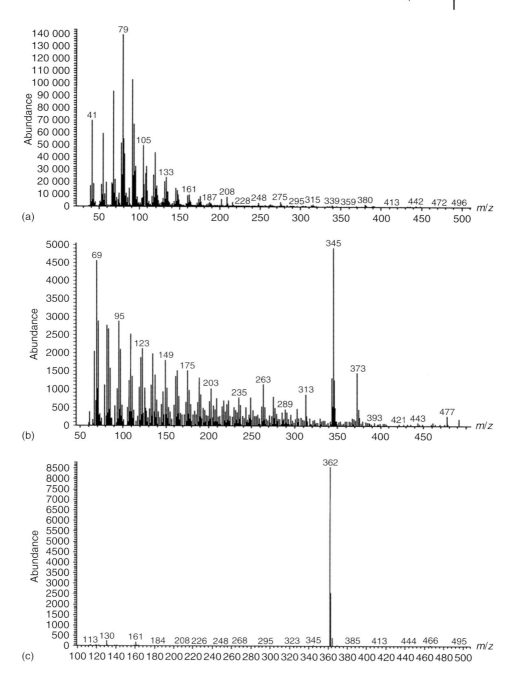

Figure 7.14 EI (a) and positive CI spectra, using methane (b) and ammonia (c) as reagent gases, of the FAME C22 : 5 eluting at 52.55 min using the RTL GC–MS method.

phases have an upper temperature limit of 250 °C and, under these conditions, the elution of cholesterol takes approximately 45 min resulting in a total run time of over 100 min. A more elegant way is to implement capillary flow technology [108] by means of a so-called quick swap (installed in between column outlet and MS inlet) allowing column back flushing at 250 °C once all FAMES have reached the MS detector. When using this, run time is only extended with 15 min.

The above-described methodology does not reveal the origin of the different fatty acids. Its use as the back-end step in a multidimensional setup can be highly valuable for the class-directed fatty acid analysis. As described earlier, lipid class separation can be obtained using thin-layer chromatography (TLC), NP-LC, or SPE. Figure 7.15 shows the typical FAME composition of CEs, PCs, and TGs enriched from human blood plasma by means of NP-LC [109–111] and of skin CERs enriched by the SPE methodology displayed in Figure 7.4. Note that the latter sample is known to contain various α- and ω-OH fatty acids that are not immediately targeted with the current methodology. These species can, however, be assessed using a GCxGC method described earlier by our group employing an apolar column in the first dimension and a polar column in the second dimension [59] or by a variant of the LC–MS methodology described above by simply adjusting the gradient program (linear gradient starting at 30% MeOH).

7.5.3
Data Processing and Analysis

As for LC–MS analysis, GC–MS data processing can be performed in a targeted or untargeted manner. As EI spectra display substantial fragmentation and quadrupole MS instruments operate at lower resolution and mass accuracy, algorithms used to extract features out of the MS data differ from those described earlier. Both automatic mass spectral deconvolution and identification system (AMDIS) [112] and XC-MS [98] have been explored in our laboratory with different rates of success. Both are freely available but substantially differ in their *modus operandi*. AMDIS is a deconvolution package that localizes and combines all covarying ions with the creation of so-called pure compound spectra. AMDIS deconvolution generates ELU files compatible with various data processing and analysis software packages. ELU files can be imported in MassProfiler Professional as peak lists composed of base peak ion, retention time, and intensity (e.g., 200@16.67 + intensity) that can subsequently be aligned to generate a data matrix accessible for statistical analysis. XC-MS, on the other hand, considers each m/z entity in an EI spectrum as unique, thereby generating dozens of features from a single compound (e.g., M200T1000 + intensity), which are subsequently aligned resulting in an extensive data matrix. We normally apply a feature reduction tool onto the data matrix to limit the number of features per retention time, before statistical analysis. Typically, one feature/retention time is maintained based on intensity. Figure 7.16 compares the precision of AMDIS and XCMS data processing strategies based on the replicate analysis of blood plasma FAMEs. Precision is higher using XC-MS. We typically apply conservative AMDIS settings resulting in the extraction of only

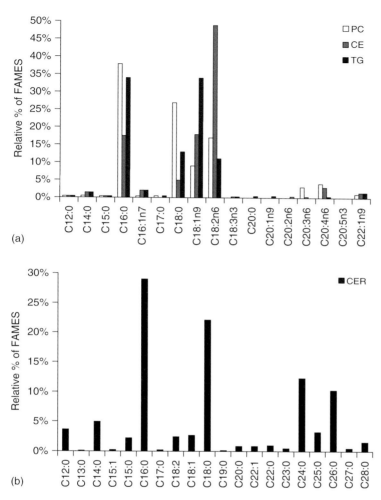

Figure 7.15 Lipid class FAME composition of blood plasma CE, PC, and TG (a) and in skin CER (b). Lipids were enrichment from human blood plasma via NP-LC with fraction collection and subjected onto the GC–MS methodology, following hydrolysis and methylation. Skin CERs were isolated via SPE as demonstrated in Figure 7.4.

a limited number of features (e.g., 50 from blood plasma samples displayed in Figure 7.13). Putting less constraints on the AMDIS settings results in substantial overdeconvolution with the generation of a lot of false positives and marginal precision. The precision of XC–MS allows the untargeted detection (t-test) of subtle differences between samples (as low as 20% concentration differences) as revealed by spiking different concentrations of fatty acids in the same blood plasma matrix.

The identification of a variety of fatty acids in different samples, combined with the use of RTL, allowed us to construct an RTL library that can be assessed in

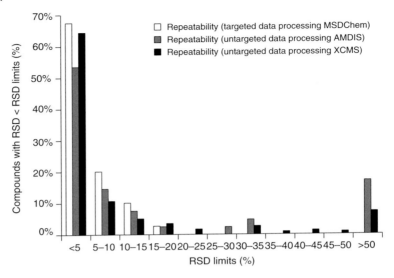

Figure 7.16 Technical repeatability of blood plasma FAME analysis: RSD% is calculated for features with 100% frequency after untargeted (AMDIS and XCMS) or targeted (MSDChem) data processing.

different manners. The output from AMDIS deconvolution can be matched onto the library generating so-called FIN files, containing identified fatty acids with their intensity (e.g., C18 : 0 + intensity). These can subsequently be imported in data processing and analysis software packages. The most precise manner to handle the data is to use the RTL library to access the fatty acids in a targeted manner making use of quantifier and qualifier ions. These ions are selectively extracted and integrated using the MSDChem software (Agilent Technologies), and a list is generated with the areas of all fatty acids detected in the sample. Integration can be corrected manually. This is more labor intensive compared to untargeted processing but can still be performed within a reasonable timescale with the current method. This is not practical with the above-described LC–MS method because of the huge amounts of data available. The precision of targeted processing is displayed in Figure 7.16 as well. As it is to be expected, targeted data processing outperforms untargeted data processing, but the latter allows the detection of species not present in the library. As is the case with the AMRT library described for LC–MS, the RTL library is continuously updated with new FAMES.

7.6
Conclusion

The field of lipidomics is in a continuous search/quest to further mine the lipidome, in a comparative manner, with the ultimate goal to widen our biological knowledge. With this chapter, describing the art and practice of lipidomics from the authors' perspective, we hope to have inspired and provided the reader with a solid basis to perform its own lipidomics experiments. Complementary, highly informative, and precise LC–MS and GC–MS methodologies have been described in detail. These methodologies are continuously being updated and improved with knowledge and experience gained from real biological experiments and in response to new evolutions in instrumentation and informatics tools.

References

1. Hanash, S. (2003) *Nature*, **422**, 226–232.
2. Van der Greef, J., Martin, S., Juhasz, P., Adourian, A., Plasterer, T., Verheij, E.R., and McBurney, R.N. (2007) *J. Proteome Res.*, **6**, 1540–1559.
3. Sreekumar, A., Poisson, L.M., Rajendiran, T.M., Khan, A.P., Cao, Q., Yu, J., Laxman, B., Mehra, R., Lonigro, R.J., Li, Y., Nyati, M.K., Ahsan, A., Kalyana-Sundaram, S., Han, B., Cao, X., Byun, J., Omenn, G.S., Ghosh, D., Pennathur, S., Alexander, D.C., Berger, A., Shuster, J.R., Wei, J.T., Varambally, S., Beecher, C., and Chinnaiyan, A.M. (2009) *Nature*, **457** (7231), 910–914.
4. Fernandis, A.Z. and Wenk, M.R. (2009) *J. Chromatogr. B*, **877**, 2830–2835.
5. Wenk, M.R. (2005) *Nat. Rev. Drug Discov.*, **4**, 594–610.
6. Morris, M. and Watkins, S.M. (2005) *Curr. Opin. Chem. Biol.*, **9**, 407–412.
7. Han, X. (2007) *Curr. Opin. Mol. Ther.*, **9**, 586–591.
8. Wolf, C. and Quinn, P.J. (2008) *Prog. Lipid Res.*, **47**, 15–36.
9. Hu, C., van der Heijden, R., Wang, M., van der Greef, J., Hankemeier, T., and Xu, G. (2009) *J. Chromatogr. B*, **877**, 2836–2846.
10. Quehenberger, O., Armando, A.M., Brown, A.H., Milne, S.B., Myers, D.S., Merrill, A.H., Bandyopadhyay, S., Jones, K.N., Kelly, S., Shaner, R.L., Sullards, C.M., Wang, E., Murphy, R.C., Barkley, R.M., Leiker, T.J., Raetz, C.R.H., Guan, Z., Laird, G.M., Six, D.A., Russell, D.W., McDonald, J.G., Subramanian, S., Fahy, E., and Dennis, E.A. (2010) *J. Lipid Res.*, **51**, 3299–3305.
11. Wang, C., Kong, H., Guan, Y., Yang, J., Gu, J., Yang, S., and Xu, G. (2005) *Anal. Chem.*, **77**, 4108–4116.
12. Han, X., Yang, J., Yang, K., Zhao, Z., Abendschein, D.R., and Gross, R.W. (2007) *Biochemistry*, **46**, 6417–6428.
13. Han, X., Rozen, S., Boyle, S.H., Hellegers, C., Cheng, H., Burke, J.R., Welsh-Bohmer, K.A., Doraiswamy, P.M., and Kaddurah-Daouk, R. (2011) *PLoS ONE*, **6** (7), e21643.
14. Ejsing, C.S., Sampaio, J.L., Surendranath, V., Duchoslav, E., Ekroos, K., Klemm, R.W., Simons, K., and Shevchenko, A. (2009) *Proc. Natl. Acad. Sci. U.S.A.*, **106** (7), 2136–2141.
15. Jentzmik, F., Stephan, C., Lein, M., Miller, K., Kamlage, B., Bethan, B., Kristiansen, G., and Jung, K. (2010) *J. Urol.*, **185** (2), 706–711.
16. Struys, E.A., Heijboer, A.C., van Moorselaar, J., Jakobs, C., and Blankenstein, M.A. (2010) *Ann. Clin. Biochem.*, **47** (3), 282.
17. Jentzmik, F., Stephan, C., Miller, K., Schrader, M., Erbersdobler, A., Kristiansen, G., Lein, M., and Jung, K. (2010) *Eur. Urol.*, **58** (1), 12–18.
18. Sandra, K., dos Santos Pereira, A., Vanhoenacker, G., David, F., and Sandra, P. (2010) *J. Chromatogr. A*, **1217**, 4087–4099.
19. t'Kindt, R., Jorge, L., Dumont, E., Couturon, P., David, F., Sandra, P., and Sandra, K. (2012) *Anal. Chem.*, **84**, 403–411.
20. Bicalho, B., David, F., Rumplel, K., Kindt, E., and Sandra, P. (2008) *J. Chromatogr. A*, **1211**, 120–128.
21. Yetukuri, L., Katajamaa, M., Medina-Gomez, G., Seppänen-Laakso, T., Vidal-Puig, A., and Oresic, M. (2007) *BMC Syst. Biol.*, **15**, 1–12.

22. Fahy, E., Subramanian, S., Brown, H.A., Glass, C.K., Merrill, A.H., Murphy, R.C., Raetz, C.R.H., Russell, D.W., Seyama, Y., Shaw, W., Shimizu, T., Spener, F., van Meer, G., VanNieuwenhze, M.S., White, S.H., Witztum, J.L., and Dennis, E.A. Jr. (2005) *J. Lipid Res.*, **46**, 839–861.

23. Fahy, E., Subramanian, S., Murphy, R.C., Nishijima, M., Raetz, C.R.H., Shimizu, T., Spener, F., van Meer, G., Wakelam, M.J.O., and Dennis, E.A. (2009) *J. Lipid Res.*, **50**, S9–S14.

24. Wishart, D.S., Tzur, D., Knox, C., Eisner, R., Guo, A.C., Young, N., Cheng, D., Jewell, K., Arndt, D., Sawhney, S., Fung, C., Nikolai, L., Lewis, M., Coutouly, M.A., Forsythe, I., Tang, P., Shrivastava, S., Jeroncic, K., Stothard, P., Amegbey, G., Block, D., Hau, D.D., Wagner, J., Miniaci, J., Clements, M., Gebremedhin, M., Guo, N., Zhang, Y., Duggan, G.E., Macinnis, G.D., Weljie, A.M., Dowlatabadi, R., Bamforth, F., Clive, D., Greiner, R., Li, L., Marrie, T., Sykes, B.D., Vogel, H.J., and Querengesser, L. (2007) *Nucleic Acids Res.*, **35**, D521–D526.

25. Han, X., Yang, K., and Gross, R.W. (2011) *Mass Spectrom. Rev.*, doi: 10.1002/mas.20342

26. Fahy, E., Cotter, D., Sud, M., and Subramaniam, S. (2011) *Biochim. Biophys. Acta*, doi: 10.1016/j.bbalip.2011.06.009

27. Roberts, L.D., McCombie, G., Titman, C.M., and Griffin, J.L. (2008) *J. Chromatogr. B*, **871**, 174–181.

28. Hou, W., Zhou, H., Elisma, F., Bennett, S.A., and Figeys, D. (2008) *Briefings Funct. Genomics*, **7** (5), 395–409.

29. Seppänen-Laakso, T. and Oresic, M. (2009) *J. Mol. Endocrinol.*, **42** (3), 185–190.

30. Navas-Iglesias, N., Carrasco-Pancorbo, A., and Cuadros-Rodriguez, L. (2009) *Trends Anal. Chem.*, **28**, 393–403.

31. Rainville, P.D., Stumpf, C.L., Shockcor, J.P., Plumb, R.S., and Nicholson, J.K. (2007) *J. Proteome Res.*, **6** (2), 552–558.

32. Pietiläinen, K.H., Sysi-Aho, M., Rissanen, A., Seppänen-Laakso, T., Yki-Järvinen, H., Kaprio, J., and Oresic, M. (2007) *PLoS ONE*, **2** (2), e218.

33. Ding, J., Sorensen, C.M., Jaitly, N., Jiang, H., Orton, D.J., Monroe, M.E., Moore, R.J., Smith, R.D., and Metz, T.O. (2008) *J. Chromatogr. B*, **871**, 243–252.

34. Hu, C., van Dommelen, J., van der Heijden, R., Spijksma, G., Reijmers, T.H., Wang, M., Slee, E., Lu, X., Xu, G., van der Greef, J., and Hankemeier, T. (2008) *J. Proteome Res.*, **7** (11), 4982–4991.

35. Stahlman, M., Ejsing, C.S., Tarasov, K., Perman, J., Boreén, J., and Ekroos, K. (2009) *J. Chromatogr. B*, **877**, 2664–2672.

36. Bird, S., Marur, V., Sniatynski, M., Greenberg, H., and Kristal, B. (2011) *Anal. Chem.*, **83** (17), 6648–6657.

37. Koulman, A., Woffendin, G., Narayana, V., Welchman, H., Crone, C., and Volmer, D. (2009) *Rapid Commun. Mass Spectrom.*, **23**, 1411–1418.

38. Taguchi, R., Houjou, T., Nakanishi, H., Yamazaki, T., Ishida, M., Imagawa, M., and Shimizu, T. (2005) *J. Chromatogr. B*, **823**, 26–36.

39. Han, X. and Gross, R.W. (2005) *Mass Spectrom. Rev.*, **24**, 367–412.

40. Cui, Z. and Thomas, M.J. (2009) *J. Chromatogr. B*, **877** (26), 2709–2715.

41. Mesaros, C., Lee, S.H., and Blair, I.A. (2009) *J. Chromatogr. B*, **877** (26), 2736–2745.

42. Deems, R., Buczynski, M., Bowers-Gentry, R., Harkewicz, R., and Dennis, E. (2007) *Methods Enzymol.*, **432**, 59–81.

43. Honda, A., Yamashita, K., Miyazaki, H., Shirai, M., Ikegami, T., Xu, G., Numazawa, M., Hara, T., and Matsuzaki, Y. (2008) *J. Lipid Res.*, **49** (9), 2063–2073.

44. Han, X. and Gross, R.W. (2005) *Expert Rev. Proteomics*, **2** (2), 253–264.

45. Lisa, M., Cifkova, E., and Holcapek, M. (2011) *J. Chromatogr. A*, **1218**, 5146–5156.

46. Houjou, T., Yamatani, K., Imagawa, M., Shimizu, T., and Taguchi, R. (2005) *Rapid Commun. Mass Spectrom.*, **19**, 654–666.

47. Sommer, U., Herscovitz, H., Welty, F.K., and Costello, C.E. (2006) *J. Lipid Res.*, **47**, 804–814.
48. Barroso, B. and Bischoff, R. (2005) *J. Chromatogr. B*, **814**, 21–28.
49. Ikeda, K., Oike, Y., Shimizu, T., and Taguchi, R. (2009) *J. Chromatogr. B*, **877** (25), 2639–2647.
50. Lisa, M., Velinska, H., and Holcapek, M. (2009) *Anal. Chem.*, **81**, 3903–3910.
51. Mondello, L. (2011) *Comprehensive Chromatography in Combination with Mass Spectrometry*, John Wiley & Sons, Inc., Hoboken, NJ.
52. Bamba, T., Shimonishi, N., Matsubara, A., Hirata, K., Nakazawa, Y., Kobayashi, A., and Fukusaki, E. (2008) *J. Biosci. Bioeng.*, **105**, 460–469.
53. François, I., Pereira, A., and Sandra, P. (2010) *J. Sep. Sci.*, **33** (10), 1504–1512.
54. Quehenberger, O., Armando, A.M., and Dennis, E.A. (2011) *Biochim. Biophys. Acta*, doi: 10.1016/j.bbalip.2011.07.006
55. Härtig, C. (2008) *J. Chromatogr. A*, **1177**, 159–169.
56. Delmonte, P., Fardin Kia, A.R., Kramer, J.K., Mossoba, M.M., Sidisky, L., and Rader, J.I. (2011) *J. Chromatogr. A*, **1218** (3), 545–554.
57. De Geus, H.J., Aidos, I., de Boer, J., Luten, J.B., and Brinkman, U.A. (2001) *J. Chromatogr. A*, **910**, 95–103.
58. Mondello, L., Casilli, A., Tranchida, P.Q., Dugo, P., and Dugo, C. (2003) *J. Chromatogr. A*, **1019**, 187–196.
59. David, F., Tienpont, B., and Sandra, P. (2008) *J. Sep. Sci.*, **31**, 3395–3403.
60. Gu, Q., David, F., Lynen, F., Vanormelingen, P., Vyverman, W., Rumpel, K., Xu, G., and Sandra, P. (2011) *J. Chromatogr. A*, **1218** (20), 3056–3063.
61. Jover, E., Adahchour, M., Bayona, J.M., Vreuls, R.J.J., and Brinkman, U.A. (2005) *J. Chromatogr. A*, **1086**, 2–11.
62. Shvartsburg, A.A., Isaac, G., Leveque, N., Smith, R.D., and Metz, T.O. (2011) *J. Am. Soc. Mass Spectrom.*, **22**, 1146–1155.
63. Kliman, M., May, J.C., and McLean, J.A. (2011) *Biochim. Biophys. Acta*, doi: 10.1016/j.bbalip.2011.05.016
64. Goto-Inoue, N., Hayasaka, T., Zaima, N., and Setou, M. (2011) *Biochim. Biophys. Acta*, doi: 10.1016/j.bbalip.2011.03.004
65. Fernández, J.A., Ochoa, B., Fresnedo, O., Giralt, M.T., and Rodríguez-Puertas, R. (2011) *Anal. Bioanal. Chem.*, **401** (1), 29–51.
66. Manicke, N., Nefliu, M., Wu, C., Woods, J., Reiser, V., Hendrickson, R., and Cooks, G. (2009) *Anal. Chem.*, **81** (21), 8758–8764.
67. Folch, J., Lees, M., and Sloane Stanley, G.H. (1957) *J. Biol. Chem.*, **226**, 497–509.
68. Bligh, E.G. and Dyer, W.J. (1959) *Can. J. Biochem. Physiol.*, **37**, 911–917.
69. Jung, H.R., Sylvanne, T., Koistinen, K.M., Tarasov, K., Kauhanen, D., and Ekroos, K. (2011) *Biochim. Biophys. Acta*, doi: 10.1016/j.bbalip.2011.06.025
70. Matyash, V., Liebisch, G., Kurzchalia, T.V., Shevchenko, A., and Schwudke, D. (2010) *J. Lipid Res.*, **49** (5), 1137–1146.
71. Ruiz-Gutierrez, V. and Perez-Camino, M.C. (2000) *J. Chromatogr. A*, **885**, 321–341.
72. Sato, Y., Nakamura, T., Aoshima, K., and Oda, Y. (2010) *Anal. Chem.*, **82** (23), 9858–9864.
73. Kaluzny, M.A., Duncan, L.A., Merritt, M.V., and Epps, D.E. (1985) *J. Lipid Res.*, **26**, 135–140.
74. Blewett, A.J., Varma, D., Gilles, T., Libonati, J.R., and Jansen, S.A. (2007) *J. Pharm. Biomed. Anal.*, **46** (4), 653–662.
75. Masukawa, Y., Narita, H., Shimizu, E., Kondo, N., Sugai, Y., Oba, T., Homma, R., Ishikawa, J., Takagi, Y., Kitahara, T., Takema, Y., and Kita, K. (2008) *J. Lipid Res.*, **49**, 1466–1476.
76. McHowat, J., Jones, J.H., and Creer, M.H. (1997) *J. Chromatogr. B*, **702**, 21–32.
77. Haworth, O. and Levy, B.D. (2007) *Eur. Respir. J.*, **30**, 980–992.
78. Koivusalo, M., Haimi, P., Heikinheimo, L., Kostiainen, R., and Somerharju, P. (2001) *J. Lipid Res.*, **42**, 663–672.
79. Cai, S.S. and Syage, J.A. (2006) *Anal. Chem.*, **78**, 1191–1199.
80. Cai, S.S., Short, L.C., Syage, J.A., Potvin, M., and Curtis, J.M. (2007) *J. Chromatogr. A*, **1173** (1-2), 88–97.

81. Short, L.C., Cai, S.S., and Syage, J.A. (2007) *J. Am. Soc. Mass Spectrom.*, **18** (4), 589–599.
82. Hsu, F. and Turk, J. (2001) *J. Am. Soc. Mass Spectrom.*, **12**, 1036–1043.
83. Hsu, F. and Turk, J. (2009) *J. Chromatogr. B*, **877**, 2673–2695.
84. Pulfer, M. and Murphy, R.C. (2003) *Mass Spectrom. Rev.*, **22**, 332–364.
85. Schwudke, D., Oegema, J., Burton, L., Entchev, E., Hannich, J.T., Ejsing, C.S., Kurzchalia, T., and Shevchenko, A. (2006) *Anal. Chem.*, **78**, 585–595.
86. Murphy, R.C. and Axelsen, P.H. (2010) *Mass Spectrom. Rev.*, **30**, 579–599.
87. Murphy, R.C., Leiker, T.J., and Barkley, R.M. (2011) *Biochim. Biophys. Acta*, doi: 10.1016/j.bbalip.2011.06.019
88. Van Smeden, J., Hoppel, L., Van der Heijden, R., Hankemeier, T., Vreeken, R.J., and Bouwstra, J.A. (2011) *J. Lipid Res.*, **52**, 1211–1221.
89. Farwanah, H., Pierstorff, B., Schmelzer, C.E., Raith, K., Neubert, R.H., Kolter, T., and Sandhoff, K. (2007) *J. Chromatogr. B*, **852**, 562–570.
90. Ann, Q. and Adams, J. (1993) *Anal. Chem.*, **65**, 7–13.
91. Lee, M.H., Lee, G.H., and Yoo, J.S. (2003) *Rapid Commun. Mass Spectrom.*, **17**, 64–75.
92. Hinder, A., Schmelzer, C.E.H., Rawlings, A.V., and Neubert, R.H.H. (2011) *Skin Pharmacol. Physiol.*, **24**, 127–135.
93. Hsu, F. and Turk, J. (2002) *J. Am. Soc. Mass Spectrom.*, **13**, 558–570.
94. Thomas, M.C., Mitchell, T.W., Harman, D.G., Deeley, J.M., Murphy, R.C., and Blanksby, S.J. (2007) *Anal. Chem.*, **79** (13), 5013–5022.
95. Brown, S.H., Mitchell, T.W., and Blanksby, S.J. (2011) *Biochim. Biophys. Acta*, doi: 10.1016/j.bbalip.2011.04.015
96. Mitchell, T.W., Pham, H., Thomas, M.C., and Blanksby, S.J. (2009) *J. Chromatogr. B*, **877**, 2722–2735.
97. Yang, K., Zhao, Z., Gross, R.W., and Han, X. (2011) *Anal. Chem.*, **83**, 4243–4250.
98. Smith, C.A., Want, E.J., O'Maille, G., Abagyan, R., and Siuzdak, G. (2006) *Anal. Chem.*, **78** (3), 779–787.
99. Katajamaa, M., Miettinen, J., and Oresic, M. (2006) *Bioinformatics*, **22** (5), 634–636.
100. Lommen, A. (2009) *Anal. Chem.*, **81** (8), 3079–3086.
101. Scheltema, R.A., Jankevics, A., Jansen, R.C., Swertz, M.A., and Breitling, R. (2011) *Anal. Chem.*, **83** (7), 2786–2793.
102. Stopforth, A., Tredoux, A., Crouch, A., van Helden, P., and Sandra, P. (2005) *J. Chromatogr. A*, **1071**, 135–139.
103. David, F., Tienpont, B., Klee, M.S., and Tripp, P. (2009) Automated Sample Preparation for Profiling Fatty Acids in Blood and Plasma Using the Agilent 7693 ALS, Agilent Technologies, Application Note 5990-4822EN.
104. Johnson, S.B. and Brown, R.E. (1992) *J. Chromatogr.*, **605**, 281–286.
105. De Jong, C. and Badings, H. (1990) *J. High Resolut. Chromatogr.*, **13**, 94–98.
106. James, A.T. and Martin, A.J.P. (1952) *Biochem. J.*, **50** (5), 679–690.
107. James, A.T. and Martin, A.J.P. (1956) *Biochem. J.*, **63** (1), 144–152.
108. David, F. and Klee, M.S. (2007) GC-MS Analysis of PCBs in Waste Oil Using the Backflush Capability of the Agilent QuickSwap Accessory, Agilent Technologies, Application note 5989-7601EN.
109. Christie, W.W. (1985) *J. Lipid Res.*, **26**, 507–512.
110. Silversand, C. and Haux, C. (1997) *J. Chromatogr. B*, **703**, 7–14.
111. De Vrieze, M. (2009) Fractionation of lipid classes in biological fluids, Master dissertation. Ghent University, Belgium.
112. Stein, S.E. (1999) *J. Am. Soc. Mass Spectrom.*, **10**, 770–781.

8
The Role of CE–MS in Metabolomics

Rawi Ramautar, Govert W. Somsen, and Gerhardus J. de Jong

Abbreviations

CE–MS	Capillary electrophoresis–mass spectrometry
EOF	Electroosmotic flow
PB-PVS	Polybrene-poly(vinyl sulfonate)
PB-DS-PB	Polybrene-dextran sulfate-polybrene
BGE	Background electrolyte
TOF–MS	*Time-of-flight*–mass spectrometry
TQ	Triple quadrupole
IT	Ion trap
MEKC	Micellar electrokinetic chromatography
APPI	Atmospheric pressure photoionization
APCI	Atmospheric pressure chemical ionization.

8.1
Introduction

The metabolome is the complete set of endogenous low-molecular-weight metabolites in cells, tissues, body fluids, and plants [1–3]. The comprehensive analysis of these metabolites in biological samples is known as *metabolomics*. Two different approaches may be distinguished in metabolomics, that is, the nontargeted and targeted approach [4, 5]. The nontargeted approach involves the profiling of as many low-molecular-weight metabolites as possible in biological samples, without having a priori knowledge on the nature and identity of the measured metabolites. This approach results, in first instance, in lists of molecular features present in the sample rather than specific metabolites. Relative changes in molecular features, for example, in response to disease, or other alterations, are revealed using multivariate techniques. Subsequent identification of molecular features responsible for sample classification might be carried out to reveal potential biomarkers for disease. The second approach is focused on the quantitative analysis of preselected (target) metabolites in a biological sample. Internal standards, each representing a class of compounds (e.g., amino acids), are often used for the quantification of metabolites. The two approaches can also be regarded as stages in metabolomics-based research where the first approach is used to discover putative biomarkers of disease, which are then validated and subsequently quantified in the second stage.

At present, advanced analytical techniques in combination with multivariate data analysis are used for global metabolic profiling and biomarker discovery. Next to NMR, efficient separation techniques such as gas chromatography (GC), liquid chromatography (LC), and capillary electrophoresis (CE) are very suitable for this goal. Especially, when these techniques are combined with mass spectrometry (MS), powerful systems for metabolomics are obtained.

This chapter deals with the possibilities of CE–MS for the analysis of the metabolome of different biological samples. CE is particularly suited for the analysis of polar and charged compounds, as compounds are separated on the basis of their charge-to-mass ratio. The separation mechanism of CE fundamentally differs from reversed-phase LC, and, therefore, CE can provide complementary or additional information on the composition of a biological sample. By incorporating additives (e.g., micelles or chiral selectors) into the background electrolyte (BGE), a variety of separation modes can be explored. For example, micellar electrokinetic chromatography (MEKC), which employs micelles as pseudo-stationary phases in the BGE, can be used for the simultaneous separation of neutral and charged metabolites. However, the coupling of MEKC with MS is problematic and often provides limited sensitivity. Because of the BGE component constraints when coupling with MS via electrospray ionization (ESI), the most often used CE separation mode is capillary zone electrophoresis (CZE). CE separations can be achieved in a fast and highly efficient way without the need for extensive sample pretreatment. Other advantages of CE include the very low – or even absence of – organic solvent consumption, the small amount of reagents needed, and the use of simple fused-silica capillaries instead of more expensive LC columns. The relatively poor concentration sensitivity of CE, as a result of the small loading volumes, can be (partly) circumvented by the incorporation of preconcentration strategies [6]. The potential of CE and CE–MS for metabolomics has been described in detail in a number of reviews [7–13]. Although CE–MS has emerged as an attractive complementary technique for metabolomics studies, the use of CE–MS in metabolomics has lagged behind GC–MS and LC–MS. This might be due to the reduced migration time reproducibility, which is occasionally observed as a result of sample-induced variations in electroosmotic flow (EOF). Furthermore, the practical expertise required to deal with the experimental variables that influence the performance of CE–MS is limited among the metabolomics community.

In this chapter we describe the most recent developments in CE–MS for metabolomics including both nontargeted and targeted approaches. Strategies that can be used to obtain reproducible metabolomics data with CE–MS are discussed. This chapter starts by providing an overview of the CE separation modes and capillary coatings used in CE–MS for metabolomics, followed by a discussion of aspects related to the coupling of CE with MS in the context

8.2 CE–MS

8.2.1
CE Separation Conditions

CE is a highly efficient separation technique providing various modes of separation. In the most common mode, CZE and normally known as *CE*, the capillary is filled with a BGE only and analyte separation is based on differences in electrophoretic mobility. Currently, the predominant way to couple CE with MS is through ESI [10–16]. In ESI-based interfaces, the CE effluent, with or without a sheath liquid, is sprayed into a plume of fine, charged droplets under the influence of a strong electrical field. As the solvent evaporates, compounds that exist as ions in solution can end up as ions in the gas phase and subsequently be detected by MS. The combination of CE and MS by ESI is highly compatible as both CE and ESI are especially suited to compounds that can form ions in solution. On the other hand, ESI is prone to analyte signal suppression by high buffer concentrations, nonvolatile constituents, and surfactants. Moreover, nonvolatile constituents may cause source contamination and high background signals. Therefore, relatively low concentrations of volatile BGEs, usually comprising formic acid, acetic acid, and/or ammonia, are typical BGEs for CE-ESI–MS. In general, low-pH (~2) and high-pH (~9) BGE conditions in conjunction with ESI(+) and ESI(−) are employed for the analysis of cationic and anionic metabolites, respectively [10–13]. The first application of CZE–MS for nontargeted metabolic profiling was demonstrated for bacterial extracts, in which more than 1600 metabolites were analyzed [17]. In this study, distinct CZE-ESI-MS methods were described for cationic and anionic metabolites. CZE-ESI-MS analysis of the cationic metabolites was performed with a bare fused-silica capillary using 1 M formic acid (pH 1.8) as BGE. The analysis of the anionic metabolites was performed with a cationic polymer-coated capillary using 50 mM ammonium acetate (pH 8.5) as BGE. As already mentioned, volatile buffers such as ammonium acetate or formic acid are often used for CE-ESI-MS [18, 19]; however, these BGEs may not give optimum CE separations. The addition of organic modifiers, such as methanol, to the BGE can improve the separation and MS detection performance of metabolites. For instance, in a CE–MS method for amino acids, a baseline separation of the isomers leucine and isoleucine was obtained using a BGE of 2 M formic acid (pH 1.8) with 20% methanol [20].

A stable CE–MS method is crucial to obtain reproducible results. For instance, the stability of analyte migration times is of utmost importance in metabolomics studies where multiple biological samples have to be profiled and compared. Furthermore, for CE–MS, a constant and appreciable EOF is often essential to achieve adequate and reproducible electrospray conditions and, therefore, efficient analyte ionization [21]. When conventional bare fused-silica capillaries are used, separation efficiencies may be compromised as a result of adverse analyte–capillary wall interactions. In addition, the analysis of biological samples with minimal sample pretreatment may lead to adsorption of matrix components to the capillary wall causing unacceptable changes of the EOF and, as a result, analyte migration times. To minimize these problems, fused-silica capillaries coated with polymers may be used, as has been demonstrated for various CE-MS-based metabolomics studies [17, 21–24]. Especially, charged capillary coatings have been used frequently as they can aid in changing the direction and magnitude of the EOF in CE–MS using normal or reversed CE polarity, providing CE systems for anions and cations, as shown in Figure 8.1a [24]. The usefulness of these noncovalently coated capillaries with layers of charged polymers was recently investigated in our laboratory for global metabolic profiling of rat urine. Capillaries were coated with a bilayer of polybrene (PB) and poly(vinyl sulfonate) (PVS) or with a triple layer of PB, dextran sulfate (DS), and PB. The bilayer and triple layer coatings were evaluated at acidic (pH 2.0) and alkaline (pH 9.0) separation conditions, thereby providing separation conditions for basic and acidic compounds. A representative metabolite mixture and spiked urine samples applying no sample pretreatment were used for the evaluation of the four CE methods. Migration time repeatability (RSD < 2%) and plate numbers (N, 100 000–400 000) were similar for the test compounds in all CE methods. The analysis of cationic compounds with the PB-DS-PB CE method at low pH (i.e., migration after the EOF time) provided a wider separation window and a larger number of separated peaks in urine compared to the analysis with the PB-PVS CE method at low pH (i.e., migration before the EOF time), as shown in Figure 8.1b. Approximately 600 molecular features were detected in rat urine by the PB-DS-PB CE-TOF-MS method, whereas only about 300 features were found with the PB-PVS CE-TOF-MS method.

8.2.2
CE–MS Coupling

8.2.2.1 Interfacing

The coupling of CE with MS is not straightforward owing to the very low flow rates (nanoliter per minute range) in CE and the need to establish a closed electrical circuit to maintain the high voltage across the capillary necessary for CE separation. Several recent reviews describe the design, performance, and application of CE–MS systems over the past two decades [25–27]. Until now, ESI is the main ionization technique used for CE–MS in metabolomics. The coupling of CE with ESI–MS can be performed via a sheath–liquid interface or a sheathless interface [28]. The sheath–liquid interface is most widely used for CE–MS. In this configuration,

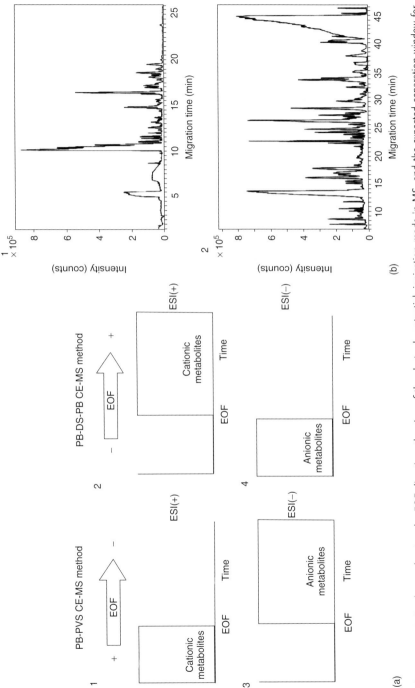

Figure 8.1 (a) A scheme showing the EOF direction, the sign of the electrode potential, ionization mode in MS, and the expected separation window for cations and anions in the PB-PVS and PB-DS-PB CE–MS methods at low and high pH separation conditions. (1) and (2) show the conditions at pH 2.0; (3) and (4) show the conditions at pH 9.0. (Source: Taken from Ref. [24].) (b) Base peak electropherograms obtained during CE–MS analysis of rat urine using (1) a PB-PVS-coated capillary and (2) a PB-DS-PB-coated capillary. Conditions: BGE, 1 M formic acid (pH 2.0); sample injection, 35 mbar for 60 s. (Source: Taken from Ref. [24].)

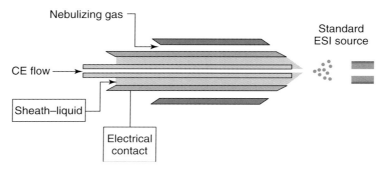

Figure 8.2 Design of a sheath–liquid interface for CE–MS in which sheath liquid and nebulizing gas are added coaxially to the CE effluent. (Source: Taken from Ref. [27].)

the separation capillary is inserted into a tube of larger diameter in a coaxial setting (Figure 8.2). The conductive sheath liquid, to which the CE terminating voltage is applied, is administered via this outer tube and merges with the CE effluent at the capillary outlet. Usually, a gas flow is applied via a third coaxial capillary in order to facilitate spray formation in the ESI source. The sheath liquid can be used to optimize the ESI process, and, therefore, the composition and flow rate of the sheath liquid is very important. For instance, for the CE–MS analysis of catecholamines, a number of organic solvent mixtures (methanol/water, acetonitrile/water, and 2-propanol/water) were tested as sheath liquid [29]. The most optimal signal-to-noise ratio for the catecholamines was obtained with a sheath liquid of methanol/water (80 : 20 v/v) containing 0.5% acetic acid using a flow rate of 6 µl min^{-1}.

In the sheathless interface configuration, the CE voltage is directly applied to the CE buffer at the capillary outlet. This can be achieved by applying a metal coating to the end of a tapered separation capillary or by connecting a metal-coated, full metal, or conductive polymeric sprayer tip to the CE outlet. Another way to make a closed circuit is by insertion of a metal microelectrode through the capillary wall into the CE buffer end or by direct introduction of a microelectrode into the end of a CE capillary [25]. A CE–MS method using a sheathless interface was used to improve the sensitivity for metabolite analysis in extracts of prokaryotes [30]. To accomplish this, the separation capillary was modified by creating a porous junction near the outlet where the electrospray voltage and cathodic voltage for CE were applied. The outlet of the capillary was pulled to a 5 µm inner diameter to form an electrospray emitter and had a frit fabricated near the exit to prevent clogging as shown in Figure 8.3. During analysis, pressure (+4 psi) was applied at the inlet of the separation column to create sufficient flow toward the detector. Table 8.1 shows the limits of detection (LODs) for 19 metabolites obtained in full scan mode with this sheathless CE–MS method using a quadrupole ion trap (QIT) mass analyzer, which ranged from 20 nM for ADP ribose to 2.5 µM for α-ketoglutarate using 40 nl injections. Recently, Busnel *et al.* [31] coupled CE with MS via a sheathless

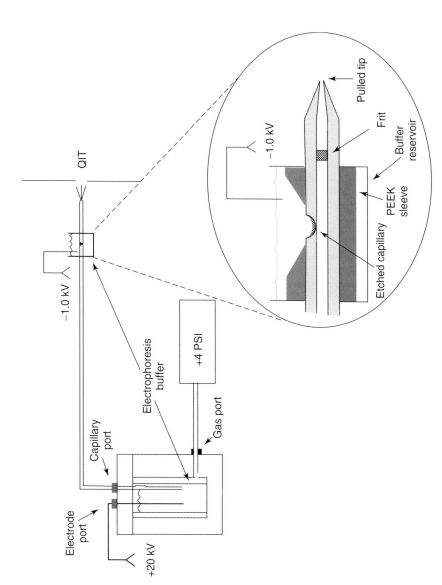

Figure 8.3 Diagram of sheathless CE interfaced to quadrupole ion trap (QIT) MS. Inlet of CE is pressurized for separation using the reservoir shown. Inset shows the etched porous junction. (Source: Taken from Ref. [30].)

Table 8.1 Limit of detection for 19 metabolite standards[a] as determined by CE–MS using a sheathless interface (shown in Figure 8.2) and an LCQ Deca XP Plus Quadrupole ion trap MS instrument.

Compound	m/z [M−H]⁻	LOD (nM) full scan
NAADP	743	100
NAD	662	40
ADP ribose	558	20
GTP	522	300
ATP	506	100
GDP	442	200
ADP	426	70
GMP	362	100
Cyclic GMP	344	70
Fructose-1,6-bisphosphate	339	400
Cyclic-AMP	328	200
Glucose-6-phosphate	259	300
Fructose-6-phosphate	259	300
Citrate	191	600
Phosphoglycerate	185	400
Phosphoenolpyruvate	167	800
α-Ketoglutarate	145	2500
Malate	133	200
Succinate	117	2200

[a] Limit of detection was determined as the concentration that generated a signal-to-noise ratio (S/N) of 3 in reconstructed ion chromatograms. Noise was calculated as the root mean square of baseline in the chromatogram, and signal was peak height. Concentrations generating an S/N of 10 or less were used in the calculation and assuming a linear response to the detection limit.
Source: Taken from Ref. [30].

interface making use of a porous tip sprayer, which was originally developed by Moini [32]. The use of this approach resulted in subnanomolar LODs for some peptides of a protein digest. Although not demonstrated yet, these very favorable LODs indicate that the sheathless CE–MS approach based on a porous tip sprayer shows a strong potential for highly sensitive metabolic profiling of biological samples.

8.2.2.2 Mass Analyzers

So far, the triple quadrupole (TQ) and ion trap (IT) have been the most commonly used mass analyzers in CE–MS for the analysis of low-molecular-weight compounds in biological samples. These MS instruments, especially the TQ, provide high sensitivity with the capability to obtain structural information on unknown compounds. However, a disadvantage of these mass analyzers, especially with respect to fast and highly efficient CE separations, is the relatively slow scanning process and low-duty cycle. These MS instruments may not be able to obtain sufficient data points across a very narrow CE peak to accurately define it. In addition, the mass resolution of these instruments is limited and often does not allow distinction of hundreds of metabolites in a single CE–MS run. Using a quadrupole mass analyzer, Soga et al. [17] limited the scan range to a window of 30 m/z for the detection of metabolites in a bacterial extract. To cover a range of m/z 70–1027, 33 runs were required resulting in an analysis time of ∼16 h. In general, TQ and IT mass analyzers are more suited for targeted metabolomic studies by CE–MS.

Time-of-flight (TOF) mass analyzers have an inherent ability for high spectral acquisition rates allowing a high number of data points (e.g., 10 spectra s^{-1}) to be collected across a narrow CE peak. TOF–MS also provides a high mass resolution and high mass accuracy with errors below 5 ppm [33, 34]. The high spectral acquisition rate and high mass resolution of TOF–MS makes this instrument very compatible with fast and efficient CE separations. Indeed, when the same analysis of metabolites in a bacterial extract as mentioned earlier was performed with a CE-TOF-MS method, three runs were sufficient to measure the metabolites in the m/z 70–1027 region [11, 17]. TOF–MS instruments are increasingly being used in nontargeted metabolomics studies. The accurate mass obtained for (unknown) metabolites also can strongly aid in their provisional identification. Despite the increased resolution of TOF analyzers, potential interferences from solvent ions, adducts, and compounds with the same nominal mass as the metabolites in the biological sample can still disturb the analysis. Therefore, efficient separations before MS analysis are of pivotal importance.

8.3
Sample Pretreatment

Sample pretreatment for metabolomic analysis depends on the goal of the study. For nontargeted metabolomic analysis, it is desirable that the biological sample is analyzed with minimal pretreatment to prevent the loss of metabolites. In order to allow proper analysis, it is essential that the low-molecular-weight metabolites are separated from large molecules (proteins, lipids, and large peptides) and salts. For nontargeted extraction of metabolites from bacteria, several procedures have been investigated, such as extraction with cold methanol, hot methanol, ethanol, lysis with chloroform–methanol, and extractions with strong acids [35]. Extraction with cold methanol showed the highest recovery for the polar metabolites, and more compounds were extracted compared to the other methods. Extraction with cold methanol was applied to the comprehensive analysis of anionic metabolites from *Bacillus subtilis* cells by CE–MS [17].

Recently, Cifuentes and coworkers [36] investigated four different metabolite purification approaches for metabolic profiling of human HT29 colon cancer cells by CE–MS. The purification methods studied were methanol deproteinization, ultrafiltration, and two solid-phase extraction (SPE) methods, namely, Isolute C18

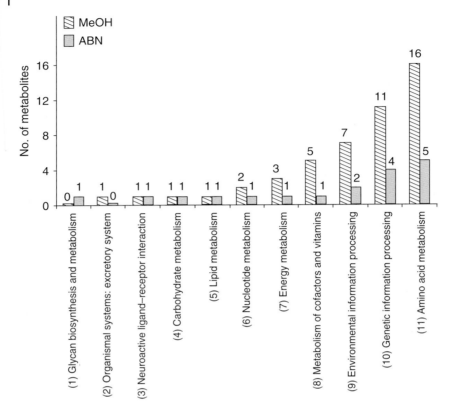

Figure 8.4 Number of compounds (y-axis) identified for some metabolic pathways (indicated in the x-axis) in metabolite extracts obtained from SPE (ABN cartridge) or methanol deproteinization. (Source: Taken from Ref. [36].)

Endcapped cartridges (100 mg) and Evolute™ acidic, basic, and neutral (ABN) columns (25 mg). In both SPE methods, 150 μl of cytosolic fraction obtained from the cell culture was diluted with 350 μl of water. After sample loading, the sorbent was washed with 1 ml of water–methanol (95 : 5 v/v). Sample elution was performed with 500 μl of methanol, which was divided in 100 μl aliquots. The 100 μl aliquots were vacuum dried, and dried extracts were dissolved in 20 μl of water for CE–MS analysis. CE-MS-based metabolome analysis revealed in some cases important differences among the studied metabolite purification procedures. For example, more compounds were observed in human HT29 colon cancer cell extracts obtained after methanol deproteinization than in extracts obtained after ultrafiltration. Figure 8.4 shows that more metabolites were detected in cell extracts deproteinized with methanol than in extracts pretreated with SPE using an ABN cartridge. As indicated in Figure 8.4, metabolites from both methanol and ABN extracts were associated to nine metabolic pathways (numbered from 3 to 11); however, the number of metabolites observed after each purification procedure was different. For instance, in metabolic pathway number 11 of Figure 8.4, 16 metabolites were found in the extract obtained after methanol deproteinization, while only five were found in the extract obtained after SPE. In general, the use of methanol deproteinization brought about a higher number of metabolites associated to known metabolic pathways and, therefore, more metabolomics information was obtained with this approach.

For both nontargeted and targeted metabolomic analysis of plasma and serum, containing relatively large amounts of proteins, deproteinization with an organic solvent is often applied to prevent the adsorption of proteins to the capillary wall during CE analysis [37]. Urine and cerebrospinal fluid (CSF) have often been analyzed with minimal sample pretreatment. CE–MS has been used for metabolic profiling of human urine with direct sample injection [20, 23]. For nontargeted metabolomics, the sample is often analyzed at both low and high pH separation conditions in order to cover as many metabolites as possible [17, 38].

In targeted metabolomics, the sample preparation procedure can be adapted to the target metabolites, because the analytes are known and internal standards or stable isotope-labeled standards can be used to optimize the reliability of the method. Sample preparation includes the separation of low-molecular-weight metabolites from large molecules by deproteinization techniques, followed by liquid–liquid extraction (LLE) or off-line SPE for the selective preconcentration of the target compounds. For instance, phenylboronic-acid-based SPE columns have been used for the selective preconcentration of nucleosides from human urine and subsequent nucleoside profiling by CE [39]. In contrast to nontargeted profiling studies, method development and optimization is relatively straightforward for targeted metabolomics as it is known to focus on which type of compounds. As the identity of the target analytes is known, the selectivity and sensitivity can be further improved using multiple reaction monitoring (MRM) with a TQ MS instrument.

8.4
Data Analysis

Metabolomic analysis generates large and complex data sets. Therefore, multivariate data analysis techniques such as principal component analysis (PCA) and partial least squares-discriminant analysis (PLS-DA) have become an essential part in CE-MS-based metabolomics studies as they can provide interpretable models for complex data. As data treatment and chemometrics are described in detail in other chapters of this book, we only focus on migration time alignment and metabolite identification strategies used in CE-MS-based metabolomics in this section.

A limitation of CE may be the poor migration time reproducibility when analyzing biological samples using minimal sample pretreatment with bare fused-silica

188 | 8 The Role of CE–MS in Metabolomics

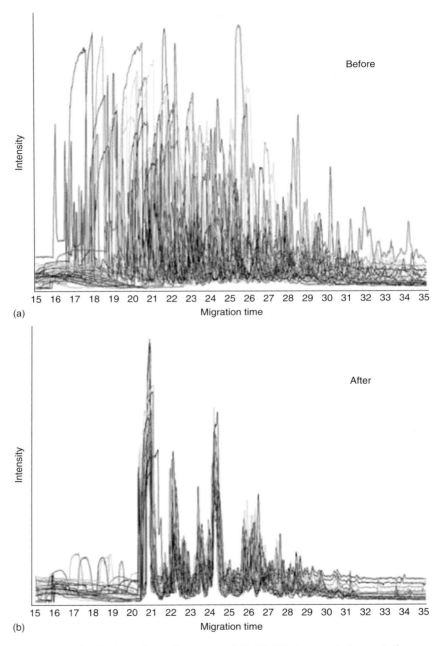

Figure 8.5 (a,b) Total ion electropherograms of 20 CE–MS data sets before and after alignment using accurate mass information. (Source: Taken from Ref. [40].)

capillaries. Therefore, a number of alignment algorithms have been developed, but not all of them can tackle the relatively large and irregular migration time shifts between CE–MS runs. For this purpose, a genetic algorithm designed for alignment of CE–MS data using accurate mass information was developed [40]. The utility of this algorithm was demonstrated for urine samples of mice analyzed by CE–MS. The new algorithm showed a significant reduction of migration time variation in the aligned data sets (Figure 8.5). Significantly improved alignments were provided for data obtained with an ultrahigh resolution TOF–MS instrument than for data obtained with a standard TOF–MS instrument emphasizing the crucial importance of mass accuracy for the performance of the algorithm.

Metabolite identification is based on m/z value using standards and databases such as the Human Metabolome Database. For this purpose, an MS instrument with a high mass resolution and accuracy such as TOF–MS is needed as the accurate mass obtained for unknown compounds considerably reduces the list of possible candidates in the database. For example, CE-TOF-MS has been used for metabolic profiling of transgenic versus conventional soybean extracts [41]. As TOF–MS was used for detection, a number of possible elemental compositions were obtained from the accurate mass of the metabolite peaks. These elemental compositions were matched against available databases using the deduced molecular formula as a search criterion. Additional information about physicochemical properties of the compounds was provided by the migration times.

Despite the good mass accuracy of TOF–MS instruments, accurate m/z values and isotope pattern may not be sufficient for reliable metabolite identification. The lack of commercially available metabolite standards further limits the identification of molecular features detected by CE–MS. To facilitate metabolite identification, Sugimoto *et al.* [42] recently developed a mathematical model using support vector regression to identify unknown peaks based on the predicted migration time ($t(m)$) and accurate m/z values. The model yielded good correlations between the predicted and measured $t(m)$ for 375 cationic metabolite standards. The inclusion of the predicted migration time significantly reduced the number of candidate metabolites for a given m/z and isotope pattern. The group of Britz-McKibbin modeled the migration behavior of charged metabolites in CE as a qualitative tool to support MS characterization based on two fundamental analyte physicochemical properties, namely, absolute mobility and acid dissociation constant [43, 44]. Computer simulations using Simul 5.0 were used to better understand the dynamics of analyte electromigration, as well as aiding *de novo* identification of unknown compounds. There was a good agreement between computer-simulated and experimental electropherograms for several classes of cationic metabolites as reflected by their relative migration times with an average error of <2.0%.

Identification of unknown compounds can also be achieved by performing further CE-MS/MS experiments using an IT, TQ, or quadrupole TOF–MS to provide fragment spectra [45, 46]. For example, to increase the confidence in identification of analytes from *E. coli* extracts, MS/MS experiments with an IT were performed to collect fragment spectra [30]. These fragmentation patterns allowed the assignment of peaks with multiple possible candidates as derived from the database based on molecular mass only.

8.5
Applications

An overview of targeted and nontargeted metabolomics studies performed by CE–MS and published between January 2006 and June 2011 is given in Tables 8.2 and 8.3, respectively. The tables provide information about the type of biological sample and compounds analyzed, the BGE, sample pretreatment procedure, the MS analyzer employed, and LOD (when provided by the authors). Representative applications are discussed in the following sections.

8.5.1
Targeted Approaches

In general, the sample preparation procedure can be adapted to the target metabolites in targeted metabolomics, because the analytes are known and internal standards or stable isotope-labeled standards can be used to optimize the reliability of the method. For example, for the analysis of short-chain carnitines in human plasma by CE–MS, plasma samples were deproteinized with cold acetonitrile [37]. Despite the resulting sample dilution, the peak intensities of the analytes were not significantly decreased due to an acetonitrile-induced stacking effect. The CE–MS analysis of the short-chain carnitines in the deproteinized plasma samples took less than 10 min, with high sensitivity and specificity, as the mass spectrometer was used in product ion scan mode. For quantification of short-chain carnitines in plasma, deuterated carnitine was used as an internal standard, which was added before deproteinization. LODs varied from 0.25 to 1 µM. The relative standard deviations (RSDs) for within-day and between-day repeatability were within 15% for low and high concentrations of short-chain carnitines in plasma.

A CE–MS method was developed by Chalcraft and Britz-McKibbin [47] for the direct analysis of amino acids, acylcarnitines, and their stereoisomers from dried blood spot (DBS) extracts without chemical derivatization. In-capillary preconcentration with desalting by CE–MS allowed for improved concentration sensitivity and as a result the detection of low-abundance metabolites in complex biological samples without ionization suppression or isomeric/isobaric interferences. Figure 8.6 shows a profile of amino acids and acylcarnitines derived from a filtered dried blood extract from a healthy volunteer. Method validation demonstrated that accurate and precise quantification could be achieved for 20 different amino acids and acylcarnitine biomarkers associated with inborn errors of metabolism when using a single internal standard.

Table 8.2 Overview of CE–MS applications in targeted metabolomics.

Sample matrix	Compounds	BGE	Sample pretreatment	MS analyzer	LOD[a]	Remarks	References
Human urine	Amino acids	2 M formic acid and 20% methanol (pH 1.8)	Direct sample injection	TOF	Mid nanomolar range	pH-mediated stacking for preconcentration	[20]
Human cerebrospinal fluid	Amino acids	1 M formic acid (pH 1.8)	Dilution with BGE (1:1)	TOF	20–215 nM	Polybrene-poly (vinyl-sulfonate)-coated capillary	[23]
RBC	Cationic and anionic- metabolites	1.0 M formic acid (pH 1.8); 50 mM ammonium acetate (pH 8.5)	See Ref. [44]	Ion trap	Low micromolar	Internal standards for quantification	[43]
Red blood for quantification	Cationic metabolites	1.4 M formic acid (pH 1.8)	RBC lyzed with ice-cold water; Ultracentrifugation with 3 kDa filter; filtered lysate diluted with ammonium acetate (1:1)	Ion trap	Low micromolar	Internal standards for quantification	[44]
Dried blood spot (DBS)	Cationic metabolites	1.4 M formic acid (pH 1.8)	Extraction with ice-cold methanol/ water (1:1)	Ion trap	0.008–5 μM	Internal standard for quantification	[47]
Bacterial extract	Organic acids and nucleotides	20 mM ammonium acetate (pH 6.8)	Extraction with ice-coldmethanol/ water (1:1, v/v)	Ion trap	ns	Polyethylene -glycol-coated capillary; quantification by comparing peak areas with standards	[48]

(continued overleaf)

Table 8.2 (Continued)

Sample matrix	Compounds	BGE	Sample pretreatment	MS analyzer	LOD[a]	Remarks	References
Mouse liver extract	Anionic metabolites	50 mM ammonium acetate (pH 8.5)	Methanol extraction	TOF	0.03–0.87 µM	Platinum ESI needle; internal standards for quantification	[49]
Human plasma	Arginine and methylated metabolites	1.0 M formic acid (pH 1.8)	Plasma deproteinized with 1% formic acid and acetonitrile	Ion trap	Low micromolar	Internal standard for quantification; field amplified stacking	[50]
Human plasma	Thiols	1 M formic acid (pH 1.8)	Ultracentrifugation with 3 kDa filter	Ion trap	Low nanomolar	Thiol-selective maleimide labeling; internal standards for quantification	[51]
Human tumor and normal tissues	Cationic and anionic metabolites	50 mM ammonium acetate (pH 8.5); 1 M formic acid (pH 1.8)	Methanol extraction	TOF	ns	Polybrene-dextran sulfate-polybrene-coated capillary for anionic metabolites; internal standards for quantification	[52]
Human urine	Amino acids	3% formic acid (pH 1.8)	Centrifugation; 490 µl urine mixed with 20 µl internal standard	Quadrupole	0.008–2.4 µg ml^{-1}	pH-mediated stacking	[53]

Mouse urine	Amino acids and related compounds	2 M formic acid (pH 1.8) + 10% MeOH	Centrifugation and dilution with BGE (1:1, v/v)	TOF	Low nanomolar	pH-mediated stacking	[54]
Bacterial extract	Cationic and anionic metabolites	50 mM ammonium acetate (pH 9.0); 1 M formic acid (pH 1.8)	Extraction with methanol, chloroform, and water	TOF and triple quadrupole	ns	Polybrene-dextran sulfate-polybrene-coated capillary for anions	[55]
Bacterial extract	Nucleotides	50 mM ammonium acetate (pH 7.5)	Extraction with methanol, chloroform, and waters	TOF	ns	Capillary treated with phosphate to prevent adsorption of phosphorylated compound	[56]
Plant extract	Cationic and anionic metabolites	50 mM ammonium acetate (pH 9.0); 1 M formic acid (pH 1.8)	Extraction with methanol, chloroform, and water	TOF	ns	Internal standards for quantification	[57]
Plant extract	Cationic and anionic metabolites	50 mM ammonium acetate (pH 9.0); 1 M formic acid (pH 1.8)	Extraction with methanol, chloroform, and water	TOF	ns	Polybrene-dextran sulfate-polybrene-coated capillary for anions	[58]
Plant extract	Anionic metabolites	50 mM ammonium acetate (pH 9.0)	Extraction with methanol, chloroform, and water; Deproteinization with 5 kDa filter	Triple quadrupole	Low micromolar	Sulfonated capillary	[59]

(continued overleaf)

Table 8.2 (Continued)

Sample matrix	Compounds	BGE	Sample pretreatment	MS analyzer	LOD[a]	Remarks	References
Plant extract	Cationic and anionic metabolites	50 mM ammonium acetate (pH 9.0); 1 M formic acid (pH 1.8)	Extraction with methanol/water (1 : 1, v/v)	Ion trap	ns	Pressure-assisted CE–MS analysis of anions; quantification performed using known concentrations of selected compounds	[60]
Bacterial extract	Amino acids	1.6 M formic acid in MeOH/H_2O (2 : 8, v/v)	Cold methanol extraction of metabolites. Salts removed by off-line SPE	FT-ICR	0.1 μM	pH-mediated stacking and tITP for preconcentration	[61]
Bacterial extract	Sugar phosphates	—	30 mM morpholine/formate (pH 9.0)	Ice-cold ethanol extraction linear-ion trap	Triple quadrupole	2.5–10 μM	[62]
Bacterial extract	Intracellular metabolites	1 M formic acid (pH 1.8)	Extraction with water, chloroform, and methanol. Methanol extract treated with 5 kDa filter and lyophilized	TOF	ns	—	[63]

Sample	Analytes	BGE	Sample preparation	MS analyzer	LOD^a	Remarks	Ref.
Bacterial extract	Nucleotides	50 mM ammonium acetate (pH 7.5)	Extraction with water, chloroform, and methanol. Methanol extract filtered with 5 kDa filter and lyophilized	Ion trap	0.5–1.7 μM	Capillary treated with phosphate to prevent adsorption of multiphosphorylated compounds	[64]
Yeast extract	Sulfur-related metabolites	1 M formic acid (pH 1.8)	Extraction with cold chloroform and cold methanol	Ion trap	ns	Matrix effects studied	[65]
Human cerebrospinal fluid	Amino acids	5 mM ammonium acetate (pH 9.7) + 5% acetonitrile	CSF diluted with water (1:5, v/v)	TOF	20–67 nM	Capillary coated with 1-(4-iodobutyl)4-aza-1-azoniabicyclo[2.2.2]octane iodide	[66]
Human urine	Amino acids	1 M formic acid (pH 1.8)	Dilution with BGE (1:1, v/v)	TOF	85–280 nM	pH-mediated stacking for preconcentration; polybrene-poly(vinyl-sulfonate)-coated capillary	[67]

^{a}LOD = limit of detection (S/N = 3); ns, not specified in paper.

Table 8.3 Overview of CE–MS applications in nontargeted metabolomics.

Sample matrix	Compounds	BGE	Sample pretreatment	MS analyzer	LOD[a]	Remarks	References
Rat urine	Cationic and anionic metabolites	1 M formic acid (pH 1.8); 20 mM ammonium acetate (pH 9.0)	Dilution with BGE (1:1)	TOF	ns	—	[24]
Bacterial extract	Anionic metabolites	20 mM ammonium acetate and 2-propanol (8:2, v/v)	Cold methanol extraction	Ion trap	0.02–2.5 µM	Sheathless interface	[30]
Plant extract	Cationic metabolites	5% formic acid (pH 1.9)	Extraction with methanol-water (1:1, v/v)	TOF	ns	—	[41]
Mouse liver extracts	Cationic and anionic metabolites	50 mM ammonium acetate (pH 9.0) and 1 M formic acid (pH 1.8)	Extraction with water and chloroform	TOF	0.1–1.7 µM	Polybrene-dextran sulfate-polybrene-coated capillary	[45]

8.5 Applications | 197

Human urine	Cationic metabolites	1 M formic acid (pH 1.8)	Dilution with BGE (1:1)	TOF	Low nM	pH-mediated stacking or preconcentration	[68]
Human urine	Positively and negatively charged metabolites	20 mM formic acid/ammonium formate (pH 3.0); 20 mM ammonium acetate (pH 9.0)	Direct sample injection	Triple quadrupole	ns	Cationic polymer polyE-323 for anions	[69]
Plant extract	Anionic metabolites	50 mM ammonium formate (pH 8.0); de-proteinization with 3 kDa 50 mM ammonium acetate (pH 10.0) containing 50% methanol	Extraction with cold methanol and cold water; filter	Ion trap	0.13–17 µM	Internal standards for quantification	[70]

(continued overleaf)

Table 8.3 (Continued)

Sample matrix	Compounds	BGE	Sample pretreatment	MS analyzer	LOD[a]	Remarks	References
Human saliva	Cationic and anionic metabolites	50 mM ammonium acetate (pH 9.0); 1 M formic acid (pH 1.8)	Methanol extraction	TOF	ns	Polybrene-dextran sulfate-polybrene -coated capillary for anionic metabolites	[71]
Red blood cell lysate	Cationic metabolites	1.4 M formic acid (pH 1.8)	RBC lyzed with ice-cold water; ultra-centrifugation with 3 kDa filter; filtered lysate diluted with ammonium acetate (1:1)	Ion trap	Low micromolar	—	[72]
Mouse blood and tissues	Cationic and anionic metabolites	50 mM ammonium acetate (pH 9.0); 1 M formic acid (pH 1.8)	Methanol extraction	TOF	ns	Polybrene-dextran sulfate-polybrene -coated capillary for anionic metabolites	[73]

Rat liver extract	Cationic and anionic metabolites	50 mM ammonium acetate (pH 9.0); 1 M formic acid (pH 1.8)	Methanol extraction	TOF	ns	Polybrene-dextran sulfate-polybrene-coated capillary for anionic metabolites	[74]
Bacterial extract	Anionic metabolites	50 mM ammonium acetate (pH 9.0)	Methanol extraction; centrifugation with 5 kDa and lyophillization	TOF	ns	Polybrene-dextran sulfate-polybrene-coated capillary for anionic metabolites	[75]
Bacterial extract	Anionic metabolites	50 mM ammonium acetate (pH 8.7)	Extraction with 80% methanol in water	TOF	0.2–2 µM	Cationic polymer polyE-323 for EOF reversal	[76]
In vitro biochemical assay	Cationic and anionic metabolites	1 M formic acid (pH 1.8)	Ultrafiltration to remove proteins	Quadrupole	0.3–11 µM	Polybrene-dextran sulfate-polybrene-coated capillary for anionic metabolites	[77]
Yeast extract	Anionic metabolites	150 mM ammonium hydrogen-carbonate/formate (pH 6.0)	Extraction with cold chloroform and cold methanol	Ion trap	ns	Pressure-assisted CE using PEEK	[78]

(continued overleaf)

Table 8.3 (Continued)

Sample matrix	Compounds	BGE	Sample pretreatment	MS analyzer	LOD[a]	Remarks	References
Plant extract	Anionic metabolites	50 mM ammonium acetate (pH 9.0)	Extraction with methanol, chloroform, and water	Triple quadrupole linear-ion trap	0.1–8.8 µM	Sulfonated coated capillary	[79]
Maize extract	Cationic metabolites	5% formic acid (pH 1.9)	Pressurized liquid extraction	TOF	ns	—	[80]
Japanese Sake (wine)	Cationic and anionic metabolites	50 mM ammonium acetate (pH 9.0); 1 M formic acid (pH 1.8)	Centrifugation with 5 kDa filter	TOF	ns	Polybrene-dextran sulfate-polybrene-coated capillary for anionic metabolites	[81]

[a] LOD = limit of detection (S/N = 3); ns, not specified in paper.

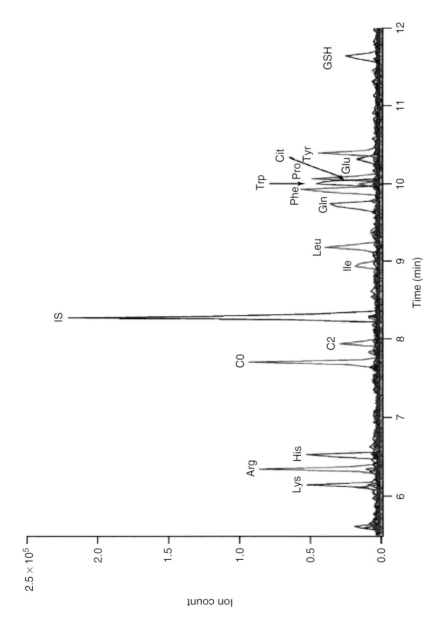

Figure 8.6 A multiple extracted ion electropherogram showing a profile of some amino acids and acylcarnitines derived from a filtered dried blood spot (3.2 mm or ≈ 3.4 µl) extract of a healthy adult volunteer that was reconstituted in 20 µl of sample buffer before CE-ESI-MS analysis. BGE, 1.4 M formic acid, 15% ACN, and pH 1.8. (Source: Taken from Ref. [47].)

CE–MS was used for the analysis of changes in the amounts of metabolites on the shift from photoautotrophic to photomixotrophic conditions in the cyanobacterium *Synechocystis* sp. PCC 6803 [48]. Fifty milliliters of cultures were harvested by centrifugation at 15 000 g and 4 °C for 2 min, and the cell pellets obtained (30–50 mg in fresh weight) were frozen in liquid nitrogen. Samples were vortexed with 200 µl of ice-cold 50% (v/v) methanol containing an internal standard (50 µM PIPES) for 10 min. The supernatant was deproteinized before CE–MS analysis. Separation of metabolites was performed on a polyethylene glycol-coated capillary using 20 mM ammonium acetate (pH 6.8) as BGE. Metabolites in the bacterial extract were identified by comparison of the migration time and m/z ratio with those of organic acid and nucleotide standards.

CE-TOF-MS was used for the global analysis and quantification of endogenous metabolites in tumor and grossly normal tissues obtained from 16 colon and 12 stomach cancer patients [52]. The use of three CE-TOF-MS methods (one for cations, one for anions, and one for nucleotides) resulted in the detection of 738 (normal) and 877 (tumor) peaks in colon and 1007 (normal) and 1142 (tumor) peaks in stomach tissues on average, after eliminating redundant peaks, such as fragments and adduct ions. Among these, 94 peaks in colon and 95 peaks in stomach were identified by matching the closest m/z values and normalized migration times and quantified with metabolite standards. Extremely low glucose and high lactate and glycolytic intermediate concentrations were found in both colon and stomach tumor tissues, which indicated enhanced glycolysis.

8.5.2
Nontargeted Approaches

For nontargeted metabolomics, the use of sample pretreatment should be preferably kept to a minimum in order to prevent the loss of metabolites. However, the use of CE–MS for the analysis of body fluids using minimal sample pretreatment is often hindered by reproducibility problems, as indicated by the occurrence of migration-time shifts between analyses. This means that system stability should be improved in order to establish the role of CE–MS for nontargeted metabolomics studies. For this purpose, Ramautar *et al.* developed CE-TOF-MS approaches based on noncovalently coated capillaries for metabolic profiling of body fluids (Section 8.2.1). For example, CE-TOF-MS using a PB-PVS-coated capillary was used for nontargeted metabolic profiling of urine from patients with complex regional pain syndrome (CRPS) [68]. The CE-TOF-MS method provided fast and stable metabolic profiles of urine samples. Migration time and peak area RSDs ($n = 10$) of various endogenous compounds in pooled urine were <2 and <9%, respectively. With the use of multivariate statistics, discrimination between urine samples from CRPS patients and controls was obtained, emphasizing differences in metabolic signatures between CRPS-diseased patients and controls. Several compounds, such as 3-methylhistidine, were responsible for discriminating the samples. Although these findings have to be validated in a further study with larger numbers of patients and age- and sex-matched controls, the results of this study demonstrated an obvious increased muscle catabolism.

A CE–MS method using fused-silica capillaries for the profiling of anionic metabolites was recently developed by Sato and Yanagisawa [70] using a combination of two different analytical modes. A high-speed mode was used to simultaneously analyze a number of major anionic metabolite classes including organic acids, sugar phosphates, nucleotides, and coenzymes. Using ammonium formate (pH 8.0) as BGE and applying pressure-assisted flow, a standard mixture including 38 compounds could be analyzed in less than 16 min. RSDs were better than 0.7% for migration times and between 1.2 and 7.0% for peak areas. Figure 8.7 shows selected ion electropherograms for the standard mixture and for an extract from moss. As the peaks of several isomers overlapped in this high-speed mode, a high-resolution CE–MS method was developed for these compounds. The high-resolution mode was achieved by suppressing the EOF through the use of a mixture of ammonium acetate (pH 10.0) and methanol (1 : 1, v/v) as BGE. These conditions allowed the separation of structurally related compounds and isomers, such as hexose phosphate isomers and pentose phosphate isomers.

A differential metabolomics strategy based on CE–MS was used by Lee *et al.* [72] to assess the efficacy of nutritional intervention to attenuate oxidative stress induced by strenuous exercise. A healthy volunteer was recruited to perform a submaximal prolonged ergometer cycling trial until volitional exhaustion with frequent blood collection over a 6 h time interval, which included pre-, during-, and postexercise periods. A follow-up study was subsequently performed by the same subject after high-dose oral intake of *N*-acetyl-L-cysteine (NAC) before performing the same exercise protocol under standardized conditions. Nontargeted metabolic profiling of filtered red blood cell lysates by CE–MS allowed for the identification of several putative early- and late-stage biomarkers that reflected oxidative stress inhibition due to nutritional intervention, including oxidized glutathione, reduced glutathione, 3-methylhistidine, L-carnitine, *O*-acetyl-L-carnitine, and creatine.

8.6
Conclusions and Perspectives

The use of CE–MS for metabolomics has increased considerably over the last few years. The studies reported so far in the literature demonstrate that CE combined with MS has the potential to provide informative metabolic profiles of biological samples. As CE is particularly useful for the separation of highly polar and charged metabolites, CE–MS offers complementary information on metabolic compositions of biological samples with respect to reversed-phase LC–MS [11, 82]. The use of both techniques would provide a broad coverage of the metabolome.

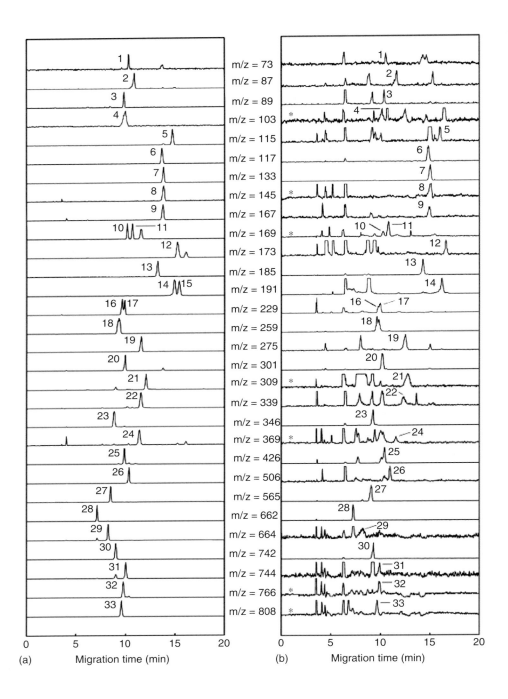

8.6 Conclusions and Perspectives

So far, the application of CE–MS for large-scale metabolic profiling studies, that is, for a big cohort of urine, plasma, or CSF samples, is still limited. In order to increase the applicability of CE–MS for biomedical and clinical metabolomics studies, the utility of CE–MS should be demonstrated for the analysis of large cohorts of clinical samples. In this regard, reproducibility of migration times and peak areas is of utmost importance for a reliable comparison of metabolic profiles and to observe small changes in sample composition. As outlined in this chapter, a promising approach to achieve reproducible CE–MS systems for metabolic profiling of body fluids is the use of noncovalently coated capillaries [24, 68].

The applicability of CE for metabolomics may be further broadened using CEC, an analytical separation technique that combines the high peak efficiency of CE with the stationary phase selectivity of LC and MEKC as additional separation modes. All CE-MS-based metabolomics studies have been performed with ESI, which is especially suitable for the analysis of (highly) polar compounds. MEKC combined with MS using APPI and/or APPI as ionization techniques would be highly suited for the analysis of relatively nonpolar compounds [11, 27].

In general, a sheath–liquid interface is used for the coupling of CE with MS. The use of a sheathless interface instead of a sheath–liquid interface for CE–MS would be very interesting to find out whether an improved coverage of metabolites in a biological sample is obtained. Recently, the sheathless interface approach of Moini was developed by Beckman Coulter into a novel front-end separation and ionization technology called *CESI*, which combines the low-flow characteristics of CE with an integrated ESI source [31, 32]. Sheathless CE–MS shows a strong potential for highly sensitive analysis and, therefore, one can envisage that the use of this approach will probably not only allow the detection of low-abundance metabolites present in biological samples but also in getting a more extended coverage of the metabolome.

Figure 8.7 Selected ion electropherograms of anionic metabolites analyzed in high-speed CE–MS mode using 50 mM ammonium acetate (pH 8.0) as BGE and 50 mbar of air pressure was applied at the CE inlet. Standard mixtures (a) and moss extracts (b) were analyzed. All compounds except for hexose phosphate isomers could be separately identified in this mode. The asterisks shown at the left side of the electropherogram (b) indicate that the mass signals were separately scanned to obtain a sufficient peak intensity of the target metabolite. Peak identification: 1, glyoxylate; 2, pyruvate; 3, lactate; 4, hydroxypyruvate; 5, fumarate; 6, succinate; 7, malate; 8, 2-oxoglutarate; 9, phosphoenolpyruvate; 10, glyceraldehyde 3-phosphate; 11, dihydroxyacetonephosphate; 12, *cis*-aconitate; 13, 3-phosphoglycerate; 14, citrate; 15, isocitrate; 16, ribose 5-phosphate; 17, ribulose 5-phosphate; 18, hexose phosphate isomers; 19, 6-phosphogluconate; 20, PIPES as an internal standard; 21, ribulose 1,5-bisphosphate; 22, fructose 1,6-bisphosphate; 23, AMP; 24, sedoheptulose 1,7-bisphosphate; 25, ADP; 26, ATP; 27, UDP-glucose; 28, NAD; 29, NADH; 30, NADP; 31, NADPH; 32, CoA; and 33, acetyl-CoA. (Source: Taken from Ref. [70].)

References

1. Fiehn, O. (2002) *Plant Mol. Biol.*, **48**, 155–171.
2. Van der Greef, J., Stroobant, P., and Van der Heijden, R. (2004) *Curr. Opin. Chem. Biol.*, **8**, 559–565.
3. Nicholson, J.K. and Wilson, I.D. (2003) *Nat. Rev. Drug Discov.*, **2**, 668–676.
4. Villas-Boas, S.G., Mas, S., Akesson, M., Smedsgaard, J., and Nielsen, J. (2005) *Mass Spectrom. Rev.*, **24**, 613–646.
5. Lenz, E.M. and Wilson, I.D. (2007) *J. Proteome Res.*, **6**, 443–458.
6. Ptolemy, A.S. and Britz-McKibbin, P. (2008) *Analyst*, **133**, 1643–1648.
7. Soga, T. (2007) *Methods Mol. Biol.*, **358**, 129–137.
8. Ramautar, R., Demirci, A., and De Jong, G.J. (2006) *Trends Anal. Chem.*, **25**, 455–466.
9. Barbas, C., Moraes, E.P., and Villaseñor, A. (2011) *J. Pharm. Biomed. Anal.*, **55**, 823–831.
10. Britz-McKibbin, P. (2011) *Methods Mol. Biol.*, **708**, 229–246.
11. Monton, M.R.N. and Soga, T. (2007) *J. Chromatogr. A*, **1168**, 237–246.
12. Ramautar, R., Somsen, G.W., and De Jong, G.J. (2009) *Electrophoresis*, **30**, 276–291.
13. Ramautar, R., Mayboroda, O.A., Somsen, G.W., and De Jong, G.J. (2011) *Electrophoresis*, **32**, 52–65.
14. Gaspar, A., Englmann, M., Fekete, A., Harir, M., and Schmitt-Kopplin, P. (2008) *Electrophoresis*, **29**, 66–79.
15. Hernandes-Borges, J., Neusüß, C., Cifuentes, A., and Pelzing, M. (2004) *Electrophoresis*, **25**, 2257–2281.
16. Staub, A., Schappler, J., Rudaz, S., and Veuthey, J.L. (2009) *Electrophoresis*, **30**, 1610–1623.
17. Soga, T., Ohashi, Y., Ueno, Y., Naraoka, H., Tomita, M., and Nishioka, T. (2003) *J. Proteome Res.*, **2**, 488–494.
18. Soga, T., Ueno, Y., Naraoka, H., Ohashi, Y., Tomita, M., and Nishioka, T. (2002) *Anal. Chem.*, **74**, 2233–2239.
19. Neusüß, C., Pelzing, M., and Macht, M. (2002) *Electrophoresis*, **18**, 3149–3159.
20. Mayboroda, O.A., Neusüß, C., Pelzing, M., Zurek, G. et al. (2007) *J. Chromatogr. A*, **1159**, 149–153.
21. Huhn, C., Ramautar, R., Wuhrer, M., and Somsen, G.W. (2010) *Anal. Bioanal. Chem.*, **396**, 297–314.
22. Johannesson, N., Olsson, L., Bäckström, D., Wetterhall, M. et al. (2007) *Electrophoresis*, **28**, 1435–1443.
23. Ramautar, R., Mayboroda, O.A., Deelder, A.M., Somsen, G.W., and De Jong, G.J. (2008) *J. Chromatogr. B*, **871**, 370–374.
24. Ramautar, R., Toraño, J.S., Somsen, G.W., and De Jong, G.J. (2010) *Electrophoresis*, **31**, 2319–2327.
25. Klampfl, C.W. (2009) *Electrophoresis*, **30**, S83–S91.
26. Maxwell, E.J. and Chen, D.D. (2008) *Anal. Chim. Acta*, **627**, 25–33.
27. Hommerson, P., Khan, A.M., De Jong, G.J., and Somsen, G.W. (2011) *Mass Spectrom. Rev.*, in press. **30**, 1096–1120.
28. Simpson, D.C. and Smith, R.D. (2005) *Electrophoresis*, **26**, 1291–1305.
29. Vuorensola, K., Kokkonen, J., Sirén, H., and Ketola, R.A. (2001) *Electrophoresis*, **22**, 4347–4354.
30. Edwards, J.L., Chisolm, C.N., Shackman, J.G., and Kennedy, R.T. (2006) *J. Chromatogr. A*, **1106**, 80–88.
31. Busnel, J.M., Schoenmaker, B., Ramautar, R., Carrasco-Pancorbo, A., Ratnayake, C., Feitelson, J.S., Chapman, J.D., Deelder, A.M., and Mayboroda, O.A. (2010) *Anal. Chem.*, **82**, 9476–9483.
32. Moini, M. (2007) *Anal. Chem.*, **79**, 4241–4246.
33. Bajad, S. and Shulaev, V. (2007) *Trends Anal. Chem.*, **26**, 625–636.
34. Lacorte, S. and Fernandez-Albaz, A.R. (2006) *Mass Spectrom. Rev.*, **25**, 866–880.
35. Maharjan, R.P. and Ferenci, T. (2003) *Anal. Biochem.*, **313**, 145–154.
36. Simó, C., Ibáñez, C., Gómez-Martínez, A., Ferragut, J.A., and Cifuentes, A. (2011) *Electrophoresis*, **32**, 1765–1777.
37. Desiderio, C., De Rossi, A., Inzitari, R., Mancinelli, A. et al. (2008) *Anal. Bioanal. Chem.*, **390**, 1637–1644.
38. Allard, E., Bäckström, D., Danielsson, R., Sjöberg, P.J.R., and Bergquist, J. (2008) *Anal. Chem.*, **80** (23), 8946–8955.

39. La, S., Cho, J., Kim, J., and Kim, K. (2003) *Anal. Chim. Acta*, **486**, 171–182.
40. Nevedomskaya, E., Derks, R., Deelder, A.M., Mayboroda, O.A., and Palmblad, M. (2009) *Anal. Bioanal. Chem.*, **395**, 2527–2533.
41. Levandi, T., Leon, C., Kaljurand, M., Garcia-Canas, V., and Cifuentes, A. (2008) *Anal. Chem.*, **80**, 6329–6335.
42. Sugimoto, M., Hirayama, A., Robert, M., Abe, S., Soga, T., and Tomita, M. (2010) *Electrophoresis*, **31**, 2311–2318.
43. Lee, R. and Britz-McKibbin, P. (2009) *Anal. Chem.*, **81**, 7047–7056.
44. Chalcraft, K.R., Lee, R., Mills, C., and Britz-McKibbin, P. (2009) *Anal. Chem.*, **81**, 2506–2515.
45. Soga, T., Baran, R., Suematsu, M., Ueno, Y. et al. (2006) *J. Biol. Chem.*, **281**, 16768–16776.
46. Rubakhin, S.S., Romanova, E.V., Nemes, P., and Sweedler, J.V. (2011) *Nat. Methods*, **8**, S20–S29.
47. Chalcraft, K.R. and Britz-McKibbin, P. (2009) *Anal. Chem.*, **81**, 307–314.
48. Takahashi, H., Uchimiya, H., and Hihara, Y. (2008) *J. Exp. Bot.*, **59**, 3009–3018.
49. Soga, T., Igarashi, K., Ito, C., Mizobuchi, K., Zimmermann, H.P., and Tomita, M. (2009) *Anal. Chem.*, **81**, 6165–6174.
50. Desiderio, C., Rossetti, D.V., Messana, I., Giardina, B., and Castagnola, M. (2010) *Electrophoresis*, **31**, 1894–1902.
51. D'Agostino, L.A., Lam, K.P., Lee, R., and Britz-McKibbin, P. (2011) *J. Proteome Res.*, **10**, 592–603.
52. Hirayama, A., Kami, K., Sugimoto, M., Sugawara, M., Toki, N., Onozuka, H., Kinoshita, T., Saito, N., Ochiai, A., Tomita, M., Esumi, H., and Soga, T. (2009) *Cancer Res.*, **69**, 4918–4925.
53. Wang, S., Yang, P., and Zhao, X. (2009) *Chromatographia*, **70**, 1479–1484.
54. Nevedomskaya, E., Ramautar, R., Derks, R., Westbroek, I., Zondag, G., Van der Pluijm, I., Deelder, A.M., and Mayboroda, O.A. (2010) *J. Proteome Res.*, **9**, 4869–4874.
55. Saito, N., Robert, M., Kochi, H., Matsuo, G., Kakazu, Y., Soga, T., and Tomita, M. (2009) *J. Biol. Chem.*, **284**, 16442–16451.
56. Ooga, T., Ohashi, Y., Kuramitsu, S., Koyama, Y., Tomita, M., Soga, T., and Masui, R. (2009) *J. Biol. Chem.*, **284**, 15549–15556.
57. Hasunuma, T., Harada, K., Miyazawa, S., Kondo, A., Fukusaki, E., and Miyake, C. (2010) *J. Exp. Bot.*, **61**, 1041–1051.
58. Urano, K., Maruyama, K., Ogata, Y., Morishita, Y., Takeda, M., Sakurai, N., Suzuki, H., Saito, K., Shibata, D., Kobayashi, M., Yamaguchi-Shinozaki, K., and Shinozaki, K. (2009) *Plant J.*, **57**, 1065–1078.
59. Jumtee, K., Okazawa, A., Harada, K., Fukusaki, E., Takano, M., and Kobayashi, A. (2009) *J. Biosci. Bioeng.*, **108**, 151–159.
60. Takahara, K., Kasajima, I., Takahashi, H., Hashida, S.N., Itami, T., Onodera, H., Toki, S., Yanagisawa, S., Kawai-Yamada, M., and Uchimiya, H. (2010) *Plant Physiol.*, **152**, 1863–1873.
61. Baidoo, E.E.K., Benke, P.I., Neusüss, C., Pelzing, M. et al. (2008) *Anal. Chem.*, **80**, 3112–3122.
62. Hui, J.P.M., Yang, J., Thorson, J.S., and Soo, E.C. (2007) *Chembiochem*, **8**, 1180–1188.
63. Toya, Y., Ishii, N., Hirasawa, T., Naba, M. et al. (2007) *J. Chromatogr. A*, **1159**, 134–141.
64. Soga, T., Ishikawa, T., Igarashi, S., Sugawara, K. et al. (2007) *J. Chromatogr. A*, **1159**, 125–133.
65. Tanaka, Y., Higashi, T., Rakwal, R., Wakida, S. et al. (2007) *J. Pharm. Biomed. Anal.*, **44**, 608–613.
66. Arvidsson, B., Johannesson, N., Citterio, A., Righetti, P.G., and Bergquist, J. (2007) *J. Chromatogr. A*, **1159**, 154–158.
67. Ramautar, R., Mayboroda, O.A., Derks, R.J.E., Van Nieuwkoop, C. et al. (2008) *Electrophoresis*, **29**, 2714–2722.
68. Ramautar, R., Van der Plas, A.A., Nevedomskaya, E., Derks, R. et al. (2009) *J. Proteome Res.*, **8**, 5559–5567.
69. Ullsten, S., Danielsson, R., Bäckström, D., Sjöberg, P., and Bergquist, J. (2006) *J. Chromatogr. A*, **1117**, 87–93.
70. Sato, S. and Yanagisawa, S. (2010) *Metabolomics*, **6**, 529–540.

71. Sugimoto, M., Wong, D.T., Hirayama, A., Soga, T., and Tomita, M. (2009) *Metabolomics*, **6**, 78–95.
72. Lee, R., West, D., Phillips, S.M., and Britz-McKibbin, P. (2010) *Anal. Chem.*, **82**, 2959–2968.
73. Kato, Y., Kubo, Y., Iwata, D., Kato, S., Sudo, T., Sugiura, T., Kagaya, T., Wakayama, T., Hirayama, A., Sugimoto, M., Sugihara, K., Kaneko, S., Soga, T., Asano, M., Tomita, M., Matsui, T., Wada, M., and Tsuji, A. (2010) *Pharm. Res.*, **27**, 832–840.
74. Sakuragawa, T., Hishiki, T., Ueno, Y., Ikeda, S., Soga, T., Yachie-Kinoshita, A., Kajimura, M., and Suematsu, M. (2010) *J. Clin. Biochem. Nutr.*, **46**, 126–134.
75. Nakahigashi, K., Toya, Y., Ishii, N., Soga, T., Hasegawa, M., Watanabe, H., Takai, Y., Honma, M., Mori, H., and Tomita, M. (2009) *Mol. Syst. Biol.*, **5**, 306.
76. Timischl, B., Dettmer, K., Kaspar, H., Thieme, M. *et al.* (2008) *Electrophoresis*, **29**, 2203–2214.
77. Saito, N., Robert, M., Kitamura, S., Baran, R. *et al.* (2006) *J. Proteome Res.*, **5**, 1979–1987.
78. Tanaka, Y., Higashi, T., Rakwal, R., Wakida, S., and Iwahashi, H. (2008) *Electrophoresis*, **29**, 2016–2023.
79. Harada, K., Ohyama, Y., Tabushi, T., Kobayashi, A. *et al.* (2008) *J. Biosci. Bioeng.*, **105**, 249–260.
80. Leon, C., Rodriguez-Meizoso, I., Lucio, M., Garcia-Cañas, V., Ibañez, E., Schmitt-Kopplin, P., and Cifuentes, A. (2009) *J. Chromatogr. A*, **1216**, 7314–7323.
81. Sugimoto, M., Koseki, T., Hirayama, A., Abe, S., Sano, T., Tomita, M., and Soga, T. (2010) *J. Agric. Food Chem.*, **58**, 374–383.
82. Ramautar, R., Nevedomskaya, E., Mayboroda, O.A., Deelder, A.M., Wilson, I.D., Gika, H.G., Theodoridis, G.A., Somsen, G.W., and De Jong, G.J. (2011) *Mol. Biosyst.*, **7**, 194–199.

9
NMR-Based Metabolomics Analysis

Andrea Lubbe*, Kashif Ali, Robert Verpoorte, and Young Hae Choi

9.1
Introduction

In analogy to the terms transcriptome and proteome, the set of metabolites synthesized by an organism represents its metabolome, and *"Metabolomics"* aims to unbiased identification and quantification of the complete set of metabolites in a cell or tissue type [1, 2]. In a strict sense, it is not technically possible to do metabolomics (qualitative and quantitative analysis of *"all"* metabolites) because so far no single or combination of analytical techniques is sensitive, selective, or comprehensive enough to measure all the metabolites [3]. The reason for this is, unlike sequencing of DNA, RNA, or proteins, the great diversity in structure, chemical properties, concentration, and stability of metabolites in a given sample [4]. Instead, *"metabolite fingerprinting"* and *"metabolite profiling"* are more realistic tasks and all the studies referred to (in this chapter as well) as *metabolomics studies* are actually either fingerprinting or profiling studies. By definition, metabolite fingerprinting is high throughput qualitative screening of metabolic composition to perform discrimination analysis and comparison of different samples, for example, two different genotypes. Metabolite fingerprinting is often followed by metabolite profiling which involves identification and quantification of selected and limited number of metabolites because of their role in sample discrimination in metabolite profiling studies.

The most striking newly emerged field in biological sciences is *systems biology*, generally defined as the study of interactions between the components of different biological systems. Among the different "omics" technologies, which are the building blocks of systems biology, metabolomics is the closest to organisms' phenotype and can lead to better understanding of different biochemical mechanisms in complex systems. After the establishment of technologies for high-throughput DNA sequencing (genomics), gene expression analysis (transcriptomics), and protein

---

*Andrea Lubbe and Kashif Ali are equally contributed.

Metabolomics in Practice: Successful Strategies to Generate and Analyze Metabolic Data, First Edition.
Edited by Michael Lämmerhofer and Wolfram Weckwerth.
© 2013 Wiley-VCH Verlag GmbH & Co. KGaA. Published 2013 by Wiley-VCH Verlag GmbH & Co. KGaA.

analysis (proteomics), the remaining functional genomics challenge is that of *metabolites analysis*.

9.2
Platforms for Metabolomics

To gain a comprehensive and complete overview of the entire metabolic complement of a sample in a single analysis is, and should be, the ultimate goal in metabolomics. Chemical analysis techniques applied to metabolomics should be unbiased, rapid, reproducible, and stable over time, while requiring only simple sample preparation [5]. Different platforms are available for metabolome analysis including high-performance liquid chromatography (HPLC), gas chromatography–mass spectrometry (GC–MS), liquid chromatography–mass spectrometry (LC–MS), capillary electrophoresis–mass spectrometry (CE–MS), fourier transform-ion cyclotron-mass spectrometry (FT-ICR-MS), and nuclear magnetic resonance (NMR). Among these, GC–MS or LC–MS and NMR are most widely used. There is always a payoff between different technologies in terms of sensitivity, high throughput, robustness, quantitation analysis, and suitability for specific chemical classes of metabolites. Nevertheless, a carefully chosen analytical method can be an excellent initial strategy for gaining a first impression of a metabolic profile that can ultimately be used to identify key biochemical leads to further or more focused studies [6, 7]. Many platforms are being used for the high-throughput analysis of metabolites, but only the major ones are briefly explained below with their advantages and limitations followed by more detailed account on NMR.

9.2.1
Mass Spectrometry (MS)

MS is considered as one of the primary detection methods of choice for metabolomics because of its high sensitivity, speed, and broad application. Many papers have shown its suitability for metabolite detection in complex matrices [8, 9]. GC or LC is commonly used for metabolite separation before MS detection. These combinations of different separation techniques with MS along with their applications have been extensively discussed by Dettmer *et al.* [10]. The following sections present brief accounts on each of these techniques and their pros and cons.

9.2.1.1 Gas Chromatography–Mass Spectrometry (GC-MS)
This technique is the most popular and widely applied method in metabolomics. This popularity is mainly due to robustness of both separation and detection along with the availability of some excellent metabolite identification tools based on databases and fragmentation pattern. This technique combines high sensitivity and resolution with a reproducible fragmentation pattern of the separated molecules.

Application of two-dimensional GC–MS has resulted in the improvement of resolution [11, 12]. GC–MS is the principal technique for separation and detection of metabolites that are naturally volatile at temperatures up to 250 °C (e.g., fatty acids, aliphatic alcohols, and esters essential oils) at the cost of thermolabile compounds. The technology can also be applied to groups of nonvolatile, polar (mainly primary) metabolites, such as amino acids, sugars, and organic acids, by converting these into volatile and thermostable compounds through chemical derivatization. These derivatized samples can then be analyzed by GC–MS and detailed information on many of the key primary metabolites (e.g., in plants) (comment: here is no specific focus on plants!) can be obtained in a single chromatographic process [13, 14]. However, the extent of derivatization or incomplete derivatization can cause the problem of more than one peak for the same compound [15]. Comparison between chromatograms of identical peaks is possible, but absolute quantitation further requires calibration curves. Another limitation of this technique is that complex plant secondary metabolites, such as phenolic glycosides, cannot be analyzed by GC.

9.2.1.2 Liquid Chromatography–Mass Spectrometry (LC–MS)

It is a very important and versatile technology in metabolomics, capable of facilitating the analysis of several large groups of secondary metabolites of plant tissues without any chemical derivatization of the metabolites. Advances in chromatographic technologies (such as ultraperformance LC) together with advances in column chemistry (such as hydrophilic interaction chromatography and long monolithic columns) resulted in a significantly improved separation potential. The technology is inherently restricted to molecules, which can be ionized, either as positively or as negatively charged ions, before moving through the MS. The wide range of analytes in terms of molecular weight and polarity along with precise molecular weight determination are certainly the strong points of LC–MS [2, 16]. An authoritative review on LC–MS in plant metabolomics has been recently published [17]. Unlike GC–MS, very few mass spectral libraries are available for LC–MS and this is a key topic being given considerable attention today.

9.2.1.3 Capillary Electrophoresis–Mass Spectrometry (CE–MS)

The recent development of CE as an alternative separation technology, particularly for charged metabolites, is growing in popularity particularly when combined with MS for extra selectivity and sensitivity [18]. High-resolution chromatographic separation and sensitive detection of water-soluble extracts make a strong combination suitable for the analysis of a diverse range of polar and water-soluble primary and secondary metabolites [19]. Derived from CE, capillary electrochromatography (CEC) is another promising separation technique. It uses LC or has monolithic stationary phases; hence, a hybrid of liquid chromatography and CE. The combination of CEC with MS, the interfaces used, and different bioanalytical applications such as analysis of proteins, peptides, amino acids, and mixture of saccharides, has been reviewed by Klampfl [20].

9.2.1.4 Fourier Transform-Ion Cyclotron Resonance-Mass Spectrometry (FT-ICR-MS)

A relatively less applied technique known as *Fourier transform-ion cyclotron-mass spectrometry*, often called as *FT–MS*, is capable of nontargeted metabolic analysis and is suitable for rapid screening of similarities and dissimilarities in large collections of biological samples, for example,, plant mutant populations [21]. After a pause following the first paper on this topic by Aharoni *et al.* [22], which explains the metabolic changes during strawberry ripening and to differentiate transgenic and nontransgenic plants, more recent applications are emerging for the phenotyping studies [6, 7, 23–25]. This technology requires specialized skills and equipment, not easily accessible to most researchers. Also, the high per sample cost and the inability to separate structural isomers, which have identical mono-isomeric masses, is still seen as a significant limitation to its application.

9.2.2
Fourier Transform–Infrared Spectroscopy (FT–IR)

This technique is considered as the basic among the spectroscopic techniques with the features such as fast, high-throughput, nondestructive, and nonselective [26]. As the technique is not as expensive as the other spectrometric and spectroscopic techniques, it is a method of choice for an initial screening [27]. Analysis of food products such as meat [28, 29] or milk [30] seems to be favored by this technique. The major disadvantages related to this method are its poor reproducibility and highly intense detection of water molecules [31].

9.2.3
Nuclear Magnetic Resonance Spectroscopy: Principles and Techniques

In NMR, any molecule with one or more atoms with nonzero magnetic moments can be detected, such as ^{1}H, ^{13}C, ^{14}N, and ^{31}P. Of these ^{1}H NMR is the most commonly used in metabolomics studies because hydrogen is part of most organic metabolites and because of the high abundance (99.98%) of the ^{1}H isotope [32]. A basic solution ^{1}H NMR experiment consists of placing a liquid sample in a glass tube between two poles of a powerful magnet. This brings the sample into equilibrium, that is, all magnetic moments become aligned in direction of the magnetic field. A radiofrequency signal is transmitted into the sample, which perturbs the magnetization vector of the sample. After the pulse the vector eventually returns to equilibrium, and this change in magnetization over time (called free induction decay, FID) is monitored and from this a signal is evolved. Fourier transformation is used to transform the signal from the time domain to the frequency domain, resulting in a spectrum of intensity versus frequency. Each signal corresponds to a specific proton in a molecule, the frequency or chemical shift (in unit parts per million) of which depends on the kinds of protons found in a molecule and how they are arranged relative to each other.

9.2.3.1 One-Dimensional Nuclear Magnetic Resonance (^{1}H and ^{13}C NMR)

A pulse program for a basic one-dimensional ^{1}H NMR experiment used in metabolomics studies is shown in Table 9.1. The integrated areas under ^{1}H NMR signals are directly proportional to molar concentrations of the corresponding protons. For very accurate quantitation by ^{1}H NMR, certain pulse program parameters may need to be further optimized [33]. In a study on metabolites in *Narcissus* bulbs, for example, a longer relaxation delay allowed more accurate quantification of the major alkaloid galantamine [34]. All ^{1}H NMR signals can be quantitated with one internal standard. Commonly used internal standards in aqueous solvents are trimethylsilylpropionic acid (TMSP) and 4,4-dimethyl-4-silapentane-1-sulfonic acid (DSS), but others may be more suitable in different NMR solvents [33].

In contrast to ^{1}H NMR, signal intensity in ^{13}C NMR is not always directly proportional to the number of carbon nuclei. This is mainly due to the long and variable relaxation times of different carbon nuclei in different molecules. This, together with the low sensitivity of ^{13}C NMR, makes standard ^{13}C experiments less suitable for accurate quantitation. These problems can be overcome by using inverse-gated ^{1}H decoupling [35]. This requires long interpulse delays resulting in often prohibitively long acquisition times. Methods have, however, been developed to make such measurements faster [36].

In many types of samples, the presence of residual water causes a large, broad signal in the ^{1}H NMR spectra, which can interfere with other metabolite signals [37]. Water suppression is commonly included in pulse sequences to overcome this problem. Presaturation is often used, where selective preirradiation of water resonances equalizes the spin populations of water protons before the data is acquired. This prevents water from contributing to the NMR spectrum. Another commonly used method in metabolomics studies combines presaturation with a one-dimensional version of a two-dimensional ^{1}H-^{1}H-NOESY sequence. This and other water suppression methods are discussed thoroughly by McKay [38] and Price [39]. Some of these methods can also be used to suppress large signals belonging to components other than water. In analysis of alcoholic beverages, for example, the ethanol signal may be very large. The WET technique (water suppression enhanced through T_1 effects) is one way to suppress this dominating signal [40]. Košir and Kidrič [41] incorporated a WET pulse sequence element into 1D and 2D NMR experiments to analyze amino acids in wine, allowing their full signal assignment.

A further potential source of interference in ^{1}H NMR spectra is peak broadening caused by macromolecules (e.g., proteins, polysaccharides). In samples where this may occur, such as biofluids, a modified pulse sequence such as the Carr-Purcell-Meiboom-Gil spin echo modification can be used. This makes acquisition more selective for low-molecular-weight compounds [42].

In a complex mixture, a ^{1}H NMR spectrum is usually a crowded plot with much signal overlap. Such a plot can be used in multivariate data analysis to compare the overall signal patterns between the samples. However, metabolite identification and quantification is hampered by overlapping signals, so additional NMR experiments, multidimensional NMR is usually needed. ^{13}C NMR produces 1D spectrum with a wider chemical shift spread and less overlap between signals. Low sensitivity and

Table 9.1 Examples of pulse programs and processing parameters for selected 1D and 2D NMR experiments.

Experiment	Acquisition parameters	Processing parameters
1D ^{1}H NMR with water presaturation	Pulse sequence comprising (relaxation delay-60°-acquire) where pulse power is set to achieve 60° flip angle, 10 kHz spectral width, and water saturation is applied during 1.5 s relaxation delay	Zero-fill to 64 k data points, apply exponential line broadening of 0.3 Hz, apply Fourier transformation, manually phase spectrum (zero- and first- order corrections), manually correct baseline, calibrate chemical shift to internal standard
J-resolved NMR	J-resolve pulse sequence, two-pulse echo sequence (relaxation delay-90°-[t1/2]-180°-[t1/2]-acquire), water presaturation during 1.5 s relaxation delay. Acquire FID using data matrix 64 × 4096 points covering 66 × 6361 Hz, with 16 scans for each increment	Zero-fill to 128 × 4096 and apply sine bell-shaped window function in both dimensions before magnitude mode 2D Fourier transformation. Tilt resulting spectra along rows by 45° relative to frequency axis and symmetrize about the central line along F2. Manually correct baseline, calibrate chemical shift to internal standard
COSY (^{1}H-^{1}H)	Use phase-sensitive/magnitude mode standard three-pulse sequence with presaturation during relaxation delay of 1 s. Acquire FID using data matrix 512 × 4096 points covering 6361 × 6361 Hz, record with eight scans for each increment	Zero-fill to 4096 × 4096 and apply sine bell-shaped window function shifted by /2 in the F1 and /4 in the F2 dimension before Stated-TPPI type 2D Fourier transformation. Manually phase all spectra, correct baseline and calibrate chemical shift to internal standard
HSQC (^{1}H-^{13}C)	Use a data matrix of 254 × 4096 points covering 27 164 × 6361 Hz with 256 scans for each increment with relaxation delay of 1 s	Qsine (SSB = 2.0) used for window function. Coupling constants optimized to 145 Hz
HMBC (^{1}H-^{13}C)	Use a data matrix of 254 × 4096 points covering 27 164 × 6361 Hz with 256 scans for each increment with relaxation delay of 1 s	Data should be linear predicted to 512 × 4096 points using 32 coefficients before magnitude type 2D Fourier transformation and apply a sine bell-shaped window function shifted by /2 in the F1 and /6 in the F2 dimension. Calibrate chemical shift to internal standard (^{1}H and ^{13}C chemical shift)

long acquisition times may prohibit the use of ^{13}C for large metabolomics studies. For certain sample types, however, ^{13}C NMR offers advantages over ^{1}H NMR. For the analysis of vegetable oils (e.g., olive oil), for example, ^{13}C NMR provides structure-specific information that cannot be obtained with ^{1}H NMR or even the more traditional GC methods [43, 44].

9.2.3.2 J-Resolved Spectroscopy (JRES)

Two-dimensional ^{1}H J-resolved (JRES) NMR spectroscopy is a homonuclear method that can be used to generate less congested spectra. Chemical shifts and spin–spin couplings are visualized along two different axes, which make it easier to tell signals apart in crowded regions of the spectra [45]. An example of a portion of a J-resolved spectrum is shown in Figure 9.1a. The J-resolved spectrum can also be projected on the chemical shift axis to generate a proton-decoupled projected one-dimensional spectrum (p-JRES) [46]. In such spectra, all ^{1}H signals appear as processed singlets, resulting in much simpler and better resolved signals more suitable for quantitation [47]. p-JRES spectra can also be used in multivariate data analysis, where simplified spectra may improve interpretability of analyses [46]. One potential problem to look out for using p-JRES is the presence of "strong coupling artifacts." They occur when strong coupling leads to additional peaks in the J-RES spectrum. When the signals are projected onto the chemical shift axis, the projection does not represent a fully decoupled 1D spectrum, since additional peaks appear. Some methods for suppression of such artifacts were proposed by Thrippleton *et al.* [48]. Even though the acquisition time per sample is about double that of standard ^{1}H NMR measurements, the advantages of using p-JRES spectra may outweigh these drawbacks.

9.2.3.3 Correlation Spectroscopy (COSY)

Correlation spectroscopy (COSY) is another homonuclear two-dimensional method that shows correlations between protons with mutual spin–spin couplings. The one-dimensional NMR spectrum runs diagonally across the plot, with spin–spin couplings (J-couplings) indicated as cross-peaks in the off-diagonal space [49]. COSY can indicate couplings between multiplets three bonds away, but long range coupling between protons four bonds away may sometimes be seen. In plant samples, this method is particularly useful for metabolite identification in the aromatic region of the spectrum (around 6.0–8.0 ppm). This is where many secondary metabolite signals occur, often with a large degree of overlap and at low signal intensity. For example, trans-phenylpropanoids in plants typically have doublet signals at 6.3–6.5 ppm ($J = 16$) with COSY correlations to doublets at 7.3–7.85 ppm ($J = 16$) [50]. With the help of COSY and J-resolved spectra, Ali *et al.* [51] assigned various phenylpropanoid and flavonoid signals in grapes, including quercetin-3-*O*-glucoside and a trans-feruloyl derivative associated with resistance to downy mildew. An example of a COSY spectrum is shown in Figure 9.1b.

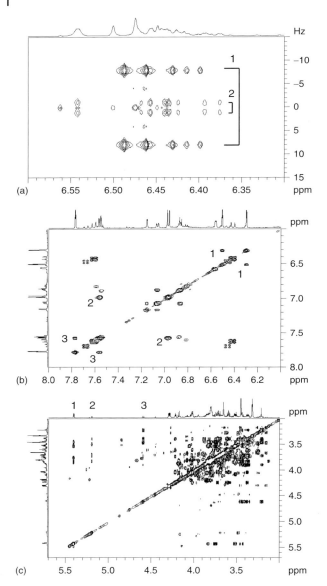

Figure 9.1 Examples of two-dimensional homonuclear ^{1}H–^{1}H experiments of various plant samples recorded in CH$_3$OH-d_4-D$_2$O (KH$_2$PO$_4$ buffer, pH 6.0). (a) ^{1}H-^{1}H-J-resolved spectrum of *Brassica rapa* leaves in the range of δ 6.3–6.6. (1) H-8 of phenylpropanoids and (2) H-6 or H-8 of 5,7-dihydroxyflavonoids. (b) COSY spectrum of *Vitis vinifera* leaves in the range of δ 6.6–8.0. (1) Correlation between H-6 and H-8 of quercetin analogs, (2) correlation between H-5′ and H-6′ of quercetin analogs, and (3) correlation between H-1′ and H-6′ of quercetin analogs. (c) TOCSY spectra of *Cannabis sativa* flowers in the range of δ 3.0–5.7. (1) sucrose, (2) α-glucose, and (3) β-glucose.

9.2.3.4 Total Correlation Spectroscopy (TOCSY)

While COSY shows correlations between geminal and vicinal protons of a molecule, two-dimensional total correlation spectroscopy (TOCSY) creates correlations between all protons in a spin system, as long as there are couplings between every intervening proton. TOCSY allows one to see which signals belong to the same molecule and is particularly useful for assigning carbohydrate and amino acid signals in the usually crowded region where they occur (Figure 9.1c). Another variant, selective TOCSY is a 1D method where a selected signal is excited and the excitation is progressively transferred along all coupled protons in the spin system. This can be done in a complex mixture to give rise to a 1D spectrum showing only the peak of interest and signals of the same spin system. Overlapping signals can be resolved to help confirm metabolite identities and to aid in quantitation of minor compounds [52].

9.2.3.5 Heteronuclear Two-Dimensional Methods

Heteronuclear two-dimensional NMR experiments such as heteronuclear single quantum coherence (HSQC), heteronuclear multiple quantum coherence (HMQC), and heteronuclear multiple bond coherence (HMBC) are used to show correlations between ^{13}C and ^{1}H. HSQC and HMQC both generate plots of the ^{13}C spectrum versus the ^{1}H spectrum, with direct correlations (J_1) indicated as cross-peaks in the plot (Figure 9.2a,b). Both experiments provide the same information, but in HMQC, broadening of resonances by homonuclear coupling can occur. For this reason, better resolution can be obtained in the carbon dimension using HSQC. Some examples where this method is useful for signal assignment include anomeric carbons of carbohydrates (δ90–110), C-6 and C-8 of flavonoids (δ95–110), and methyl groups of terpenoids (δ10–25) [53]. Similar to p-JRES, an F_1 projected HSQC spectrum can be produced to generate new one-dimensional variables containing additional information from ^{13}C for use in multivariate data analysis [54]. HMBC shows long-range correlations (J_2, J_3) between ^{13}C and ^{1}H. This is very useful in assigning quaternary carbon signals, as was illustrated for progroitin and other secondary metabolites in *Brassica rapa* leaves [55]. In addition, it is also a valuable tool to confirm structures of molecules.

9.2.3.6 Combined Two-Dimensional Methods

Two-dimensional NMR experiments can be combined to help resolve overlapping signals. An example is the 2D HSQC-TOCSY experiment, where a TOCSY mixing sequence is added after an initial HSQC pulse sequence. This extends the original proton–carbon correlation peak onto neighboring protons within the same spin system to produce a ^{13}C-dispersed TOCSY spectrum (Figure 9.2c). Cross peaks in such a spectrum will indicate correlations between all J-coupled protons and all carbons in that spin system. Similar information can be obtained by combining the HMQC and COSY experiments [56]. A good example of the application of combined 2D experiments for metabolite identification in a complex mixture was described by Leiss *et al.* [57]. HSQC-TOCSY, together with other 2D NMR experiments, was used

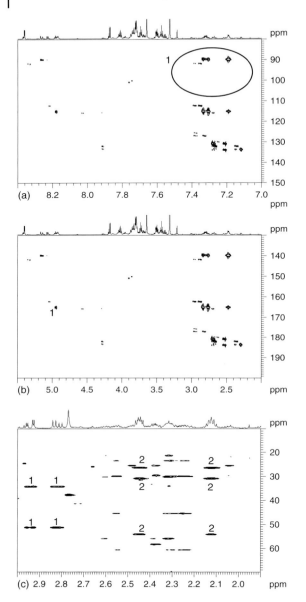

Figure 9.2 Examples of two-dimensional heteronuclear ^{13}C-^{1}H experiments of various plant samples recorded in CH$_3$OH-d_4-D$_2$O (KH$_2$PO$_4$ buffer, pH 6.0). (a) HMQC spectrum of *Genista tenera* leaves in the range of δ 1.9–3.0 (^{1}H) and δ 80–160 (^{13}C). (1) C-6 or C-8 of 5,7-dihydroxyflavonoids. (b) HMBC spectrum of *Brassica rapa* leaves in the range of δ 2.0–5.5 (^{1}H) and δ 130–200 (^{13}C). (1) Correlation between anomeric proton and carbon of C=N in glucosinoates. (c) HSQC-TOCSY spectrum of *Nicotiana plumbaginifolia* cell lines in the range of δ 6.0–8.3 (^{1}H) and δ 10–70 (^{13}C). (1) Carbons of aspartic acid and (2) carbons of glutamic acid.

to find candidate metabolites important for the difference between thrips-resistant and susceptible *Senecio* hybrid plants.

9.3 NMR for Metabolomics

NMR spectroscopy has been widely applied in plant metabolomics and is the first choice for medical metabonomics. NMR analysis is a favored choice for the major metabolites [58–60]. However, NMR application in metabolomics studies is often criticized because of its poor sensitivity [61]. Sensitivity and resolution increase with increasing magnetic field strength [62]. The highest magnetic field is 1 GHz to date but most metabolomics work is currently carried out on 500 or 600 MHz NMR instruments. In the past, a metabonomics study compared spectra of the same urine samples acquired with 250, 400, 500, and 800 MHz NMR instruments to assess the effect of magnetic field strength on the information contained in such data sets [63]. In the multivariate data analysis, the same biomarkers were identified in all the NMR spectra measured in different magnetic fields. The predictive performance of the model increased with the higher field strength measurements, however. The sensitivity of NMR is also determined by the time of accumulation of the spectra. With a standard 500 MHz NMR, a spectrum of an extract of 50 mg dry weight plant material can be obtained in about 10 min (128 scans). Higher field strength, cryoprobes, and microprobes have greatly contributed to the shorter time needed to record a spectrum in NMR spectroscopy. Even in terms of sensitivity NMR with cryo- or microprobe greatly improves the sensitivity making it comparable with most chromatographic methods. Also, the amount needed in most type of experiments is not limiting. So sensitivity is more related to the dynamic range and the overlapping of signals.

NMR has some unique advantages over chromatography- and MS-based methods. First, it is a more uniform detection system and can directly be used to identify and quantify metabolites, even *in vivo*. The most promising features of NMR are its nondestructive nature, simple sample preparation in relative short time or even direct measurement of samples, for example, urine. Another major advantage of NMR is that quantification is easy for all compounds as with a single internal standard all the detected metabolites can be quantified without the need of calibration curves for each single compound as signal intensity is only dependent on molar concentration of the compound. Since nearly no sample pretreatment is required in NMR spectroscopy, the inherent properties of the sample are well kept. The nonselectiveness of NMR makes it an ideal tool for the profiling of a broad range of metabolites [64]. NMR has been already demonstrated to be a robust method and unaffected by instrumental and experimental factors as is the case in other analytical methods. Continuous improvements in instrumentation design may lead to increasing popularity of this approach, and a full overview of the current potentials and limitations of NMR has been provided by Ratcliffe and Shachar-Hill [65] and recently by Schripsema [66].

9.3.1
Sample Preparation

Most NMR-based metabolomics studies are performed with solution NMR. For biofluids (e.g., urine, plasma), minimal processing is needed as samples are already in liquid form [67]. In plant studies, the low-molecular-weight compounds making up the metabolome need to be transferred from the plant matrix to a deuterated solvent for analysis by NMR. This sample preparation usually involves the following steps: harvesting, drying, extraction, and preparation for analysis [68]. Sample preparation is a very critical part of a metabolomics experiment, as it will determine the quality of the results obtained. For each step in the process, a number of practical considerations should be taken into account.

If plant material is to be harvested on different days, it should ideally be done at the same time of day. It is well known that levels of certain plant metabolites fluctuate throughout the day [69]. The plant part to be harvested should be chosen carefully. If possible organs should be analyzed separately, as large differences in metabolic profiles may be seen between them. Also, organs of the same type but of different ages may differ considerably in metabolite levels. As plant cells are often very specialized, one should bear in mind that the metabolic profile of, for example, a leaf is in fact a mixture of many different cell metabolomes.

While removing the fresh plant material from the original plant, the aim is to keep all metabolites in their original state. Handling and wounding can break cell compartments, and unwanted enzymatic and chemical reactions can lead to undesired changes or degradation of metabolites. The best way to avoid this is the rapid freezing of fresh plant material in liquid nitrogen. This stops any enzymatic activity and allows further processing or storage (at $-80\,^\circ$C) without any metabolic changes occurring as long as it is frozen. At this point, plant material may be ground in liquid nitrogen to homogenize tissue for ease of handling and improved extraction.

In some well-established methods, extraction with perchloric acid ($HClO_4$) proceeds from this point, leading to denaturation and precipitation of unwanted proteins and extraction of polar metabolites [70]. An alternative is to first dry the plant material, which helps stabilization by removing the matrix in which enzymatic reactions take place. Removing the water also reduces the interfering water signal in NMR spectra, as well as variable chemical shifts caused by differences in pH. In addition, drying the samples allows for more accurate quantitation of metabolites. Various methods can be used for drying plant material, but freeze-drying is the most commonly used. Except for some volatile compounds, most compounds are well preserved during freeze-drying, which is mild in comparison to other methods [68]. Typically, 50–100 mg of dry plant material is extracted per sample but less than that can also be used. For smaller amounts, it is very important that the material is homogenously ground. The aim of solvent extraction in sample preparation for metabolomics is to obtain an accurate snap shot of the metabolome. This is not a simple task as many factors such as solvent, time, temperature, pH, energy, solubility, and dissolution rate influence the extraction process. The huge diversity in structure, concentration range, and polarities of the low-molecular-weight metabolites in any given tissue also means that no single solvent can extract all low-molecular-weight metabolites. Many extraction methods have been employed in the past, and each one has advantages and disadvantages. The method of Kruger *et al.* [70] is very good for polar metabolites but excludes hydrophobic compounds and may also cause degradation of acid-labile metabolites. Organic solvents in mixtures with other organic solvents or water are often used to increase the range of metabolites extracted. A two-phase solvent system comprised of chloroform, methanol, and water (2 : 1 : 1, v/v) is one such combination. This method produces two extracts per sample, one with more polar and the other with nonpolar compounds. More information can be obtained from the analysis of the two phases, and it has been shown to be useful for plant materials where hydrophilic secondary metabolites are causing discrimination between samples [59, 71]. A drawback of this method is that it requires time-consuming separation and evaporation steps, as well as the need to reconstitute the extracts in NMR solvents. Extraction can be simplified by directly performing it in deuterated NMR solvents. This narrows the choice of solvents somewhat since not all solvents are available in deuterated form. The combination of deuterated water and methanol (1 : 1, v/v) was found to be a good general purpose solvent for the extraction of a range of primary and secondary metabolites. Replacing the water with a phosphate buffer (pH 6.0) avoids signal shift because of pH variation. This extraction solvent (CH_3OH-d_4-KH_2PO_4 buffer in D_2O, pH 6.0) has now become a well-established method in our group for metabolomics studies on a range of plants (Figure 9.3) [72].

Apart from the solubility of compounds in a chosen extraction solvent, the efficiency of an extraction also depends on the dissolution rate. This can be increased by increasing the extraction time or by adding energy to the system (through heat or using methods such as ultrasonication or microwave extraction). Care should be taken however, as using such methods to improve the extraction yields also increase the risk of artifact formation [73].

9.3.2
Metabolite Identification

Even with all the 1- and 2D NMR experiments available, structure elucidation of metabolites can be very challenging and time consuming, especially in a complex mixture. The rapid acquisition of many ^1H NMR spectra for a metabolomics study is often offset by the long time needed to assign metabolite signals in complex mixtures. Comparing signals to spectral data of reference compounds measured in the same conditions is the most straightforward way of compound identification. Availability of reference compounds may be a problem, especially with secondary plant metabolites. Even with reference spectra, manually comparing signals is a time-consuming task.

Statistical methods can be used to aid metabolite identification. One such method is statistical total correlation spectroscopy (STOCSY), which identifies

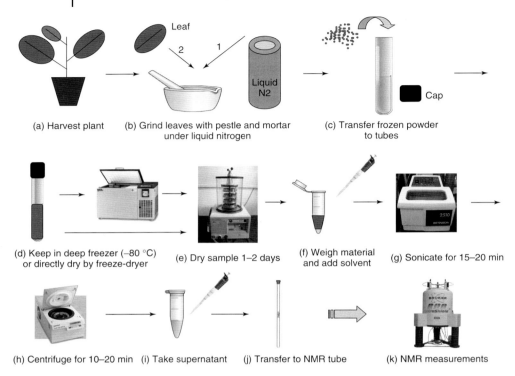

Figure 9.3 (a–k) Schematic representation of the experimental procedures for sample preparation. (Source: Adapted from Kim et al. [72].)

peaks belonging to the same molecule in a mixture by them being highly correlated [74]. Many spectra are analyzed simultaneously, and in addition to information about intramolecular connectivity, negative or low correlations between signals can indicate intermolecular connections via a metabolic pathway or common regulatory mechanisms [75]. This is a useful tool in metabolic studies; however, care should be taken in interpretation as results may differ depending on how many spectra are being compared.

Various computational methods have been developed to automatically or semi-automatically identify and quantify metabolites in NMR spectra. The ChenomX NMR Suite is an example of a software package that semiautomatically identifies compounds from ^{1}H NMR spectra [76]. A spectral library of about 260 compounds is used for the peak fitting process. While this method speeds up the identification process, the manual nature of the fitting and analysis may lead to inconsistent interpretations when used by different individuals. A more automated method developed by Zheng et al. [77] uses linear mixed modeling with Bayesian model selection on local regions of ^{1}H NMR spectra for metabolite identification and quantitation. This method simultaneously models the entire collection of spectra, and unlike other models perform identification and quantitation at the same time to give improved results compared to similar methods. MetaboMiner was developed to identify metabolites in biofluids based on 2D NMR spectra [78]. Prior knowledge on biofluid type is used to achieve this semiautomatically with very high accuracy. Unlike some of the aforementioned methods, this program does not provide quantitative analysis of the identified metabolites. A method developed by Xi *et al.* [49] also used a 2D NMR (COSY)-based method to identify metabolites in complex mixtures through comparison with a library of precollected spectra. The method gave good accurate results but may have limited practicality because the spectral library contained only 19 recorded spectra.

These studies show that it is possible to automatically assign a signal or provide a list of potential assignments. Ideally, these methods should be linked to large spectral databases with many entries. Many of the aforementioned methods were linked to small databases containing only a few tens or hundreds of metabolites, which limits its usefulness. Several large online databases exist with NMR spectra of small molecules (Table 9.2). In some cases (e.g., HMDB, MMCD, and BMRB), they can be queried directly with peak lists from a compound mixture.

Standardization of acquisition conditions is essential for such databases to be useful, particularly for ^{1}H NMR spectra, where solvent effects and pH can cause variation in chemical shift. ^{13}C-NMR chemical shifts are less sensitive to small differences in the surrounding environment, and it has therefore been suggested that 2D ^{13}C-^{1}H spectra (e.g., HSQC) should always be included in spectral databases [79].

9.3.3
Data Analysis: Turning Data into Information, Possibly Knowledge

Although different in many ways, all the platforms for metabolomics studies produce huge amounts of data and suffer from some machine inaccuracies. In order to generate information and eventually knowledge, this raw data need considerable processing and then suitable statistical analysis. The success of any metabolomics-based study relies on how efficiently the data is processed and analyzed. As metabolomics data can rarely be handled manually, the processing and analysis require specialized bioinformatics tools for the *in silico* multivariate analysis of this data set. Multivariate statistics is a large discipline and this chapter briefly explains a few of the most common multivariate data analysis methods, but before that a brief account on the preprocessing of NMR spectra is presented.

9.3.3.1 Data Preprocessing
The preparing of data for further analysis is often referred to as *data preprocessing (or pretreatment)*. In NMR-based metabolomics studies, the spectroscopic data needs to be processed in order to proceed with the multivariate data analysis. Several methods to process the NMR data, such as peak alignment, normalization, scaling, and bucketing (binning) are available and explained briefly in this chapter.

Resonances in the NMR spectra of different samples (replicates of same treatment) can be shifted and cannot be uniquely defined for each metabolite. This

Table 9.2 Online databases for NMR spectroscopy.

Name	URL	Developed/maintained by	Spectra	Access
Biological Magnetic Resonance Data Bank (BMRB)	www.bmrb.wisc.edu	University of Wisconsin-Madison	NMR spectra of proteins, peptides, and nucleic acids	Free
NMRshiftDB	http://nmrshiftdb.org	University of Mainz, Max Planck Institute Chemical Ecology in Jena, European Bioinformatics Institute in Cambridge	More than 48 000 measured NMR spectra (^{13}C, ^{1}H, ^{15}N, ^{11}B, ^{19}F, ^{29}S, ^{31}P)	Free
Madison Metabolomics Consortium Database (MMCD)	http://mmcd.nmrfam.wisc.edu/	University of Wisconsin-Madison	More than 20 000 small molecules, ^{1}H and ^{13}C spectra, 1D and 2D	Free
Human Metabolome Database (HMDB)	www.hmdb.ca	University of Alberta	More than 7900 human metabolite NMR spectra (^{13}C and ^{1}H)	Free
Spectral Database for Organic Compounds (SDBS)	http://riodb01.ibase.aist.go.jp/sdbs/cgi-bin/cre_index.cgi	Japanese National Institute of Advanced Industrial Science and Technology	More than 27 000 NMR spectra (^{13}C and ^{1}H)	Free
PRIMe	http://prime.psc.riken.jp	RIKEN Plant Science Center, Yokohama	NMR spectra of standard compounds, also other tools for signal assignment and integration of -omics data	Free
NMR metabolomics database of Linkoping	http://www.liu.se/hu/mdl/main/	University of Linkoping, Sweden	NMR spectra of mostly primary and some secondary metabolites	Free
Advanced Chemistry Development	www.acdlabs.com	ACDLabs	NMR spectral matching with databases and predictor tools (^{13}C, ^{1}H, ^{15}N, ^{31}P)	Purchase or lease
SpecInfo on the Internet	http://onlinelibrary.wiley.com/book/10.1002/9780471692294	John Wiley & Sons	NMR spectral databases (^{13}C, ^{1}H, ^{19}F, ^{31}P, ^{29}Si)	Purchase or lease
The Aldrich FT-NMR Library	http://www.sigmaaldrich.com/analytical-chromatography/spectroscopy/learning-center/nmr-spectroscopy/spectral-viewer.html	Sigma-Aldrich	^{13}C and ^{1}H spectra of more than 11 800 metabolites	Purchase or lease

usually happens because of instrument variations or possible interferents in the analytes. These shifts need to be corrected in the process called *peak alignment*. Another type of variation, which is obvious in metabolite analysis, is the variation in the concentrations of the metabolites. This difference in concentration hinders the quantification (absolute or relative) of metabolites and is addressed by a process of normalization. For normalizing the signals, a stable standard is added to the sample with known concentration and the rest of the spectrum normalizes with respect to that internal standard. The signal of NMR solvent can also be used for normalization. This data preprocessing is essential for the correct quantification of the identified metabolites.

If large metabolic differences existed among the analyzed samples, scaling is inevitable in order to reduce the influence of highly inconsistent signals (or variables). There are many ways to scale the data before putting it into the analysis. The most basic is mean centering, which is done with each variable across all the samples. In mean centering, the mean value for each variable is calculated and then subtracted from the data. Another widely applied scaling method is unit variance (UV) in which each variable is divided by its standard deviation to give each variable an equal variance. This method of scaling is recommended if no prior information of the data is available. Sometimes "no scaling" is appropriate, especially if the data is in the same unit, for example, spectroscopic data. Another development in scaling methods is the Pareto scaling in which unlike UV scaling, the scaling factor is square root of standard deviation, which means each variable is divided by the root of its standard deviation. In this way, Pareto scaling scales each variable according to its initial standard deviation and is intermediate between no scaling and UV scaling.

Bucketing or binning is another type of data processing, which divides the NMR spectrum into a number of buckets of desired width and calculates the peak areas within that segment of the spectrum. Bucketing can help to reduce the effects of pH-induced variations in the chemical shifts by ensuring the measurement of same resonance across the samples. As bucketing calculates the peak areas, it is also helpful to quantify the metabolites, as the peak intensities in NMR spectrum are proportional to the molar concentrations of the metabolites. Considering the resolution, pH-dependent chemical shift variations, and splitting of NMR signals, a bucket of 0.04 pm widths is the optimum but not the only one that should be used.

9.3.3.2 Principal Component Analysis (PCA)

Principal component analysis (PCA) is the most common multivariate data analysis method, which is used to reduce the dimensionality of a multivariate dataset by data decomposition. This method is designed to extract the X-data matrix and maximum variation. As this modeling is done solely on the explanatory variables and without any prior information of samples, PCA is an unsupervised and hence unbiased method. This method generates score and loading vectors and can be represented in a graphical form known as a *score plot* and *loadings plot*, respectively. The score plot can be used to identify the differences or similarities among the

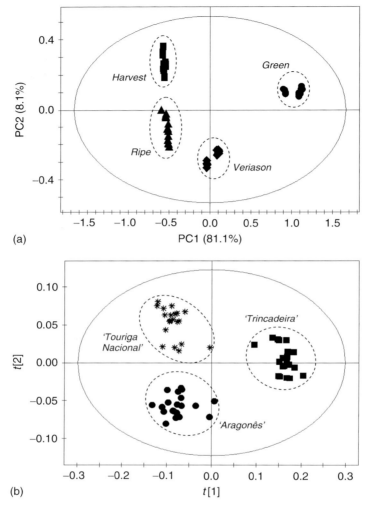

Figure 9.4 Score plots of different multivariate data analysis. (a) Score plot of PCA shows the clustering of samples according to the developmental stages of grape berries. (b) Score plot of PLS-DA shows the classification of grape berries based on the cultivars. (Source: Adapted from Ali et al. [78].)

samples with easy identification of an outlier (Figure 9.4a). The loadings plot can be used to identify the spectral signals responsible for the grouping or separation among the samples, which ultimately resulted in the identification of metabolites responsible for the separation on the score plot.

9.3.3.3 Partial Least Squares (PLS) Projections to Latent Structures
As a regression extension of PCA, partial least squares (PLS)-based modeling is the basis of supervised multivariate data analysis algorithms. When sample-specific

information is available as Y-data matrix (bioactivity data, for instance), this method often does more efficient data decomposition than PCA and can connect the X-matrix (descriptor) and the Y-matrix (response) to each other. Similar to PCA, score and loadings plots are used to visualize the respective score and loadings vectors. Another PLS-based technique, known as partial least squares-discriminant analysis (PLS-DA), is capable of separating "tight" (with least internal variation) classes of observations based on their X-data matrix (Figure 9.4b). In PLS-DA, the Y-matrix represents class membership by a set of "dummy" variables. This method separates the observations according to the class membership as a discriminant plane is found by fitting a PLS model between and X- and artificial Y-matrices. The low number of classes (not more than 5 or 6) and high degree of tightness are the basic requirement for this model to work efficiently.

9.3.3.4 Bidirectional Orthogonal-PLS (O2PLS)

Methods such as PLS regression can cause systematic variation because of structured noise present in the data matrices. Another algorithm, such as O2PLS, is an extension of PLS regression with an integrated OSC (orthogonal signal correction) filter. Briefly, OSC filter, initially developed to preprocess spectral data by Wold *et al.* [80], can identify the systematic variation in the X-matrix by employing information regarding the Y-matrix. Depending on the study, this information can be removed or retained. Similarly, analyses such as O2PLS (and O2PLS-DA for discriminant analysis) are multivariate projection methods that remove the structured noise by extracting linear relationships from independent and dependent data blocks, in a bidirectional way, and result in the decomposition of systematic variation into two model parts: the predictive or parallel part and the orthogonal part [81, 82]. These methods are really effective in correlating two data types (spectral and activity data, for instance) or for the integration of different omics as illustrated in the applications section of this chapter.

9.3.3.5 Validation

In metabolomics studies, one can use as advanced and fancy multivariate data analysis methods as one likes by following the most basic and critical rule in statistics: validation. The methods explained above cannot always be trusted as they are sensitive to false correlation or the risk of over fitting is quite obvious. Conclusions on the basis of false-correlated or over-fitted models can be disastrous, especially in the case of medical based studies. Different validation tools, such as cross-validation, permutation tests, Jack-knifing model parameters, and test-set validation, are available to overcome this problem [83]. In this chapter, we briefly explain the three most widely applied validation methods in metabolomics, that is, cross-validation, permutation tests, and test-set validation.

Cross-validation is a technique that divides data into equally sized blocks. The idea is to use the available data blocks minus some blocks to fit the model and the left out data blocks to test the model. As the fitting and testing of the model is done on the same original data, this method is biased, and to further test the model, a separate data set is required. This problem can be overcome by using double cross-validation. The significance of classification can also be evaluated by a permutation test. In permutation testing, permutation of the class assignment is done several times and a model between the data and the class assignment is built for each permutation. Also, the discrimination of the model based on original classification is compared with the discrimination between the classes of the model based on the permutated-class assignment. The test-set validation involves a separate data set (treated in the same way as the training data) to check the validity and fitness of the statistical analysis. If the test data is kept separated, the reproducibility and accuracy of the produced model will be discovered by validation. It is also possible to divide a large data set into bigger training data set and smaller test data set, in case a separate test data set is not available.

9.4
Applications of NMR-Based Metabolomics

Metabolite profiling or fingerprinting has a potentially broad field of applications as reflected by the variety in metabolite analysis-based studies. As a valuable tool for fundamental science and particularly systems biology type of studies, examples are also emerging where metabolomics is being used in applied situations concerning, for example, quality characteristics [84] or identifying potential biochemical markers to detect product contamination and adulteration [85]. It has been shown that metabolomics contributes to our understanding of plant metabolism and its role for plant survival. Recent advancement of NMR-based metabolomics applications is discussed in the following sections.

9.4.1
Understanding Stress Response

Metabolomics is currently providing and being increasingly used as the best platform to understand cellular phenotypes and to study their response under various types of biotic stresses. Although NMR- and chemometrics-based fingerprinting studies acquired a unique place in studying stress response of plants against broad range of pathogens, surprisingly very few of the NMR-based metabolomics studies have focused on the interaction between two organisms. Choi *et al.* [59] employed an NMR-based approach to characterize phytoplasma-infected *Catharanthus roseus* and reported the association of phenylpropanoids and terpenoid indole alkaloids with the plant response toward the infection. tobacco mosaic virus (TMV) infection to tobacco plants has also been studied by the same group [47]. This study revealed that on infection, the host triggered programmed cell death to restrict the spreading of infection. This was followed by the synthesis of signaling molecules, such as salicylic acid, to initiate systemic acquired resistance in the host. Many metabolites known to be involved in plant resistance, such as caffeoyl quinic acid, sesquiterpenoids, and diterpenoids, were also found elicited in the infected plant.

More recent work on tomato infection using NMR in combination with PCA and PLS-DA has been published by Lopez-Gresa *et al.* [86]. The report showed the response generated by the tomato plants on infection with a viroid (citrus exocortis viroid, CEVd) and a bacterium (*Pseudomonas syringae* pv. *tomato*). It was observed that the host responds differently against different type of pathogens. In the case of bacterial infection, phenylpropanoids and flavonoids (rutin) were the main inducible metabolites, while glycosylated gentisic acid was found critical in viroid infection. The NMR-PCA combination was also effective in highlighting the metabolic response of tomato against thrips (*Frankliniella occidentalis*), which is characterized by elevated levels of acylsugars in the resistant tomato species. These acylsugars are well known for their negative effect on herbivores [87].

Studies on *Brassica rapa* interaction with bacterial [50] and fungal [88] pathogen also revealed the underline mechanism of this plant to deal with biotic stress (Figure 9.5). Metabolites such as phenylpropanoids conjugated with malic acid and flavonoids were found in higher levels after infection. NMR-based metabolomics was recently applied to grapevine in order to characterize the metabolic response against Esca disease suggesting the increased production of phenolics with reduction in carbohydrates [89]. Similar observations were made by one of our studies where, on inoculation with a fungal pathogen, the resistant grapevine cultivar showed accumulation of caffeoyl tartaric acid, quercetin, and alanine, within 24 h after inoculation (unpublished data).

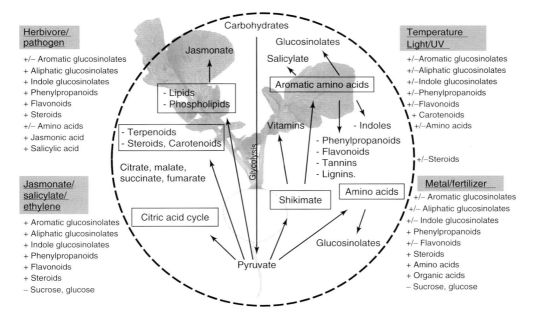

Figure 9.5 Biosynthetic pathways for the stress-induced metabolites. Different types of stresses are shown by the boxes out side the main circle with the induced metabolites. (Source: Adapted from Jahangir *et al.* [88].)

Plant physiology, dependent on multiple factors, is a complex phenomenon in relation to tissue or organ type, developmental stage, cultivation practice, and so on. In unsuitable growth conditions, plants must initiate a defensive response at cell organelle level and move upward. As illustrated by the above examples, NMR-based metabolomics is a prime tool if we wish to understand these complex processes and how strategies to make plants with enhanced resistance might be designed.

9.4.2
Application to Bioactivity Screening

The chemicals present in plants generally have an *in vivo* function related to their structure and are often known to have some bioactivities. These bioactivities can be of great relevance regarding how plants are important for humans (e.g., Chinese medicines, plant infusions with anticancer activity, antioxidant activity, etc.). NMR spectroscopy and chemometrics-based metabolomics is an emerging field that promises to provide information about the active ingredients of a crude plant extracts. Supervised multivariate data analysis methods such as PLS or O2PLS are the integral part of such studies. Roos *et al.* [90] evaluated 24 different extracts of four different accessions of St. John's Wort (*Hypericum perforatum*) extracted by six distinct solvents. PLS analysis was used as a regression model and proved effective to identify the resonances in the ^{1}H NMR spectrum correlated with the activity. The same NMR-PLS-based approach was used to predict the antiplasmodial activity in different *Artemisia annua* extracts and allowed their classification based on this activity [91].

Another successful attempt was made on Mexican anxiolytic and sedative plant, *Galphimia glauca*. This chapter discusses the use of NMR metabolomics to characterize the chemical profile of this plant collected from six different locations along with the discrimination of active and nonactive samples using PLS-DA modeling. The signals related to activity were found associated with a specific class of metabolites known as *galphimines* [92]. More recently, Cho *et al.* [93] successfully reported the use of NMR and PLS modeling for the classification and prediction of free radical scavenging activity in the fruit of *Citrus grandis* plant at different stages of its development.

In one of the recent publications by our group, a comprehensive extraction of *Orthosiphon stamineus* was combined with NMR-based metabolomics for the screening of adenosine A1 receptor binding compounds [94]. The method used PLS- and OPLS-based modeling to correlate the metabolomics and bioactivity data. Our laboratory also determined the TNFα inhibition activity in three grape cultivars at different stages of their development from the 2007 and 2008 vintages. We used NMR spectroscopy and chemometric methods (PLS and O2PLS) to identify the active ingredients responsible for the activity. Several phenylpropanoids and flavonoids, which were previously reported to have antioxidant and anti-inflammatory activity, were found positively correlated with anti-TNFα activity. Similar observations were made when different red wines were assessed for the same activity and the proposed methodology was found effective in terms of correlating the activity and metabolomics data and also to identify the active ingredients in the crude wine extracts (submitted to *Metabolomics* for publication).

Using the NMR- and chemometrics-based approach, the analysis of NMR shifts in relation to pharmacological activity can provide an idea about what part of the NMR spectrum correlates with the activity and gives information about the active ingredients in crude extracts of medicinal plants. The presented approach proved to be an effective tool to short-list a large set of crude extracts based on bioactivity. The effect of different variables on the activity of a sample can also be measured. Another advantage of this method is that compounds related to activity can be identified without extensive and elaborate chromatographic separation and thus allows rapid identification of extracts with biological activity. It is also important to know that this method is, and should be, considered as preliminary and final conclusions on metabolite–activity relationship should be drawn on the basis of activity data on the pure compound.

9.4.3
Quality Control of Herbal Medicines

Quality control of herbal medicinal material is an increasingly important issue all over the world. Regulatory authorities demand better standardization of material in terms of active components and more objective and reproducible analytical methods to achieve this [95]. The quantitative nature of NMR analysis makes it a very suitable method for accurate analysis of active substances in medicinal plant material. Choi *et al.* [96] developed a simple, rapid ^{1}H NMR method for the quantitation of the active compounds in *Ginkgo biloba* leaves and commercial herbal products. Quantitation was performed using the H-12 signal of bilobalide and ginkgolide A, B, and C, while separation of flavonoid and other terpenoid signals was also achieved.

Herbal medicines are usually consumed in a crude form (crude extract or powder) so it is often not clear which compounds are responsible for the observed biological activities. Not only does it make quality control more difficult but variable chemical composition also hampers studies into the pharmacological efficacy of the material. A good example is ginseng, which is well known for its adaptogenic properties. The roots of the plant contain compounds of many chemical classes, such as triterpene saponins and many others. In a study by Yang *et al.* [97], three commercial ginseng preparations and four kinds of ginseng root were analyzed by NMR-based methods (^{1}H NMR and J-resolved) together with multivariate data analysis. The different samples were clearly discriminated based on the content of various primary and secondary metabolites. This study showed the feasibility of using NMR as a quick method to characterize the overall chemical profiles of ginseng root material.

The ability to analyze many diverse compound types simultaneously makes NMR a suitable method to address purity, another quality aspect of medicinal plant material. It is often difficult to assess the purity of the material in terms of the desired plant part or assess adulteration with other plant species. Wang et al. [95] analyzed chamomile flower head samples containing different proportions of stalks by ^{1}H NMR. PCA classified samples based on content of stalk material, with a linear relationship between the average scores and percentages of stalks. NMR-based metabolic analysis can be used for discriminating medicinal plant species of the same genus. This was illustrated well in a study on *Ephedra* species by Kim et al. [98]. Of the more than 50 species, *Ephedra sinica* is the main species used in medicine. The presence of different species may affect the efficacy of the medicine, as species vary in their active alkaloid profiles. Three *Ephedra* species were analyzed by NMR, and after PCA was shown to be discriminated by alkaloid as well as benzoic acid derivative signals. Further analysis of commercially available *Ephedra* products revealed that most were composed of *E. sinica*, while some were composed of mixtures of species. In a study by van der Kooy et al. [99], ^{1}H NMR and multivariate data analysis was used to characterize a commercially available herbal antimalarial remedy. Capsules were claimed to contain leaves of *Artemisia afra* and contain the antimalarial compound artemisinin. Artemisinin has not been reported in *A. afra* but is well known to be present in the related species *A. annua*. Analysis of leaves of *A. afra* and *A. annua*, as well as the capsules showed a clear differentiation between the two species, with artemisinin an important marker for the discrimination. The herbal capsules were clearly grouped with the *A. afra* samples, and further targeted analysis confirmed the absence of artemisinin.

9.4.4
Chemotaxonomy

The ability to distinguish plant species chemically is not only useful for quality control of medicinal plants but has also led to the use of NMR-based metabolic analysis in chemotaxonomic applications. A study on 11 species of *Ilex* grown from seed under the same conditions revealed four groups of similar chemical profiles [54]. PCA and PLS analysis grouped the samples according to their NMR profiles, while hierarchical cluster analysis was applied to these data to show the closeness between species and groups. The results corresponded well with a phylogenetic study of *Ilex* species using DNA fingerprinting (amplified fragment length polymorphisms).

Safer et al. [100] used an NMR metabolomics approach together with LC–MS analysis to study relationships between some members of genus *Leontopodium*. Apart from *Leontopodium alpinum* (Alpine Edelweiss), not much is known about the other species, especially those occurring in Asia. Twenty-three species were analyzed, and with the help of PCA and PLS-DA, an unidentified species was shown to be closely related to two others, suggesting that hybridization occurred between them. With the help of DNA fingerprinting, two other species were unambiguously classified, which helped to clear up confusion caused by their very similar morphological characteristics.

9.4.5
Agricultural Applications

While chemotaxonomic studies often keep environmental conditions constant to assess the effects of genetics on plant metabolite profiles, other studies grow the same planting stock in different conditions to see how these affect metabolite profiles. Lubbe *et al.* [101] conducted a field experiment with *Narcissus pseudonarcissus* cultivated as a medicinal crop to determine the effect of fertilizers on the levels of galanthamine in the bulbs. Quantitative ^{1}H NMR analysis showed that application of different levels of fertilizers changed galanthamine concentration as compared to a control. PCA highlighted differences in metabolite patterns that helped to explain the observed changes in galanthamine in terms of biosynthetic precursors and other metabolites. A study on mandarin oranges in California assessed the effects of various growth conditions on metabolic profiles by ^{1}H NMR and multivariate data analysis [102]. Factors such as rootstock, soil composition, and elevation were found to influence the nutrient composition, which in turn also influence taste profiles of the fruit. The authors concluded that such NMR-based analysis could be very useful in the development of agricultural practices to obtain mandarins with optimized tastes.

Another area of agricultural research to which NMR-based metabolomics has been applied is the study of genetically modified (GM) crops. New genes may be inserted into crop plants to, for example, improve resistance to pesticides or environmental stress conditions. Alternatively, the aim of genetic modification may be to alter the content of specific metabolites related to nutritional or other quality aspects. In the first case, the new trait should be present without other unexpected changes in the metabolite profile. In the second case, the desired effect is to see only a change in the targeted chemical components, without unexpected pleiotropic effects. The ability of NMR-based methods to detect a wide range of chemical compounds together with the unbiased pattern-recognition abilities of multivariate data analysis make NMR metabolomics useful for seeing unexpected metabolic changes. Some examples of this application in the literature include studies comparing wild-type and transgenic maize [103], tomato [104], tobacco [105], and wheat [106]. These studies show the potential of such techniques for assessing the effects of genetic modification of plants, but for a more complete understanding of the consequences, results should ideally be integrated with other complementary techniques [107].

9.5
Future Prospects and Conclusions

NMR is a powerful analytical tool for the characterization of complex metabolite mixtures. It has been applied to many different sample types and has found use in diverse applications. From the increasing amounts of publications in the field, it seems that NMR-based methods will continue to make important contributions to the field of metabolomics. Complementing NMR-based methods with other platforms such as MS-based techniques will vastly improve the amount of information obtained. Integrating metabolite data with that of the gene, transcript, and protein level also had great potential to increase understanding of organisms as a whole.

The technical strong points of NMR are accurate quantitation and good reproducibility. In contrast, sensitivity and resolution remain major challenges in metabolite mixture analysis. The development of stronger magnets is one way to improve sensitivity. Sensitivity can also be improved by the use of cryoprobes, where the NMR probe electronics is cooled to reduce thermal noise. Microcoil probes allow measurement of small amounts of samples, down to micro- or nanoliter volumes. Signal-to-noise ratio can further be improved with the use of microcoils with solenoidal instead of Helmholz geometry [108]. Apart from technical advances, experimental techniques may be used to enhance sensitivity. *In vivo* isotope labeling is one such strategy. Cells can be incubated on labeled medium, and plants or invertebrate animals can be fed labeled nutrients so that samples enriched in ^{13}C or ^{15}N can be obtained [109, 110]. Signals of these normally low abundance nuclei can be enhanced for more rapid analysis by 2D NMR methods.

Even with more sensitive NMR measurements, signal overlap remains an obstacle to data analysis. As described in this chapter, 2D NMR experiments can help resolve signals allowing easier identification and quantitation [45, 49, 56]. More methods to shorten acquisition times for 2D experiments (e.g., ASAP HMQC, [111]) will be valuable for metabolomics studies with great sample numbers. Further development and improvement of automated computational tools for signal deconvolution and assignment may also help speed up the process of data analysis. Availability of large metabolite databases with 1- and 2D NMR spectra to freely access and link to computational tools will contribute to advances in the field. A number of large databases exist, but there is still a need for more comprehensive plant metabolite libraries.

To improve reproducibility of metabolomics results and facilitate data exchange between working groups, the Metabolomics Standards Initiative was developed [112]. This is coordinated by the Metabolomics Society (*www.metabolomicssociety.org/mstandards.html*) and is an ongoing project to develop guidelines for reporting of results and metadata of experiments. Initially, the focus was on mass spectrometry as analytical platform, but guidelines for NMR-based studies have also been incorporated [113].

References

1. Oliver, S.G., Winson, M.K., Kell, D.B., and Baganz, F. (1998) *Trends Biotechnol.*, **16**, 373–378.
2. Sumner, L.W., Mendes, P., and Dixon, R.A. (2003) *Phytochemistry*, **62**, 817–836.
3. Weckwerth, W. (2003) *Annu. Rev. Plant Biol.*, **54**, 669–689.
4. Verpoorte, R., Choi, Y.H., Mustafa, N.R., and Kim, H.K. (2008) *Phytochem. Rev.*, **7**, 525–537.
5. Verpoorte, R. and Kim, H.K. (2010) *Flavour Fragr. J.*, **25**, 128–131.
6. Hirai, M.Y., Yano, M., Goodenowe, D.B., Kanaya, S., Kimura, T., Awazuhara, M., Arita, M., Fujiwara, T.,

and Saito, K. (2004) *Proc. Natl. Acad. Sci. U.S.A.*, **101**, 10205–10210.
7. Hirai, M.Y., Klein, M., Fujikawa, Y., Yano, M., Goodenowe, D.B., Yamazaki, Y., Kanaya, S., Nakamura, Y., Kitayama, M., Suzuki, H., Sakurai, N., Shibata, D., Tokuhisa, J., Reichelt, M., Gershenzon, J., Papenbrock, J., and Saito, K. (2005) *J. Biol. Chem.*, **280**, 25590–25595.
8. Fiehn, O. (2002) *Plant Mol. Biol.*, **48**, 155–171.
9. Fiehn, O., Kopka, J., Dormann, P., Altmann, T., Trethewey, R.N., and Willmitzer, L. (2000) *Nat. Biotechnol.*, **18**, 1157–1161.
10. Dettmer, K., Aranov, P.A., and Hammock, B.D. (2007) *Mass Spectrom. Rev.*, **26**, 51–78.
11. Dallüge, J., Beens, J., and Brinkman, U.A.T. (2003) *J. Chromatogr. A*, **1000**, 69–108.
12. Vial, J., Pezous, B., Thiébaut, D., Sassiat, P., Teillet, B., Cahours, X., and Rivals, I. (2011) *Talanta*, **83**, 1295–1301.
13. Roessner, U., Luedemann, A., Brust, D., Fiehn, O., Linke, T., Willmitzer, L., and Fernie, A.R. (2001) *Plant Cell Online*, **13**, 11–29.
14. Desbrosses, G.G., Kopka, J., and Udvardi, M.K. (2005) *Plant Physiol.*, **137**, 1302–1318.
15. Ryan, D. and Robards, K. (2006) *Anal. Chem.*, **78**, 7954–7958.
16. Gobey, J., Cole, M., Janiszewski, J., Covey, T., Chau, T., Kovarik, P., and Corr, J. (2005) *Anal. Chem.*, **77**, 5643–5654.
17. Allwood, W. and Goodacre, R. (2010) *Phytochem. Anal.*, **21**, 33–47.
18. Soga, T., Ohashi, Y., Ueno, Y., Naraoka, H., Tomita, M., and Nishioka, T. (2003) *J. Proteome Res.*, **2**, 488–494.
19. Sato, S., Soga, T., Nishioka, T., and Tomita, M. (2004) *Plant J.*, **40**, 151–163.
20. Christian, W.K. (2004) *J. Chromatogr. A*, **1044**, 131–144.
21. Cooper, H.J. and Marshall, A.G. (2001) *J. Agric. Food Chem.*, **49**, 5710–5718.
22. Aharoni, A., de Vos, C.H.R., Verhoeven, H.A., Maliepaard, C.A., Kruppa, G., Bino, R.J., and Goodenowe, D.B. (2002) *OMICS*, **6**, 217–234.
23. Murch, S.J., Rupasinghe, H.P.V., Goodenowe, D.B., and Saxena, P.K. (2004) *Plant Cell Rep.*, **23**, 419–425.
24. Tohge, T., Nishiyama, Y., Hirai, M.Y., Yano, M., Nakajima, J., Awazuhara, M., Inoue, E., Takahashi, H., Goodenowe, D.B., Kitayama, M., Noji, M., Yamazaki, M., and Saito, K. (2005) *Plant J.*, **42**, 218–235.
25. Brown, S.C., Kruppa, G., and Dasseux, J.L. (2005) *Mass Spectrom. Rev.*, **24**, 223–231.
26. Ellis, D.I., Dunn, W.B., Griffin, J.L., Allwood, J.W., and Goodacre, R. (2007) *Pharmacogenomics*, **8**, 1243–1266.
27. Allwood, J.W., Ellis, D.I., Heald, J.K., and Mur, L.A.J. (2006) *Plant J.*, **46**, 351–368.
28. Ellis, D.I., Broadhurst, D., Kell, D.B., Rowland, J.J., and Goodacre, R. (2002) *Appl. Environ. Microbiol.*, **68**, 2822–2828.
29. Ellis, D.I., Broadhurst, D., Clarke, S.J., and Goodacre, R. (2005) *Analyst*, **130**, 1648–1654.
30. Aernouts, B., Polshin, E., Saeys, W., and Lammertyn, J. (2011) *Anal. Chim. Acta*, **705**, 88–97.
31. Allwood, J.W., Ellis, D.I., and Goodacre, R. (2007) *Physiol. Plant.*, **132**, 117–135.
32. Krishnan, P., Kruger, N.J., and Ratcliffe, R.G. (2005) *J. Exp. Bot.*, **56**, 255–265.
33. Pauli, G.F., Jaki, B.U., and Lankin, D.C. (2004) *J. Nat. Prod.*, **68**, 133–149.
34. Lubbe, A., Pomahacova, B., Choi, Y.H., and Verpoorte, R. (2010) *Phytochem. Anal.*, **21**, 66–72.
35. Freeman, R., Hill, H.D., and Kaptein, R. (1972) *J. Magn. Reson. (1969)*, **7**, 327–329.
36. Giraudeau, P. and Baguet, E. (2006) *J. Magn. Reson.*, **180**, 110–117.
37. Gottlieb, H.E., Kotlyar, V., and Nudelman, A. (1997) *J. Org. Chem.*, **62**, 7512–7515.
38. McKay Ryan, T. (2009) *Recent Advances in Solvent Suppression for Solution NMR: A Practical Reference*, Chapter 2, Academic Press, pp. 33–76.

39. William, S.P. (1999) *Water Signal Suppression in NMR Spectroscopy*, Academic Press, pp. 289–354.
40. Ogg, R.J., Kingsley, R.B., and Taylor, J.S. (1994) *J. Magn. Reson., B*, **104**, 1–10.
41. Košir, I.J. and Kidrič, J. (2001) *J. Agric. Food Chem.*, **49**, 50–56.
42. Viant, M.R. (2007) *Metabolomics: Methods and Protocols*, Humana Press, pp. 229–246.
43. Sacchi, R., Addeo, F., and Paolillo, L. (1998) *Magn. Reson. Chem.*, **35**, S133–S145.
44. Vlahov, G. (2006) *Anal. Chim. Acta*, **577**, 281–287.
45. Viant, M.R. and Ludwig, C. (2010) *Phytochem. Anal.*, **21**, 22–32.
46. Viant, M.R. (2003) *Biochem. Biophys. Res. Commun.*, **310**, 943–948.
47. Choi, Y.H., Kim, H.K., Linthorst, H.J.M., Hollander, J.G., Lefeber, A.W.M., Erkelens, C., Nuzillard, J.-M., and Verpoorte, R. (2006) *J. Nat. Prod.*, **69**, 742–748.
48. Thrippleton, M.J., Edden, R.A.E., and Keeler, J. (2005) *J. Magn. Reson.*, **174**, 97–109.
49. Xi, Y., de Ropp, J.S., Viant, M.R., Woodruff, D.L., and Yu, P. *Metabolomics*, **2**, 221–233.
50. Jahangir, M., Kim, H.K., Choi, Y.H., and Verpoorte, R. (2008) *Food Chem.*, **107**, 362–368.
51. Ali, K., Maltese, F., Zyprian, E., Rex, M., Choi, Y.H., and Verpoorte, R. (2009) *J. Agric. Food Chem.*, **57**, 9599–9606.
52. Sandusky, P. and Raftery, D. (2005) *Anal. Chem.*, **77**, 2455–2463.
53. Kim, H., Choi, Y., and Verpoorte, R. (2006) *Plant Metabolomics*, Springer, Leipzig, pp. 261–276.
54. Kim, H.K., Saifullah Khan, S., Wilson, E.G., Kricun, S.D.P., Meissner, A., Goraler, S., Deelder, A.M., Choi, Y.H., and Verpoorte, R. (2010) *Phytochemistry*, **71**, 773–784.
55. Abdel-Farid, I.B., Kim, H.K., Choi, Y.H., and Verpoorte, R. (2007) *J. Agric. Food Chem.*, **55**, 7936–7943.
56. Hu, K., Westler, W.M., and Markley, J.L. (2011) *J. Biomol. NMR*, **49**, 291–296.
57. Leiss, K.A., Choi, Y.H., Abdel-Farid, I.B., Verpoorte, R., and Klinkhamer, P.G.L. (2009) *J. Chem. Ecol.*, **35**, 219–229.
58. Defernez, M., Gunning, Y.M., Parr, A.J., Shepherd, L.V.T., Davies, H.V., and Colquhoun, I.J. (2004) *J. Agric. Food Chem.*, **52**, 6075–6085.
59. Choi, Y.H., Tapias, E.C., Kim, H.K., Lefeber, A.W.M., Erkelens, C., Verhoeven, J.T.J., Brzin, J., Zel, J., and Verpoorte, R. (2004) *Plant Physiol.*, **135**, 2398–2410.
60. Liang, Y.-S., Kim, H.K., Lefeber, A.W.M., Erkelens, C., Choi, Y.H., and Verpoorte, R. (2006) *J. Chromatogr. A*, **1112**, 148–155.
61. Kaddurah-Daouk, R., Beecher, C., Kristal, B.S., Matson, W.R., Bogdanov, M., and Asa, D.J. (2004) *Pharmagenomics*, **4**, 46–52.
62. Kupce, E. (2001) *Chem. Heterocycl. Compd.*, **37**, 1429–1438.
63. Bertram, H.C., Malmendal, A., Petersen, B.O., Madsen, J.C., Pedersen, H., Nielsen, N.C., Hoppe, C., Mølgaard, C., Michaelsen, K.F., and Duus, J.Ø. (2007) *Anal. Chem.*, **79**, 7110–7115.
64. Dixon, R.A., Gang, D.R., Charlton, A.J., Fiehn, O., Kuiper, H.A., Reynolds, T.L., Tjeerdema, R.S., Jeffery, E.H., German, J.B., Ridley, W.P., and Seiber, J.N. (2006) *J. Agric. Food Chem.*, **54**, 8984–8994.
65. Ratcliffe, R.G. and Shachar-Hill, Y. (2005) *Biol. Rev.*, **80**, 27–43.
66. Schripsema, J. (2010) *Phytochem. Anal.*, **21**, 14–21.
67. Beckonert, O., Keun, H.C., Ebbels, T.M.D., Bundy, J., Holmes, E., Lindon, J.C., and Nicholson, J.K. (2007) *Nat. Protoc.*, **2**, 2692–2703.
68. Kim, H.K. and Verpoorte, R. (2010) *Phytochem. Anal.*, **21**, 4–13.
69. Queiroz, O. (1974) *Annu. Rev. Plant Physiol.*, **25**, 115–134.
70. Kruger, N.J., Troncoso-Ponce, M.A., and Ratcliffe, R.G. (2008) *Nat. Protoc.*, **3**, 1001–1012.
71. Choi, Y.H., Kim, H.K., Hazekamp, A., Erkelens, C., Lefeber, A.W.M., and Verpoorte, R. (2004) *J. Nat. Prod.*, **67**, 953–957.

72. Kim, H.K., Choi, Y.H., and Verpoorte, R. (2010) *Nat. Protoc.*, **5**, 536–549.
73. Maltese, F., van der Kooy, F., and Verpoorte, R. (2009) *Nat. Prod. Commun.*, **4**, 447–454.
74. Cloarec, O., Dumas, M.-E., Craig, A., Barton, R.H., Trygg, J., Hudson, J., Blancher, C., Gauguier, D., Lindon, J.C., Holmes, E., and Nicholson, J. (2005) *Anal. Chem.*, **77**, 1282–1289.
75. Couto Alves, A., Rantalainen, M., Holmes, E., Nicholson, J.K., and Ebbels, T.M.D. (2009) *Anal. Chem.*, **81**, 2075–2084.
76. Holmes, E. and Antti, H. (2002) *Analyst*, **127**, 1549–1557.
77. Zheng, C., Zhang, S., Ragg, S., Raftery, D., and Vitek, O. (2011) *Bioinformatics*, **27**, 1637–1644.
78. Xia, J., Bjorndahl, T.C., Tang, P., and Wishart, D.S. *BMC Bioinformatics*, **9**, 507.
79. Kikuchi, J. and Hirayama, T. (2006) Hetero-nuclear NMR-based Metabolomics, *Plant Metabolomics: Biotechnology in Agriculture and Forestry*, Springer.
80. Wold, S., Antti, H., Lindgren, F., and Öhman, J. (1998) *Chemom. Intell. Lab. Syst.*, **44**, 175–185.
81. Trygg, J. and Wold, S. (2002) *J. Chemom.*, **16**, 119–128.
82. Trygg, J. and World, S. (2003) *J. Chemom.*, **17**, 53–64.
83. Rubingh, C.M., Bijlsma, S., Derks, E.P.P.A., Bobeldijk, I., Verheij, E.R., Kochhar, S., and Smilde, A.K. (2006) *Metabolomics*, **2**, 53–61.
84. Ali, K., Maltese, F., Toepfer, R., Choi, Y.H., and Verpoorte, R. (2011) *J. Biomol. NMR*, **49**, 255–266.
85. van der Kooy, F., Maltese, F., Hae Choi, Y., Kyong Kim, H., and Verpoorte, R. (2009) *Planta Med.*, **75**, 763, 775.
86. Lopez-Gresa, M.P., Maltese, F., Belles, J.M., Conejero, V., Kim, H.K., Choi, Y.H., and Verpoorte, R. (2010) *Phytochem. Anal.*, **21**, 89–94.
87. Mirnezhad, M., Romero-Gonzalez, R.R., Leiss, K.A., Choi, Y.H., and Verpoorte, R. (2010) *Phytochem. Anal.*, **21**, 110–117.
88. Abdel-Farid, I.B., Jahangir, M., van den Hondel, C.A.M.J.J., Kim, H.K., Choi, Y.H., and Verpoorte, R. (2009) *Plant Sci.*, **176**, 608–615.
89. Lima, M.R.M., Felgueiras, M.L., Graça, G., Rodrigues, J.E.A., Barros, A., Gil, A.M., and Dias, A.C.P. (2010) *J. Exp. Bot.*, **61**, 4033–4042.
90. Roos, G., Röseler, C., Büter, K.B., and Simmen, U. (2004) *Planta Med.*, **70**, 771, 777.
91. Bailey, N.J.C., Wang, Y., Sampson, J., Davis, W., Whitcombe, I., Hylands, P.J., Croft, S.L., and Holmes, E. (2004) *J. Pharm. Biomed. Anal.*, **35**, 117–126.
92. Cardoso-Taketa, A.T., Pereda-Miranda, R., Choi, Y.H., Verpoorte, R., and Villarreal, M.L. (2008) *Planta Med.*, **74**, 1295, 1301.
93. Cho, S.K., Yang, S.-O., Kim, S.-H., Kim, H., Ko, J.S., Riu, K.Z., Lee, H.-Y., and Choi, H.-K. (2009) *J. Pharm. Biomed. Anal.*, **49**, 567–571.
94. Yuliana, N.D., Khatib, A., Verpoorte, R., and Choi, Y.H. (2011) *Anal. Chem.*, **83**, 6902–6906.
95. Wang, Y., Tang, H., Nicholson, J.K., Hylands, P.J., Sampson, J., Whitcombe, I., Stewart, C.G., Caiger, S., Oru, I., and Holmes, E. (2004) *Planta Med.*, **70**, 250, 255.
96. Choi, Y.H., Choi, H.-K., Hazekamp, A., Bermejo, P., Schilder, Y., Erkelens, C., and Verpoorte, R. (2003) *Chem. Pharm. Bull.*, **51**, 158–161.
97. Yang, S.Y., Kim, H.K., Lefeber, A.W.M., Erkelens, C., Angelova, N., Choi, Y.H., and Verpoorte, R. (2006) *Planta Med.*, **72**, 364, 369.
98. Kim, H.K., Choi, Y.H., Chang, W.-T., and Verpoorte, R. (2003) *Chem. Pharm. Bull.*, **51**, 1382–1385.
99. Van der Kooy, F., Verpoorte, R., and Marion Meyer, J.J. (2008) *S. Afr. J. Bot.*, **74**, 186–189.
100. Safer, S., Tremetsberger, K., Guo, Y.P., Kohl, G., Samuel, M.R., Stuessy, T.F., and Stuppner, H. (2011) *Bot. J. Linn. Soc.*, **165**, 364–377.
101. Lubbe, A., Choi, Y.H., Vreeburg, P., and Verpoorte, R. (2011) *J. Agric. Food Chem.*, **59**, 3155–3161.

102. Zhang, X., Breksa, A.P., Mishchuk, D.O., and Slupsky, C.M. (2011) *J. Agric. Food Chem.*, **59**, 2672–2679.
103. Manetti, C., Bianchetti, C., Casciani, L., Castro, C., Di Cocco, M.E., Miccheli, A., Motto, M., and Conti, F. (2006) *J. Exp. Bot.*, **57**, 2613–2625.
104. Le Gall, G., Colquhoun, I.J., Davis, A.L., Collins, G.J., and Verhoeyen, M.E. (2003) *J. Agric. Food Chem.*, **51**, 2447–2456.
105. Choi, H.-K., Choi, Y.H., Verberne, M., Lefeber, A.W.M., Erkelens, C., and Verpoorte, R. (2004) *Phytochemistry*, **65**, 857–864.
106. Baker, J.M., Hawkins, N.D., Ward, J.L., Lovegrove, A., Napier, J.A., Shewry, P.R., and Beale, M.H. (2006) *Plant Biotechnol. J.*, **4**, 381–392.
107. Rischer, H. and Oksman-Caldentey, K.-M. (2006) *Trends Biotechnol.*, **24**, 102–104.
108. Zhang, S., Nagana Gowda, G.A., Ye, T., and Raftery, D. (2010) *Analyst*, **135**, 1490–1498.
109. Chikayama, E., Suto, M., Nishihara, T., Shinozaki, K., Hirayama, T., and Kikuchi, J. (2008) *PLoS ONE*, **3**, e3805.
110. Lundberg, P. and Lundquist, P.O. (2004) *Planta*, **219**, 661–672.
111. Kupce, E. and Freeman, R. (2007) *Magn. Reson. Chem.*, **45**, 2–4.
112. Fiehn, O., Robertson, D., Griffin, J., van der Werf, M., Nikolau, B., Morrison, N., Sumner, L., Goodacre, R., Hardy, N.W., Taylor, C., Fostel, J., Kristal, B., Kaddurah-Daouk, R., Mendes, P., van Ommen, B., Lindon, J.C., and Sansone, S.A. (2007) *Metabolomics*, **3**, 175–178.
113. Sumner, L.W., Amberg, A., Barrett, D., Beale, M.H., Beger, R., Daykin, C.A., Fan, T.W.M., Fiehn, O., Goodacre, R., Griffin, J.L., Henkemeier, T., Hardy, N., Higashi, R., Kopka, J., Lane, A.N., Lindon, J.C., Marriott, P., Nicholls, A.W., Reily, M.D., Thaden, J.J., and Viant, M.R. (2007) *Metabolomics*, **3**, 211–221.

10
Potential of Microfluidics and Single Cell Analysis in Metabolomics (Micrometabolomics)

Meghan M. Mensack, Ryan E. Holcomb, and Charles S. Henry

10.1
Introduction

The field of metabolomics has experienced rapid growth since its inception over a decade ago. It has expanded to include a variety of different approaches and subdisciplines that range from broad-based strategies such as metabolomics, metabonomics, and metabolic fingerprinting, to specific monitoring methods such as metabolite profiling and targeted analysis [1, 2]. Data acquired from these studies are analyzed using a variety of techniques, highlighting aspects of the biochemistry occurring within the cell or organism. The diversity and sheer number of analytes to be measured in a metabolomics study are daunting, and thus have led to the development of many different analytical methods to address these challenges. While the most common tools in metabolomics are nuclear magnetic resonance (NMR) and mass spectrometry (MS), a new class of techniques based on microfluidics is being applied to metabolomic analysis. Microfluidics itself is also a relatively young field, but has the ability to contribute in a number of unique ways to metabolomics as an advanced analytical technique capable of handling very small samples and multiplexing chemical analysis [3]. Methods for handling, delivering, and analyzing volumes of fluids in the microliter to submicroliter regime fall under the general heading of microfluidics. For the purpose of this chapter, the definition of microfluidics will be further constrained to devices that are made using micromachining methods (as opposed to solutions flowing through small bore tubing) as this represents the field of microfluidics most accurately. The physical characteristics of microfluidic systems have the potential to drastically impact the process flow of a given metabolomics analysis. For example, time- and labor-intensive sample preparation techniques consisting of multiple extractions and derivatization reactions are frequently used in metabolomics studies [4]. If performed in a microfluidic format, a significant reduction in analysis time coupled with an increase in efficiency could be obtained. At present, a number of microfluidic systems couple on-chip sample preparation and selective separations with off-chip detection techniques such as NMR and MS for metabolite determination [5]. In addition, a number of microfluidic systems have been

Metabolomics in Practice: Successful Strategies to Generate and Analyze Metabolic Data, First Edition.
Edited by Michael Lämmerhofer and Wolfram Weckwerth.
© 2013 Wiley-VCH Verlag GmbH & Co. KGaA. Published 2013 by Wiley-VCH Verlag GmbH & Co. KGaA.

developed that integrate biological systems (cells) onto a microfluidic platform for metabolomic studies [6–10]. The goal of this chapter is to summarize past contributions and highlight potential uses of microfluidics as a novel tool for conducting metabolomics studies of both traditional biological samples and single cells.

10.2
Sample Processing for Metabolomics

The use of microfluidics for metabolic analysis is an exciting prospect, as it has the potential to streamline process flow and increase sample throughput. Much of this potential lies in the ability to incorporate multiple functionalities into microfluidic systems in a high density format [11]. Indeed, an entire metabolomics analysis from sample preprocessing to metabolite measurement could be conducted using a properly designed system. To date, use of fully integrated systems such as these for metabolic analyses have been limited [3]; however, the functionality of microfluidic technologies that comprise these systems has been demonstrated with great success [3, 6, 12, 13]. For sample preprocessing, this includes stand-alone or integrated microfluidic components to minimize hands-on sample preparation. Not only does this save time, but it also eliminates error that can be introduced during the many steps involved in the sample preparation. The most commonly used microfluidic sample preprocessing techniques for metabolomics applications are discussed in the following sections.

10.2.1
Solid Phase Extraction

One common form of sample preprocessing is solid phase extraction (SPE). With SPE, analytes are isolated on a stationary phase material and eluted with an appropriate solvent system. The resulting concentrated analyte fraction can be directly analyzed using a suitable measurement technique [14, 15]. Currently, the most common SPE format consists of single use disposable cartridges. Using specially designed vacuum manifolds, multiple SPE cartridges can be run simultaneously in a high throughput manner. Owing to the high-density fluidic architectures and inexpensive disposable nature of many microfluidic platforms [11], microfluidic SPE can be conducted in much the same way making it a viable alternative to currently used SPE technologies.

Microfluidic SPE systems are generally constructed by polymerizing an appropriate extraction material in a microfluidic channel [16, 17] or by packing the channel with a commercially available sorbent [18]. Many times the SPE portion is integrated alongside other preprocessing functionalities upstream of a detection step. One example from the literature describing a system such as this involves on-chip sample/reagent mixing and SPE before in-line MS detection [16]. In this example, P450 drug metabolism was studied by diffusively mixing human liver microsomes and several antidepressant drugs (imipramine, doxepin, and amitriptyline) on-chip with laminar flow mixing, a technique commonly employed with microfluidics [11, 16]. The SPE sorbent acted to retain the drugs and their resulting metabolites, as well as desalt the sample before MS analysis. The lifetime of this device was contingent upon the lifetime of the sorbent. To increase lifetime and throughput, multiple SPE elements can be integrated into a single device, and this was successfully demonstrated in a subsequent study [17]. While microfluidic SPE is not as highly developed as conventional SPE, it is clear that future advances in device engineering will enable performance comparable to the current state-of-the-art for conventional systems. Indeed, microfluidic SPE will have the advantage of decreased sample/reagent consumption, decreased analysis times, disposability of SPE material to prevent carryover, and the ability to integrate in-line with an appropriate measurement technique such as MS [16, 19], something which is neither possible nor practical with conventional SPE systems.

10.2.2
Laminar Diffusion

Another microfluidic sample pretreatment technique used in conjunction with metabolite analysis involves diffusional separation of biofluid components at a laminar flow interface [20, 21]. These interfaces form when two fluid streams with low Reynolds numbers are brought together [22]. Low Reynolds numbers are normal for microfluidic devices and form readily even when two solutions are highly miscible (e.g., two different aqueous buffers). A visual representation of this phenomenon is shown in Figure 10.1a, where two buffer streams that differ only by addition of fluorescein converge at a Y-mixer in a microfluidic device; Figure 10.1b shows the flow profile for a single fluid stream. Starting at the

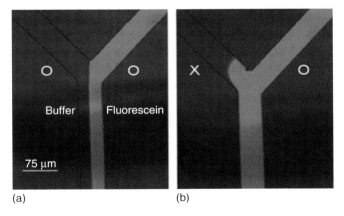

Figure 10.1 (a) Microscope image showing laminar flow of buffer (left) and fluorescein solution (right) in a microfluidic Y-mixer. The "O" denotes fluid flow is occurring in both incoming channels. (b) The same image, but with no incoming fluid flow ("X") from the channel on the left (buffer). (Images courtesy of Lucas J. Mason.)

convergence point, the two streams begin to mix slowly by diffusion. At some point downstream from the convergence, the streams become fully mixed. The distance downstream is dependent on the linear flow rate and diffusion coefficients of the molecules involved; thus, laminar flow provides a means of separation based on either diffusion or selective interactions between molecules in the two different streams. Laminar fluid flow is commonly used in microfluidics for a number of preprocessing tasks [11, 16]. An example which exploits this phenomenon describes use of a laminar fluid diffusion interface for sample cleanup before profiling of serum creatinine [20, 21]. In this example, the flow interface was used to separate low and high molecular weight serum components into separate sample streams before analysis. The flow interface accommodates input from both a sample (serum) and receiver stream (buffer) that converge in a common channel, in a manner similar to that shown in Figure 10.1a. Owing to faster diffusion rates of lower molecular weight components, they are preferentially collected in the receiver stream while the majority of high molecular weight components remain in the original biofluid stream. The two streams are split downstream, the target stream collected and dried, and the metabolite profile obtained using infrared spectroscopy.

10.2.3
Fluidic Pumping for On-Chip Mixing

A final preprocessing application well suited for microfluidics is in-line sample mixing and labeling. Many times, microfluidic mixing involves either use of the laminar diffusion techniques described earlier [16] or alternatively more elaborate schemes which utilize integrated valving to deliver and mix sample and assay reagents on-chip. With the latter, one of the most popular schemes for in-line fluid manipulation is pneumatic valving (Quake style) [23]. This type of valving system utilizes a gas to compress microfluidic channels fabricated in elastomeric substrates (e.g., poly(dimethylsiloxane), PDMS), and can be used for both channel sealing and fluid pumping [23]. These systems are comprised of multilayer microfluidic architectures with one layer containing the valves (in the form of dead-end microfluidic channels) and the other, the channel network in which sample and reagents are transported and mixed. Valve actuation allows for discrete metering of fluidic volumes as well as sample isolation for quantitative analysis. A relevant metabolomics example which utilizes this approach for sample preprocessing involves an integrated microfluidic system to study embryo metabolism [24]. With this system, sample/reagent delivery, mixing, and transport to a measurement site were facilitated by pneumatic valves. This whole process, including data acquisition and analysis, could be conducted in an autonomous manner once the sample and reagents were loaded onto the microfluidic device. The ability to automate these processes is beneficial because it reduces analysis time by making it less labor intensive. Qualities such as these make integrated microfluidic systems attractive alternatives to currently used analytical methodologies.

10.3
Microfluidic Separations for Metabolic Analysis

Modern metabolic analyses are based largely on separation techniques such as gas and liquid chromatography. Unfortunately, the time required to complete separations with these techniques is long. Additionally, the high instrumentation cost has the potential to limit the number of scientists who can actively contribute to the field. Microfluidic devices excel at chemical separations, providing speed, resolution, and low overall cost compared to traditional systems. The most common mode of microchip separation is microchip capillary electrophoresis (CE), largely because it is simple to implement. Microchip liquid and gas chromatography methods have been gaining ground on microchip electrophoresis [25] and are poised to make strong contributions to the field of metabolomics as well.

10.3.1
Microchip Capillary Electrophoresis

CE is commonly used for metabolomics applications [1, 5, 26] largely because of its excellent resolving power and relatively fast separation times [27, 28]. Additionally, analysis of biofluids, such as serum and urine, can often be conducted with little or no sample preparation [1]. Miniaturized CE, or microchip capillary electrophoresis (MCE) [29, 30], functions on the same principles as CE and is capable of many of the same types of analyses. While relatively new to the field of metabolomics, MCE has the potential to be a useful analytical tool because of the shorter analysis times, smaller reagent/sample consumption, portability, and lower cost compared to currently used instrumentation [3]. The aim of this section is to familiarize the reader with this technique and formats that are currently being used for targeted metabolic analysis. In addition, practical considerations for conducting metabolic analyses with MCE are discussed.

10.3.1.1 MCE Systems

Implicit in the name, MCE relies on electrophoresis for fluid manipulation and analyte separation. MCE was the first widely used microfluidic separation technique because of its instrumental and operational simplicity [31]. In CE, bulk fluid flow is achieved using electroosmotic flow (EOF) while resolution of the analytes is achieved using the inherent electrophoretic mobility of the analytes and/or selective partitioning interactions. Most MCE systems utilize microchannels with critical dimensions ranging from 10 to 75 µm; electrokinetic fluid flow in these channels is achieved by applying voltages to electrodes integrated into the device [3]. While MCE devices vary significantly in architecture, most contain the basic features shown in Figure 10.2a. The operational features are also essentially the same, and are presented here as such for simplicity. In most MCE systems, there are at least four solution reservoirs that hold the sample, electrophoresis buffer, sample waste, and electrophoresis waste. Samples are injected at the intersection of the sample, buffer, and sample waste channels and detected at a distance down

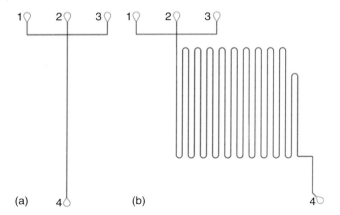

Figure 10.2 Schematic of generic MCE devices. (a) Straight channel (b) serpentine channel. Injected samples are separated in a microfluidic channel that terminates at the buffer waste reservoir. Device dimensions vary depending on the application; most have length and width dimensions of multiple centimeters (∼2–5 cm). Reservoirs are labeled as follows: (1) sample reservoir, (2) buffer reservoir, (3) sample waste, and (4) buffer waste. (Image courtesy of Jason Emory.)

the separation channel. Typically, analytes are detected toward the end of the separation channel to improve resolution, but can be detected anywhere along the channel, provided the detector is movable. If more resolution is desired, a longer serpentine channel can be employed at the cost of longer separation times (Figure 10.2b) [32].

10.3.1.2 Sample Injection

With MCE, there are two commonly used forms of injection, pinched [32] and gated [33]. Both have their associated benefits and limitations, and choice of one or the other depends on the overall aims of the study and metabolites to be measured. Gated injection allows for sample preconcentration using field amplified sample injection (FASI), but does not give a true concentration profile because of electrokinetic biasing [33]. For a more representative concentration profile, a pinched injection scheme can be employed. In addition to the type of injection scheme used, the analyst needs to be aware of how the physical state of the sample will affect injection. Ionic strength and viscosity mismatches between the sample and background electrolyte (BGE) have been shown to cause sample destacking and other EOF (bulk flow) effects with electrokinetic injections that result in a loss of resolution and/or sensitivity [34].

10.3.1.3 Electrophoretic Separations

With MCE, several different electrophoretic separation strategies may be employed depending on the analyte's chemistry. The majority of targeted metabolic analyses using MCE are conducted using normal polarity capillary zone electrophoresis (CZE) [3]. In normal polarity CZE, the microchannel is filled with BGE and charged analytes separated based on their differential migration velocities in the applied electrical field. CZE relies on the presence of bulk fluid flow, termed electroosmotic flow (EOF), to move analytes (both charged and neutral) toward the detector. When using this technique, it is important to maintain a constant EOF to ensure reproducible separations. Alterations in the EOF can be expected as a result of adsorption of analytes or matrix components to the capillary wall when analyzing biofluids. To counteract this, surfactants such as sodium dodecyl sulfate (SDS) [35] or polyethylene glycol (PEG) [36] can be added to the BGE to act as a dynamic coating to suppress surface adsorption.

A drawback of CZE is that neutral species cannot be separated. When neutral analytes are to be measured, other modes of CE separation can be used including micellar electrokinetic chromatography (MEKC) and capillary electrochromatography (CEC). In MEKC, surfactant is added to the BGE above its critical micelle concentration to form a pseudostationary phase. A pseudostationary phase is formed when the separation phase (micelles in this case) are free in solution and not bound to a surface. Neutral compounds are separated based on their differential partitioning with the micelle while charged species can also be retained via hydrophobic and/or electrostatic interactions. The utility of MEKC for the separation of metabolites has been demonstrated extensively in metabolic fingerprinting studies [1, 37, 38].

CEC is an alternative to MEKC for separation of neutral compounds that uses traditional chromatographic stationary phases (or slight modifications thereof) to resolve analytes. In CEC, EOF is used to drive fluid through a packed bed and elute compounds. The major difference between CEC and traditional chromatographic techniques is the higher separation efficiencies (often in the hundreds of thousands of theoretical plates [39]) provided by electrophoresis compared to those provided the pressure driven flows of liquid chromatography (often measured in tens of thousands of plates).

10.3.2
Analyte Detection

The type of detection mode used for a given metabolomics study using MCE is largely dependent on the physical properties of the analytes of interest. Although many detection modes can be used, optical and electrochemical techniques are the most common with MCE [40]. A description of these detection techniques, considerations for their use, and current applications in the field of metabolomics are discussed below.

10.3.2.1 Optical Detection
Fluorescence is frequently used with MCE because of its high sensitivity, low limits of detection (LODs), and relative ease of implementation with MCE systems [41]. The most common instrumental approaches using this technique are laser-induced fluorescence (LIF) and fluorescence microscopy. Fluorescence detection requires analytes to exhibit native fluorescence or have functional groups, which enable

conjugation with an appropriate fluorophore. Of these, the latter is most common although there are some instances were direct analysis of natively fluorescent metabolites has been pursued [42, 43]. While labeling of amines, sulfhydryls, and carboxylates is a viable method for metabolite analysis, the technique can be limited by low coupling efficiencies and short fluorophore lifetimes (photobleaching). Regardless, the technique is one of the most sensitive and remains a top choice for measuring low concentration metabolites; LODs commonly obtainable using this technique are in the nanomolar regime although femtomolar levels have been reached [41]. To date, MCE coupled with fluorescence detection has been used for the targeted profiling of intracellular amines, aromatic molecules, flavin metabolites, and glycans [44–47].

Another form of optical detection used for MCE metabolomics applications is absorbance. While not as sensitive as fluorescence, this technique does not require a labeling step, thus decreasing the time and complexity of the analysis. Demonstrated applications of UV–vis detection include targeted analysis of drug metabolites [48], nitrate and nitrite in serum and saliva [49–51], antimicrobial metabolites [52], and plant metabolites [53]. While not explored using MCE, an additional application for this detection technique is clinical monitoring of disease state and treatment efficacy using metabolic fingerprinting. Recently, the suitability of conventional CE for these types of analyses was demonstrated [1, 37, 38], and the same type of analyses should be possible using MCE as well. With these fingerprinting studies, data sets (peaks in an electropherogram) are globally compared to evaluate whether differences exist in the metabolite makeup (i.e., fingerprint). These data sets are usually compared using multivariate techniques such as partial least squares (PLS), random forest, or principal component analysis (PCA) to discern what differences, if any, exist in the data [1, 54]. One benefit of using this approach is the ability to distinguish a disease state without having to target and analyze specific disease biomarkers. Therefore, a nonspecific measurement technique such as UV–vis absorbance is well suited for this type of analysis. Integration with a rapid technique such as MCE is beneficial, in that it would act to increase the throughput of these measurements.

10.3.2.2 Electrochemical Detection

The second type of detection commonly employed with MCE is electrochemical detection (ECD) [55, 56]. ECD is an attractive alternative to optical techniques as it can be easily miniaturized without loss of performance. Additionally, it allows for a range of detection options including direct detection using amperometric methods and general detection using conductivity detection. Often, target metabolites can be measured without derivatization, and selectivity provided by differences in the oxidation/reduction potentials of the analytes of interest.

While various types of ECD can be employed with MCE, the most useful for metabolic analyses are DC amperometry and pulsed electrochemical detection (PED). DC amperometry is the most common because of its simplicity and sensitivity [55, 56]. For signal generation, a constant potential is applied to the working electrode and the current measured as analytes undergo either oxidation or reduction. DC amperometry is routinely used for measuring catecholamines, phenols, and other related compounds as these species are easily oxidized or reduced. To date, several successful profiling analyses of neurotransmitters have been conducted using DC amperometry [3, 57]. Besides neurotransmitters, DC amperometry is also effective for detecting pharmaceuticals and other xenobiotics present in biofluids [58–60], making this technique relevant for metabonomics studies. PED differs from DC amperometry, in that it utilizes a pulsed waveform during detection for *in situ* regeneration of the working electrode surface [61]. This technique is commonly employed for analytes which cannot be easily measured using DC amperometry including aliphatic amines, thiols, and carbohydrates [61]. These analytes, which are common in biological samples, tend to rapidly foul the working electrode, and thus limit its effective lifetime and sensitivity [62].

One of the benefits of using ECD is the ability to conduct multipotential analyses with electrode arrays. Using this approach, multiple working electrodes can be individually biased to different detection potentials to improve detection selectivity and increase the total number of measurable metabolites [63–65]. The utility of this technique for metabolic profiling has been demonstrated previously for liquid chromatography systems where simultaneous profiling of hundreds of electrochemically active compounds have been performed [65]. Recently, the same concept was applied with MCE in the form of an eight working electrode array detector [66]; a schematic of this microchip system is shown in Figure 10.3. Using this device, selective detection of biologically relevant metabolites and xenobiotics was performed using potential control. The ability to selectively detect and electrochemically resolve analytes in a system such as this should find many applications in metabolomics studies because of the chemical complexity of the samples involved. In another example, Ewing's group designed a carbon–fiber microarray with seven individually addressable electrodes to electrochemically image neurochemical secretions from single PC12 cells. Both spatial and temporal resolution were obtained using a combination of fast scan cyclic voltammetry (FCSV) and amperometry [67]. In the future, high-density electrochemical arrays integrated with MCE could prove to be a useful tool for conducting rapid metabolic profiling analyses.

10.4
Microfluidics for Cellular Analysis

10.4.1
Requirements for Single Cell Metabolomics

Obtaining a complete cellular metabolite profile is difficult because of the limited amount of material within a single cell. Single cell metabolomics requires efficient cell lysis and analyte extraction with minimal dilution. In addition, the complexity of the cellular metabolome requires these extraction techniques be coupled with high

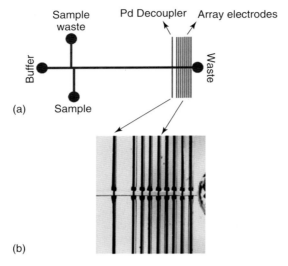

Figure 10.3 (a) Schematic of a MCE-ECD array device. (b) Brightfield image of array electrodes showing a palladium decoupler and eight gold working electrodes. All electrodes are 25 m in diameter while the separation channel is 50 μm in height. Color and contrast have been modified to aid in visualization. (Reprinted from Ref. [66] with permission from the Analyst 2009.)

resolution separation techniques. The ability of microfluidic devices to deliver small volumes of liquid and the small confined geometries make microfluidic devices well suited for cell growth and single cell analysis. Miniaturized devices also offer decreased reaction times partnered with increased sensitivity. The performance of a microfluidic device for cellular analysis is highly dependent on the internal structure of the device. Traditional *in vitro* cell culture often alters cellular properties, which results in proliferation being the only observed cellular process [68]. In contrast, the small, complex structures fabricated within a microfluidic device can often mimic *in vivo* environments.

10.4.2
Types of Microfluidic Instrumentation

Microfluidic instrumentation for cellular analysis can be grouped into two major categories: finite volume devices and flow analysis systems. Finite volume devices are akin to batch reactors and include microchambers and droplet microfluidics. Media conditions in finite volume devices are highly dependent on cell growth and metabolic activity of the cells within the device. Of these finite volume devices, microchambers are, in general, the simplest design for chemical and biological analysis of single cells. These types of devices allow for easy control of well-defined cell types. Individual reaction chambers such as those described by Heo *et al.* [69] are also useful for studying ion balance, metabolism, and cell growth rate resulting from evaporative osmolality shifts.

Droplet microfluidic designs enable the segregation of one to a few cells in each reactor (droplet). This is especially advantageous as proteins and metabolites from adjacent cells can easily influence the metabolite profile of the cells of interest. Each microdroplet acts as a spatially separated, independent reactor. These picoliter to nanoliter microdroplets are typically aqueous and isolated from other droplets by an immiscible oil phase. The use of droplet microfluidics has been shown for fast kinetic enzymatic assays [70], cellular protein expression [71], and cell-based assays [72, 73]. In 2009, Huebner *et al.* [74] introduced a single layer PDMS microdroplet array that was capable of generating, storing, incubating, monitoring, isolating, and recovering droplets for analysis without electrical or optical actuators.

Flow channel systems are similar to perfusion reactors and other continuous cultivation systems. Unlike fixed volume reactors, cellular activity does not greatly influence the medium conditions as a constant flux of media continually removes cellular products. Dam structures, notches, microwells, and other physical structures within flow analysis systems allow for control over cell position. For instance, a cell array cytometer consisting of 440 micromechanical traps which hydrodynamically capture nonadherent cells was introduced by Wlodkowic *et al.* [75] in 2009 to investigate the pharmacokinetic (PK) effects of anticancer drugs in hematopoetic cells. A similar device was reported by Wang's group in 2010 for rapidly screening antiproliferation effects of the anticancer drugs mitomycin-C and tamoxifen on Michigan Cancer Foundation-7 (MCF-7) breast cancer cells [76]. Flow systems can also be designed for parallel analysis of several single cells in separate microfluidic channels within a single device. An example of this can be seen in Figure 10.4 from the Kennedy group. The microfluidic device shown here is designed to monitor insulin secretion from 15 single pancreatic islet cells simultaneously [77].

10.4.3
Biological Questions

10.4.3.1 Monitoring Metabolic Response to Stimulation and Cell-to-Cell Signaling

Integrated microfluidic devices are capable of exposing or stimulating cells as well as monitoring the metabolic response as a result. For example, Zhang and Roper developed a device capable of monitoring [Ca^{2+}] secretions from single islets of Langerhans following glucose stimulation [78]. The three-layer glass/polymer hybrid device consisted of two pneumatic pumps, a 12 cm mixing channel, and a 0.21 cm cell chamber. Online mixing and glucose dilutions were conducted using the two pumps. Figure 10.5 shows a schematic of this perfusion system.

Cooper's group produced a microfluidic device capable of monitoring cell-to-cell signaling between pairs of cardiomyocytes; previously, these events were primarily examined in bulk solution or with cultured cells on-chip [79]. They looked at electrical and mechanical coupling between primary heart cells and monitored the movement of Ca^{2+} waves between the cells under physiological conditions. Cells were electrically stimulated using microelectrodes embedded in the PDMS

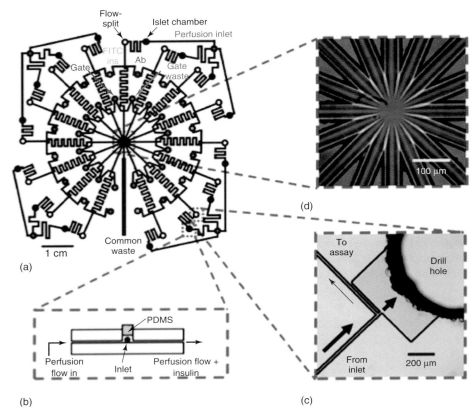

Figure 10.4 Design and photomicrographs of images detailing portions of a microfluidic chip for monitoring insulin secretion from 15 independent pancreatic islets. (a) Chip schematic illustrating the channel network where the colored reservoirs correspond to different solutions. (b) Side view of an islet perfusion chamber. Islets are loaded in phosphate buffer and the chamber sealed with a piece of PDMS. (c) Image of on-chip flow-splitter that allows for compatibility between the fast flowing insulin sampling stream and the slower EOF-driven immunoassay reagents. (d) Brightfield image of the detection area in the center of the chip. Waste is removed from the chip with a single waste channel at the bottom of the detection area. (Reprinted from Ref. [77] with permission from the Journal of Analytical Chemistry.)

device. The cells were also challenged with pharmaceutically relevant compounds. The Ca^{2+} movement between the cells for both physiological and nonphysiological stimulation was monitored using a fluorescent tracer for Ca^{2+}. A second device designed for real time monitoring of metabolic flux in individual cardiac cells has been reported by Cheng et al. [80]. Specifically, levels of lactate release were measured under continuous field stimulation using enzyme modified platinum working electrodes. A detection limit of 4.8 fmol (7.4 µM) was obtained using this technique.

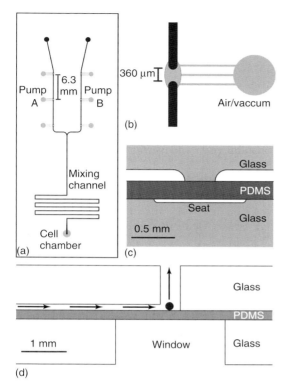

Figure 10.5 Microfluidic perfusion system. (a) Schematic of microfluidic design. Pumps A and B were operated at various flow rates to deliver controlled amounts of stimulant and diluent at a constant flow rate into the mixing channel. (b) Zoomed-in image of a single valve. (c) Side view of a single valve illustrating the three layers within the device. (d) Side view of cell chamber. Perfusion direction given by arrows. The window below the islet (black dot) allowed for fluorescence monitoring of Ca^{2+} concentration. (Adapted from Ref. [78] with permission from the Journal of Analytical Chemistry.)

Yoo et al. [81] developed a microfluidic chip for general toxicology screening using bioluminescent bacteria. Two different strains of bacteria were examined. The first, GC2, is a useful marker of general toxicity while the second, DK1, shows specificity for toxicity relating to oxidative stress. A dose dependent response was seen in the immobilized bioluminescent GC2 cells upon phenol exposure. In the case of the DK1 cells, a 10-fold increase in bioluminescence was observed upon exposure to 0.88 mM peroxide.

A continuous flow microfluidic system designed to monitor adipocyte metabolism, and in particular glycerol secretion, was developed by the Kennedy group [82]. The online enzyme assay had a detection limit of 4 μM glycerol and temporal resolution for changes in glycerol concentrations of 90 s. Pharmacological dosing of isoproterenol, a B-adrenergic agonist, triggered a threefold increase in glycerol secretion as a result of lipolysis.

10.4.3.2 Pharmacokinetics/Pharmacodynamics

PK refers to how substances such as pharmaceutical agents, hormones, nutrients, and toxins are metabolized by the body. In contrast, pharmacodynamics (PD) examines the effect of these substances on the body. Traditionally, PK/PD studies are monitored using liquid chromatography coupled to MS. Microfluidic devices can be useful for conducting these studies as well, and to date have been successfully implemented in various capacities for drug toxicity and PK/PD studies. For instance, Prot *et al.* [83] developed a device in which hepatocyte cells were cultured and exposed to a cocktail of pharmaceutically relevant small molecules such as acetaminophen. The xenobiotic metabolism and PK were monitored over the course of several hours. Metabolite detection and identification was carried out using LC-MS/MS. A second PDMS device consisting of cell culture reservoirs and tapered microchannels packed with polymeric beads for metabolite extraction was introduced by Gao *et al.* in 2010 [84]. The metabolism of vitamin E in human lung epithelial cells was studied. The total sample pretreatment took only 15 min and the solvent volume required for metabolite extraction was a mere 100 µl. A diagram of this system is shown in Figure 10.6. A microfluidic *in situ* microreactor was introduced to monitor phase I and phase II metabolism of candidate drugs

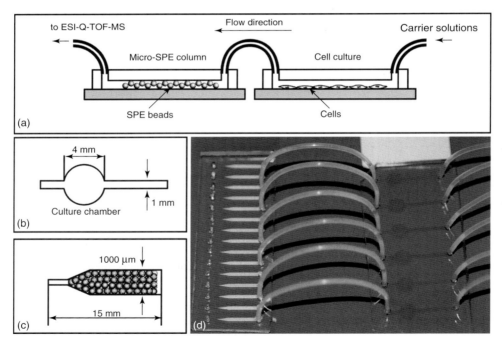

Figure 10.6 Microfluidic device for cell culture and monitoring of cellular metabolism. (a) Schematic of entire microfluidic device, (b) schematic of culture chamber portion of device, (c) schematic of micro-SPE portion of device, and (d) photomicrograph of device. (Adapted from Ref. [84] with permission from the Journal of Analytical Chemistry.)

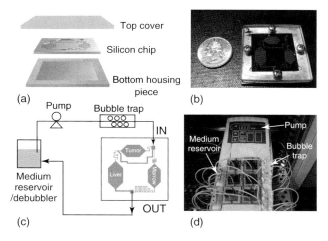

Figure 10.7 Micro cell culture analog (μCCA) device. (a) Assembly of the μCCA device. (b) Photomicrograph of single mCCA device. Channels have been filled with dye for visualization. (c) Schematic of operational setup for single μCCA device. (d) Photomicrograph of μCCA system with multiple chips. (Adapted from Ref. [86] with permission from RSC.)

in microsomes (phase I) and hepatocytes (phase I and II) encapsulated in PEG hydrogels [85]. No loss of enzymatic activity (as determined by P450 activity) was seen following photopolymerization. The mesh size of the PEG hydrogel can be used to determine diffusivity/mass transfer of the candidate drug. Arrays of microsome and hepatocyte microreactors were patterned in PDMS devices by filling microfluidic chambers with hydrogels containing either microsomes or hepatocytes and photopolymerizing the microreactors using a patterned chrome mask. These arrays help to facilitate sample delivery.

A microfluidic device has been developed to test the cytotoxicity of anticancer drugs on 3D hydrogel cell cultures. This device, shown in Figure 10.7, has been designed to closely mimic the PK/PD effects expected in human drug trials by embedding colon cancer cells, hepatoma cells, and myeloblasts in separate, interconnected hydrogel regions within a single microfluidic platform [86]. Toxicity studies of tegafur, an oral prodrug of Fluorouracil 5-(FU), were carried out in this device and produced results that were consistent with previous clinical results.

Leclerc's group designed a device consisting of microfluidic chambers interconnected by a microfluidic network intended to closely mimic the kidney *in vitro* [87]. The fluidic network allows for continuous cell feeding as well as waste removal. Madin Darby Canine Kidney (MDCK) cells were cultured on PDMS microchips coated with fibronectin. Chronic toxicity as a result of ammonium chloride (0–10 mM) exposure was monitored via glucose consumption and NH_3 production using enzymatic reactions and direct and indirect absorbance measurements, respectively. The results indicated that at higher concentrations of ammonium chloride, glucose consumption was increased while NH_3 production decreased.

The effect of flow rate on glucose consumption and ammonia production was also determined for the MDCK culture.

10.4.3.3 Clinical Diagnostics

Another application for microfluidic devices is disease diagnosis in a clinical setting. Several applications have been published to this effect. For example, a device consisting of parallel channels was used to monitor C-peptide resistance in type II diabetes by quantifying glutathione (GSH) levels in red blood cells of diseased and control patients [88]. Mohammed and coworkers developed and characterized a device, designed to screen the functionality of pancreatic islets of Langerhans, before transplantation into the organ recipient by fluorescently imaging the mitochondrial membrane potential and intracellular calcium as well as performing an Enzyme-linked immunosorbent assay (ELISA) to measure secreted insulin [89]. Satoh et al. [90] describe a PDMS/glass hybrid microfluidic device for hepatocyte culturing in 1 ml volume using electrowetting-based liquid handling to deliver nanoliter solution volumes (Figure 10.8). Ammonia metabolism was measured in primary hepatocytes using an integrated sensor consisting of an iridium oxide pH electrode and an Ag/AgCl reference electrode. A microfluidic device for multiplexed isolation of CD4 and CD8 T-cells from RBC depleted blood capable of monitoring secreted cytokines was developed for monitoring and diagnosing HIV and other immunological diseases [91].

10.5
A Look Forward

Microfluidics and MCE are emerging technologies, which show promise as useful analytical tools in the field of metabolomics. The ability to integrate sample preprocessing and measurement capabilities onto a single platform is exciting, in that it has the potential to increase efficiency and streamline workflow. Additionally, the rapid analysis times inherent to most microfluidic systems will allow for increased sample throughput, thus decreasing the total time needed to conduct a metabolomics study. Currently, the use of integrated microfluidic systems such as these has been limited, but the effectiveness of the components comprising these systems has been successfully demonstrated. Preprocessing components are largely used for sample cleanup, preconcentration, and automated on-chip mixing of assay reagents. In regards to analysis, microfluidics has largely been utilized in the form of MCE for separation of target metabolites. Continuing advances in microchip LC and GC technologies may soon make them viable alternatives to MCE as separation techniques for micrometabolomics studies.

Innovations in microscale-LC analysis include chip-based nanospray devices such as those developed by Hop for PK and metabolite identification studies [92] and Wickremsinhe et al. [93] for monitoring circulating levels of the antidepressant Reboxetine in dogs. The combination of microchip LC with MS allows for much higher sample throughput as well as identification of unknowns. Furthermore, the

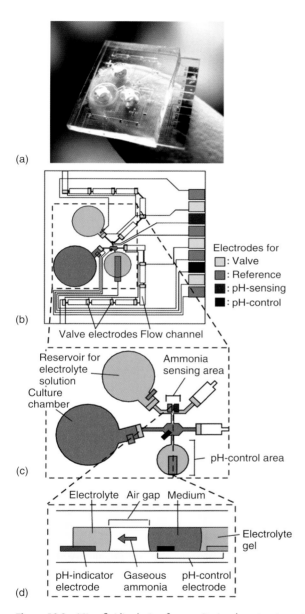

Figure 10.8 Microfluidic device for monitoring hepatocyte metabolism of ammonia. (a) Assembled device, (b) schematic of microfluidic components, (c) ammonia sensing area of chip, and (d) cross-section of ammonia sensor. (Reprinted from Ref. [90] with permission from RSC.)

cost of microchip LC columns is less than that of traditional columns. In 2008, Lin et al. [94] introduced a nanoSpliter LC-MS coupled to a microdroplet NMR for the identification of trace natural products in cyanbacterial crude extract. All of these examples point toward the capabilities of microchip LC as an attractive emerging platform for metabolomics analysis.

Microfluidic systems that are able to detect and measure temporal and spatial metabolic changes in response to stimulus or chemical exposure will serve as useful tools for PK/PD studies and clinical diagnosis. Perhaps the biggest challenge for single cell metabolomics within a microfluidic device is the integration of function into microfluidic devices and analysis of true biological samples. Examples have been given of organ systems on a chip; however, these devices still require off-chip detection methods. Spatially resolved imaging of metabolites including phospholipids within tissues can provide valuable insight into underlying biological processes. This area is relatively new with most literature being published on entire mouse organs or large histological biopsy samples. For example, Murphy's group has been developing the ability to define regional distributions of lipids within rat brain tissues using matrix-assisted laser desorption ionization (MALDI) imaging [95]. Imaging MS has also been used by Caprioli's group to examine the distribution of therapeutic drugs and their metabolites from dosed tissue [96]. In another report, Jun et al. [97] report high spatial, high mass resolution of surface metabolites of the plant *Arabidopsis thaliana* by reducing the MALDI laser spot size. While these techniques provided important new information, MALDI imaging is not routine and the overall measurements could benefit from a simplified microfluidic approach.

The advent of MCE-MS [19] should open up new doors for more comprehensive metabolite studies using microfluidics. With continued advances in microfluidic and total analysis system engineering, it may not be long before these systems are routinely used for conducting complex metabolomics studies. Furthermore, as the field of single cell metabolomics continues to advance, these technologies may soon be used to image a small group of metabolites of interest produced by individual cells (or small groups of cells) immobilized on a microfluidic platform.

References

1. Garcia-Perez, I., Vallejo, M., Garcia, A., Legido-Quigley, C., and Barbas, C. (2008) *J. Chromatogr. A*, **1204** (2), 130–139.
2. Dunn, W.B. and Ellis, D.I. (2005) *Trac-Trends Anal. Chem.*, **24** (4), 285–294.
3. Kraly, J.R., Holcomb, R.E., Guan, Q., and Henry, C.S. (2009) *Anal. Chim. Acta*, **653** (1), 23–35.
4. Jonsson, P., Gullberg, J., Nordstrom, A., Kusano, M., Kowalczyk, M., Sjostrom, M., and Moritz, T. (2004) *Anal. Chem.*, **76** (6), 1738–1745.
5. Ramautar, R., Somsen, G.W., and de Jong, G.J. (2009) *Electrophoresis*, **30** (1), 276–291.
6. Merten, C.A., Vyawahare, S., and Griffiths, A.D. (2010) *Chem. Biol.*, **17** (10), 1052–1065.
7. Wlodkowic, D. and Cooper, J.M. (2010) *Curr. Opin. Chem. Biol.*, **14** (5), 556–567.

8. Wang, D.J. and Bodovitz, S. (2010) *Trends Biotechnol.*, **28** (6), 281–290.
9. Lenshof, A. and Laurell, T. (2010) *Chem. Soc. Rev.*, **39** (3), 1203–1217.
10. Beebe, D.J. and Young, E.W.K. (2010) *Chem. Soc. Rev.*, **39** (3), 1036–1048.
11. West, J., Becker, M., Tombrink, S., and Manz, A. (2008) *Anal. Chem.*, **80** (12), 4403–4419.
12. Kim, T., Vinuselvi, P., Park, S., Kim, M., Park, J.M., and Lee, S.K. (2011) *Int. J. Mol. Sci.*, **12** (6), 3576–3593.
13. Zeng, A.P., Wurm, M., Schopke, B., Lutz, D., and Muller, J. (2010) *J. Biotechnol.*, **149** (1-2), 33–51.
14. Marchiarullo, D.J., Lim, J.Y., Vaksman, Z., Ferrance, J.P., Putcha, L., and Landers, J.P. (2008) *J. Chromatogr. A*, **1200** (2), 198–203.
15. Yang, Y., Li, C., Kameoka, J., Lee, K.H., and Craighead, H.G. (2005) *Lab Chip*, **5** (8), 869–876.
16. Benetton, S., Kameoka, J., Tan, A.M., Wachs, T., Craighead, H., and Henion, J.D. (2003) *Anal. Chem.*, **75** (23), 6430–6436.
17. Tan, A.M., Benetton, S., and Henion, J.D. (2003) *Anal. Chem.*, **75** (20), 5504–5511.
18. Marchiarullo, D.J., Lim, J.Y., Vaksman, Z., Ferrance, J.P., Putcha, L., and Landers, J.P. (2008) *J. Chromatogr. A*, **1200** (2), 198–203.
19. Zhang, B., Liu, H., Karger, B.L., and Foret, F. (1999) *Anal. Chem.*, **71** (15), 3258–3264.
20. Mansfield, C.D., Man, A., and Shaw, R.A. (2006) *IEE Proc. Nanobiotechnol.*, **153** (4), 74–80.
21. Shaw, R.A., Rigatto, C., Reslerova, M., Ying, S.L., Man, A., Schattka, B., Battrell, C.F., Matthewson, J., and Mansfield, C. (2009) *Analyst*, **134** (6), 1224–1231.
22. Beebe, D.J., Mensing, G.A., and Walker, G.M. (2002) *Annu. Rev. Biomed. Eng.*, **4**, 261–286.
23. Thorsen, T., Maerkl, S.J., and Quake, S.R. (2002) *Science*, **298** (5593), 580–584.
24. Urbanski, J.P., Johnson, M.T., Craig, D.D., Potter, D.L., Gardner, D.K., and Thorsen, T. (2008) *Anal. Chem.*, **80** (17), 6500–6507.
25. Faure, K. (2010) *Electrophoresis*, **31** (15), 2499–2511.
26. Monton, M.R.N. and Soga, T. (2007) *J. Chromatogr. A*, **1168** (1-2), 237–246.
27. Jorgenson, J.W. and Lukacs, K.D. (1981) *Anal. Chem.*, **53** (8), 1298–1302.
28. Landers, J.P. (2008) *Handbook of Capillary and Microchip Electrophoresis and Associated Microtechniques*, CRC Press, Boca Raton, FL.
29. Manz, A., Harrison, D.J., Verpoorte, E.M.J., Fettinger, J.C., Paulus, A., Ludi, H., and Widmer, H.M. (1992) *J. Chromatogr.*, **593** (1-2), 253–258.
30. Harrison, D.J., Manz, A., Fan, Z.H., Ludi, H., and Widmer, H.M. (1992) *Anal. Chem.*, **64** (17), 1926–1932.
31. Ewing, A.G., Wallingford, R.A., and Olefirowicz, T.M. (1989) *Anal. Chem.*, **61** (4), 292A–294A, 296A, 298A, 300A–303A.
32. Jacobson, S.C., Hergenroder, R., Koutny, L.B., Warmack, R.J., and Ramsey, J.M. (1994) *Anal. Chem.*, **66** (7), 1107–1113.
33. Jacobson, S.C., Koutny, L.B., Hergenroder, R., Moore, A.W., and Ramsey, J.M. (1994) *Anal. Chem.*, **66** (20), 3472–3476.
34. Shultz-Lockyear, L.L., Colyer, C.L., Fan, Z.H., Roy, K.I., and Harrison, D.J. (1999) *Electrophoresis*, **20** (3), 529–538.
35. Garcia, C.D., Dressen, B.M., Henderson, A., and Henry, C.S. (2005) *Electrophoresis*, **26** (3), 703–709.
36. Horvath, J. and Dolnik, V. (2001) *Electrophoresis*, **22** (4), 644–655.
37. Barbas, C., Vallejo, M., Garcia, A., Barlow, D., and Hanna-Brown, A. (2008) *J. Pharm. Biomed. Anal.*, **47** (2), 388–398.
38. Garcia-Perez, I., Whitfield, P., Bartlett, A., Angulo, S., Legido-Quigley, C., Hanna-Brown, M., and Barbas, C. (2008) *Electrophoresis*, **29** (15), 3201–3206.
39. Yang, Y., Boysen, R.I., Matyska, M.T., Pesek, J.J., and Hearn, M.T. (2007) *Anal. Chem.*, **79** (13), 4942–4949.
40. Ohno, K.-I., Tachikawa, K., and Manz, A. (2008) *Electrophoresis*, **29** (22), 4443–4453.
41. Johnson, M.E. and Landers, J.P. (2004) *Electrophoresis*, **25** (21-22), 3513–3527.

42. Liu, B.F., Hisamoto, H., and Terabe, S. (2003) *J. Chromatogr. A*, **1021** (1-2), 201–207.
43. Schulze, P., Ludwig, M., Kohler, F., and Belder, D. (2005) *Anal. Chem.*, **77** (5), 1325–1329.
44. Allen, P.B., Doepker, B.R., and Chiu, D.T. (2009) *Anal. Chem.*, **81** (10), 3784–3791.
45. Schulze, P., Ludwig, M., Kohler, F., and Belder, D. (2005) *Anal. Chem.*, **77** (5), 1325–1329.
46. Liu, B.-F., Hisamoto, H., and Terabe, S. (2003) *J. Chromatogr. A*, **1021** (1-2), 201–207.
47. Callewaert, N., Contreras, R., Mitnik-Gankin, L., Carey, L., Matsudaira, P., and Ehrlich, D. (2004) *Electrophoresis*, **25** (18-19), 3128–3131.
48. Ma, B., Zhang, G.H., Qin, J.H., and Lin, B.C. (2009) *Lab Chip*, **9** (2), 232–238.
49. Miyado, T., Tanaka, Y., Nagai, H., Takeda, S., Saito, K., Fukushi, K., Yoshida, Y., Wakida, S., and Niki, E. (2004) *J. Chromatogr. A*, **1051** (1-2), 185–191.
50. Miyado, T., Tanaka, Y., Nagai, H., Takeda, S., Saito, K., Fukushi, K., Yoshida, Y., Wakida, S., and Niki, E. (2006) *J. Chromatogr. A*, **1109** (2), 174–178.
51. Miyado, T., Wakida, S., Aizawa, H., Shibutani, Y., Kanie, T., Katayama, M., Nose, K., and Shimouchi, A. (2008) *J. Chromatogr. A*, **1206** (1), 41–44.
52. Guihen, E. and Glennon, J.D. (2005) *J. Chromatogr. A*, **1071** (1-2), 223–228.
53. Fouad, M., Jabasini, M., Kaji, N., Terasaka, K., Tokeshi, M., Mizukami, H., and Baba, Y. (2008) *Electrophoresis*, **29** (11), 2280–2287.
54. Szymanska, E., Markuszewski, M.J., Capron, X., van Nederkassel, A.M., Vander Heyden, Y., Markuszewski, M., Krajka, K., and Kaliszan, R. (2007) *J. Pharm. Biomed. Anal.*, **43** (2), 413–420.
55. Vandaveer, W.R., Pasas-Farmer, S.A., Fischer, D.J., Frankenfeld, C.N., and Lunte, S.M. (2004) *Electrophoresis*, **25** (21-22), 3528–3549.
56. Chen, R.S., Cheng, H., Wu, W.Z., Ai, X.O., Huang, W.H., Wang, Z.L., and Cheng, J.K. (2007) *Electrophoresis*, **28** (19), 3347–3361.
57. Vlckova, M. and Schwarz, M.A. (2007) *J. Chromatogr. A*, **1142** (2), 214–221.
58. Chen, C.M., Ho, Y.H., Wu, S.M., Chang, G.L., and Lin, C.H. (2009) *J. Nanosci. Nanotechnol.*, **9** (2), 718–722.
59. Zhang, Q.L., Lian, H.Z., Wang, W.H., and Chen, H.Y. (2005) *J. Chromatogr. A*, **1098** (1-2), 172–176.
60. Zhang, Q.L., Xu, J.J., Li, X.Y., Lian, H.Z., and Chen, H.Y. (2007) *J. Pharm. Biomed. Anal.*, **43** (1), 237–242.
61. LaCourse, W.R. (1997) *Pulsed Electrochemical Detection in High-Performance Liquid Chromatography*, John Wiley & Sons, Inc., New York, NY.
62. Evrovski, J., Callaghan, M., and Cole, D.E.C. (1995) *Clin. Chem.*, **41** (5), 757–758.
63. Gamache, P., Ryan, E., Svendsen, C., Murayama, K., and Acworth, I.N. (1993) *J. Chromatogr. B*, **614** (2), 213–220.
64. Gamache, P.H., Meyer, D.F., Granger, M.C., and Acworth, I.N. (2004) *J. Am. Soc. Mass Spectrom.*, **15** (12), 1717–1726.
65. Vigneau-Callahan, K.E., Shestopalov, A.I., Milbury, P.E., Matson, W.R., and Kristal, B.S. (2001) *J. Nutr.*, **131** (3), 924S–9932.
66. Holcomb, R.E., Kraly, J.R., and Henry, C.S. (2009) *Analyst*, **134** (3), 486–492.
67. Zhang, B., Heien, M.L., Santillo, M.F., Mellander, L., and Ewing, A.G. (2011) *Anal. Chem.*, **83** (2), 571–577.
68. Yeon, J.H. and Park, J.K. (2007) *Biochip J.*, **1** (1), 17–27.
69. Heo, Y., Cabrera, L., Song, J., Futai, N., Tung, Y., Smith, G., and Takayama, S. (2007) *Anal. Chem.*, **79** (3), 1126–1134.
70. Song, H. and Ismagilov, R.F. (2003) *J. Am. Chem. Soc.*, **125** (47), 14613–14619.
71. Huebner, A., Srisa-Art, M., Holt, D., Abell, C., Hollfelder, F., DeMello, A.J., and Edel, J.B. (2007) *Chem. Commun.*, (12), 1218–1220.
72. Chiu, D.T. and Lorenz, R.M. (2009) *Acc. Chem. Res.*, **42** (5), 649–658.
73. Hettiarachchi, K. and Lee, A.P. (2010) *J. Colloid Interface Sci.*, **344** (2), 521–527.

74. Huebner, A., Bratton, D., Whyte, G., Yang, M., Demello, A.J., Abell, C., and Hollfelder, F. (2009) *Lab Chip*, **9** (5), 692.
75. Wlodkowic, D., Faley, S., Zagnoni, M., Wikswo, J.P., and Cooper, J.M. (2009) *Anal. Chem.*, **81** (13), 5517–5523.
76. Song, H., Chen, T., Zhang, B., Ma, Y., and Wang, Z. (2010) *Biomicrofluidics*, **4** (4), 044104.
77. Dishinger, J.F., Reid, K.R., and Kennedy, R.T. (2009) *Anal. Chem.*, **81** (8), 3119–3127.
78. Zhang, X. and Roper, M. (2009) *Anal. Chem.*, **81** (3), 1162–1168.
79. Klauke, N., Smith, G., and Cooper, J. (2007) *Lab Chip*, **7** (6), 731–739.
80. Cheng, W., Klauke, N., Sedgwick, H., Smith, G., and Cooper, J. (2006) *Lab Chip*, **6** (11), 1424–1431.
81. Yoo, S., Lee, J., Yun, S., Gu, M., and Lee, J. (2007) *Biosens. Bioelectron.*, **22** (8), 1586–1592.
82. Clark, A., Sousa, K., Jennings, C., MacDougald, O., and Kennedy, R. (2009) *Anal. Chem.*, **81** (6), 2350–2356.
83. Prot, J.-M., Videau, O., Brochot, C., Legallais, C., Bénech, H., and Leclerc, E. (2011) *Int. J. Pharm.*, **408** (1-2), 67–75.
84. Gao, D., Wei, H., Guo, G.S., and Lin, J.M. (2010) *Anal. Chem.*, **82** (13), 5679–5685.
85. Zguris, J., Itle, L., Hayes, D., and Pishko, M. (2005) *Biomed. Microdevices*, **7** (2), 117–125.
86. Sung, J.H. and Shuler, M.L. (2009) *Lab Chip*, **9** (10), 1385.
87. Baudoin, R., Griscom, L., Monge, M., Legallais, C., and Leclerc, E. (2007) *Biotechnol. Prog.*, **23** (5), 1245–1253.
88. Oblak, T., Meyer, J., and Spence, D. (2009) *Analyst*, **134** (1), 188–193.
89. Mohammed, J.S., Wang, Y., Harvat, T.A., Oberholzer, J., and Eddington, D.T. (2009) *Lab Chip*, **9** (1), 97–106.
90. Satoh, W., Takahashi, S., Sassa, F., Fukuda, J., and Suzuki, H. (2009) *Lab Chip*, **9** (1), 35–37.
91. Zhu, H., Stybayeva, G., Macal, M., Ramanculov, E., George, M., Dandekar, S., and Revzin, A. (2008) *Lab Chip*, **8** (12), 2197–2205.
92. Hop, C.E. (2006) *Curr. Drug Metab.*, **7** (5), 557–563.
93. Wickremsinhe, E.R., Ackermann, B.L., and Chaudhary, A.K. (2005) *Rapid Commun. Mass Spectrom.*, **19** (1), 47–56.
94. Lin, Y., Schiavo, S., Orjala, J., Vouros, P., and Kautz, R. (2008) *Anal. Chem.*, **80** (21), 8045–8054.
95. Murphy, R.C. and Gaskell, S.J. (2011) *J. Biol. Chem.*, **286** (29), 25427–25433.
96. Cornett, D.S., Frappier, S.L., and Caprioli, R.M. (2008) *Anal. Chem.*, **80** (14), 5648–5653.
97. Jun, J.H., Song, Z., Liu, Z., Nikolau, B.J., Yeung, E.S., and Lee, Y.J. (2010) *Anal. Chem.*, **82** (8), 3255–3265.

11
Data Processing in Metabolomics
Age K. Smilde, Margriet M.W.B. Hendriks, Johan A. Westerhuis, and Huub C.J. Hoefsloot

11.1
Introduction and Scope

To set the scene for this chapter, Figure 11.1 shows the metabolomics pipeline. This is a general pipeline for a metabolomics study showing the different steps. In many of these steps, data analysis methods are needed [1]. At the start of the study, the design of experiments methods have to be used to set up a proper study design. Such a study design should be accompanied by a proper measurement design to assure high-quality data without confounding. The measurement design is especially important when large series of samples have to be measured [2]. After the measurements have been performed, the resulting raw data should be preprocessed to arrive at clean data, suitable for the subsequent processing. This is usually a very time-consuming effort with its own difficulties. Depending on the type of preprocessing, the resulting clean data can be in the form of standardized intensities or relative concentrations. Only in rare cases (with a lot of instrumental effort), absolute concentrations are obtained. We do not discuss preprocessing in this chapter but focus on the processing of clean data. The aim is to draw statistical conclusions from the clean data, which then have to be interpreted in terms of the substantive biology. In this chapter, we discuss such data processing methods for a wide range of metabolomics applications.

11.2
Characteristics of Metabolomics Data

11.2.1
Correlation Structure of Metabolomics Data

Measured metabolite concentrations are correlated. These correlations and the structure of the resulting correlation matrix depend crucially on the type of experiment, the design of the experiment, the biological system under study, and the measurement error of the instrument.

Metabolomics in Practice: Successful Strategies to Generate and Analyze Metabolic Data, First Edition.
Edited by Michael Lämmerhofer and Wolfram Weckwerth.
© 2013 Wiley-VCH Verlag GmbH & Co. KGaA. Published 2013 by Wiley-VCH Verlag GmbH & Co. KGaA.

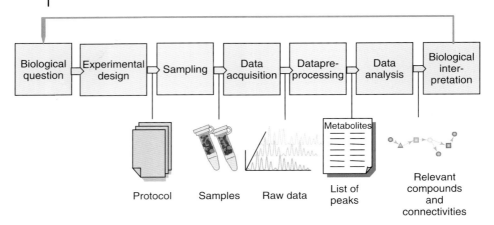

Figure 11.1 The metabolomics pipeline.

Two examples of the interpretation of correlations are the following. First, if measurements are made directly on cellular systems (e.g., microorganisms) whereby the differences between samples are pure biological variation, then it can be expected that the correlations between the metabolite concentrations can be linked to the metabolic network. This relationship, however, is not at all simple and unequivocal [3]. Second, another type of correlation arises when considering measured metabolite levels in plasma or serum for a number of subjects. As blood collects metabolites from different tissues in the body, it is not immediately clear how these correlations can be interpreted in biochemical pathways of the different cell types in the human body.

In almost all metabolomics cases, the interpretation of correlations is even more problematic. In many studies, there is a design underlying the samples, for example, subjects that underwent a treatment. Calculating correlations across samples mixes up two types of variation: group or treatment effects and biological variation. This requires even more careful analysis and interpretation of correlations (Sections 11.3.3). In general, the interpretation depends on the conceptual model underlying the data generation and requires substantial knowledge of the relevant biology [4].

11.2.2
Informative versus Noninformative Variation

In most cases, variation in metabolomics data can be divided into informative and noninformative variation. An obvious example is biomarker studies where the informative variation is present in the biomarkers and the other metabolites do not carry information about the disease. Separating informative from noninformative variation translates then into the statistical problem of variable selection (Sections 11.5.2, 11.5.4, and 11.5.5).

For exploratory purposes, it is useful to separate informative from noninformative variation because the important patterns become more clear than without such a separation. Usually, the informative parts can have clear and strong correlation structures, whereas the noninformative part has a more "loose" correlation structure. Simplivariate models are based on this idea and try to (i) separate the informative from the noninformative parts and (ii) model the informative parts with simple models [5].

11.2.3
Low Samples-to-Variables Ratio

A general characteristic of metabolomics data is its low samples-to-variables ratio. This holds also true for other functional genomics data such as transcriptomics and proteomics data. Depending on the type of metabolomics application, the number of samples ranges from tens to hundreds and the number of variables from hundreds to about a thousand. This has repercussions for the processing of metabolomics data, as the risk of overfitting is high and special methods have to be used to overcome the curse of dimensionality (Section 11.4.2). Careful validation is necessary (Section 11.4) before reporting results.

11.2.4
Measurement Error

In almost all methods currently in use for analyzing metabolomics data, the underlying assumption is that the data are identically and independently distributed. This is a very famous assumption in statistics (called the *iid assumption*), and it does not hold for metabolomics data. This has been shown at several instances [6, 7] but has hardly resulted in an adaptation of the standard methods. Yet, measurement error structure can be assessed by a proper experimental design using repeated measurements. Subsequently, maximum likelihood methods adjusting for heterogeneous error structure can be used [8]. Admittedly, these methods are more difficult to use and are not yet available in the standard toolbox. However, they may proof to be much better in the long run in giving answers to the questions raised.

11.2.5
Dynamics

A growing number of metabolomics applications contain dynamic or time-resolved data. The reason is that dynamic data usually gives more information about the biological system than static data, for example, in assessing homeostatic resilience, onset of toxicity, and development of diseases. The number of time points at which the samples are taken is typically low, and this restricts the range of methods that can be used [9].

11.2.6
Nonlinear Relations

Almost certainly, there exist nonlinear relationships between measured metabolites of a system. This is most clear for cellular metabolism governed by (nonlinear) kinetics. Depending on the type of perturbation and the type of sampling, there will certainly be nonlinear relationships between the concentrations of the metabolites. For whole body samples (plasma, serum, and urine), such relationships will also exist but are more difficult to interpret. Most of the data processing methods in metabolomics are linear by nature and will approximate these nonlinear relationships to a certain extent. Nonlinear methods can be useful, but the type of nonlinearity has to be chosen and using these methods usually requires many samples.

11.3
Types of Biological Questions Asked

11.3.1
Methods Should Follow the Questions

There is a multitude of biological questions that are addressed by metabolomics. It is important to retain a close match between the question asked, the data collected, and the data analysis method employed. In this section, a very brief overview is given of the potential biological questions together with a quick preview of a possible suitable data analysis methods.

11.3.2
Biomarkers

The largest number of applications in metabolomics concerns biomarkers. A biomarker "marks" a biological process and this already indicates the versatility of the concept. Biomarkers can be used as diagnostics for discovering early stages of disease, to follow disease progression or recovery, to follow the impact of toxic insults, and so on. Many of such applications can be cast in the framework of discriminating between controls and diseased which calls for discriminant analysis methods with variable selection (Section 11.5.5).

11.3.3
Treatment Effects

Treatments can come in different ways. In nutritional research, treatments are usually in the form of diets or special food ingredients. In drug studies, treatments are usually in the form of administering a certain drug. Other treatments can be toxic compounds or other ways of manipulating a system to study its resilience.

The treatment effects can differ greatly in size. In nutritional studies, the effects are usually small, superimposed on between-individual variation, and can also differ in nature. This calls for special designs (cross-over designs) and, subsequently, special data analysis methods (Multilevel methods, see Section 11.5.6). Contrary, in toxicological studies or drug-related studies, the treatment effect is usually larger and can be seen with more standard methods such as OPLSDA or PCA. Even in those cases it might pay to use more advanced methods such as ASCA and PARAFASCA to explore the data and find underlying patterns which go otherwise unnoticed.

11.3.4
Networks and Mechanistic Insight

An upcoming area in metabolomics is to collect metabolite concentration data to infer metabolic networks. This proves to be a difficult task but it is completely focused at obtaining results directly interpretable in terms of biochemistry [10, 11]. In some cases, an association network is built from the data not directly reflecting the biochemistry but relationships at a higher aggregation level, for example, correlation networks. Methods to perform this task are discussed in Section 11.5.7.

11.4
Validation

11.4.1
Several Levels of Validation

There are different levels of validation; these come down to distinguishing different levels of validity [12].

On the lowest level, a numerical algorithm should be correct: it should do what it is supposed to do. In other words, the outcome should be stable from a numerical point of view. In commercial software (e.g., SAS or SIMCA), the algorithms are tested extensively and only allowed to be incorporated once they have proved to be stable. Such a rigorous testing is usually not done for freeware; hence, this software should be used with some care.

Once the result of an algorithm becomes available, the next question is whether this is statistically a sound result. In other words, is there a reasonable probability that the result is not a chance result? This question proves hard to tackle and is discussed in more depth in this section.

The ultimate validation is the biology. Is the result also biologically valid? The best way to test this is to run a follow-up experiment based on this result and specifically testing this result in a well-defined system. Unfortunately, this is not done often probably because of high costs and sociological/psychological factors as being not perceived as original work.

11.4.2
Curse of Dimensionality

An essential property of most metabolomics studies is that the number of samples is (much) lower than the numbers of variables (i.e., measured metabolites, channels of the NMR instruments, number of peaks in LC–MS, etc.). This difference becomes even more pronounced when considering the trend of fusing different analytical platforms. This property has repercussions for the data processing, both for univariate and multivariate analysis.

For univariate analysis, care has to be taken when testing multiple hypotheses. Suppose that we want to distinguish between two groups (case vs control) and we have measured 100 metabolites. Then, simple t-tests (or nonparametric versions) can be run on these 100 metabolites at a level of significance $\alpha = 0.05$. If there is no biological effect, then purely by chance we can expect five significant findings (Section 11.5.2).

Although the problem of multiple testing is avoided in multivariate approaches (since all variables are considered simultaneously), another problem is introduced in the multivariate arena: the curse of dimensionality. This problem is illustrated in Figures 11.2 and 11.3.

Figure 11.2 illustrates the ideal case for multivariate analysis: many samples and a limited number of variables. For sake of clarity, only two variables are considered but this should be seen as illustrative for the many samples — limited variables scenario. Most classical multivariate analysis methods work well in this situation, and if grouping is present (indicated by the colored dots), then it is relatively easy to find a line separating the groups. Figure 11.3 shows the opposite: many variables (four) — few samples (three). This is exaggerated in the figure to make a clear case: in practice, the number of samples may be 50 and the number of metabolites, for example, 200. The grouping present in the data is only carried by a small number of samples. Indeed, many random separation lines (broken lines) can be drawn with an equal performance as the true one (drawn line)

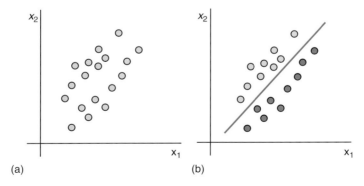

Figure 11.2 The oversampled case: (a) many samples and two variables; (b) a grouping structure can easily be assessed (red and blue dots indicate two groups, and the green line is a separation line).

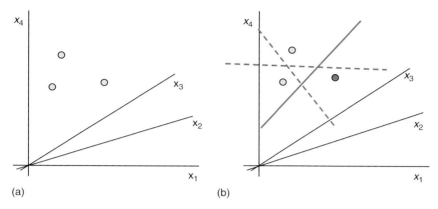

Figure 11.3 The undersampled case: (a) many variables and three samples; (b) a grouping structure cannot be assessed (red and blue dots indicate two groups, the drawn green line is the true separation line, and the broken lines are equally valid alternatives).

Clearly, avoiding spurious conclusions when using multivariate analysis methods requires careful validation.

11.4.3
Cross-Validation and Permutations

Generalizability is the central question regarding whether a result of metabolomics data processing is relevant: can the finding be reproduced in a completely new data set thereby giving credibility as to its general validity? One way of testing generalizability is to collect new data, repeat the whole analysis including (pre-)processing of the data and check whether the result is the same. This strategy can be implemented, for example, by holding out test samples, which are not used in the subsequent data processing and after obtaining a model based on the remaining samples (the training set) to check whether using the model on the test data gives good results. In almost all cases, there are too few samples to follow this strategy and the so-called resampling methods are used. One of those resampling methods is called *cross-validation*, the idea of which already goes back a long time [13].

We explain cross-validation with a much used application in mind, namely, finding biomarkers for a disease. The typical setup of such a biomarker study is to have cases and controls, perform metabolomics experiments (e.g., on the plasma, serum, or urine), and use discrimination methods to find the biomarkers. Technically, the problem comes down to finding a stable classifier and the variables that contribute most to this classification. Note that finding the good classifier is not the main goal of such an analysis, but finding good variables (i.e., biomarkers) is. However, the selected variables will depend on the model, and therefore (the stability of), the classification performance is used as a proxy for biomarker stability.

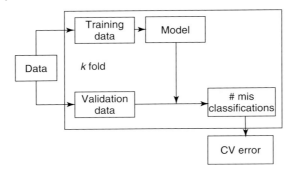

Figure 11.4 The basic layout of a cross-validation (CV) scheme.

The basic layout of a cross-validation scheme is given in Figure 11.4. The available data is split into two parts: the training data and the validation data. The discrimination model is then trained on the training data and used to classify the samples of the validation set. As the true class memberships of the samples in the validation set are known, the number of misclassifications can be recorded. This process is repeated a number of cycles where in each cycle a different split is made of the data in training and validation data in such a way that each object serves once and only once in the validation set. The crucial step underlying cross-validation is that a left out sample (i.e., a sample in the validation set) is not used in fitting the model. The end result is a cross-validation error. There are different ways of implementing this scheme: by leaving out a single sample or groups of samples. There is a growing consensus to use the leave-more-out cross-validation option, where in each cycle about 10% of the data is left out (k-fold cross-validation; [14]).

Although the idea of cross-validation is simple, the implementation and interpretation is not always straightforward. When the data set is structured (e.g., extra treatments are underlying the samples or some kind of stratification is present in the sampling direction), then the data split should reflect this. The resulting cross-validatory assessment should not be trusted on face value. When the number of samples is very low, then the sampling is likely to be not representative [15] and the cross-validation result should be interpreted with great care.

A problem often encountered while performing cross-validation is that hidden in the whole procedure still the validation samples are used, for example, for model selection. This is a well-known phenomenon (called *selection bias*; [16]), but it can be obscured in the procedure. An example is the use of cross-validation in establishing the value of a so-called meta-parameters (MP; such as the number of latent variables in a PLSDA or PCDA method). One way of setting up cross-validation for this purpose is depicted in Figure 11.5.

A regular cross-validation is ran for different values of the MP. A plot is made of the MP value (x-axis) *versus* the cross-validation error (y-axis), and the MP minimizing the cross-validation error is then selected. The critical part of this procedure is that all the validation samples contribute to making such a

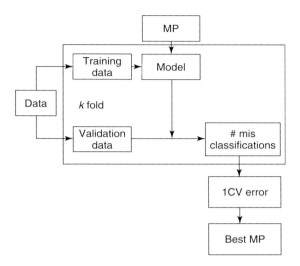

Figure 11.5 Cross-validation used for meta-parameter (MP) optimization.

performance plot and thereby affect the model. Stated otherwise, the validation samples are not completely independent of the selected model. One way to circumvent this problem is by using a double-cross-validation scheme as depicted in Figure 11.6. Although the scheme might seem a bit complicated, the basic idea is to nest two cross-validation schemes, thereby assuring that the test data results are independent of the model. This kind of schemes goes under the name of double cross-validation [17] or cross-model-validation [18], although the basic idea was stated already much earlier [13]. Note that in the biomarker example discussed

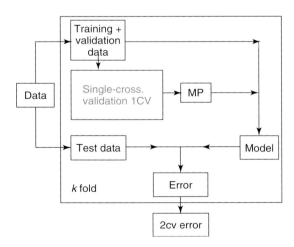

Figure 11.6 Double cross-validation used for metaparameter (MP) optimization and independent statistical assessment.

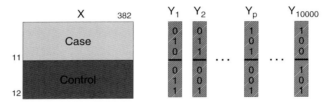

Figure 11.7 Setup of a case-control biomarker study including permutations.

earlier the variable selection should also be subjected to a double cross-validation to ensure unbiasedness.

Another basic notion to validate results obtained from data processing is by asking the simple question: could such a result have been obtained by randomness alone? This is not a superfluous question, as is illustrated in Section 11.4.2. An already very old way of answering this question is by providing the used method with random data. There are different ways of doing this, but permutations are very convenient in this respect because they stay close to the original data (structure). The idea of permutations will be explained for the prototypical example of metabolomics: finding biomarkers of a disease in a case-control study using a classifier (Section 11.5.5).

The setup of the study is that 11 cases are available and 12controls of which the same 382 metabolites are measured. This table makes up the matrix labeled as "X" in Figure 11.7. The technical way of performing a classification is by using a class label y that has values of 0 for the case and 1 for the control. Then, discriminant analysis techniques can be used to build a model between X and y (Section 11.5.5). New samples can be classified by using the model and new measured metabolites, which results in a predicted class label. This setting can also be used in cross-validation by leaving out samples, predicting the class labels, and comparing those predictions with the real labels. The statistic "number of misclassifications" can then be used to summarize the results: the lower the number of misclassifications, the better the classifier is. This number of misclassifications can be calculated using the fit (i.e., the model is built using all samples and then the same samples are "predicted" by the model) or using a single- or double-cross-validation scheme. The idea of permutations is now to built so-called nonsense models by permutating the class labels (leaving X intact!) and repeating the whole modeling procedure using these permutated class labels. This is performed a large number of times for different permuted labels, and all the results are by definition random results as the class labels do not correspond with the rows of the X matrix.

The result is shown in Figure 11.8 in the form of histograms. The cross-validations were performed multiple times with different leave-more-out schemes, resulting in a distribution of misclassifications (indicated by the red bars). These cross-validations were also used to estimate the MP (i.e., the number of latent variables in the PLSDA model, see Section 11.5.5), and since a different value of this MP may be obtained, also for the fit values, some variability in misclassification error is present. The blue bars are the result of the permuted class labels

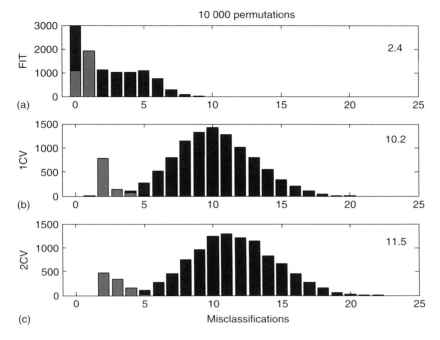

Figure 11.8 (a–c) Results of a permutation test.

and produce a distribution of misclassifications under the null hypothesis that the classes are equal in terms of the measured metabolites. Different observations can now be made. First, the fit results are far too optimistic, as expected: the average misclassification error of the permutation distribution is only 2.4 and obviously much too low. The single cross-validation performs better in this respect and gives an average of 10.2, which is still a bit too optimistic. The average misclassification of the permutation distribution for the double cross-validation is 11.5, which is exactly the value expected for a pure random assignment in groups of size 11 and 12. This is an illustration of the points made earlier that using (single) cross-validation both for MP optimization and for prediction is biased upward (i.e., too optimistic). Double cross-validation is correcting this. Second, the distribution of the "true" results (red bars) can be compared to the permutation distribution to get an idea of the significance of the result. If needed, a p-value associated with the quality of the classifier can be calculated. In this case, the classifier performs good as is evident from Figure 11.8c.

The beauty of permutation testing is its versatility. It can be used in virtually all kinds of classification and regression settings with a minimum of reprogramming of the software. In classification studies, the only requirement is to run the software multiple times with permuted class labels and collecting the end results in a meaningful manner. Another useful application is permuting the columns of X to check on the importance of a metabolite associated with that column.

11.5
Overview of Methods

11.5.1
Exploratory Analysis

One of the workhorses in metabolomics data processing is principal component analysis (PCA). This method already goes back a long time [19], and good books exist about the method [20]. The basic idea behind PCA is simple and explained in Figure 11.9.

When the objects are arranged in a table with rows containing these objects and columns containing the measured metabolites M_1, M_2, and M_3, then each sample in such a row can be represented as a point in the three-dimensional space. Hence, all samples in such a table make up a cloud of points (Figure 11.9a). Not all dimensions in this space are equally occupied and this property is utilized by PCA: it finds the directions in space that are well populated and these directions are shown by the arrows in Figure 11.9a. These two arrows — also called principal components — define a plane, and the points in the original three-dimensional space can be projected onto this two-dimensional plane. These projected points are then visualized in Figure 11.9b; a score plot. This score plot shows the relative positions of the original samples and can thus be interpreted in terms of (dis)similarities between these samples. It is important to realize that even if we would have measured a hundred metabolites and thus would be working in the hundred-dimensional space, this projection on a lower-dimensional space would still be possible and would give a tremendous dimensional reduction.

The arrows in Figure 11.9a point in a certain direction, and this direction gives information about the relative importance of the original metabolites for the representation of Figure 11.9b. Such directions can also be plotted in the so-called

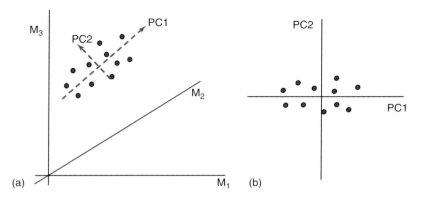

Figure 11.9 Principal component analysis: (a) points in the original space and (b) in the reduced space. For an explanation, see text.

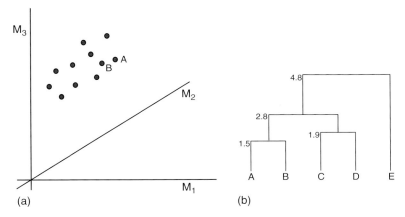

Figure 11.10 (a,b) Cluster analysis result: the dendrogram. For an explanation, see text.

loading plots. Of course, the projection is not perfect: the original points do not fall exactly onto the plane. The distance between the original and the projected point is called the (size of the) *residual*. The plane is not arbitrary: it minimizes the sum of those distances that correspond to a well-defined problem in data analysis with a simple solution.

Another very popular technique to visualize high-dimensional functional genomics data (especially in gene expression or transcriptomics) is cluster analysis. Starting again from Figure 11.10a, distances can be defined between the points. Those points that are close to each other are clustered in the same group (e.g., A and B). Then, new points are added to these groups and on the highest level the groups are clustered. The typical outcome of such an exercise is a dendrogram (Figure 11.10b). The points A and B are close together (distance = 1.5). and this is indicated in the dendrogram.

There are clear differences between cluster analysis and PCA. While PCA is a dimension reduction technique, it has the property of reducing the noise in the data. This property is not shared by cluster analysis. On the other hand, for an overview of all the data including the dimensions containing less variability, cluster analysis is convenient. Although cluster analysis does not provide an analog of a loading plot, more advance versions of cluster analysis do provide such measures, for example, in the case of Clustering Objects on Subsets of Attributes (COSA) [21].

11.5.2
ANOVA and Other Univariate Methods

Traditionally, assessing the difference between two groups of subjects or samples for a certain outcome (e.g., control vs treatment effect on triglyceride levels), Student's *t*-test is often used [22]. This *t*-test determines whether on average the value of a readout is different between two groups (unpaired *t*-test) or between two different situations for the same group (paired *t*-test, for example, comparison of a readout before and after a treatment). The *t*-test assumes that the outcomes for each group originate from normal distributions. If this assumption does not hold for the studied outcome, alternatives exist. One of these so-called distribution-free tests for testing differences between two groups is the Mann–Whitney or the Wilcoxon test [22]. This test ranks the outcomes of both groups, ranking the highest as 1 and so on. The sum of the rank numbers of one of the groups is then tested against the distribution of the sum of rank numbers of random outcome data.

ANOVA (analysis of variance) is a generalized version of the *t*-test [23]. Instead of testing the differences in outcome between two groups, it allows for testing of differences between multiple groups. Moreover, it allows for testing of multiple (nested) groupings or treatment types. An example of the latter is a study where a drug is administered to one group, while the other group did not receive the treatment (control group). If the groups consist of men and women and also the difference in outcome for each of these subgroups is of interest, the so-called two-factor ANOVA is to be used. Also, ANOVA assumes normality of the data.

In metabolomics, usually not a single outcome is of interest, but a large set of outcomes, the set of measured metabolites. How to deal with multivariate data sets has been discussed partly in the previous paragraph. Many researchers, however, are still very much attached to investing every outcome (e.g., metabolite levels) by itself instead of in relation to other metabolites. This results in multiple tests because for every outcome an ANOVA or *t*-test is performed. The drawback of this repeated univariate testing was discussed in Section 11.4.2. Especially in the exploratory phase of a study often very, many hypotheses are tested. As outlined in [24], chance findings may disturbingly influence the results and conclusions. Not only testing multiple outcomes may lead to erroneous conclusions but also when testing multiple subgroupings, predictors, or time points.

In the literature, many solutions to this problem induced by multiple testing can be found [25]. All solutions are based on adjusting the significance level for the number of tests performed. One of the oldest adjustments, Bonferroni correction, most often gives too conservative results for metabolomics data, because of the often high number of tests that are performed. A too conservative correction leads to loss of power: an increase of false negatives. Less conservative approaches take the correlation between the individual tests into account. Subsets of measured metabolites are likely to be correlated, making the assumption of independence of tests invalid. Newer methods take these relations into account and try to balance type I and type II errors (the number of false positives vs the number of false negatives). A popular one among these newer methods controls the false discovery rate (FDR), which is the proportion of false positives among the number of positives, for example, the number of tests for which the null hypothesis was rejected [26, 27].

The comprehensive profiling approach that is essential in metabolomics studies necessarily results in measuring informative as well as noninformative metabolites. Variable selection therefore is an important strategy in metabolomics and aims at finding a subset (which may have size 1) from the whole set of measured variables. Variable selection can be used for the purpose of model building, with the aim to create better and more parsimonious models, for enhancing insight, or for biomarker selection. Usually, variable selection techniques are divided into three classes [28]. The first class consists of filter techniques that act on the intrinsic properties of the data itself, ignoring the classification or prediction algorithm that is subsequently used. Examples are using *t*-tests, ANOVA, or rank products to select important variables. The second class consists of wrapper methods, where the variable selection is included in the prediction algorithm. These methods are more prone to overfitting and are computationally more intensive. Less computationally intensive alternatives are embedded techniques, where variable selection is built into the prediction algorithm, for instance, Random Forests [14] or the LASSO [29].

11.5.3
Advanced Exploratory Analysis

PCA and cluster analysis are powerful visualization tools, but they are less convincing when there is structure in the data. One such a structure is shown in Figure 11.11 where rats are subjected to a dose of a toxicant and their urine is collected for metabolomics analysis. For simplicity, only two doses ($d = 1$ and $d = 2$) and two time points ($t = 1$ and $t = 2$) are considered for five rats per group. The J metabolites measured per occasion can be arranged in a matrix (Figure 11.11), and this matrix can be subjected to a PCA or a cluster analysis. This is, however, not very insightful.

The reason for the lack of insight when using standard methods is that the data set contains different types of variation. There is variation due to treatment, time, and biological variation on the level of the rats per group. Both cluster analysis and

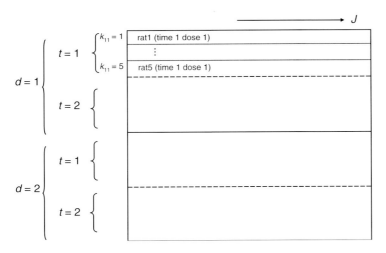

Figure 11.11 Structured metabolomics data from a rat toxicity study. For an explanation, see text.

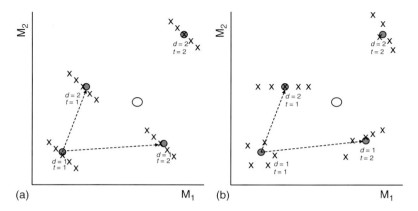

Figure 11.12 Data shown for the simple example for only two metabolites: (a) with equal between group correlation; (b) with different between group correlations. For an explanation, see text.

PCA completely disregard these levels of variation. A simple plot of the resulting data is shown in Figure 11.12 for only two metabolites. Each cross in the figure is a row in the data table and represents the measurement performed on one rat. The open dot is the overall mean of the data (usually subtracted from the data before a PCA), and the filled dots are the means per group. Figure 11.12a shows the case of equal within-group correlation, whereas Figure 11.12b shows the case of unequal within-group correlation. The different types of variation can be distinguished easily in this example, but for higher-dimensional data, similar configurations can be present.

A way to proceed visualizing such data is by using ASCA [29–31] or SMART [33], which explicitly take into account the underlying design. ASCA calculates the group means (filled dots) and visualizes those (Figure 11.13a) which immediately shows the effects of dose and time (and their interaction). The within-group variation can also be visualized (Figures 11.13b and c) and can be subjected to a PCA to study within-group variation. ASCA is a combination of ANOVA and PCA, whereby the initial ANOVA step is used to separate sources of variation and, subsequently, PCA is used to visualize these sources of variation.

Taking it one step further, ASCA can also be combined with PARAFAC; a generalization of PCA for analyzing three-way data [34]. The resulting method is called *PARAFASCA* and can be used for certain types of metabolomics data, namely, repeated measurements performed on rats in time, which are subjected to a treatment, for example, a dose of a toxicant. In such a study, rats were randomly assigned in three groups (control, low, and high dose) and their urine was collected at regular time points (two of which predose) [35]. The resulting NMR data can be analyzed by PCA (Figure 11.14), showing all data in one principal component arranged according to the time points. There are clearly patterns visible, such as treatment effects and within-group variability in the high-dose group.

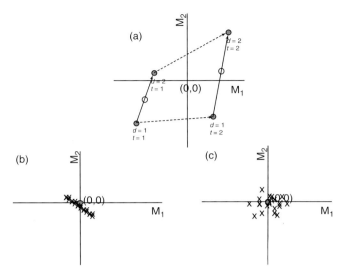

Figure 11.13 ASCA decomposition of the data and explanation: (a) the mean levels, (b) the within variation of Figure 11.12a, and (c) the within variation of Figure 11.12b. For an explanation, see text.

Subjecting the same data to an ASCA analysis and then continuing by decomposing the time-treatment-means (the dots of Figure 11.13a) further by PARAFAC gives the plots of Figure 11.15. There is a short-term effect (Figure 11.15a) contrasting control versus treatment (see loadings Figure 11.15b), and a long-term lasting effect (Figure 11.15c) contrasting control and low control versus high levels of toxicant. These figures are much more informative than a PCA on the whole data.

11.5.4
Regression Methods

Regression is the construction of a model that describes the connection of an outcome variable (denominated the dependent or response variable) with a set of other variables (denominated independent or predictor variables). Most researchers are familiar with ordinary multiple linear regression (MLR) with which linear associations between predictor and response variables can be assessed. Figure 11.16a shows MLR that can be used in the case of many samples (the rows of **X**), a few variables (the columns of **X**), and a single response **y**. The regression coefficient of a predictor variable in the regression model is a measure of the association with the response variable in the context of the other predictor variables.

Often, the set of potential predictor variables is large, and it is important to select a subset of relevant variables to explain variation in the response variable, thereby obtaining more robust and interpretable models. All kinds of variable subset selection methods exist for MLR. For information on all-subset, forward

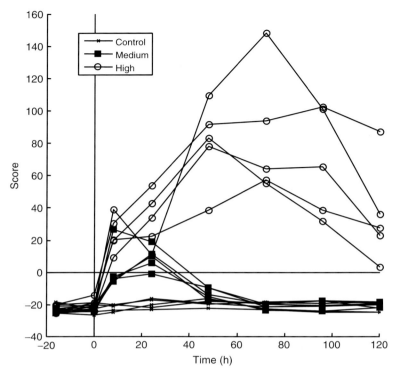

Figure 11.14 PCA of the rats in the PARAFASCA example. Each dot represents an individual rat. For an explanation, see text.

selection, backward elimination, and stepwise variable selection methods, we refer to [36]. In metabolomics, the predictor variables are usually the abundances of metabolites, while the response variable is some other extraneous variable, for example, phenylalanine production [37].

When the number of predictor variables exceeds the number of subjects or samples in the data set, MLR cannot be used anymore. Under this condition, there are many values for each of the regression coefficients that equally well exactly describe the response variable, so no unique solution exists. As this condition occurs very often in metabolomics studies, we have to resolve to other multivariate analysis techniques. These other approaches in the context of linear regression all aim at restricting the freedom in the estimation of the regression coefficients to obtain unique coefficient values. Principal component regression (PCR) [38] does this by using the first principal components of the predictor variables, those explaining the highest variation, of a PCA as a new set of predictor variables in an MLR. The number of principal components to be included in the regression model can be estimated using cross-validation or an independent test set. Partial least squares (PLS) regression resembles PCR, but the principal components (now called *latent vectors*) are not solely based on describing maximum variation in the

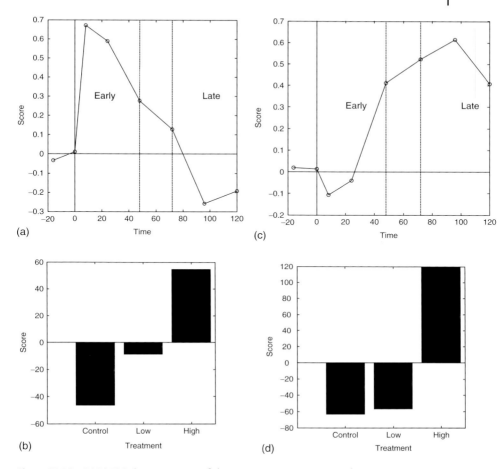

Figure 11.15 PARAFAC decomposition of the time-treatment-means in the rat toxicity example: (a) and (b) are the time and treatment loadings, respectively, of the short-term effect and (c) and (d) the corresponding ones for the long-term effect.

set of predictor variables (metabolite levels) but also on explaining variation in the response variable [39]. Figure 11.16b shows how PLS works. A linear combination (denoted by the vector **w**) of the variables produces a score vector **t**; this score vector **t** summarizes the variability across the samples in **X** and is used as a regressor for **y**.

Variable selection to come to more parsimonious, robust, and informative models in multivariate regression comes in many different shapes. For PLS, the most commonly used techniques are based on Variable Importance on Projection (VIP) [40], PLS weights [41], or regression coefficients [42].

If we represent metabolomics data in a table or matrix (such as **X**), then this has a two-way structure. In the first direction (or way) of the table, subjects or samples are the identifiers; for the second way, metabolites are the identifiers. Each entry

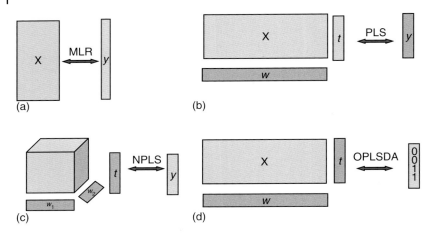

Figure 11.16 Regression and discrimination methods: (a) MLR, (b) PLS, (c) NPLS, and (d) OPLSDA. For an explanation, see text.

in the table, an abundance measure, has two identifiers: the sample in which and the metabolite for which it was measured. In certain circumstances, extra ways are added. Consider the case where for a set of subjects metabolite levels are measured at different moments in time. A time point now can be seen as a third identifier, and the abundance data can be represented in a three-way structure or data cube (Figure 11.16c). In cases where the data has a three (or multi)-way structure, it can be beneficial to use explicitly this structure in the data analysis. In Section 11.5.3, we have already seen an example of such a methodology for exploratory analysis (PARAFASCA). To relate a multiway data set of metabolite abundance data to an outcome variable, multilinear PLS (NPLS) is one of the proved methodologies [37, 43, 44].

11.5.5
Discriminant Analysis

An important class of methods in metabolomics is discriminant analysis methods. These methods model the relation between a set of predictor variables and a categorical response variable. A categorical variable is a variable that indicates to what class or group a certain sample or subject belongs. Linear discriminant analysis (LDA) is a method that can separate two or more classes based on a set of variables. The method uses a linear combination of the variables to create a rule by which sample classes can be separated. Unfortunately, this method is not applicable when the number of variables exceeds the number of samples. To overcome this problem, other methods have come available, two of which are described in this paragraph.

Principal component discriminant analysis (PCDA) [17] combines the methodology of PCA with LDA, by constructing an LDA model on a set of PCA score vectors. The number of principal components to be used in the LDA model can be estimated by a cross-validation procedure (Section 11.4.3).

Orthogonal partial least squares (OPLS) regression is an extension of PLS, where only the variation in the set of predictor variables (metabolite abundances) is retained that correlates with the response variable. If this response variable is a categorical variable, this automatically transforms to OPLSDA [45]. Figure 11.16d shows how this works: a PLS model is built between **X** and a categorical **y**. All the machinery for variable selection can be used also in the context of PCDA and OPLSDA. Special care has to be taken to validate the models properly because overfitting can occur easily [46].

11.5.6
Multilevel Approaches

One of the first steps to take when processing metabolomics data is to investigate the types and levels of variability in the data. Usually, many types of variability are present; some are wanted, while others are not (nuisance variation). An example of nuisance variation is variation between individuals in trials where treatment effects are considered. Especially, in nutritional studies, this interindividual variation can be problematic and obscures the (small) effects of dietary interventions [47].

A way of dealing with unwanted interindividual variation is by using the properties of paired data. The basic idea of pairing of data for univariate analysis is explained in Figure 11.17. Two groups of subjects are randomly assigned to a treatment group or a control group. Their metabolite levels are measured and indicated with blue (control) or red (treatment) dots in t Figure 11.17a. There is interindividual variation present in each group, and this is visualized by the normal

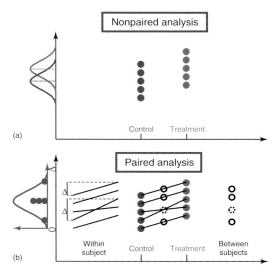

Figure 11.17 (a,b) The idea of paired analysis. For an explanation, see the text.

distributions on the y-axis. To test whether there is a treatment effect, the means of both distributions have to be compared with their standard deviations (a *t*-test). This test does not give a significant result.

When using the fact that each subject underwent both treatments (separated in time, with a washout period and randomized), the same numbers can be arranged in a different way. For each subject, the average of both treatment levels can be calculated. These averages are indicated by black dots and happen to be the same for two of the subjects. These black dots (on the right side of the lower panel) represent interindividual variation. The difference per subject between the control and treatment period is indicated by a Δ. These Δ values are now shown as a distribution on the y-axis, and there is a clear treatment effect. The reason for the improved significance is the removal of the unwanted interindividual effect.

This idea can be generalized for multivariate analysis. It can be used in the context of PCA, for example, when time-resolved urine NMR measurements are available for several monkeys, then the data can be separated in between monkey differences and within monkey changes giving rise to multilevel component analysis (MSCA; [6]). It can also be used in the context of analyzing glucose challenge test data from repeated measurements on the same subject, giving rise to multilevel NPLSDA [48]. Another field of application is in nutritional studies as mentioned earlier [49]. This type of multilevel analysis can give markedly better results than the standard methods [50].

11.5.7
Network Inference

Network inference methods are powerful tools to show relations within sets of measured metabolites. The product of network inference is a visual representation of these relations that can enhance biological interpretation. Network inference is based on the understanding of metabolites being related to each other through pathways and/or other types of connecting mechanisms such as regulation. These mechanisms and pathways can be seen as a network of associations resulting from direct and also indirect relations. The latter poses a problem. Take, for example, the following small pathway: metabolite A produces B and metabolite B produces C. The association between A and C will be high, although there is no direct link between the two. The result is a network that has a lot of spurious edges, but fortunately, this can be solved by a number of techniques. One of these is using Gaussian graphical models (GGM). The basic idea behind GGMs is the use of partial correlation coefficients describing the correlation between two nodes (metabolites) conditional on all the other metabolites in the network [51].

The system of complex relations between metabolites in a cell creates a metabolic network, which may be inferred from data [11]. Often, however, the information on metabolite levels is not measured at the level of cells, but on a higher aggregation level, for example, plasma. In this case, one has to refer to the construction of association networks. The visual representation of a network for metabolites consists of nodes, representing the metabolites and edges, representing the relations between the metabolites. Estimation of the edges (do they exist or not between two metabolites) is the main challenge of network inference. To this purpose, many different inference methods were developed [10, 52–53].

Construction of an association network consists of estimating all pairwise associations between metabolites, after which some assessment is performed to determine which associations are relevant. As many tests (each association between two metabolites is tested) are involved, multiple testing corrections have to be included.

References

1. M.M.W.B., Hendriks, van F.A., van Eeuwijk, R.H., Jellema, J.A., Westerhuis, T., Reijmers, H.C.J., Hoefsloot, and A.K. Smilde (2011) Quantitative strategies for metabolomics studies *Trends in Analytical Chemistry (TrAC)*, **30**, 1685–1698.
2. Bijlsma, S., Bobeldijk, I., Verheij, E.R., Ramaker, R., Kochhar, S., Macdonald, I.A., van Ommen, B., and Smilde, A.K. (2006) *Anal. Chem.*, **78**, 567–574.
3. Steuer, R., Kurths, J., Fiehn, O., and Weckwerth, W. (2003) *Bioinformatics*, **19**, 1019–1026.
4. Van Mechelen, I. and Smilde, A.K. (2011) *Chemom. Intell. Lab. Syst.*, **106**, 2–11.
5. Hageman, J.A., Hendriks, M.M.W.B., Westerhuis, J.A., van der Werf, M.J., Berger, R., and Smilde, A.K. (2008) *Plos One*, **3**, e3259.
6. Jansen, J.J., Hoefsloot, H.C.J., van der Greef, J., Timmerman, M.E., and Smilde, A.K. (2005) *Anal. Chim. Acta*, **530**, 173–183.
7. Van Batenburg, M.F., Coullier, L., van Eeuwijk, F.A., Smilde, A.K., and Westerhuis, J.A. (2011) *Anal. Chem.*, **83**, 3267–3274.
8. Wentzell, P.D., Andrews, D.T., Hamilton, D.C., Faber, K., and Kowalski, B.R. (1997) *J. Chemom.*, **11**, 339–366.
9. Smilde, A.K., Westerhuis, J.A., Hoefsloot, H.C.J., Bijlsma, S., Rubingh, C.M., Vis, D.J., Jellema, R.H., Pijl, H., Roelfsema, F., and van der Greef, J. (2010) *Metabolomics*, **6**, 3–17.
10. Cakir, T., Hendriks, M.M.W.B., Westerhuis, J.A., and Smilde, A.K. (2009) *Metabolomics*, **5**, 318–329.
11. Hendrickx, D.M., Hendriks, M.M.W.B., Eilers, P.H.C., Smilde, A.K., and Hoefsloot, H.C.J. (2011) *Mol. Biosyst.*, **7**, 511–520.
12. Harshman, R.A. and Lundy, M.E. (1984) The PARAFAC model for three-way factor analysis and multidimensional scaling, in *Research Methods for Multimode Data Analysis* (eds H.G. Law, C.W. Snyder, J.A Hattie, and R.P. McDonald), Praeger, New York, pp. 122–215.
13. Stone, M. (1974) *J. R. Stat. Soc. Ser. B Stat. Methodol.*, **36**, 111–148.
14. Hastie, T., Tibshirani, R., and Friedman, J. (2001) *The Elements of Statistical Learning*, Springer Verlag.
15. Rubingh, C.M., Bijlsma, S., Derks, E.P.P.A., Bobeldijk, I., Verheij, E.R., Kochhar, S., and Smilde, A.K. (2006) *Metabolomics*, **2**, 53–61.
16. Ambroise, C. and McLachlan, G.J. (2002) *Proc. Natl. Acad. Sci. U.S.A.*, **99**, (10), 6562–6566.
17. Smit, S., van Breemen, M.J., Hoefsloot, H.C.J., Smilde, A.K., Aerts, J.M.F.G., and de Koster, C.G. (2007) *Anal. Chim. Acta*, **592**, 210–217.
18. Anderssen, E., Dyrstad, K., Westad, F., and Martens, H. (2006) *Chemom. Intell. Lab. Syst.*, **84**, 69–74.
19. Pearson, K. (1901) *Philos. Mag.*, **2**, 559–572.
20. Jolliffe, I.T. (1986) *Principal Component Analysis*, Springer-Verlag, Berlin.
21. Damian, D., Oresic, M., Verheij, E., Meulman, J., Friedman, J., Adourian, A., Morel, N., Smilde, A., and van der Greef, J. (2007) *Metabolomics*, **3**, 69–77.
22. Armitage, P., Berry, G., and Matthews, J.N.S. (2002) *Statistical Methods in Medical Research*, Blackwell Publishing.
23. Searle, S.R. (1971) *Linear Models*, John Wiley & Sons, Inc.

24. Sainani, K.L. (2009) *PM R*, **1**, 1098–1103.
25. Dudoit, S. and van der Laan, M.J. (2008) *Multiple Testing Procedures with Applications to Genomics*, Springer.
26. Benjamini, Y. and Hochberg, Y. (1995) *J. R. Stat. Soc. Ser. B Stat. Methodol.*, **57**, 289–300.
27. Storey, J.D. and Tibshirani, R. (2003) *Methods Mol. Biol.*, **224**, 149–157.
28. Saeys, Y., Inza, I., and Larranaga, P. (2007) *Bioinformatics*, **23**, 2507–2517.
29. Tibshirani, R. (1996) *J. R. Stat. Soc. Ser. B Stat. Methodol.*, **58**, 267–288.
30. Smilde, A.K., Jansen, J.J., Hoefsloot, H.C.J., Lamers, R.A.N., van der Greef, J., and Timmerman, M.E. (2005) *Bioinformatics*, **21**, (13), 3043–3048.
31. Jansen, J.J., Hoefsloot, H.C.J., van der Greef, J., Timmerman, M.E., Westerhuis, J.A., and Smilde, A.K. (2005) *J. Chemom.*, **19**, 469–481.
32. Smilde, A.K., Hoefsloot, H.C.J., and Westerhuis, J.A. (2008) *J. Chemom.*, **22**, (7-8), 464–471.
33. Keun, H.C., Ebbels, T.M., Bollard, M.E., Beckonert, O., Antti, H., Holmes, E., Lindon, J.C., and Nicholson, J.K. (2004) *Chem. Res. Toxicol.*, **17**, (5), 579–587.
34. Smilde, A.K., Bro, R., and Geladi, P. (2004) *Multiway Analysis: Applications in the Chemical Sciences*, John Wiley & Sons, Inc., New York.
35. Jansen, J.J., Bro, R., Hoefsloot, H.C.J., van den Berg, F.W.J., Westerhuis, J.A., and Smilde, A.K. (2008) *J. Chemom.*, **22**, (1-2), 114–121.
36. Draper, N.R. and Smith, H. (1981) *Applied Regression Analysis*, John Wiley & Sons, Inc.
37. Rubingh, C.M., Bijlsma, S., Jellema, R.H., Overkamp, K.M., van der Werf, M.J., and Smilde, A.K. (2009) *J. Proteome Res.*, **8**, 4319–4327.
38. Massart, D.L. Vandeginste, B.G.M., Lewi, P.J., Smeyers-Verbeke, J., Buydens, L.M.C., and de Jong, S. (1998) *Handbook of Chemometrics and Qualimetrics: Part A*, Elsevier.
39. Hoskuldson, A. (1988) *J. Chemom.*, **2**, 211–228.
40. Chong, I. and Jun, C. (2005) *Chemom. Intell. Lab. Syst.*, **78**, 103–112.
41. Hoskuldson, A. (2001) *Chemom. Intell. Lab. Syst.*, **55**, 23–38.
42. Centner, V., Massart, D.L., de Noord, O.E., de Jong, S., Vandeginste, B.M., and Sterna, C. (1996) *Anal. Chem.*, **68**, 3851–3858.
43. Bro, R. (1996) *J. Chemom.*, **10**, 47–61.
44. Bro, R., Smilde, A.K., and de Jong, S. (2001) *Chemom. Intell. Lab. Syst.*, **58**, 3–13.
45. Bylesjø, M., Rantalainen, M., Cloarec, O., Nicholson, J.K., Holmes, E., and Trygg, J. (2006) *J. Chemom.*, **20**, 341–351.
46. Westerhuis, J.A., Hoefsloot, H.C.J., Smit, S., Vis, D.J., Smilde, A.K., van Velzen, E.J.J., van Duijnhoven, J.P.M., and van Dorsten, F.A. (2008) *Metabolomics*, **4**, 81–89.
47. van Duynhoven, J., Vaughan, E.E., Jacobs, D.M., Kemperman, R.A., van Velzen, E.J.J., Gross, G., Roger, L.C., Possemiers, S., Smilde, A.K., Dore, J., Westerhuis, J.A., and Van de Wiele, T. (2011) *Proc. Natl. Acad. Sci. U.S.A.*, **108**, 4531–4538.
48. Rubingh, C.M., van Erk, M.J., Wopereis, S., van Vliet, T., Verheij, E.R., Cnubben, N.H.P., van Ommen, B., van der Greef, J., Hendriks, H.F.J., and Smilde, A.K. (2011) *Chemom. Intell. Lab. Syst.*, **106**, 108–114.
49. van Velzen, E.J.J., Westerhuis, J.A., and van Duynhoven, J.P.M., van Dorsten, F.I.A., Hoefsloot, H.C.J., Jacobs, D.M., Smit, S., Draijer, R., Kroner, C.I., and Smilde, A.K. (2008) *J. Proteome Res.*, **7**, 4483–4491.
50. Westerhuis, J.A., van Velzen, E.J.J., Hoefsloot, H.C.J., and Smilde, A.K. (2010) *Metabolomics*, **6**, 119–128.
51. Schafer, J. and Strimmer, K. (2005) *Stat. Appl. Genet. Mol. Biol.*, **4**, Article 32.
52. Steuer, R., Gross, T., Selbig, J., and Blasius, B. (2006) *Proc. Natl. Acad. Sci. U.S.A*, **103**, 11868–11873.
53. Nemenman, I., Escola, G.S., Hlavacek, W.S., Unkefer, P.J., Unkefer, C.J., and Wall, M.E. (2007) *Ann. N. Y. Acad. Sci.*, **1115**, 102–115.

12
Metabolic Flux Analysis

Christoph Wittmann and Jean-Charles Portais

12.1
Introduction

Recent years have seen impressive achievements toward the understanding of biological systems. Metabolomics and fluxomics are among the huge technological their developments for a new, fascinating era of research – analysis and engineering of metabolic and regulatory properties on a global scale [1]. Metabolism comprises a complex network of hundreds of biochemical reactions for the various functions required for growth or product formation. Metabolic flux analysis aims at the quantification of the molecular fluxes (the fluxome) through these reactions, that is, of the *in vivo* activities of intracellular enzymes and pathways. The most relevant and most intensively studied part of metabolism concerning fluxes is a highly connected network of about 100 reactions that comprises the high carbon flux part of metabolism [2]. Meanwhile, a large set of prominent pathways is routinely accessible by metabolic flux analysis, among which are the glycolysis, the gluconeogenesis, the pentose phosphate pathway (PPP), the Entner–Doudoroff pathway, the tricarboxylic acid (TCA) cycle, and the glyoxylate shunt or the reactions at the pyruvate node. Fluxes are of central importance for systems level understanding, because – most closely linked to the cellular phenotype they reflect the integrated flow of material and information between the cellular components, that is, the genes, transcripts, proteins, and metabolites (Figure 12.1). Meanwhile applicable, the analysis of metabolic fluxes has developed into a valuable toolbox in functional genomics, metabolic engineering, and systems biology. This contribution provides an overview on methods and concepts for the quantification of metabolic fluxes. It gives important prerequisites and practical guidelines for a wide range of flux experiments including rather simple isotope studies on single pathways, first model-based network investigations, and highly advanced state-of-art fluxomics. In addition, the enormous potential of metabolic flux analysis is illustrated by selected examples from recent literature.

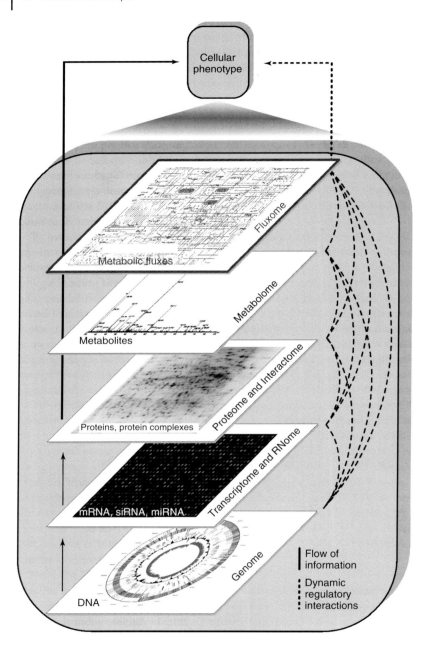

Figure 12.1 Systems level view on a cellular system. The fluxome, that is, the set of molecular fluxes in and out of the cell and within the cellular network, integrates the interactions among and within the different functional layers. The solid lines denote flow of information; the dashed lines indicate interactions between molecular species. (The figure is adapted from a recent review [1].)

12.2
Prerequisites for Flux Studies

12.2.1
Network Topology and Cellular Composition

Metabolic flux studies require knowledge about the pathway repertoire of the investigated organism, that is, the topology of the underlying metabolic network. The corresponding level of information for the various organisms of interest is continuously increasing with the advent of fast sequencing and bioinformatics tools for gene annotation and network reconstruction. As example, the network information on the carbon core metabolism of many microorganisms is rather detailed. For *Escherichia coli, Corynebacterium glutamicum*, or *Bacillus subtilis* it can be extracted from metabolic models up to genome scale [3–5] or databases such as KEGG (*http://www.genome.jp/kegg/*). For less-studied microbial species or higher organisms such as eukaryotic microorganisms, plants, or mammals, with an even more complex, and compartmented metabolism, simplifying assumptions, or supporting measurements can help to derive a suitable network, as described later. For flux studies with growing cells, the demand for metabolic intermediates as anabolic precursors has to be additionally considered, as anabolic reactions carry flux during growth and withdraw carbon from the central routes [6]. The basic organic biomass constituents are proteins, carbohydrates, lipids, DNA, and RNA. To obtain about 80 different building blocks for these polymeric fractions (including e.g., amino acids, sugars, fatty acids, or nucleotides), the cell typically recruits 12 central precursor metabolites: glucose 6-phosphate, fructose 6-phosphate, ribose 5-phosphate, erythrose 4-phosphate, glyceraldehyde 3-phosphate, 3-phosphoglycerate, pyruvate, phosphoenolpyruvate, acetyl-CoA, 2-oxoglutarate, succinyl-CoA, and oxaloacetate [7]. Detailed information on the cellular composition and the exact stoichiometric demand for anabolism is available for selected microorganisms such as the above-mentioned bacteria *E. coli* [8], *C. glutamicum* [6], or *B. subtilis* [9]. In flux studies, where the cellular composition for the investigated system is not known, this information is sometimes adapted from other microorganisms or growth conditions. This, however, might falsify the determined fluxes, because the cellular composition can vary significantly between different species and with a change of nutrient level or growth conditions [10].

12.2.2
Network Formulation and Condensation

A condensation of the genome scale network representation from about 1000 and more reactions to a compact, yet highly informative size is an important entry step into metabolic flux analysis. For most of the flux studies, which are carried out in steady state (Chapter 13), it is straightforward to lump subsequent reaction steps in a linear pathway into a single reaction without a loss of information. Similarly, biosynthetic reactions at the network periphery originating from the same precursor can be condensed into a single reaction to only account for the total consumption of this precursor from the central routes. Moreover, inaccessible fine structures of the metabolic network can be condensed. In purely stoichiometric flux analysis (see below), these might be parallel pathways with the same overall stoichiometry, cyclic, or bidirectional reactions, which cannot be resolved by this approach and have to be represented as a single net reaction. Further assumptions may include (i) the consideration of reactions with a large standard free energy term ($\Delta G'_0$) as irreversible *in vivo*, (ii) the assumption of rapid equilibrium for certain highly reversible enzymatic conversions, catalyzed, for example, by isomerases or epimerases in the PPP, (iii) the assumption that energy stoichiometry is known or (iv) that all reactions consuming or producing NADPH or NADH are known and can be balanced. Such assumptions are optional in the framework to be described and should be carefully documented to achieve transparency of the performed study and reproducibility of the results [11]. For eukaryotic cells, compartmentation has to be taken into account [12]. If particular information on the network is not available, one can test for the presence of certain reactions or pathways by *in vitro* activity measurement of the corresponding enzymes or consider data from other omics analyses, for example, present proteins or metabolic pathway intermediates as indicators. Alternatively, one can also limit the flux study to certain aspects of the metabolic network. The final size of the network strongly depends on the aim of the performed flux study, as well as on the possibilities given by the utilized flux approach. For state-of-art flux analysis, a typical network contains about 50–70 reactions.

12.2.3
Metabolic and Isotopic Steady State

It should be noticed that the experimental flux approaches mostly rely on metabolic steady state. Metabolic steady state is given, when the concentrations of all intracellular metabolites are constant. It displays an important prerequisite, as only under these conditions can one neglect *in vivo* enzyme kinetics, information which is still mostly unavailable and generally would lead to much more involved models for flux estimation. In the case of labeling studies, routine flux analysis typically also involves isotopic steady state. During isotopic steady state, the labeling patterns, that is, the relative fractions of metabolite isotopomers, are constant over time. Exceptions are transient studies which allow for shorter experiments [13] or the use of carbon dioxide as tracer substrate [14]. Steady state can be practically achieved by a continuous culture. For cultures of single cells, a continuously operated bioreactor is the preferred system to ensure constant conditions. However, its use is linked to long lasting experiments, high labeling costs, and only limited relevance for biotechnological processes, typically not continuous. A large fraction of labeling studies is therefore performed in batch culture. In batch cultivations the steady state assumption is generally questionable, but, with a thorough experimental setup, intracellular metabolite concentrations can indeed be constant even in inherently transient systems so that steady state assumptions can be basically applied [15]. In all cases, experimental evidence should be provided to be sure about the validity of assuming metabolic and isotopic steady state. For validation, intracellular concentrations, ^{13}C labeling data, specific rates of substrate uptake, growth, product formation or respiration, or yield coefficients can be checked for constancy over the inspected time period [16].

12.2.4
Definition of Isotope Labeling Patterns

Concerning the nomenclature of isotope labeling there are no strict guidelines. Basically, molecules are characterized by the number and position(s) of the corresponding isotopes in their constitutive atoms. The following definitions are exemplified for the most commonly used carbon isotopes ^{12}C and ^{13}C, but can be generally applied for other isotope labels as well (Figure 12.2). *Positional isotopomers* have an exactly determined labeling pattern with a specific number of ^{13}C atoms in specific positions of the molecule. For example, [1-^{13}C] alanine and [2-^{13}C] alanine are different positional isotopomers, carrying the ^{13}C label at carbons C1 and C2. Generally, a compound with n carbons has 2^n possible positional isotopomers. For alanine, this results in eight different variants. It should be noted that the number of positional isotopomers of symmetrical molecules such as succinate is

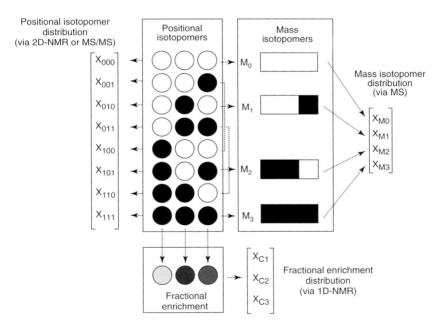

Figure 12.2 Positional isotopomers, mass isotopomers, and fractional enrichment exemplified for a molecule with three carbon atoms either ^{12}C (white) or ^{13}C (black). The arrows indicate the grouping of positional isotopomers into different mass isotopomer fractions or the contribution to the fractional enrichment of the single carbons, respectively.

only $2^n/2$. *Mass isotopomers* differ only by the number of heavy isotopes, resulting in different molecular weights. They do not differ by the position of the label. For example, [$^{12}C_3$] alanine, [$^{13}C - ^{12}C_2$] alanine, [$^{13}C_2^{12}C$] alanine, and [$^{13}C_3$] alanine are different mass isotopomers with different molecular weights. Usually, the lightest atoms are not specified and the mass isotopomers are classified as alanine, [^{13}C] alanine, [$^{13}C_2$] alanine, and [$^{13}C_3$] alanine, in agreement with the International Union of Chemistry. The corresponding mass isotopomer fractions are often given as M_0, M_1, M_2, and M_3 or as m, $m+1$, $m+2$, and $m+3$. For a compound with n atoms, the number of possible mass isotopomers is $(n+1)$. Most mass isotopomers, except m and $m+n$, include multiple positional isotopomers. As example, [1-^{13}C] alanine and [2-^{13}C] alanine both belong to the M_1 or $m+1$ mass isotopomer fraction. The ^{13}C labeling state of a molecule with n carbons can be also expressed as *molar enrichment* (ME) [17] or as *summed fractional labeling* (SFL) [18] describing the weighted sum of mass isotopomer fractions. The *fractional enrichment* denotes the relative enrichment of a specific carbon atom with ^{13}C.

12.3
Stoichiometric Flux Analysis

The classical way to perform metabolic flux analysis is based on direct measurement of extracellular fluxes and a stoichiometric model of the metabolic network that allows determining the remaining intracellular fluxes via metabolite mass balances around intracellular pools [19]. Accordingly, this approach for flux analysis is also named metabolite balancing or stoichiometric balancing. It is easy applicable because the measurement of only a few extracellular fluxes, for example, uptake of substrate, formation of biomass, secreted products, or carbon dioxide is necessary to estimate the intracellular flux distribution. Hereby, the measured formation of biomass together with the cellular composition can be utilized to calculate the stoichiometric demand, that is, the relative fluxes, from all precursors toward anabolism. The main assumption that has to be considered is metabolic steady state of the culture (Chapter 11). Under these conditions, the intracellular concentration of metabolites is constant, so that the sum of the ingoing fluxes into a metabolite pool equals the sum of fluxes out of that pool. This assumption facilitates the mathematical description of the metabolic network into a system of linear equations that can be easily solved by taking the measured extracellular and known anabolic fluxes for biomass formation into account. The procedure is illustrated in the following example. The network described here comprises a small set of reactions (Figure 12.3). For the two intracellular metabolites, B and D, two mass balances can be formulated.

$$v_1 - v_2 - v_3 = 0 \tag{12.1}$$

$$v_2 - v_4 - v_5 = 0 \tag{12.2}$$

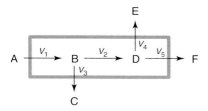

Figure 12.3 Example network for illustration of metabolite balancing. It contains extracellular and intracellular metabolites (M) and interconverting reactions (v).

With three direct extracellular flux measurements, for example, on v_1, v_4, and v_5, the remaining two unknown fluxes in the network can be calculated via the two mass balances (Eqs. (12.1) and (12.2)). Exactly the same principle is also applied to larger networks. Here, the number of metabolites, reactions, and mass balances is substantially larger, but matrix calculus allows easy handling of the equation system. For this purpose, the network stoichiometry is written in matrix form. The stoichiometric matrix **S** has m columns representing all reactions and n rows representing all balanced or internal metabolites. The stoichiometric matrix **S** for the small example network above reads as follows:

$$\begin{array}{c|ccccc} & v_1 & v_2 & v_3 & v_4 & v_5 \\ \hline A & -1 & 0 & 0 & 0 & 0 \\ B & 1 & -1 & -1 & 0 & 0 \\ C & 0 & 0 & 1 & 0 & 0 \\ D & 0 & 1 & 0 & -1 & -1 \\ E & 0 & 0 & 0 & 1 & 0 \\ F & 0 & 0 & 0 & 0 & 1 \end{array} = S \quad (12.3)$$

At steady state the set of mass balance equations is generally given by Eq. (12.4), with v as the vector of fluxes.

$$S \times v = 0 \qquad (12.4)$$

This equation is the basis for metabolite balancing. Separating the measurable (m) and nonmeasurable (nm) reactions allows calculating the unknown fluxes (v_{nm}) via the measured fluxes (v_m) and the stoichiometric information. If the matrix S_{nm} is square and full rank, the nonmeasurable fluxes are accessible by:

$$v_{nm} = S_{nm}^{-1} \times (-v_m \times S_m) \qquad (12.5)$$

If the system is overdetermined, the calculation involves pseudo-inversion via the transpose of S_{nm} denoted here by S_{nm}^T.

$$v_{nm} = \left(S_{nm}^T \times S_{nm}\right)^{-1} \times S_{nm}^T \times (-v_m \times S_m) \qquad (12.6)$$

The above calculations are easily performed using mathematical software platforms such as MATLAB (The Mathworks, Natick, MA) or MATHCAD (PTC, Needham, MA). For large networks with various intracellular branch points, it can sometimes be difficult to arrive at a fully determined system because the number of possible measurements might be lower than the degrees of freedom. One possibility to overcome this problem is the simplification of the network by lumping reactions or assuming scenarios that lack specific pathways [20]. In other cases, the number of mass balances is increased by additional constraints and balancing of energy and reduction equivalents. This is legitimate as metabolic pathways are linked via these components, but their exact balancing is often impossible as not all reactions linked to a specific cofactor are accessible or futile cycles occur [21, 22]. Moreover, the stoichiometric coefficients which have to be included in the balances, for example, on energy stoichiometry, may be uncertain and are therefore controversially discussed. Alternatively, one can incorporate growth or product formation as optimization criterion and obtain at least predictions for the intracellular fluxes within certain boundaries via minimization or maximization of the objective function (flux balance analysis) [23, 24].

In the past, metabolite balancing provided important information on metabolic fluxes especially in microbial systems. Pioneering studies investigated citrate production in yeast [20], anaerobic bacteria [25], and lysine producing *C. glutamicum* with a series of remarkable contributions [26–29]. As example, flux analysis in different stages of a batch cultivation revealed changes of the intracellular flux distribution of *C. glutamicum* during the transition from growth to lysine production. Other studies focused on *Escherichia coli* as a production host for amino acids and recombinant proteins [30–33]. As shown, the basic inputs required for the determination of intracellular fluxes by stoichiometric balancing are measured extracellular fluxes. Despite this approach has been successfully applied to various microbial systems, one should be aware that important intracellular fluxes are accessible only to a limited extent (see above). Examples are parallel pathways, metabolic cycles, or bidirectional pathways, which are either not accessible or rely on uncertain assumptions on reduction or energy balances. These limitations can be overcome by the use of stable isotopes, providing complementary information on intracellular fluxes of interest.

12.4
Labeling Studies Using Isotopes

The use of labeled compounds to estimate fluxes has a long tradition in biochemistry and quantitative physiology and significantly contributed to our understanding of metabolic functioning and control. Basically, such approaches can be differentiated into flux estimation from transient tracer studies and from metabolic and isotopic steady state experiments. The overall concept is that the compounds, specifically labeled with radioactive or stable isotopes are metabolized by the studied organism leading to distribution of the label throughout the biological network and a specific "labeling fingerprint" among the metabolites involved. The most prominent stable isotope to assess carbon fluxes is the stable isotope ^{13}C, which replaced the radioactive ^{14}C with the advent of mass spectrometry (MS) and nuclear magnetic resonance (NMR). For specific questions, tracer substrates labeled with isotopes such as ^{18}O [22], ^{15}N [34], or ^{2}H [35, 36] can be applied. For the analysis of stable isotope labeling patterns, we have a number of efficient analytical methods at hand that generate detailed and precise information. The labeling measurement is a crucial part of ^{13}C flux analysis, as it has to provide data sets with high information content and high precision for the quantified labeling patterns. MS [37–39] and NMR spectroscopy [40, 41] display excellent instrumental solutions. MS allows distinguishing different mass isotopomers of a compound which differ by the number of labeled atoms in the molecule (Figure 12.2). 1D-NMR spectroscopy provides information on the relative degree of labeling at a certain atom position (fractional enrichment) [42]. Advanced measurement techniques such as high resolution 2D-NMR or extended fragment analysis by MS can be applied to further unravel the labeling status ending up in full resolution of positional isotopomer pools. The links between labeling patterns and fluxes are the very well known and defined transitions of the single atoms in biochemical reactions [43, 44]. The atom transition of each reaction is hereby fixed. One should note that this may hold even for chemically symmetrical molecules, which are biologically asymmetrical because of the configuration of the active site of enzymes (e.g., the citrate synthase) [45]. It should be further mentioned here that phenomena of metabolite channeling, proposed for enzymes of the TCA cycle and the PPP, may influence the distribution of carbon atoms through the reaction network [46–48]. The basic principle of how flux is related to labeling patterns can be nicely illustrated from simple example networks. Let us consider the prominent case of glucose catabolism to visualize this. Depending on the pathway involved, glucose carbons are subjected to different transfer pattern (Figure 12.4). The carbon atom C_1 of glucose is transferred to the pyruvate C_3 by the Entner–Doudoroff pathway [49], and to the C_1 by the glycolysis. In contrast, it is cleaved off as CO_2 in the oxidative PPP [50]. It is obvious that depending on the substrate used, each pathway is linked to a specific *labeling fingerprint*. Accordingly, the resulting labeling patterns in the cellular metabolites are originating from the relative use of different pathways and thus can be taken as a measure for the estimation of the corresponding relative fluxes. By coupling this to extracellular rate measurements, absolute fluxes can be determined. We will return later to explore the above example in more detail.

12.4.1
Radiolabeled Isotopes

Transient intensity measurements of radiolabeled compounds can be applied to probe metabolic pathways. Hereby, the radioactivity of a specific compound, isolated from the experiment, is quantified over time as an estimate for the flux of interest. The design of the experiment hereby depends on the question to be answered. Let us take the example of estimating flux through a decarboxylation reaction via transient $^{14}CO_2$ measurement. The network, shown in Figure 12.5, comprises a subset of reactions from glycolysis, PPP, and TCA cycle and displays a simplified

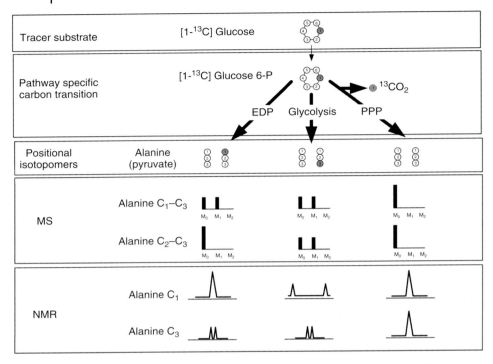

Figure 12.4 Quantification of the flux partitioning between the Entner–Doudoroff pathway (EDP), the glycolysis, and the pentose phosphate pathway (PPP) via isotope labeling studies. The strategy is based on [1-^{13}C] glucose as tracer substrate combined with labeling analysis of alanine formed by mass spectrometry or nuclear magnetic resonance spectroscopy. The MS signals illustrate the mass isotopomer distributions of ion clusters obtained by GC-MS from t-butyl-dimethyl-silyl derivatized alanine at m/z 260 containing carbons C_1–C_3 and at m/z 232 containing carbons C_2–C_3 [51]. The NMR fingerprints reflect the labeling of carbon C1 and C3 of alanine accessible via the proton signatures of H2 and H3, respectively, in 1D-^{1}H-NMR. For the sake of clarity, the proton coupling is neglected in the illustrations.

form of carbon core metabolism. As can be seen, CO_2 is formed by different decarboxylating reactions during aerobic growth on glucose. A close inspection of the underlying biochemistry shows that these reactions, with certain simplifying assumptions (not further discussed here) release only specific carbon atoms from the original glucose substrate as CO_2 (Figure 12.5a). This can be exploited to assess the relative flux through these pathways by introducing different tracer substrates. A pulse with [3,4-$^{14}C_2$] glucose yields $^{14}CO_2$, which can be linked to the pyruvate dehydrogenase flux. Similar relations can be seen for [2-^{14}C] glucose (TCA cycle) and [1-^{14}C] glucose (PPP plus TCA cycle). Parallel pulse experiments with all three tracer substrates now allow a direct flux estimate of the three pathways, whereby the relative area of the $^{14}CO_2$ signal is directly proportional to the corresponding relative flux in each experiment (Figure 12.5b). Wang et al. [52] have utilized this

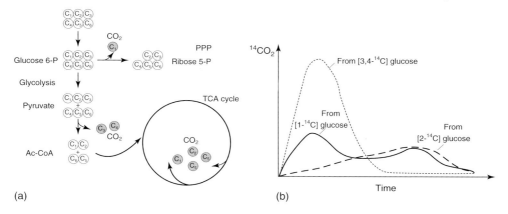

Figure 12.5 Metabolic flux analysis of central catabolic pathways in *E. coli* based on transient experiments with ^{14}C-glucose. The straightforward approach is based on a simplified metabolic network, assuming release of radiolabeled carbon dioxide specifically linked to certain input labeling of the substrate (a). (b) The resulting transient $^{14}CO_2$ responses from parallel tracer studies are adapted from previous work [52].

approach to quantify fluxes through central metabolism in *E. coli* – more than 50 years ago. Subsequently, this technique was applied to follow fluxes during glucose mineralization in other microorganisms as well as in complex habitats such as soil [53, 54]. In an even more reduced form, transient measurements can also be applied to determine a local flux through a distinct metabolite (M) as depicted previously [55]. Here, the corresponding metabolite is added in radiolabeled form as a pulse to the system. Assuming steady state, radioactivity of M inside the cell will decrease as a function of pool exchange by the synthesizing and consuming reactions, that is, the flux through the pathway. From the time dependent decrease of the radioactivity of purified M and its intracellular concentration, the pathway flux can be directly obtained. Admittedly, this requires a delicate experimental handling as perturbation during the experiment has to be minimal and cells have to be in steady state.

12.4.2
Stable Isotopes

The advent of MS and NMR as analytical techniques for labeling measurements has strongly pushed the use of stable isotopes to quantify intracellular fluxes. Aside the much risk less use of stable isotopes instead of experiments contaminated with radioactivity, it is especially the increased, position specific resolution of labeling patterns achievable by MS or NMR, which gives a benefit for flux analysis from stable isotope experiments. The power of such studies can be inferred for the three major pathways of glucose utilization. The relative flux can be resolved by simply analyzing the labeling pattern of alanine deriving its carbon backbone directly from pyruvate, where the three pathways, converge (Figure 12.4). Feeding [1-^{13}C] glucose leads to a unique mass isotopomer distribution for PPP-derived alanine C1–C3 and its C2–C3 fragment which can be easily obtained by GC-MS. The relative contribution of the different pathways to the formation of alanine (pyruvate) is accessible directly from the detected abundance of the different mass isotopomers by probabilistic equations [49]. Similarly, NMR can assess the position specific enrichment of the ^{13}C to resolve the fluxes as becomes evident from the fingerprints resulting in ^{1}H-NMR (Figure 12.4). Application of stable isotopes in combination with analytical equations for flux estimation provided interesting information on different microbial systems. A fairly advanced example is a study of fluxes through, for example, PEP carboxylase, citric acid cycle, and glyoxylate shunt in glutamate producing *M. ammoniaphilum* [56]. Other studies investigated certain aspects of the amino acid producer *C. glutamicum*, including glycolysis and PPP [57], the pyruvate node [58] or the two alternative pathways in lysine biosynthesis [59]. The use of stable isotopes for flux analysis was subsequently further improved into a straightforward, systematic framework (flux ratio approach) of probabilistic equations that relate GC-MS or NMR derived labeling patterns in proteinogenic amino acids to 10–15 ratios of fluxes through converging pathways and reactions in the central carbon metabolism [43, 60]. Inherently limited to selected flux ratios in relatively small and simplified networks accessible and high efforts to derive the case specific analytical equations, the isotope approaches have been extensively developed further. In order to extend the determinable parameters from single flux ratios to *in vivo* enzyme and pathway activities in the form of absolute rates, the following development included the concept of stoichiometric models and extracellular rate measurements. The derived experimental and computational frameworks are generally applicable and allow the complete resolution of fluxes in metabolic networks with all relevant biochemical fine structures. In addition, the isotopic fingerprints can be also used directly. This does a priori not provide fluxes, but can lead to valuable information for the relative comparison and discrimination of microorganisms, and metabolism-based screening [61].

12.5
State-of-Art ^{13}C Flux Analysis

The different pioneering studies for flux analysis utilizing either metabolite balancing or isotope labeling served as a strategic basis for the development of the current state-of-art methodologies. The comprehensive approaches available at present combine isotope labeling analysis with metabolite balancing and extracellular flux measurement. For flux calculation, experimental design and statistics modeling tools and software packages are available, which display a highly efficient framework for flux studies (Table 12.1). Usually, they comprise different tool boxes for model generation, experimental design, parameter estimation, and statistical analysis (Figure 12.6).

Table 12.1 Overview on publically available software platforms for ^{13}C metabolic flux analysis.

	OpenFLUX	^{13}CFLUX	FiatFLUX	IMMFLUX
Basic features				
Implementation	MATLAB	C++ (Linux)	MATLAB	MATLAB
Calculus for carbon transition	Elementary mode units	Cumomers	Probabilistic equations	Isotopomers
Flux estimation	Global minimization	Global minimization	Local flux ratios	Global minimization
Computation speed	++	+	+	+
Flexibility				
Support of MS data	+	+	+	+
Support of NMR data	−	+	−	+
Implementation of parallel tracer studies	−	−	+	+
Source code available	+	−	+	+
Consideration of variable carbon sources or mixtures	+	+	−	+
Toolboxes				
Experimental design	+	+	−	+
Statistical flux evaluation	+	+	+	+
Flux calculation	+	+	+	+
Natural isotope correction	+	+	+	+
Reference	Quek et al. [62]	Wiechert et al. [63]	Zamboni et al. [64]	Wittmann and Heinzle [39]

"−" not supported
"+" supported
"++" strongly supported

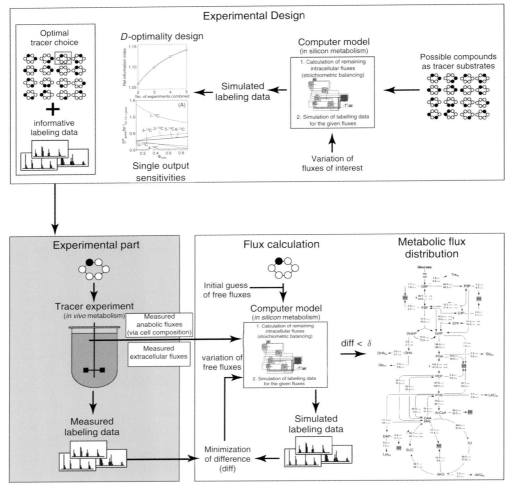

Figure 12.6 Schematic work flow for ^{13}C metabolic flux analysis with (i) experimental design, (ii) isotope tracer experiments including measurement of ^{13}C labeling patterns and of extracellular fluxes and biosynthetic requirements, and (iii) computational calculation of fluxes. This basic concept can be varied in several aspects depending on the focus of the flux study. (The figure is adapted from Ref. [1].)

12.5.1
Modeling of Carbon Transitions

To resolve the high complexity of the often nonlinear metabolic networks, manually derived analytical equations for flux calculation have only limited value. Most suitable here are modeling frameworks with a systematic and general protocol for the quantitative description of the transfer of labeled ^{13}C atoms in metabolic networks [21, 22]. An elegant way makes use of vector notation for the labeling state of metabolites and matrix notation for the carbon transition through the biochemical reactions [65]. One possibility to mathematically formulate this is based on atom mapping matrices [66]. Regarding the reaction of pyruvate dehydrogenase, the carbons C_2 and C_3 from pyruvate are transferred into carbon positions C_1 and C_2 of acetyl-CoA. The fractional enrichment of the acetyl-CoA carbons can be calculated form the corresponding pyruvate labeling and the atom mapping matrix for the reaction.

$$\begin{bmatrix} 010 \\ 001 \end{bmatrix} \times \begin{pmatrix} x_{C1,pyr} \\ x_{C2,pyr} \\ x_{C3,pyr} \end{pmatrix} = \begin{pmatrix} x_{C1,aca} \\ x_{C2,aca} \end{pmatrix} \qquad (12.7)$$

This concept is rather straightforward. The atom mapping matrices can be easily generated from biochemical knowledge available in textbooks. It is, however, restricted to the tracing of the fractional enrichment of single carbons. High information flux studies, which make use of experimental mass isotopomer or positional isotopomer distributions require the more extended simulation and solving of isotopomer balances. For this purpose, different formal approaches have been established in a generally applicable way. The isotopomer mapping matrix approach uses isotopomer mapping matrices, which can be automatically generated from the much simpler atom mapping matrices, and isotopomer distribution vectors to describe the transfer of carbon [67]. The different positional isotopomers of a compound are arranged in the isotopomer distribution vector according to their labeling state encoded as binary code with 0 for ^{12}C and 1 for ^{13}C (Figure 12.2). The transition matrix approach is based on a transformation of the isotopomer balances into cumomer balances exhibiting greater simplicity and being capable to solve the nonlinear isotopomer balances analytically [68]. Another state variable used is based on the bondomer concept resulting in considerably smaller models than corresponding cumomer models, thus saving computational time, allowing easier identifiability analysis, and yielding new insights in the information content of labeling data [69]. A further significant reduction of the number of necessary isotopomer balances with the benefit of reduced model complexity and increased simulation speed was recently achieved by the concept of elementary mode units [70]. These types of modeling frameworks represent clear, systematic, and general approaches for the quantitative description of the transfer of labeled ^{13}C atoms in metabolic networks.

12.5.2
Experimental Design

The success of a flux study strongly depends on the ^{13}C-labeling strategy. The resolution of a particular flux depends on the network topology and also the substrate utilized, so general guidelines cannot be given. For flux studies with glucose, [1-^{13}C] glucose as well as a mixture of [^{13}C$_6$] glucose and unlabeled glucose has proven suitable to determine fluxes in the central metabolism of different organisms. A mixture of [^{13}C$_6$] glucose and unlabeled glucose is particularly useful to resolve fluxes downstream of PEP and some exchange fluxes that result in C–C bond cleavages, whereas the use of [1-^{13}C] glucose is valuable for resolving the upper part of metabolism, in particular the oxidative PPP, glycolysis and the Entner–Doudoroff pathway [71]. Combining [U-^{13}C] glucose and [1-^{13}C] glucose, either in two separate experiments [16, 39, 71, 72] or as a substrate mixture [15] may lead to an even better estimate. Other combinations of labeled substrates can be applied to resolve specific network topologies [73, 74]. Questions concerning the determinability and the predicted accuracy of a certain flux together with an optimal experimental approach can be effectively answered by computer based experimental design [50, 63, 68]. Here, various possible scenarios can be tested by simulations (Figure 12.6). By the calculation of single output sensitivities, the optimum setup, for example, the optimum tracer substrate and the key labeling measurements, can be found for specific flux parameters [50], whereas D-optimality design allows comparing the information content for different strategies in a quantitative manner [75].

12.5.3
Flux Calculation and Statistical Evaluation of Flux Data

For flux calculation, the measured ^{13}C labeling data are usually utilized to globally fit the unknown fluxes (Figure 12.6). Starting with a random initial guess for the free fluxes, the model first computes all remaining dependent fluxes via stoichiometric mass balances. The set of fluxes is then handed over to the isotopomer modeling part, where the ^{13}C labeling patterns of all compounds in the network are calculated for the given fluxes and, after correction for natural isotopes [76, 77], are compared with the experimental labeling data. The experimental data can be mass isotopomer fractions [39], fractional enrichments [78], positional isotopomer fractions [43], or combination of different types [79] depending on the system of interest. By iterative variation of the free fluxes, these steps are repeated, whereby an optimization algorithm is applied which minimizes the weighed deviation of the labeling data between simulation ($r_{i,exp}$) and experiment ($r_{i,calc}$) until optimum fit is obtained. Owing to the extended labeling data sets of, for example, multiple mass isotopomers, this usually benefits from a high redundancy for the flux calculation. Different optimization functions such as gradient or adaptive random search functions are available. To increase the probability of identifying the global solution optimum, the convergence is usually tested for different initial guesses. As minimization criterion a weighted sum of least squares (SLS) is typically considered

(Eq. (12.8)).

$$\text{SLS} = \sum_i \frac{\left(\dfrac{r_{i,\text{exp}} - r_{i,\text{calc}}}{r_{i,\text{exp}}}\right)^2}{s_{r,i}^2} \tag{12.8}$$

In advanced studies involving thorough analysis and validation, an excellent agreement can be achieved resulting in high confidence in the flux data [16]. In combination with experimentally determined extracellular fluxes, absolute carbon fluxes throughout the network are obtained. As an alternative to global flux fitting based on computer models of varying complexity, absolute fluxes can also be obtained via the flux ratio method, which provides direct evidence for the relative *in vivo* activity of a set of reactions. As a drawback, these are only about 10–15 selected fluxes and the probabilistic equations do hold only for a restricted selection of substrates. Recently, the method has been extended toward estimation of absolute fluxes [71, 80] and can also be seen as a complementary extension to model-based flux analysis [81, 82]. Statistical analysis of flux data is of central importance to verify whether differences in intracellular flux distributions can be really attributed to strain or condition specific differences. Such analysis can be performed by Monte Carlo simulations, including multiple parameter estimation runs, with statistical variation of the experimental data. These simulations result in mean values and standard deviations for the intracellular fluxes, from which confidence limits can be calculated [83, 84].

12.5.4
Labeling Analysis by Mass Spectrometry

The most routinely applied MS method is GC-MS because of the outstanding characteristics concerning resolution, sensitivity, accuracy, and robustness [44]. As promising alternative toward flux dynamics, LC-MS [85, 86], and collision energy (CE)-MS [87] for direct labeling analysis of intracellular intermediates have recently come into focus. Other studies have utilized matrix-assisted laser desorption-time of flight (MALDI-TOF) MS allowing fast analysis from extremely small sample amounts toward high throughput flux screening [57]. An interesting extension is the recent introduction of GC isotope ratio (ir) MS to flux analysis. It provides highly precise estimates at ultralow labeling enrichment and thus seems promising to study large systems such as environmental mesocosms or industrial bioreactors [88, 89]. Additionally, membrane inlet MS has been utilized for flux analysis with CO_2-based labeling measurement, which is valuable to study nongrowing cells [90, 91]. Flux studies typically utilize labeling analysis of cellular constituents formed during growth of the examined cells. Their labeling pattern reflects the whole growth process up to the point of harvest, so care has to be taken to ensure metabolic and isotopic steady state (see above). The most prominent compounds utilized in flux analysis are amino acids stemming from the cell protein [9, 92], which are rich in flux information and fairly abundant in relatively high concentration. Other cell constituents considered are nucleotides from DNA [93] or monomers from glycogen [94]. In situations where cells are not growing such as in biotechnological production processes, secreted products such as sugars or amino acids [72] as well as CO_2 [90] can serve for labeling analysis. Fluxes in selected phases of batch or fed batch processes can be elucidated via measurement of the labeling of intracellular compounds [15].

12.5.5
Labeling Analysis by Nuclear Magnetic Resonance Spectroscopy

NMR has been applied since about 40 years for the purpose of isotopic studies of metabolism. It is unique in providing direct access to the position labeled carbons in metabolites thus straightforward monitor the fate of specifically labeled substrates (Figure 12.4). This has provided high resolution labeling data for flux studies in microorganisms [74], plants [95], eukaryotic cells, and tissues [47]. From the technological point of view, different NMR measurement types can be applied to collect the isotopic data [96]. Labeling information can be extracted from the analysis of $^{13}C-^{13}C$ couplings in one-dimensional (1D) ^{13}C or two-dimensional (2D) $^1H-^{13}C$ NMR experiments. 2D experiments are usually preferred, when complex mixtures of ^{13}C-labeled metabolites – such as cell extracts or biomass hydrolysates – are considered, and therefore, they are well suited for the investigation of large metabolic networks. Besides developments in the NMR technology itself, which have increased the sensitivity of the spectrometers, novel NMR measurement algorithms have been recently proposed to extend the type and number of accessible isotopic information [97–99]. A promising approach to overcome previous problems of long experiment duration (several hours) is ultrafast NMR, which allows a complete 2D correlation to be obtained within single scan [100]. Two ultrafast sequences were recently developed for the specific purpose of isotopomer analysis [101], opening new perspectives in the field of fluxomics.

Mass spectrometry and NMR are most often applied separately but combinations of the two techniques are increasingly used for ^{13}C-labeling studies of metabolism. Such combinations allow first to collect increased numbers of isotopomer data for the purpose of flux analysis [102]. They allow also the design of elegant combinations of labeling experiments, combining the sensitivity of MS with short-term labeling for dynamic information, with the detailed positional information obtained by NMR for direct identification and quantification of metabolic pathways. This was recently used to demonstrate the operation of the ethylmalonyl-CoA pathway, a novel metabolic pathway that represents an alternative to the glyoxylate cycle for the biosynthesis of glyoxylate [103]. Specific developments aiming at a better integration of the two analytical techniques will provide powerful tools for resolving the topology of complex metabolic networks in the future.

12.6
Application of Metabolic Flux Analysis

During recent years, ^{13}C-based flux analyses have revealed fascinating insights on microbial systems. This includes simultaneous flux through glycolytic and gluconeogenic reactions around the C_3/C_4 node as widespread, but still not yet fully understood the characteristic of bacterial metabolism [16, 57, 58, 81, 104, 105]. Further highlights are the discovery of novel pathways such as the PEP-glyoxylate cycle in *E. coli* [60] or the homolanthionine route for isoleucine biosynthesis [106]. The following studies illustrate the application of metabolic flux analysis to the two key areas of metabolic engineering and systems biology. The few examples selected do not aim to give a full overview on all interesting studies available, but should highlight the potential of metabolic flux analysis toward understanding and optimizing biological systems and biotechnological processes. For a more extended reading, the reader is addressed to recent review articles [1, 2, 107–110].

12.6.1
Improvement of Industrial Production Strains

The need for a rational basis in selecting the right targets for metabolic engineering was one of the most important stimulants of recent flux approaches. As metabolic engineering particularly is an engineering of flux, knowledge about the actual carbon flux distribution in an organism turned out early on to be the key information for successful improvement of production strains [111]. At present, metabolic flux analysis clearly is a fundamental tool for metabolic engineering [1, 108, 112]. A remarkable example of flux-based metabolic engineering has led to substantial improvement of lysine production in *C. glutamicum*. Comparative flux studies on different carbon sources suggested insufficient NADPH supply through the pentose phosphate limiting lysine production [72, 113] as shown for glucose and fructose (Figure 12.7a,b). This led to the novel strategy of redirecting flux toward the PPP via over expression of glucose-repressed fructose 1,6-bisphosphatase (Figure 12.7c,d). As shown, the genetic modification substantially increased PPP flux, NADPH supply, and lysine yield [114]. This flux-based strategy was successfully extended by additional over expression of glucose 6-phosphate dehydrogenase [115] and the entire operon of PPP genes [116]. Beyond the prediction of product supporting pathways, ^{13}C flux analysis also allowed to identify competing reactions as candidates for attenuation or deletion. The TCA cycle, resulting in high loss of carbon via decarboxylation, is a promising example (Figure 12.7a,b). Predicted from flux analysis, targeted attenuation of the expression of isocitrate dehydrogenase [117] and pyruvate dehydrogenase [118] significantly improved lysine production. Another interesting example was the discovery of substantial flux through PEP carboxykinase withdrawing the lysine precursor oxaloacetate in *C. glutamicum* [58]. Subsequent deletion of this activity was beneficial for lysine production [119]. Beyond these single target studies, global flux redirection was recently achieved by identifying a multitarget combination of genes in the central metabolism of *C.*

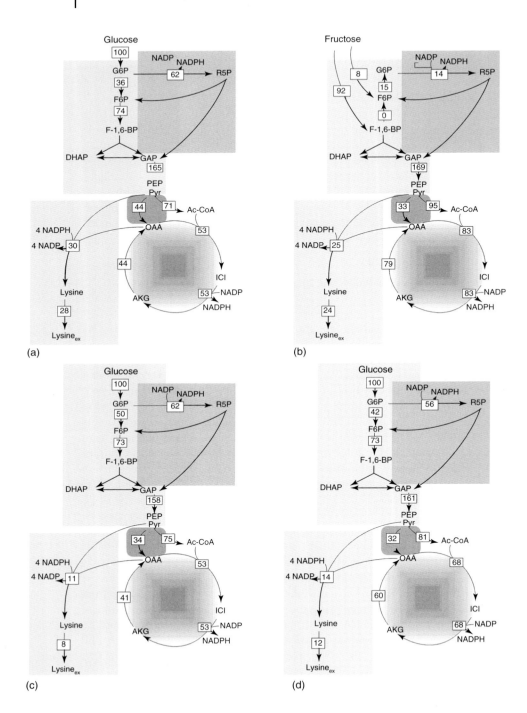

glutamicum [116]. The combination was predicted from integrating *in vivo* and *in silico* pathway fluxes. Subsequent genome-based genetic engineering provided a highly efficient lysine hyperproducer on the basis of the wild type *C. glutamicum* ATCC 13032. Including only 12 genomic traits, this tailor-made production strain is the best known lysine producer and even exceeds the performance of classically derived strains optimized for almost six decades. It allows production of 120 g l^{-1} of lysine in 30 h, a conversion yield of 55% and a productivity of 4 g l^{-1} h^{-1} [116]. This work is an outstanding example on how metabolic flux analysis can drive the development of tailor-made designer bugs. Also for glutamate production in *C. glutamicum*, flux analysis served as rational basis for metabolic engineering. A limiting flux at the level of the 2-oxoglutarate dehydrogenase complex was hereby identified as major control point of the flux at the 2-oxoglutarate node [120]. This stimulated metabolic engineering of the production strain toward increased glutamate production by attenuation of the flux through 2-oxoglutarate dehydrogenase [121]. Other studies did not directly provide nonobvious key targets, but have proven valuable for description of network responses to verify success of genetic manipulations or monitor flux changes during production processes. Among these are flux analyses of *C. glutamicum* identifying pyruvate carboxylase as major anaplerotic pathway during glutamate production [122], in genealogy of production strains unraveling flux changes linked to classical optimization strains [39], monitoring flux changes during production of phenylalanine [123]. Interesting insights into microbial metabolism were also obtained from studies of riboflavine producing *B. subtilis*, underlining the importance of the nonoxidative PPP for supply of riboflavin [124]. Production related flux changes suggesting potential targets were also observed in studies of *E. coli* producing the recombinant protein h-FGF2 [125], in penicillin producing *P. chrysogenum* [126], or in succinate producing *Actinobacillus succinogenes* [127]. More recently, ^{13}C metabolic flux analysis has also been applied to characterize evolutionary engineered *Saccharomyces cerevisiae* with attractive properties for wine production such as increased fermentation rates and aroma compound production, which could be of interest for future wine production [128].

Figure 12.7 Metabolic flux analysis as core technology in metabolic engineering. Impact of the carbon sources glucose (a) and fructose (b) on metabolic fluxes in lysine producing *C. glutamicum*, predicting fructose bisphosphatase (*fbp*) with zero flux on fructose as bottleneck for NADPH supply. Metabolic flux analysis of a rational lysine producing strain (c) and the derived *fbp* mutant exhibiting increased lysine production and PPP flux (d). The values represent relative fluxes which are normalized to the specific substrate rate set to 100%. (The data are taken from previous work [72, 114].)

12.6.2
Integration into Systems Biology Approaches

The central carbon metabolism assessed by metabolic flux analysis displays a highly interconnected network of reactions, which catalyze the materials flow through the cell.

Superimposed on the metabolic network are multiple layers of control that ensure optimal pathway usage [2]. The understanding, modeling, and prediction of function and interaction between these layers of metabolic and regulatory networks are in the heart of systems biology. Beyond doubt, metabolic flux analysis is a key technology toward this goal [129]. High-throughput flux studies have recently shown that only a few regulatory proteins in fact directly control flux, which seems a general design principle in metabolic networks [71, 80]. First examples have integrated metabolic flux analysis with other profiling tools to gain understanding of the link between selected fluxes and gene expression, that is, control of pathway flux which indicates that fluxes are rarely regulated on the genome level [15, 130]. Recent systems-oriented studies have provided substantially large data sets from various omics technologies including fluxes in order to unravel and quantify regulation mechanisms upon genetic or environmental perturbation [87]. To fully exploit such data, metabolic models are needed that quantitatively integrate the different data types. Interesting approaches in this direction are based on application of probabilistic graphical models [131] or stoichiometric, second generation genome-scale models to study links between reactions and transcription [132]. Despite detailed knowledge on network topology, the capacity to predict metabolic responses to environmental stimuli or genetic perturbations is still very limited. The major burden is the lack of direct data on *in vivo* network operation and mathematical model that infer how metabolic networks and regulatory networks cooperate in the biological response. Metabolic systems biology provides a conceptual and methodological framework for integrating flux analysis with all other information relevant to metabolism (metabolites, enzymes, reactions, etc.) into scalable and mechanistic mathematical models of metabolic networks. They can be used for getting comprehensive and quantitative understanding of the operation and regulation of metabolism at the system level, or for getting a predictive capability as regard to the behavior of the metabolic system in response to natural or nonnatural perturbations. First examples have integrated metabolic flux analysis with other profiling tools to gain understanding of the link between selected fluxes and gene expression, that is, control of pathway flux which indicates that fluxes are rarely regulated on the genome level [15, 130]. Recent systems-oriented studies have provided substantially large data sets from various omics technologies including fluxes in order to unravel and quantify regulation mechanisms upon genetic or environmental perturbation [87, 133]. To fully exploit such data toward quantitative, metabolic models are needed that quantitatively integrate the different data types. Interesting approaches in this direction are based on application of probabilistic graphical models [131], stoichiometric, second generation genome-scale models to study links between reactions and transcription [132], and large-scale kinetics models [134].

12.7
Recent Advances in the Field

12.7.1
High-Throughput Flux Screening

Several developments have recently significantly expanded the possibilities of microbial flux analysis and should be mentioned here. These include the extension of flux analysis to high-throughput, allowing screening of a large number of mutants in short time [61, 135]. The key of this technology is miniaturized and parallelized cultivation in microliter scale in conventional 96 well microtiter plates or specifically designed deep well plates with increased oxygen transfer. These are combined with highly sensitive MS labeling analysis [51, 57, 136] requiring only minimal sample amounts. Cultivation in microtiter plates can be coupled to online monitoring of dissolved oxygen and pH, thus providing important physiological data [137]. Comparative studies have proven high similarity of flux data from this small scale to shaken flasks [135] and lab scale bioreactors [71]. Flux screening has successfully been applied to study deletion mutants of *B. subtilis* [80] or yeast [138, 139] and unraveled interesting information about gene function and regulatory properties of the metabolism [140]. The concept of fluxome profiling was also introduced where flux analysis focuses on targeted flux ratios rather than on absolute values, providing model independent comparison of various strains or organisms [141].

12.7.2
Flux Dynamics

As stated earlier, a pre-requisite for current flux analysis is metabolic steady state and isotopic steady state. For the usually analyzed proteinogenic pool of amino acids it takes multiple cell divisions, before constant labeling patterns are achieved. This time period can be significantly shortened, if labeling patterns in free intracellular intermediates are considered, reaching isotopic steady state much faster [15, 142]. Labeling experiments may thus be shortened from several hours to minutes, although exchange of label with large pools of unlabeled intracellular macromolecules may extend the labeling period to $1-2$ h [2]. A further reduction to ultrafast labeling experiments, a recent concept is based on the kinetic information of label distribution during the first minute of labeling at metabolic steady state, but isotopic instationarity [143]. This approach is demanding, as it requires the accurate measurement of both labeling patterns and pool sizes, and needs extremely high computer power for flux calculation [143]. Such approaches will be of great value to monitor flux changes in dynamic systems as highlighted by first successful examples [144–146].

12.8
Concluding Remarks

Now we begin to understand that metabolism is far more than just the conversion of molecules, and that it plays a central role in the integration of cellular components to sense, communicate, or interact – and thus live. What might be more meaningful than just take a direct view? Fluxomics, providing access to the quantitative analysis of molecular pathway fluxes, has proven as a key tool toward the understanding of metabolism. Without doubt, the coming years will see an even more massive application of this technology toward a routine tool in systems biology and metabolic engineering.

Acknowledgments

Christoph Wittmann acknowledges support by the German Federal Ministry of Education and Research (BMBF) through the grants "Systems Biology of Microorganisms 2" (Grant **0313978G9**), and "Medical Infectious Genomics" (**Grant 0315833D**) and by the Institut National de Recherche Applique (INSA).

References

1. Kohlstedt, M., Becker, J., and Wittmann, C. (2010) *Appl. Microbiol. Biotechnol.*, **88**, 1065–1075.
2. Sauer, U. (2006) *Mol. Syst. Biol.*, **2**, 62.
3. Feist, A.M. and Palsson, B.O. (2008) *Nat. Biotechnol.*, **26**, 659–667.
4. Kjeldsen, K.R. and Nielsen, J. (2009) *Biotechnol. Bioeng.*, **102**, 583–597.
5. Oh, Y.K., Palsson, B.O., Park, S.M., Schilling, C.H., and Mahadevan, R. (2007) *J. Biol. Chem.*, **282**, 28791–28799.
6. Wittmann, C. and de Graaf, A. (2005) in *Handbook of Corynebacterium glutamicum* (eds L. Eggeling and M. Bott), CRC Press, Boca Raton, FL, pp. 277–304.
7. Neidhardt, F.C., Ingraham, J.L., and Schaechter, M. (1990) *Physiology of the Bacterial Cell - A Molecular Approach*, Sinauer Associates Inc., Sunderland, MA.
8. Ingraham, J.L., Maaloe, O., and Neidhardt, F.C. (1983) *Growth of the Bacterial Cell*, Sinauer Associates Inc.
9. Dauner, M. and Sauer, U. (2001) *Biotechnol. Bioeng.*, **76**, 132–143.
10. Carlson, R. and Srienc, F. (2004) *Biotechnol. Bioeng.*, **85**, 1–19.
11. Wiechert, W. and de Graaf, A. (1997) *Biotechnol. Bioeng.*, **55** (1), 102–117.
12. Frick, O. and Wittmann, C. (2005) *Microb. Cell Fact.*, **4**, 30.
13. Nöh, K. and Wiechert, W. (2011) *Appl. Microbiol. Biotechnol.* **91**, 1247–1265.
14. Shastri, A.A. and Morgan, J.A. (2007) *Phytochemistry*, **68**, 2302–2312.
15. Krömer, J.O., Sorgenfrei, O., Klopprogge, K., Heinzle, E., and Wittmann, C. (2004) *J. Bacteriol.*, **186**, 1769–1784.
16. Becker, J., Klopprogge, C., and Wittmann, C. (2008) *Microb. Cell Fact.*, **7**, 8.

17. Kelleher, J.K. (1999) *Am. J. Physiol.*, **277**, E395–E400.
18. Christensen, B., Thykaer, J., and Nielsen, J. (2000) *Appl. Microbiol. Biotechnol.*, **54**, 212–217.
19. Stephanopoulos, G. (1998) *Biotechnol. Bioeng.*, **58**, 119–120.
20. Kimata, K., Tanaka, Y., Inada, T., and Aiba, H. (2001) *EMBO J.*, **20**, 3587–3595.
21. Wiechert, W. (2001) *Metab. Eng.*, **3**, 195–206.
22. Wittmann, C. (2002) *Adv. Biochem. Eng. Biotechnol.*, **74**, 39–64.
23. Edwards, J.S. and Palsson, B.O. (2000) *BMC Bioinformatics*, **1**, 1.
24. Schilling, C.H., Edwards, J.S., Letscher, D., and Palsson, B.O. (2000) *Biotechnol. Bioeng.*, **71**, 286–306.
25. Papoutsakis, E.T. and Meyer, C.L. (1985) *Biotechnol. Bioeng.*, **27**, 67–80.
26. Kiss, R.D. and Stephanopoulos, G. (1992) *Biotechnol. Bioeng.*, **39**, 565–574.
27. Vallino, J.J. and Stephanopoulos, G. (1993) *Biotechnol. Bioeng.*, **41**, 633–646.
28. Vallino, J.J. and Stephanopoulos, G. (1994a) *Biotechnol. Prog.*, **10**, 327–334.
29. Vallino, J.J. and Stephanopoulos, G. (1994b) *Biotechnol. Prog.*, **10**, 320–326.
30. Holms, W.H. (1986) *Curr. Top. Cell. Regul.*, **28**, 69–105.
31. Varma, A., Boesch, B.W., and Palsson, B.O. (1993a) *Biotechnol. Bioeng.*, **42**, 59–73.
32. Varma, A., Boesch, B.W., and Palsson, B.O. (1993b) *Appl. Environ. Microbiol.*, **59**, 2465–2473.
33. Varma, A. and Palsson, B.O. (1994) *Appl. Environ. Microbiol.*, **60**, 3724–3731.
34. Tesch, M., de Graaf, A.A., and Sahm, H. (1999) *Appl. Environ. Microbiol.*, **65**, 1099–1109.
35. Ben-Yoseph, O., Kingsley, P.B., and Ross, B.D. (1994) *Magn. Reson. Med.*, **32**, 405–409.
36. Ross, B.D., Kingsley, P.B., and Ben-Yoseph, O. (1994) *Biochem. J.*, **302** (Pt 1), 31–38.
37. Gombert, A.K., Moreira dos Santos, M., Christensen, B., and Nielsen, J. (2001) *J. Bacteriol.*, **183**, 1441–1451.
38. Klapa, M.I., Aon, J.C., and Stephanopoulos, G. (2003) *Eur. J. Biochem.*, **270**, 3525–3542.
39. Wittmann, C., Hans, M., and Heinzle, E. (2002) *Anal. Biochem.*, **307**, 379–382.
40. Marx, A., de Graaf, A., Wiechert, W., Eggeling, L., and Sahm, H. (1996) *Biotechnol. Bioeng.*, **49** (2), 111–129.
41. Portais, J.C. and Delort, A.M. (2002) *FEMS Microbiol. Rev.*, **26**, 375–402.
42. Martin, M., Portais, J.C., Voisin, P., Rousse, N., Canioni, P., and Merle, M. (1995) *Eur. J. Biochem.*, **231**, 697–703.
43. Szyperski, T. (1995) *Eur. J. Biochem.*, **232**, 433–448.
44. Wittmann, C. (2007) *Microb. Cell Fact.*, **6**, 6.
45. Brunengraber, H., Kelleher, J.K., and Des Rosiers, C. (1997) *Annu. Rev. Nutr.*, **17**, 559–596.
46. Debnam, P.M., Shearer, G., Blackwood, L., and Kohl, D.H. (1997) *Eur. J. Biochem.*, **246**, 283–290.
47. Portais, J.C., Schuster, R., Merle, M., and Canioni, P. (1993) *Eur. J. Biochem.*, **217**, 457–468.
48. Sherry, A.D., Sumegi, B., Miller, B., Cottam, G.L., Gavva, S., Jones, J.G., and Malloy, C.R. (1994) *Biochemistry*, **33**, 6268–6275.
49. Shiio, I., Otsuka, S.I., and Tsunoda, T. (1960) *J. Biochem.*, **47**, 414–421.
50. Wittmann, C. and Heinzle, E. (2001c) *Metab. Eng.*, **3**, 173–191.
51. Wittmann, C. and Heinzle, E. (2002) *Appl. Environ. Microbiol.*, **68**, 5843–5859.
52. Wang, C.H., Stern, I., Gilmour, C.M., Klungsoyr, S., Reed, D.J., Bialy, J.J., Christensen, B.E., and Cheldelin, V.H. (1958) *J. Bacteriol.*, **76**, 207–216.
53. Fluckiger, J. and Ettlinger, L. (1977) *Arch. Microbiol.*, **114**, 183–187.
54. Mayaudon, J. (1968) *Ann. Inst. Pasteur (Paris)*, **115**, 710–730.
55. Stephanopoulos, G., Aristidou, A.A., and Nielsen, J. (1998) *Metabolic Engineering – Principles and Methodologies*, Academic Press, San Diego, CA.
56. Walker, T.E., Han, C.H., Kollman, V.H., London, R.E., and Matwiyoff, N.A. (1982) *J. Biol. Chem.*, **257**, 1189–1195.

57. Wittmann, C. and Heinzle, E. (2001a) *Eur. J. Biochem.*, **268**, 2441–2455.
58. Petersen, S., de Graaf, A.A., Eggeling, L., Möllney, M., Wiechert, W., and Sahm, H. (2000) *J. Biol. Chem.*, **275**, 35932–35941.
59. Sonntag, K., Eggeling, L., De Graaf, A.A., and Sahm, H. (1993) *Eur. J. Biochem.*, **213**, 1325–1331.
60. Fischer, E. and Sauer, U. (2003) *Eur. J. Biochem.*, **270**, 880–891.
61. Sauer, U. (2004) *Curr. Opin. Biotechnol.*, **15**, 58–63.
62. Quek, L.E., Wittmann, C., Nielsen, L.K., and Krömer, J.O. (2009) *Microb. Cell Fact.*, **8**, 25.
63. Wiechert, W., Möllney, M., Petersen, S., and de Graaf, A.A. (2001) *Metab. Eng.*, **3**, 265–283.
64. Zamboni, N., Fischer, E., and Sauer, U. (2005b) *BMC Bioinformatics*, **6**, 209.
65. Wiechert, W. and de Graaf, A.A. (1996) *Adv. Biochem. Eng. Biotechnol.*, **54**, 109–154.
66. Vogt, J.A., Yarmush, D.M., Yu, Y.M., Zupke, C., Fischman, A.J., Tompkins, R.G., and Burke, J.F. (1997) *Am. J. Physiol.*, **272**, C2049–C2062.
67. Schmidt, K., Carlsen, M., Nielsen, J., and Villadsen, J. (1997) *Biotechnol. Bioeng.*, **55** (6), 831–840.
68. Wiechert, W., Möllney, M., Isermann, N., Wurzel, M., and de Graaf, A.A. (1999) *Biotechnol. Bioeng.*, **66**, 69–85.
69. van Winden, W.A., Heijnen, J.J., and Verheijen, P.J. (2002a) *Biotechnol. Bioeng.*, **80**, 731–745.
70. Antoniewicz, M.R., Kelleher, J.K., and Stephanopoulos, G. (2007) *Metab. Eng.*, **9**, 68–86.
71. Fischer, E., Zamboni, N., and Sauer, U. (2004) *Anal. Biochem.*, **325**, 308–316.
72. Kiefer, P., Heinzle, E., Zelder, O., and Wittmann, C. (2004) *Appl. Environ. Microbiol.*, **70**, 229–239.
73. Gosselin, I., Wattraint, O., Riboul, D., Barbotin, J., and Portais, J. (2001) *FEBS Lett.*, **499**, 45–49.
74. Portais, J.C., Tavernier, P., Gosselin, I., and Barbotin, J.N. (1999) *Eur. J. Biochem.*, **265**, 473–480.
75. Yang, T.H., Heinzle, E., and Wittmann, C. (2005) *Comput. Biol. Chem.*, **29**, 121–133.
76. van Winden, W.A., Wittmann, C., Heinzle, E., and Heijnen, J.J. (2002b) *Biotechnol. Bioeng.*, **80**, 477–479.
77. Wittmann, C. and Heinzle, E. (1999) *Biotechnol. Bioeng.*, **62**, 739–750.
78. Wiechert, W. (1995) Metabolic flux determination by stationary ^{13}-C tracer experiments: analysis of sensitivity, identifiability and redundancy. 17th IFIP TC7 Conference on System Modelling and Optimization, Prague, Czech Republic, p. 8.
79. Yang, C., Hua, Q., and Shimizu, K. (2002) *J. Biosci. Bioeng.*, **93**, 78–87.
80. Fischer, E. and Sauer, U. (2005) *Nat. Genet.*, **37**, 636–640.
81. Emmerling, M., Dauner, M., Ponti, A., Fiaux, J., Hochuli, M., Szyperski, T., Wuthrich, K., Bailey, J.E., and Sauer, U. (2002) *J. Bacteriol.*, **184**, 152–164.
82. Hua, Q., Yang, C., Baba, T., Mori, H., and Shimizu, K. (2003) *J. Bacteriol.*, **185**, 7053–7067.
83. Möllney, M., Wiechert, W., Kownatzki, D., and de Graaf, A.A. (1999) *Biotechnol. Bioeng.*, **66**, 86–103.
84. Schmidt, K., Norregaard, L.C., Pedersen, B., Meissner, A., Duus, J.O., Nielsen, J.O., and Villadsen, J. (1999) *Metab. Eng.*, **1**, 166–179.
85. Iwatani, S., Van Dien, S., Shimbo, K., Kubota, K., Kageyama, N., Iwahata, R., Miyano, H., Hirayama, K., Usuda, Y., Shimizu, K., and Matsui, K. (2007) *J. Biotechnol.*, **128**, 93–111.
86. Kiefer, P., Nicolas, C., Letisse, F., and Portais, J.C. (2007) *Anal. Biochem.*, **360**, 182–188.
87. Toya, Y., Ishii, N., Hirasawa, T., Naba, M., Hirai, K., Sugawara, K., Igarashi, S., Shimizu, K., Tomita, M., and Soga, T. (2007) *J. Chromatogr. A*, **1159**, 134–141.
88. Heinzle, E., Yuan, Y., Kumar, S., Wittmann, C., Gehre, M., Richnow, H.H., Wehrung, P., Adam, P., and Albrecht, P. (2008) *Anal. Biochem.*, **380**, 202–210.
89. Yuan, Y., Yang, T.H., and Heinzle, E. (2010) *Metab. Eng.*, **12**, 392–400.

90. Yang, T.H., Wittmann, C., and Heinzle, E. (2006a) *Metab. Eng.*, **8**, 432–446.
91. Yang, T.H., Wittmann, C., and Heinzle, E. (2006b) *Metab. Eng.*, **8**, 417–431.
92. Christensen, Nielsen, and (1998) *Metab Eng 1*, **29**, 282–290.
93. Derek, C. Macallan, Catherine, A. Fullerton, Richard, A. Neese, Katherine Haddock, Sunny, S. Park, and Marc, K. Hellerstein (1998) *Measurement of cell proliferation by labeling of DNA with stable isotope-labeled glucose: Studies in vitro, in animals, and in humans.* PNAS. **95**, 708–713.
94. Shulman, RG, and Rothman, DL (2001) *Annu Rev Physiol.*, 13C NMR of intermediary metabolism: implications for systemic physiology, **63**, 15–48.
95. Troufflard, S., Roscher, A., Thomasset, B., Barbotin, J.N., Rawsthorne, S., and Portais, J.C. (2007) *Phytochemistry*, **68**, 2341–2350.
96. Szyperski, T. (1998) *Q. Rev. Biophys.*, **31**, 41–106.
97. Lewis, I.A., Karsten, R.H., Norton, M.E., Tonelli, M., Westler, W.M., and Markley, J.L. (2010) *Anal. Chem.*, **82**, 4558–4563.
98. Massou, S., Nicolas, C., Letisse, F., and Portais, J.C. (2007a) *Metab. Eng.*, **9**, 252–257.
99. Massou, S., Nicolas, C., Letisse, F., and Portais, J.C. (2007b) *Phytochemistry*, **68**, 2330–2340.
100. Frydman, L., Scherf, T., and Lupulescu, A. (2002) *Proc. Natl. Acad. Sci. U.S.A.*, **99**, 15858–15862.
101. Giraudeau, P., Massou, S., Robin, Y., Cahoreau, E., Portais, J.C., and Akoka, S. (2011) *Anal. Chem.*, **83**, 3112–3119.
102. Nicolas, C., Kiefer, P., Letisse, F., Krömer, J., Massou, S., Soucaille, P., Wittmann, C., Lindley, N.D., and Portais, J.C. (2007) *FEBS Lett.*, **581**, 3771–3776.
103. Peyraud, R., Kiefer, P., Christen, P., Massou, S., Portais, J.C., and Vorholt, J.A. (2009) *Proc. Natl. Acad. Sci. U.S.A.*, **106**, 4846–4851.
104. Sauer, U., Hatzimanikatis, V., Bailey, J.E., Hochuli, M., Szyperski, T., and Wuthrich, K. (1997) *Nat. Biotechnol.*, **15**, 448–452.
105. Yang, C., Hua, Q., Baba, T., Mori, H., and Shimizu, K. (2003) *Biotechnol. Bioeng.*, **84**, 129–144.
106. Krömer, J.O., Heinzle, E., Schröder, H., and Wittmann, C. (2006) *J. Bacteriol.*, **188**, 609–618.
107. Blank, L.M. and Kuepfer, L. (2010) *Appl. Microbiol. Biotechnol.*, **86**, 1243–1255.
108. Otero, J.M. and Nielsen, J. (2010) *Biotechnol. Bioeng.*, **105**, 439–460.
109. Shimizu, K. (2004) *Adv. Biochem. Eng. Biotechnol.*, **91**, 1–49.
110. Zamboni, N. (2011) *Curr. Opin. Biotechnol.*, **22**, 103–108.
111. Bailey, J.E. (1991) *Science*, **252**, 1668–1675.
112. Stephanopoulos, G. (1999) *Metab. Eng.*, **1**, 1–11.
113. Wittmann, C., Kiefer, P., and Zelder, O. (2004a) *Appl. Environ. Microbiol.*, **70**, 7277–7287.
114. Becker, J. et al. (2005).
115. Becker, J., Klopprogge, C., Herold, A., Zelder, O., Bolten, C.J., and Wittmann, C. (2007) *J. Biotechnol.*, **132**, 99–109.
116. Becker, J., Zelder, O., Häfner, S., Schröder, H., and Wittmann, C. (2011) *Metab. Eng.*, **13**, 159–168.
117. Becker, J., Klopprogge, C., Schröder, H., and Wittmann, C. (2009) *Appl. Environ. Microbiol.*, **75**, 7866–7869.
118. Becker, J., Buschke, N., Bücker, R., and Wittmann, C. (2010) *Eng. Life Sci.*, **10**, 430–438.
119. Petersen, S., Mack, C., de Graaf, A.A., Riedel, C., Eikmanns, B.J., and Sahm, H. (2001) *Metab. Eng.*, **3**, 344–361.
120. Shirai, T., Nakato, A., Izutani, N., Nagahisa, K., Shioya, S., Kimura, E., Kawarabayasi, Y., Yamagishi, A., Gojobori, T., and Shimizu, H. (2005) *Metab. Eng.*, **7**, 59–69.
121. Kim, J., Hirasawa, T., Sato, Y., Nagahisa, K., Furusawa, C., and Shimizu, H. (2008) *Appl. Microbiol. Biotechnol.*
122. Shirai, T., Fujimura, K., Furusawa, C., Nagahisa, K., Shioya, S., and Shimizu, H. (2007) *Microb. Cell Fact.*, **6**, 19.

123. Wahl, A., El Massaoudi, M., Schipper, D., Wiechert, W., and Takors, R. (2004) *Biotechnol. Prog.*, **20**, 706–714.
124. Zamboni, N., Fischer, E., Muffler, A., Wyss, M., Hohmann, H.P., and Sauer, U. (2005a) *Biotechnol. Bioeng.*, **89**, 219–232.
125. Wittmann, C., Weber, J., Betiku, E., Krömer, J., Bohm, D., and Rinas, U. (2007) *J. Biotechnol.*, **132**, 375–384.
126. Kleijn, R.J., van Winden, W.A., Ras, C., van Gulik, W.M., Schipper, D., and Heijnen, J.J. (2006) *Appl. Environ. Microbiol.*, **72**, 4743–4754.
127. McKinlay, J.B., Shachar-Hill, Y., Zeikus, J.G., and Vieille, C. (2007) *Metab. Eng.*, **9**, 177–192.
128. Cadiere, A., Ortiz-Julien, A., Camarasa, C., and Dequin, S. (2011) *Metab. Eng.*, **13**, 263–271.
129. Cascante, M. and Marin, S. (2008) *Essays Biochem.*, **45**, 67–81.
130. Schilling, O., Frick, O., Herzberg, C., Ehrenreich, A., Heinzle, E., Wittmann, C., and Stülke, J. (2007) *Appl. Environ. Microbiol.*, **73**, 499–507.
131. Yeang, C.H. and Vingron, M. (2006) *BMC Bioinformatics*, **7**, 332.
132. Covert, M.W., Knight, E.M., Reed, J.L., Herrgard, M.J., and Palsson, B.O. (2004) *Nature*, **429**, 92–96.
133. Moxley, J.F., Jewett, M.C., Antoniewicz, M.R., Villas-Boas, S.G., Alper, H., Wheeler, R.T., Tong, L., Hinnebusch, A.G., Ideker, T., Nielsen, J., and Stephanopoulos, G. (2009) *Proc. Natl. Acad. Sci. U.S.A.*, **106**, 6477–6482.
134. Jamshidi, N. and Palsson, B.O. (2008) *Mol. Syst. Biol.*, **4**, 171.
135. Wittmann, C., Kim, H.M., and Heinzle, E. (2004b) *Biotechnol. Bioeng.*, **87**, 1–6.
136. Wittmann, C. and Heinzle, E. (2001b) *Biotechnol. Bioeng.*, **72**, 642–647.
137. John, G.T., Klimant, I., Wittmann, C., and Heinzle, E. (2003) *Biotechnol. Bioeng.*, **81**, 829–836.
138. Blank, L.M., Kuepfer, L., and Sauer, U. (2005) *Genome Biol.*, **6**, R49.
139. Hollemeyer, K., Velagapudi, V.R., Wittmann, C., and Heinzle, E. (2007) *Rapid Commun. Mass Spectrom.*, **21**, 336–342.
140. Haverkorn van Rijsewijk, B.R., Nanchen, A., Nallet, S., Kleijn, R.J., and Sauer, U. (2011) *Mol. Syst. Biol.*, **7**, 477.
141. Zamboni, N. and Sauer, U. (2004) *Genome Biol.*, **5**, R99.
142. van Winden, W.A., van Dam, J.C., Ras, C., Kleijn, R.J., Vinke, J.L., van Gulik, W.M., and Heijnen, J.J. (2005) *FEMS Yeast Res.*, **5**, 559–568.
143. Nöh, K., Wahl, A., and Wiechert, W. (2006) *Metab. Eng.*, **8**, 554–577.
144. Costenoble, R., Muller, D., Barl, T., van Gulik, W.M., van Winden, W.A., Reuss, M., and Heijnen, J.J. (2007) *FEMS Yeast Res.*, **7**, 511–526.
145. Schaub, J., Mauch, K., and Reuss, M. (2008) *Biotechnol. Bioeng.*, **99**, 1170–1185.
146. Wahl, S.A., Nöh, K., and Wiechert, W. (2008) *BMC Bioinformatics*, **9**, 152.

13
Metabolomics: Application in Plant Sciences

Gaëtan Glauser, Julien Boccard, Jean-Luc Wolfender, and Serge Rudaz

13.1
Introduction

Since its early days, plant science has relied heavily on observable features such as morphological characteristics, growth, and reproductive rates. Genomics and proteomics have greatly changed this trend by their ability to highlight so-called "silent phenotypes" that appear identical to wild types when only morphology is considered. Although an acceptable understanding of the link between genes and proteins has been established, much remains unknown about the connection between proteome and phenotype [1]. Metabolites reflect proteins' catabolic and anabolic activities. Similar to transcripts and proteins, metabolite profiles may be regarded as a snapshot of the current state of a biological sample at a particular moment. As metabolites are the end products of cellular functions, their concentration levels allow an overall inspection of the biochemical status of an organism at a functional level in response to environmental or genetic manipulation. The relationship between phenotypic characteristics and their related genes is a central issue in modern plant science, and many properties beneficial to human health can be attributed to metabolites [2] (e.g., polyphenols). Metabolomics constitutes a growing field of investigation in plant research [3, 4] and involves new analytical challenges regarding the simultaneous monitoring of large numbers of low-molecular-weight compounds and the subsequent interpretation of the vast amount of data generated.

Estimates of the sizes of metabolomes are highly dependent on the studied organism. Verpoorte *et al.* [5] suggested a comparable number of metabolites and genes. However, it must be noted that there is no direct link between every gene involved in metabolism and a given metabolite. Other estimates propose up to 15 000 distinct compounds within a given plant species [6, 7]. Overall, the plant kingdom is known to produce a wide diversity of chemical molecules estimated at 200 000 primary and secondary metabolites [8]. Except for primary metabolites directly involved in growth, development, and reproduction, most of those metabolites remain unknown. The chemical properties of this array of compounds are greatly variable, and this diversity is a critical aspect to consider.

Metabolomics in Practice: Successful Strategies to Generate and Analyze Metabolic Data, First Edition.
Edited by Michael Lämmerhofer and Wolfram Weckwerth.
© 2013 Wiley-VCH Verlag GmbH & Co. KGaA. Published 2013 by Wiley-VCH Verlag GmbH & Co. KGaA.

The variability in molecular weight, polarity, and solubility and the wide dynamic range of concentrations (from femtomoles to millimoles) constitute an additional issue when analyzing metabolites, as no amplification process is available.

Over 20 plant genomes have been sequenced, including *Arabidopsis thaliana* (thale cress) [9] and *Oryza sativa* (rice) [10], and metabolomics represents a promising tool for the post-genomic study of plant models with respect to variations induced by perturbations including environmental changes and physical, biotic, abiotic, or nutritional stress [11]. The fingerprinting of plants by high-throughput methodologies has been intensively used to characterize these phenomena and better understand plant functions. These approaches also constitute potent tools for plant molecular biotechnology and may be used to study the effects of plant breeding, mutation, and other manipulations such as the expression of engineered genes [12].

Finding relevant answers to biological questions using a metabolomic approach is the result of a sequence of challenges, from the choice of a specific sample preparation and an adapted analytical platform to the data processing and modeling procedures. Each step of the process has to be carefully planned. This chapter discusses a broad overview of the challenges related to plant metabolomics, starting with experimental setup, sample preparation, and analytical issues to the downstream handling of data to extract biological knowledge.

13.2
Sample Preparation

Sample preparation is an essential step in the chemical analytical process as it strongly affects the quality of the measured data [13]. In metabolomics, a large number of samples are generally handled simultaneously, and therefore, simple, fast, and generic sample preparation procedures are preferred. Such procedures should be adapted to the kind of metabolites that are to be analyzed and compatible with the analytical technique that will be further employed. The sample preparation of plants represents a particular case because metabolites must be extracted from a solid matrix using an appropriate solvent. In principle, the preparation of a plant sample for metabolomics requires the following steps: culture and/or harvesting, storage and drying, and extraction. Each of these steps is discussed in the following sections, and practical guidelines for the proper preparation of plant samples are given.

13.2.1
Culture and Harvesting

Plants may be either cultivated in an environment controlled for light, temperature, and humidity or grown and/or harvested in the field. Under controlled conditions (e.g., growth chambers such as phytotrons), the biological variability of samples is obviously reduced, which may be of importance for the accuracy of metabolomic data. Plant treatments with insecticides or fungicides should be avoided because they are likely to be detected during analysis and confounded with plant constituents. Once plants are mature enough, they must be harvested. During this step, the only creed is "rapidity": the part to be harvested should be quickly separated from the rest of the plant and immediately frozen, for example, by freeze-clamping or in liquid nitrogen to quench plant metabolism. Indeed, harvesting may be perceived as a wounding of the plant, which reacts within seconds by producing signaling molecules [14]. Harvesting may be more or less difficult depending on which organs are collected. Leaf, root, and flower collection is easily achieved, while pollens, nectars, and exudates may be challenging to collect and require special methods. Generally, it is advisable to separate the different organs because the metabolite content can vary from organ to organ or within the same organ. For example, significant differences are found between primary and crown roots of maize [15]. Young and old leaves of the same plant can also exhibit differences [16]. Shroff *et al.* [17] even found that the concentration in glucosinolates spatially changes within a single leaf in Arabidopsis. The moment at which plants are harvested is another aspect to consider. Indeed, it is known that the concentration of plant constituents varies during the day [18, 19]. All these points should thus be taken into account before starting a metabolomic study of a given plant.

13.2.2
Storage and Drying

After flash-freezing samples to quickly stop enzymatic reactions, samples are usually kept at low temperature (-80 to $-196\,°C$) before use. Samples are generally dried before extraction but during this step, metabolic processes may restart [20], especially with air-drying, oven-drying, or silica-based drying. A more suitable technique is freeze-drying, which prevents sample thawing. The absence of water much reduces enzymatic activity [21]. However, the freeze-drying of tissue may lead to the irreversible adsorption of metabolites on cell walls and membranes [1]. After freeze-drying, samples should be stored in a dry environment because plant tissues are hygroscopic and may absorb sufficient amounts of water to allow for the activation of certain enzymes. Another solution is to extract fresh frozen tissues directly using a solvent or a technique that will inhibit enzyme reactivation (Section 13.2.3).

13.2.3
Extraction

To increase extraction yields, plants are usually ground to a fine powder. When fresh material is used, great care should be taken to always keep the tissues frozen during grinding. This can be done with a mortar and a pestle using liquid nitrogen or using a mixer mill system. There are various parameters that must be considered when performing extractions. First the solvent choice: metabolomics aims to cover the widest possible range of metabolites and there is no single solvent that is able to extract and dissolve all plant compounds. To obtain an exhaustive view of the metabolome, several solvents of increasing polarity should be successively employed, for example, hexane-chloroform-methanol-water. Combinations of various solvents can also be used, for example, a multiple-phase solvent system composed of a mixture of chloroform-methanol-water [22]. Furthermore, the selected solvents should be compatible with the analytical technique used in order to avoid time-consuming evaporation–redissolution steps. For nuclear magnetic resonance (NMR), the use of deuterated solvents after tissue freeze-drying is a straightforward solution. In GC–MS (gas chromatography–mass spectrometry), water should be avoided, while in LC–MS (liquid chromatography), solvents of elution strength preferably lower or equal to that of the mobile phase should be injected. The next parameter that must be considered is the extraction technique. Conventional extraction by maceration and smooth agitation is slow and poorly adapted to numerous micro-samples. To accelerate the extraction process, additional energy can be applied by directly heating the system or by other techniques such as microwave extraction [23], ultrasonic extraction [24], ball mill extraction [25], pressurized liquid extraction [26], and supercritical fluid extraction [27]. Using such methods, extraction times of less than 5 min can be attained.

During extraction, it is also possible that enzymes are reactivated, even in mildly polar solvents such as methanol or isopropanol. This is especially true with fresh material. To avoid this, solvents may either be heated to 70–80 °C, or acids may be added (formic or hydrochloric acid). In both cases, some labile metabolites may be degraded. An interesting example of this phenomenon is described in the thesis of Glauser [28]. The goal of that study was to profile oxylipin-containing galactolipids, including lysogalactolipids, after wounding in Arabidopsis. Unwounded rosettes and rosettes harvested 3 min after wounding were extracted in isopropanol at room temperature and in boiling isopropanol at 75 °C. Using boiling isopropanol, a strong increase in galactolipid levels was measured after wounding. However, at room temperature, no difference could be observed between control and wounded plants. This was attributed to the fact that at room temperature, enzymes were active and generated large quantities of oxylipin-containing galactolipids during extraction, overshadowing the real effect of wounding (Figure 13.1).

Although the goal of metabolomics is to obtain the broadest possible view of the plant metabolome, further prepurification steps may be required depending on the analytical technique used. Derivatization protocols are commonly employed before GC to enhance the volatility of polar metabolites. Liquid–liquid extraction or solid-phase extraction (SPE) procedures can remove interfering or undesired compounds, as well as concentrate samples. For details on these preanalytical procedures, readers may refer to Chapter 2.

13.3
Analytical Methods

As mentioned earlier, the size of plant metabolomes is huge and probably still underestimated. The physicochemical properties and the concentrations of the metabolites present in any given plant are so variable that no single analytical technique is able to cover entire plant metabolomes. Currently, two main techniques

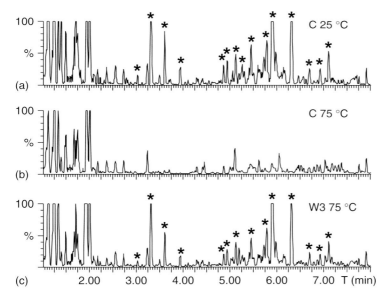

Figure 13.1 LC–MS chromatograms of (a) unwounded *Arabidopsis* extracted in isopropanol at room temperature. (b) Unwounded *Arabidopsis* extracted in boiling isopropanol. (c) Wounded *Arabidopsis* extracted in boiling isopropanol. Peaks labeled with an asterisk correspond to oxylipin-containing galactolipids.

are used for the broadscale detection of plant compounds, namely NMR and MS. Both have well-known advantages and limitations that are briefly discussed in this chapter. Other techniques are sometimes employed for studying general chemical changes in plant tissues, for example, Fourier transform-infrared (FT-IR) spectroscopy [29].

13.3.1
NMR-Based Methods

NMR is a well-established method in metabolomics and has been widely applied to record the metabolite fingerprint of the main constituents of body fluids. In plant science, the use of direct NMR methods for fingerprinting complex extracts is constantly increasing in different fields such as functional genomics, crop improvement, development, nutrition, and ecology.

13.3.1.1 Direct NMR Fingerprinting
NMR fingerprinting is used for the direct analysis of crude plant extracts without the need for prior chromatographic separation [30]. The method is simple, rapid, and does not require specific sample preparation [31, 32]. Another advantage of NMR analysis is the possibility to obtain quantitative information, as the proton signal intensity is only determined by the molar concentration [33]. Standard

protocols exist for the extraction and analysis of various plant tissues that may be shared between laboratories [34].

Concerning the extraction procedure, single-solvent systems have been used in some studies (e.g., deuterated methanol, CD_3OD). However, a combination of CD_3OD and D_2O in various ratios is sometimes preferable, because it has the capacity to extract more diverse metabolites. The use of CD_3OD and KH_2PO_4 buffer in D_2O (pH 6.0) represents a well-established protocol [34], which extracts a wide range of metabolites including amino acids, carbohydrates, fatty acids, organic acids, phenolics, and terpenoids in a single step. The use of a buffer in the solvent avoids possible fluctuation of chemical shifts of signals in the NMR spectra. For more targeted analyses, for example, in the quality control of herbal drugs, other solvent systems can be used for the preferential extraction of a given set of constituents [35]. For data analysis, NMR spectra are digitalized to numeric values for further statistical analysis. In this procedure, the NMR spectrum is divided into a series of small bins (usually of 0.02 or 0.04 ppm). The sum of the signal intensities in each bin is calculated. Scaling to the sum of the total spectrum intensity helps minimize the effect of variation between samples regarding the amount of tissue extracted.

NMR metabolomics is extremely useful when comparisons are made based on the main constituents of a plant extract and when samples have to be compared in long-term studies. Indeed, NMR fingerprints are very reproducible, independently of when analyses are performed, contrary to MS-based approaches where group clustering may be biased by instrumental noise related to contamination.

The main drawback is that, compared to MS, NMR lacks sensitivity, although constant improvement is being made with new generations of magnets and probes. Furthermore, the identification of single compounds in extracts may be hindered by overlapping signals [34]. The identification of biomarkers relies mainly on a comparison of NMR shifts of plant metabolites acquired under the same solvent conditions, while *de novo* identification can be partly achieved by the acquisition of complementary 2D NMR data [36].

13.3.1.2 Applications

NMR fingerprinting techniques have received particular attention in recent years and the number of applications in plant sciences has increased significantly over the past decade. NMR-based metabolomics has been used in various fields of plant science research.

Characterizing Genetically Modified Crops NMR-based metabolomics has been used to assess the consequence of gene modification on the primary and secondary metabolite composition of mutant plants or genetically modified (GM) crops used for consumption. For substantial equivalence, regulatory bodies are placing much emphasis on the identification of unintended effects of genetic modification [37]. As major crop metabolome modifications must be surveyed [38], NMR fingerprinting, combined with multivariate data analysis (MVA), has proved useful, for example, in the identification and classification of maize seeds obtained from transgenic plants into different classes according to changes in metabolites [39]. In another study, MVA of the NMR data collected from extracts of flour milled from three transgenic wheat seeds at two different sites for three years showed that site and year had a stronger influence on metabolite composition than genotype [40]. This revealed that the environment in which a plant grows has a significant effect on the metabolome. Metabolome variation in crop composition has been assessed in various plants and recently summarized in several reviews [38, 41–43].

Chemotaxonomy At present, only a few chemotaxonomic studies have been based on NMR fingerprinting, because this type of investigation is more classically performed with HPLC profiling methods [44]. Recently, metabolic classification of 11 South American Ilex species was performed by NMR-based metabolomics for chemotaxonomic purposes [45]. ^{1}H-NMR, combined with principal component analysis (PCA), partial least squares-discriminant analysis (PLS-DA), and hierarchical cluster analysis (HCA), showed a clear separation between species and resulted in four groups based on metabolomic similarities. The signal congestion of NMR spectra was overcome by the implementation of two-dimensional NMR experiments. One group of Ilex was characterized by higher amounts of xanthines and phenolics, the second group by oleanane-type saponins, the third by arbutin and dicaffeoylquinic acids, and the fourth by ursane-type saponins. Similarly, in another chemotaxonomic study, the relationship among 11 varieties of cannabis grown under the same environmental conditions was assessed [46]. Altogether, these results demonstrate that NMR fingerprinting may be used for rapid chemotaxonomic classification based on the main metabolites of plant species.

Developmental Changes It is well known that plant metabolite levels vary according to tissue and growth stage. NMR profiling can be used to monitor metabolome changes occurring during the development of a given plant or fruit. For example, metabolic changes in different developmental stages of *Vanilla planifolia* pods were recently assessed. MVA of the ^{1}H-NMR spectra of green pods between three and eight months after pollination revealed that older pods had a higher content of glucovanillin, vanillin, *p*-hydroxybenzaldehyde glucoside, and *p*-hydroxybenzaldehyde. Quantification of compounds based on both LC–MS and NMR analyses showed that free vanillin can reach 24% of the total vanillin content after eight months of development in the green vanilla pods [47]. Wine has also been the model of many metabolomic studies. Biochemical changes occurring during grape berry development in Portuguese cultivars were monitored using NMR fingerprinting after SPE enrichment of the extracts. Besides significant differences among cultivars, the initial stages showed comparatively high phenolics and organic acid contents such as caftaric and malic acids, while the later stages exhibited higher glucose and fructose levels. Veraison was found to be a metabolically critical stage of berry development [48]. The monitoring of bioactive ingredients is also important in the case of medicinal plants. In this respect, the metabolic alterations at different developmental stages of *Pilocarpus microphyllus* were studied with special attention to the imidazole alkaloid pilocarpine, which is used for glaucoma treatment, and pilosine, which has no recognized pharmacological use. ^{1}H NMR and electrospray (ESI)–MS profiles indicated that pilosine is produced exclusively at the mature developmental stage, and juvenile plant material seems more appropriate for further studies on pilocarpine biosynthesis [49].

Plant Nutrition Environmental factors such as the availability of essential nutrients may affect plant metabolomes. In this respect, the effect of fertilizers on the metabolic content of certain medicinal plants was assessed. The NMR fingerprints of Narcissus bulbs revealed that an application of nitrogen fertilizer at twice the standard amount resulted in the production of more amino acids and citric acid cycle intermediates, but not more galanthamine that is the bioactive ingredient used in the treatment of Alzeihmer's disease [50].

Plant–Environment Interactions and Host Plant Resistance Plants produce an immense number of secondary compounds that interact with beneficial or harmful organisms. These compounds mainly act as signal compounds and in chemical defense. Plant defense chemicals repel, restrain, or kill plant enemies [51]. Different studies discussing stress/interaction and genotype/taxonomy issues have been reported using *Brassica rapa* as a model. The metabolic alterations of *B. rapa* leaves attacked by larvae of *Plutella xylostella* and *Spodoptera exigua* were investigated by J-resolved NMR spectroscopy, followed by MVA. PCA revealed that the major signals contributing to the discrimination between uninfested and infested leaves were alanine, threonine, glucose, sucrose, feruloyl malate, sinapoyl malate, and gluconapin [52]. In the same plant, investigation of the metabolome changes induced by metal-ion accumulation revealed that glucosinolates and hydroxycinnamic acids conjugated with malates were the discriminating metabolites, as were primary metabolites such as carbohydrates and amino acids [53]. A more recent study on glucosinolate profiling of *B. rapa* cultivars after infection by *Leptosphaeria maculans* and *Fusarium oxysporum* indicated that the interaction between fungi and *B. rapa* may be both cultivar and fungal species specific [54].

In the context of our research on plant response to herbivory, NMR fingerprinting was used to assess the main metabolome variation that occurs in maize seedlings on herbivory after *Spodoptera littoralis* attack. Twelve-day-old maize seedlings were infested with *S. littoralis* larvae or left herbivore-free. After 48 h, the leaf tissue was harvested and immediately frozen in liquid nitrogen and extracted with isopropanol. The extracts were further enriched on SPE and analyzed in deuterated DMSO. The NMR fingerprints obtained for two representative samples of the whole series analyzed are displayed in Figure 13.2. Significant differences were observed and are highlighted in the zoom of the aromatic region (Figure 13.2). The induced NMR signals after herbivore attack are characteristic of glycosylated benzoxazinone derivatives and confirm results obtained from an MS-based metabolomic study [55].

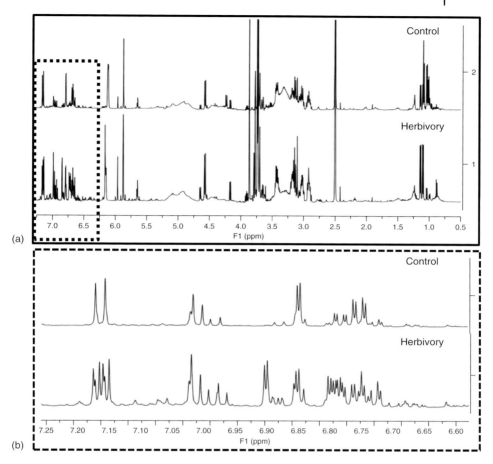

Figure 13.2 (a,b) ^{1}H-NMR spectra of noninfested (control) maize plants and *Spodoptera littoralis*-infested (herbivory) maize plants 48 h after attack. Inset: zoom on the aromatic region of the spectra highlighting the main differences.

Other applications of NMR-based metabolomics intended to identify secondary plant compounds involved in host plant resistance have been summarized in a recent review with a focus on resistance to western flower thrips (*Frankliniella occidentalis*) [51].

Natural Product Discovery and Effect of Complex Phytopreparation The investigation of bioactive natural products has historically been performed based on bioactivity-guided fractionation procedures [56]. This reductionist approach (one molecule acts on a single target) has been successfully employed in drug discovery; it does not, however, account for possible prodrugs or synergistic effects and may thus lead to the loss of activity when trying to simplify original mixtures [57].

Recently, new strategies that favor a holistic approach as opposed to the traditional reductionist method have been introduced with the purpose of overcoming the bottlenecks of natural product research [58]. Indeed, metabolomics allows a systematic study of complex mixtures such as herbal drugs or phytopharmaceuticals, which can be linked to observations obtained through biological testing systems without the need for isolating active principles [59]. NMR-based metabolomics has been used to assess the effect of complex herbal drugs *in vivo*. For example, a metabonomic strategy has been applied to the study of human biological responses to chamomile tea ingestion. Although strong intersubject variation in metabolite profiles was observed, clear differentiation between the samples collected before and after chamomile ingestion was achieved on the basis of increased urinary excretion of hippurate and glycine and depleted creatinine concentration [60]. In another study, an evaluation of the antiaging effects of the total flavone content of *Herba Epimedii*, a Chinese herbal medicine used to treat some age-related diseases, was conducted. Ten metabolites were identified as aging markers in rat urine; after treating rats with *Herba Epimedii*, the urine profile of 24 month old rats shifted toward that of 18 month old rats [61].

Metabolomics has also proved to be effective for assessing bioactivity. Crude extracts from wild plant populations from six different locations of *Galphimia glauca*, a plant popularly employed in Mexico for the treatment of central nervous system disorders, were investigated to differentiate their chemical profile. Pharmacological and phytochemical studies led to the identification of anxiolytic and sedative compounds consisting of a mixture of nor-secofriedelanes referred to as the *galphimine series*. PCA of the ^{1}H-NMR spectra of 39 crude extracts revealed clear differences among the populations; two of the populations out of the six studied expressed differences related to the presence of galphimines. This information was consistently correlated with the corresponding HPLC analysis and with neuropharmacological activity [62]. The strategy applied in this study opens new possibilities for the correct selection of plant populations with suitable metabolic and pharmacologic profiles for the development of standardized herbal medicines.

Quality Control of Phytopharmaceuticals While HPLC is routinely used for quality control of herbal medicines [63], NMR fingerprinting is increasingly employed for this purpose. NMR is appropriate for this type of study because it provides an unbiased direct ratio of principal constituents. Applications in this field have been recently reviewed [35, 64]. The classification and identification of adulterated plant products are among the major interests [65]. Direct NMR analysis has been applied to the comparison of a number of plant extracts used in phytotherapy, such as *Hypericum perforatum* [66], *Ginkgo biloba* [67], *Rhodiola rosea* [68], and *Panax ginseng* [69]. The application of NMR fingerprinting also extends to the analysis of food products [64]. An NMR-based metabolomic approach showed metabolic differences among ginseng products of different ages (four, five, and six years old) and which underwent different types of processing (white and red) [70]. Other studies also demonstrated that this approach may be used for the quality control of fresh ginseng roots of different ages [71] and different origins [69]. Recently, other metabolomic studies have evaluated the quality of traditional Chinese medicine (TCM), which often consists of multiple herbal mixtures. In this respect, NMR fingerprinting was used for the quality assessment of *Polygonum cuspidatum* and *P. multiflorum* [72] and for *Angelica acutiloba* Kitagawa roots [73].

13.3.1.3 Hyphenation of NMR to Separating Techniques

As this is the case for MS (see below), NMR can be hyphenated to HPLC. LC–NMR is a powerful means of providing structural information on single constituents of complex mixtures [74]. However, metabolomic data obtained by LC–NMR has not been used directly mainly because of its lack of sensitivity, cost, and low levels of throughput.

Instead of direct LC–NMR hyphenation, the use of HPLC fractionation combined at-line with micro-NMR is a powerful way to identify *de novo* biomarkers highlighted by LC-MS-based metabolomics. Indeed, in most cases, MS-based metabolomic strategies will not provide definitive peak annotations for given biomarkers unless a high-quality match with a given database spectrum is obtained (Section 13.4). For new or rare biomarkers, *de novo* structure determination represents a critical step that is often the bottleneck of many metabolomic studies. Thus, for definitive identification, biomarkers need to be either synthesized and compared with the detected compounds or isolated and their structure fully assigned by NMR.

Using microflow-NMR methods [75] such as CapNMR, which provide excellent sensitivity, the biomarkers only need to be dissolved in a minimal amount of deuterated solvent (5 µL). High concentrations are obtained with an optimum filling factor and high-quality 1D and 2D NMR spectra can be measured on a few micrograms of pure product [74].

13.3.1.4 Future Trends

NMR-based metabolomics is well established in various fields of plant science. However, as plants contain very different metabolites, it will be important to create a public database [76] and establish standardized protocols to facilitate the exchange of information from different laboratories, in order to render the approach more efficient and profit from the information obtained. It is only at that condition that a genuine "omic" dimension will be reached, as it has been possible for the analysis of body fluids with the so-called "metabonomics" approach [77].

13.3.2
MS-Based Methods

The ability of MS to detect multiple components with high sensitivity makes it a powerful tool for plant metabolomics, complementary to NMR. Contrary to NMR, MS can be used to detect minor compounds among complex plant extracts. It also identifies chemical groups that NMR cannot, for example, sulfate. Typically, two approaches can be employed: either direct MS methods or MS hyphenated to separation techniques.

13.3.2.1 Direct MS Methods

The coupling of chromatographic or electrophoretic methods to MS is a powerful and widespread technique for the detection and identification of metabolites within complex biological samples. Analysis, however, takes relatively long and may prevent high-throughput fingerprints of numerous samples in a reduced period of time. To increase the throughput, direct injection mass spectrometry (DIMS) can be employed. Although low-resolution analyzers can be used in this configuration, high-resolution mass spectrometers are preferred because of their increased separative power. Furthermore, new techniques such as the coupling of MS to ion mobility can separate isomers [78]. DIMS is usually employed with atmospheric pressure ionization (API) sources such as ESI or atmospheric pressure chemical ionization (APCI), and several papers have reported its use in plant metabolomics. Oikawa *et al.* [79] used direct infusion FT-ICR-MS to profile herbicide-treated and herbicide-free Arabidopsis seedlings. McDougall and Matinussen developed a rapid DIMS method for assessing the polyphenol and anthocyanin composition of different berries [80]. Luthria *et al.* [81] have discriminated cultivars and growing treatments of broccoli using MS fingerprints followed by multivariate analysis. DIMS has also been combined with ^{13}C isotope labeling of entire plant metabolomes to facilitate the identification of compounds [82]. Although DIMS is considered the method of choice for rapid MS fingerprints, it has some drawbacks: the capacity of metabolite identification is limited and often necessitates subsequent dedicated experiments. Owing to possible suppression effects, in particular with ESI, low-abundant metabolites can be overlooked and precise quantification remains problematic.

Another direct technique for metabolite fingerprinting that is complementary to API-based DIMS is matrix-assisted laser desorption/ionization (MALDI). MALDI is primarily adapted to the analysis of biomolecules (proteins and nucleotides). Its application to metabolomics has so far been limited, mainly because the conventional matrices required for ionization generate high background noise that interferes with biological molecules in the low-mass region of the spectrum. Ion-free matrices or rational protocols for matrix selection, however, may overcome this problem. Shroff *et al.* [83] have demonstrated the applicability of selected MALDI matrices to plant metabolomics. An interesting feature of MALDI is the possibility to directly analyze tissue sections and study the spatial distribution of metabolites within plant tissues. Sample preparation can be drastically reduced which decreases the risk of artifact formation. MALDI imaging metabolomics is an emerging field in plant science, and only a few methodological studies have been reported to date [84–86]. In the future, this technique will certainly play an increasing role in the progression of plant metabolomics.

13.3.2.2 Hyphenation of MS to Separating Techniques

By combining MS with separation methods, data become multidimensional (e.g., three dimensional with retention properties, mass-to-charge ratio, and intensity).

Aside from a significant reduction of potential ion suppression, this allows for higher resolution and improves the detection, identification, and quantification of metabolites. Three separation techniques are commonly coupled to MS, namely, GC, LC, and capillary electrophoresis (CE).

Gas Chromatography–Mass Spectrometry Owing to the nature of the techniques, the coupling of GC to MS is much more straightforward than that of LC or CE. GC–MS was employed for the first human metabolite profiles [87] and was used in plant analysis beginning in the early 1980s [88]. Modern capillary GC columns provide extreme peak capacity for the high-resolution separation of complex extracts. The application domain of GC is mainly dedicated to volatile compounds such as mono- or sesquiterpenes. Small polar metabolites (<400–500 Da) such as sugar hexoses, amino acids, and fatty acids can be monitored after chemical derivatization. Well-established derivatization protocols are available for polar and nonpolar metabolites, but they are time consuming and may affect the reproducibility of results [89]. In GC–MS, the most frequent ionization technique is electron ionization (EI), which generates intense and highly reproducible fragmentation patterns of molecular ions. This has permitted the creation of commercial libraries containing tens of thousands of mass spectra. By comparing experimental mass spectral data with these libraries, the identification of known metabolites is possible with a high degree of confidence. This represents a key advantage over LC–MS for which only limited databases are available.

The first GC–MS metabolomic applications appeared in 2000 with the publication of two pioneering studies [90, 91]. Roessner *et al.* detected more than 300 metabolites in a potato extract. Fiehn *et al.* detected 326 compounds in different Arabidopsis genotypes and could assign a chemical structure to approximately half of them. Since then, GC–MS metabolomics has been used for numerous applications in various domains (see below), and new analytical advances such as GCxGC–MS have further improved the selectivity and separation power of the technique.

Liquid Chromatography–Mass Spectrometry Although it is a more recent technique, LC–MS has begun to supplant GC–MS in metabolomic studies for several reasons. First, sample preparation can often be reduced and tedious derivatization steps are generally not required. Second, LC–MS can accommodate a broad range of natural compounds, that is, polar and nonpolar, low- and high-molecular-weight molecules. Reverse-phase (RP) chromatography is used in more than 80% of the separations, generally using octadecyl (C18) columns. Normal-phase (NP) chromatography can be employed for especially nonpolar molecules and in specific issues (e.g., phospholipids). APCI is often chosen for coupling NP–LC to MS because nonpolar solvents required for NP–LC, such as hexane, are not compatible with ESI. Hydrophilic interaction liquid chromatography (HILIC) is of interest for very polar compounds that are not retained by RP–LC. An alternative to HILIC is the use of MS-compatible ion-pairing agents that can increase the retention of polar ionic metabolites in RP–LC. LC–MS is generally more rapid than GC–MS, especially since the advent of ultrahigh-pressure liquid chromatography (UHPLC) systems. Using UHPLC, metabolite fingerprints can be obtained in 5–10 min while preserving all the advantages of chromatography over DIMS.

LC and MS are nowadays almost exclusively interfaced with API sources, ESI, APCI, or atmospheric pressure photoionization (APPI). API sources are soft techniques and mainly generate ions of the molecular species. Various adducts may also be produced, depending on the solvents and additives employed. As a general rule of thumb, one should use at least ESI and APCI in positive and negative ion modes to obtain the widest possible view of plant metabolomes. Extensive fragmentation can be obtained in LC–MS either by in-source collision-induced dissociation (CID) or in dedicated collision cells of tandem mass spectrometers. Contrary to GC-EI-MS, such fragmentation processes strongly depend on the spectrometer and experimental conditions used. This complicates the creation of spectra libraries, and to date, only limited databases are available (Section 13.4). Another drawback of ionization in LC–MS is its susceptibility to ion suppression effects (particularly in ESI), which can nevertheless be minimized by appropriate sample preparation procedures and improved chromatographic separation.

Capillary Electrophoresis–Mass Spectrometry Compared to GC–MS and LC–MS, CE–MS has been used to a lesser extent for plant metabolomics. CE–MS has a significant potential for the metabolomic analysis of polar or charged metabolites. In its simplest form, namely Capillary Zone Electrophoresis (CZE), the mechanism of separation is different from LC and GC and involves the separation of compounds according to their charge-to-size ratio. In CE, typical injection volumes are in the nanoliter range. While this can sometimes be an advantage (when a very limited amount of sample is available), it may also lead to low sensitivity. Until quite recently, the coupling of CE to MS was a challenging task, which may also explain why the technique has not been commonly used. However, the recent progress made in CE–MS instrumentation makes it a true alternative to HILIC for the analysis of polar metabolites such as amino acids, nucleotides, or organic acids.

13.3.2.3 Applications

In recent years, hyphenated MS-based metabolomics has been extensively used in several fields of plant science. Similar to the NMR-based applications presented above, a selection of pertinent examples are reported in this section.

Characterization of Mutants and Metabolic Pathways Metabolomics is useful for differentiating genotypes at the metabolite level [12]. It is well known that changes in DNA do not always cause changes in proteins and that protein modifications do not always lead to changes in activity. Measuring the metabolome is therefore a more direct route to linking gene modification to metabolic changes than, for example, the use of microarrays. The comparison of mutant or transgenic plants with their wild-type counterparts using MS-based metabolomic approaches can reveal biomarkers that confirm genetic mutations. Metabolomic analysis has also a strong potential to predict novel metabolic pathways [1]. Von Roepenack-Lahaye *et al.* [92] demonstrated the power of LC-QTOF-MS for the detection of subtle changes present in wild-type and mutant plants. Bottcher *et al.* [93] used capillary LC coupled to MS to analyze mutants of Arabidopsis and identify pathway-dependent metabolites. Using GC-TOF-MS, the loss of metabolic stability and generation of backup pathways has been described in *mto1* and *tt4* Arabidopsis mutants [94]. CE–MS was used to compare three lines of GM maize [95]. Differentiation between cultivars has been made on several other crops such as tomato [96], soybean [97], or potato [98]. Recently, a review was published on the use of MS-based analytical methodologies to characterize GM crops, including proteomic and metabolomic approaches [99]. While mRNA profiling remains the main tool in plant functional genomics, metabolomics is becoming an important method that should not be ignored.

Taxonomy and Phylogeny Metabolite profiles of different plant species or genera can be compared to find common or distinct traits. Wu *et al.* [100] have tested the applicability of metabolite profiling to facilitate accession identification of various Echinaceae taxa. HPLC–MS was used to compare fungal genera and deduce evolutionary lineages [101]. The technique was also employed to classify *Trichoderma* species, and results were compared to internal transcribed spacer (ITS)-based phylogenetic tree [102]. Although it may be unrealistic to attempt the characterization of taxa by metabolite profiling only, this methodology can certainly be used for the purpose of classification in combination with conventional methods [103]. Larsen *et al.* [104] even suggest that metabolomics will be an essential part of future identification and classification methods, especially in the field of drug discovery from natural sources.

Developmental Changes GC–MS was employed to study time-dependant metabolic changes that occur in the course of the germination of rice [105]. Fruit development and maturation were characterized by untargeted (GC–MS) as well as targeted (HPLC) profiling [106]. In the domain of medicinal plants, *Artemisia annua* was profiled at five developmental stages by GC–MS followed by multivariate analysis [107].

Plant Nutrition MS-based metabolomics was used to assess the impact of sulfur and nitrogen deficiency to better understand the assimilation of these essential elements by plants [108, 109]. In a seminal study, Hirai *et al.* [110] reported a generic strategy for the investigation of the plant response to nutritional stress, with integration of transcriptomics and metabolomics. Further studies have shown that nitrogen or sulfur deficiency has an impact on several metabolic pathways and considerable effort will be needed to understand the underlying phenomena implicated in nutrient stress [111, 112].

Plant–Environment Interactions Plant–environment interactions are a large field that encompasses various research topics such as plant responses to abiotic and biotic stresses, allelopathy, symbiotic interactions, and other external factors affecting plant growth (climate, light, soil, etc.). Metabolomics undoubtedly has the potential to study the chemical signals that are involved in the interactions of plants with their environment on a broad scale. In the field of abiotic stresses, hyphenated MS-based metabolomics has been applied to the study of the response of plants to dehydration [113], salt treatment [114], temperature changes [115], and light stress [116]. It is also increasingly used to understand plant responses to biotic stresses caused by insects or pathogens. Plants produce a vast array of defensive secondary metabolites in response to biotic stresses, and metabolomics may help identify the metabolites that confer resistance to plants. In this respect, Kuzina et al. [117] found a correlation between the production of saponines by winter cress and its resistance to flea beetle larvae. The authors used LC–MS to identify potential bioactive compounds involved in plant defense. Jansen et al. [118] used a similar nontargeted approach to detect induced metabolites in cabbage following jasmonic acid treatment. Interestingly, elevated levels of these compounds were also found in caterpillars that fed on induced plants. However, in both studies, the potential bioactivity of detected compounds was not confirmed by herbivore performance assays using artificial diet supplemented with pure molecules. For future studies, biological tests will be essential to establish and confirm the effect of plant metabolites on herbivore growth. Different groups have also studied the interactions between plants and pathogens using MS-based metabolomic approaches [119, 120]. Bollina et al. [121] performed extensive metabolomic work to identify and assay selected metabolites that may be involved in the defense of barley against *Gibberella zeae*. It must be noted that, when studying the interactions between plant and pathogen, the question of whether detected changes are due to host- or pathogen-produced metabolites should be carefully addressed. This may be possible by "dual metabolomics," an approach that should facilitate the modeling of reciprocal responses between plants and pathogens [122]. In conclusion, it appears that global metabolite profiling is an efficient tool for the identification of markers that are characteristic of given stresses and may lead to a better understanding of the involved phenomena. Its use in applied research will certainly increase in the future, for example, in the selection of crop varieties that exhibit elevated stress tolerance or in the development of new pest control strategies.

Quality Control of Phytopharmaceuticals MS-based metabolomics has also been used in the quality control of medicinal herbs. For instance, raw and steamed extracts of *Panax notoginseng* have been profiled using UHPLC-TOF-MS and discriminant markers that may be used for quality control were found [123]. The same technique was used to assess the quality of Japanese green tea samples [124]. GC–MS was employed to evaluate differences in primary metabolites in *Angelica acutiloba* roots according to various cultivation areas [125]. However, it must be noted that MS may not be the most appropriate tool for long-term quality control because of its relatively poor reproducibility. In this context, NMR is certainly the method of choice.

Natural Product Discovery In natural product drug discovery, redundancy is a constant issue. Extensive metabolite profiling can assist the dereplication of complex plant samples and permit the rapid identification of known molecules by combining different advanced analytical techniques such as LC–MS and LC–NMR. In this respect, the extension of spectral databases will be an essential condition for straightforward and reliable dereplication.

13.3.3
Combined Approaches

Modern analytical platforms have proved their worth by providing insight into complex networks of metabolites. However, numerous authors have demonstrated that a single analytical method is often not sufficient to grasp the chemical complexity of plant metabolomes. Therefore, the combination of information obtained from the same samples with multiple experimental platforms (e.g., NMR, GC–MS, CE–MS, or LC–MS) is expected to extend the coverage of the metabolites that characterize a biological system [126, 127]. The fields to which such methods may be applied are numerous and include the profiling of cultivars [128], the monitoring of fruit ripening and quality assessment [129, 130], and the study of plant–pathogen interactions on the basis of metabolic profiling [119]. New chemometric methodologies are therefore needed to cope with the challenges associated with the fusion, the integration, and the comparison of data generated using various analytical techniques.

The most straightforward solution is to process and analyze the data sets independently with appropriate tools and then compare the results manually using *high-level* data integration [131]. Suitable preprocessing and scaling methods are selected according to the characteristics of each analytical platform. Such an approach, however, is unable to uncover the possible interrelations between metabolites, which are measured independently. In addition, a detrimental redundancy can occur as some classes of analytes may be detected on several platforms and have a greater impact on the subsequent models. Finally, independent results may be ambiguous or divergent and biological mechanisms may be difficult to unravel.

Computing correlation profiles between the two sets of variables constitutes another simple way to evaluate the relationship between multiple data sets. On the basis that the same samples are characterized by different data sources, the sample mode constitutes a common dimension across the tables, that is, the matrices possess the same number of rows. This simple approach aims to link relevant metabolites highlighted in one of the data tables with related signals in the other matrices based on correlations, even if these signals would not have been detected when analyzing the other tables separately. This approach can be particularly well suited to searching for biomarker candidates [126]. On the other hand, correlation does not necessarily imply causation. It must also be noted that compounds present at lower concentrations are more difficult to measure and therefore more altered by the analytical noise. The statistical heterospectroscopy (SHY) method was proposed to combine NMR and LC–MS data sets based on the visual analysis of Pearson correlation profiles associated with highly statistically significant indices [132]. A similar approach was recently applied to associate data obtained with ^{1}H-NMR and GC–MS in the context of plant metabolomics [119].

As an independent analysis of an individual data set may be limited, other solutions are offered to integrate information from multiple sources, such as the horizontal concatenation of data matrices. Such *low-level* data fusion is performed by mixing variables of different origins to build a summary table so that multivariate classical methods, either unsupervised, for example, PCA, or supervised, for example, PLS-DA, can be applied to the joint data [130, 131]. However, such an approach aggravates the curse of dimensionality as it generates data structures with a prohibitive number of variables. In this context, applying variable selection or dimensionality reduction before concatenation can be useful to restrain the size of each data table [133].

Finally, more advanced chemometric methods, including multiblock and hierarchical component models, are attractive solutions for the simultaneous analysis of multiple data tables [134] but are beyond the scope of this chapter. Two levels of analysis are usually provided, that is, an individual level characterized by score and loading vectors related to each data table separately and a consensus level combining the information contained in all tables. This not only provides an overview of the major trends common to all tables but also highlights differences between them. Such advanced approaches were used in recent metabolomic studies to model multiple data sets in the context of microbial fermentation [135] and rat toxicity [136, 137] but are still uncommon in the context of plant metabolomics [138].

13.4
Metabolite Identification

13.4.1
Interpreting Mass Spectra

The identification of relevant discriminant biomarkers is probably the trickiest task and the bottleneck of metabolomic analyses. While the concept of *sequence* is a key point for the identification of proteins or nucleic acids in untargeted analyses, metabolites lack this characteristic. Alternatives must be used and MS constitutes one of the key technologies for the identification of such small molecules. It is usually achieved by comparing experimental results with standards to relate measured m/z signals to chemical entities. Computational MS has recently emerged to provide new solutions to deal with the colossal amount of data generated by modern MS devices and to relate peaks in an MS output with metabolites. Several pieces of information can be compared to reference entries in metabolite databases. They include accurate molecular mass, putative elemental composition, and specific mass spectrum.

As a starting point of compound identification, automated procedures provide quick indications about elemental composition from mass measurements. In principle, accurate values are required to determine the elemental formula of monoisotopic ions. While the information provided by an accurate mass value may be unambiguous when considering very small molecules, the number of possibilities quickly increases with ion mass. As a consequence, accurate mass measurement is often not sufficient to determine a unique elemental composition, even with very high mass resolution [139].

Rules limiting this number have been developed on the basis of physicochemical criteria. The seven golden rules proposed by Kind and Fiehn [140] to achieve an automatic exclusion of wrong or unlikely molecular formulas are among the most interesting ones. The removal of inappropriate solutions can be performed by applying the nitrogen rule, analyzing isotope ratio, or assessing valence states. The analysis of isotopic patterns is particularly useful for restraining the number of possibilities for such compounds [141]. In this respect, ultrahigh resolution instruments, such as FT-ICR or Orbitrap [142, 143], constitute attractive solutions to resolve peaks within isotope clusters. The isotopic contribution of each element can then be assessed separately. In addition, information about the charge state z can be obtained by assessing the m/z shift between isotopic homologs [144, 145]. Nevertheless, the assessment of isotopic patterns still suffers from limitations due to the intrinsic difficulties associated with the measurement of ion abundances [146].

Besides the information provided by the analysis of molecular ions with high-resolution analyzers, the identification process may be greatly enhanced by the use of additional structural information. Fragmentation patterns can be obtained either by taking advantage of in-source fragmentation or with tandem MS by performing targeted CID. Both the molecular structure and the collision energy determine the degree of fragmentation. However, a combined spectrum can be obtained in a single acquisition by applying a continuously increasing energy ramp (e.g., from 5 to 70 eV) or several discrete values. Algorithms have been recently developed to take advantage of multiple-stage mass spectrometry (MS^n) and the fragmental information of MS^n spectra [147–149]. They apply rules of possible fragmentation and link signals from precursor and product ions to assign fragment structures to the measured peaks and filter elemental composition proposals. Additional factors inherent to the ionization and fragmentation techniques themselves may lead to misinterpretation of mass spectral data and must be taken into account during the identification process. Indeed, several signals originating from the same molecule may be observed, depending mainly on the ionization mode. In API modes, a frequent cause of these additional signals is the formation of adduct ions from salt or solvent backgrounds, that is, either cations in the positive ion mode (e.g., Na^+, K^+, and $NH4^+$ adducts) or anions in the negative ion mode (e.g., $HCOO^-$ and Cl^- adducts). Another origin of these signals is the presence of many ionization products that are the result of in-source fragmentation of the molecular ion. Finally, homo- or heterodimerization processes can produce new signal from precursor ions. These phenomena induce a supplementary complexity and therefore prevent the direct interpretation of spectra. As the distinction between molecular and adduct ions is not trivial, looking for specific mass differences corresponding to known typical adducts can be helpful.

Fortunately, deconvolution approaches have been developed to assign different ion features to the same metabolite [150]. In addition, these methods are also useful for extracting single component signals in case of incomplete chromatographic separation such as overlapped peaks [151]. These algorithms often rely on the fact that fragments or adducts from the same compound present an identical retention time. Moreover, the relative abundance of molecular ions and fragments or adduct ions is expected to be stable over a given peak.

13.4.2
Databases

Extracting biological information from experimental spectra is a key point of metabolomic experiments and different types of databases have been developed to make sense of this type of data. Although great efforts have been made to create databases for researchers, these repositories remain incomplete in plant science, particularly in terms of secondary metabolites. Moreover, a large fraction of the metabolites included in complex plant extracts is still unknown and the identification of signals in the context of untargeted metabolomics experiments remains limited. The main goal of these repositories is to provide solutions for rapid data interpretation and compound identification, but some databases also collect experimental data from metabolomics experiments for public access. Two main categories can be defined, namely, compound-dedicated and pathway-oriented databases.

The former consists of a library of reference spectra measured from authentic standards (e.g., NIST [152], METLIN [153], MassBank [154]). The fragmentation mechanisms of EI during GC–MS are well described [155], and spectral data obtained from GC–MS instruments in standardized conditions exhibit high reproducibility. Unfortunately, this is not true for LC–MS data, where significant variability between instruments regarding ionization and CID fragmentation because of the difficulty of applying standardized instrumental conditions constitutes a major drawback. A score (similarity or distance function) is computed between the query spectrum and reference database entries to extract putative hits. To ensure true positive matches, similar analytical platforms must be employed. Careful attention must also be paid to a possible over- or underrepresentation of a specific class of compound in the database. Such bias may lead to misleading multiple hits in the first case or wrong unique solutions in the second [156]. The number of positive matches (true or false) for a given query is therefore highly dependent on the metabolic coverage of the database. In addition, spectral data of either individual collision energies or energy ramps cannot be compared and a harmonization of the libraries standards is highly desirable.

The second type of databases provides information about annotated metabolic pathways (e.g., Kyoto Encyclopedia of Genes and Genomes (KEGG) [157], MetaCyc

[158], Reactome [159]). They are usually manually curated and therefore constitute key tools to extract biological information from metabolomic data, by linking observed signals to existing knowledge [160]. They provide information on enzymatic reactions with respect to the proteins or genes related to a given metabolic pathway. Some of these databases are related to a specific metabolomics research project or consortium (e.g., LipidMaps or species-specific databases such as human metabolome database (HMDB)), which provide chemical structures and general information about metabolites. As they are context-specific and include biologically relevant information, they are particularly useful for the evaluation and visualization of metabolic networks. Among these, plant-specific databases including PlantCyc [161] and KNApSAcK [162] have been developed. A nonexhaustive list of databases related to the interpretation of metabolomic data is given in Table 13.1.

In a remarkable review on computational MS for metabolomics [156], Neumann and Böcker recently reported the association of mass spectral and metabolic databases to assess two arbitrarily chosen plant compounds, namely, kaempferol and reserpine. By comparing information such as molecular formula and tandem mass spectra obtained from experimental data with that of several spectral and metabolite databases, they reached a level 2 of identification (see levels of metabolite identification [163]) with a high degree of confidence for these well-known molecules. In many cases, true identification is not possible due to the fact that mass spectral databases are still limited, in particular, for LC–MS data. As a result, one often finds several hits for a calculated molecular formula but no reference

Table 13.1 Metabolomics-related databases.

Database name	Entries	Subject
NIST	MS	Generic
METLIN	MS	Generic
MassBank	MS	Generic
Madison Metabolomics Consortium Database	NMR/MS	Generic
Golm Metabolome Database (GMD)	GC–MS	Plants
DrugBank	MS	Drugs
LipidMaps	MS	Lipids
The Human Metabolome Database (HMDB)	MS	Human
KNApSAcK	MS	Plants
RIKEN MSn spectral database for phytochemicals (ReSpect)	MS	Plants
Metabolic database for Tomato (MoTo)	MS	Tomato
Kyoto Encyclopedia of Genes and Genomes (KEGG)	Pathways	Generic
Reactome	Pathways	Generic
MetaCyc	Pathways	Generic
HumanCyc	Pathways	Human
PlantCyc	Pathways	Plants
AraCyc	Pathways	*A. thaliana*

fragmentation spectrum to compare them with. A careful examination of the obtained fragments can narrow the number of possible hits but with no available standard, definite identification is often hampered.

13.5
Structural Elucidation of Novel Metabolites and Validation of Model

As mentioned earlier, the unambiguous identification of unknown or novel metabolites remains the limiting factor in nontargeted metabolomic studies. When a relevant biomarker does not positively match any spectrum from available databases or authentic standard, a possible option is to purify it in sufficient amounts from crude plant extracts and perform selected NMR experiments on the isolated compound. Such processes are usually long and tedious and may become prohibitively time consuming when numerous markers need to be identified. In general, several prepurification (e.g., SPE, LLE) and chromatographic steps from large amounts of tissue are needed to ensure that high purity is obtained. It becomes even more difficult when minor metabolites are targeted because low proportions of major compounds can easily contaminate the samples. Therefore, it is not surprising that few plant metabolomic studies have actually gone that far in the identification process of unknown relevant biomarkers. In recent years, Wolfender and coworkers have successfully used a sequential MS-NMR-based metabolomic approach to study the rapid response of *Arabidopsis thaliana* to mechanical wounding. Using a fast UHPLC–TOFMS fingerprinting method [164] combined with multivariate analysis, relevant wound markers were highlighted. Some of them were identified by comparing their chromatographic retention and mass spectral characteristics with authentic standards using a high-resolution profiling method that allowed for the precise localization of the markers and distinction between isomers. However, some minor chemical entities remained unidentified or partially identified. A three-step strategy was thus developed to isolate sufficient amounts of these relevant molecules to allow for analysis by micro-NMR [165]. Two novel and several known jasmonate derivatives were fully characterized [14, 166]. This example shows that, although time consuming, a fully integrated approach involving MS-based metabolomics followed by purification and NMR identification of markers can provide successful results.

Another issue in metabolomics studies is the difficulty of confirming the biological activity of identified markers. In the absence of commercially available standards, purification (or synthesis) is essential for validating metabolomics results with specific biological assays. The combination of metabolomics and cross-validation by bioassays has been used by Glauser and Erb to study the interaction between plants and insect herbivores [55]. In this study, maize leaves were infested with either the generalist caterpillar *Spodoptera littoralis* or the specialist *Spodoptera frugiperda*. Leaves were extracted and analyzed by UHPLC–TOFMS to evaluate the effect of herbivory on the plant metabolome. PCA revealed significant differences between control and infested plants (Figure 13.3a). A particular class of molecules, the so-called 1,4-benzoxazin-3-ones (BXDs), appeared to react strongly to herbivore attack. Forty-eight hours after infestation, concentrations of DIMBOA-Glc, the major BXD in healthy leaves, decreased slightly, while concentrations of its methylated derivative, HDMBOA-Glc, strongly increased, becoming the dominant BXD (Figure 13.3b). In disrupted tissues, both glycosides were rapidly converted to their corresponding aglucones by β-glucosidases. To evaluate their effect on insect growth, these two molecules were purified and added to artificial diets. The aglucone DIMBOA proved efficient in reducing the growth of the generalist *Spodoptera littoralis* but showed no effect on the growth of the specialist *Spodoptera frugiperda* (Figure 13.4a). HDMBOA, on the other hand, acted as a strong deterrent against both herbivores (Figure 13.4b). These results suggested that HDMBOA may have evolved from DIMBOA to provide an efficient weapon even against specialized insects that have adapted to DIMBOA. At the time this book was edited, molecular approaches were being developed to manipulate BXDs

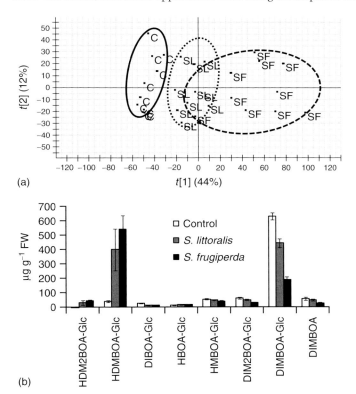

Figure 13.3 (a) Principal component analysis of metabolite profiles of control maize plants (C), *Spodoptera littoralis*-infested plants (SL), and *Spodoptera frugiperda*-infested plants (SF) 48 h after attack. (b) 1,4-Benzoxazin-3-one profiles of control plants (white bars) and plants attacked by *Spodoptera littoralis* larvae (gray bars) and *Spodoptera frugiperda* larvae (black bars) for 24 h ($n = 6$). FW, fresh weight. For abbreviations of 1,4-benzoxazin-3-ones, see Ref. [55].

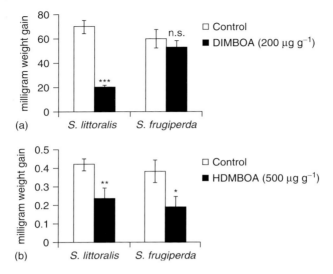

Figure 13.4 Average larval weight gain (mg ± SE) of *Spodoptera littoralis* and *Spodoptera frugiperda* on diet containing (a) DIMBOA at a concentration of 200 μg g^{-1} and (b) HDMBOA at a concentration of 500 μg g^{-1}. Asterisks indicate a significant difference between treatments within species (*$P < 0.05$, **$P < 0.01$, ***$P < 0.001$). n.s., nonsignificant.

in vivo. Linking metabolomics with molecular biology, biochemistry, and ecology will undoubtedly help unravel plant–insect interactions and, potentially, assist biological control efforts.

13.6
Conclusion and Perspectives

By providing functional information about the chemical state of a plant system, metabolomics has become an essential tool for plant science and engineering. The discovery of relevant biomarkers is expected to allow a better understanding of complex interactions occurring at the metabolite level and of the relationships between pathways and biochemical processes. Advances toward comprehensive measurements based on MS and NMR are expected to provide in-depth functional information about plant metabolism. However, improvements are still needed to address the analytical challenges related to the global monitoring of plant metabolomes and the automated processing of large metabolomics data sets. The multivariate nature of data obtained from metabolomic experiments requires specific approaches to extract relevant information. The combination of multiple data sources will undoubtedly provide a more comprehensive vision of metabolomics by extracting common traits from different data sets obtained from the same plant samples. In addition, the development and harmonization of public databases will constitute a cornerstone of metabolomics in the near future. Through this chapter, the utility of metabolomics has been demonstrated using examples from different fields of plant science. It seems likely that the range of applications will continue to widen and that metabolomics will become a routine method for the exploration of complex biological phenomena such as development, nutrition, and stress.

Acknowledgments

Part of the metabolomic studies presented has been supported financially by the Swiss National Science Foundation (Grant no. 205320-124667/1 to J-L.W. and S.R.). The work was also supported by the National Centre of Competence in Research (NCCR) Plant Survival, a research program of the Swiss National Science Foundation, as well as by an IPP project from the SystemsX.ch initiative. Dr Guillaume Marti (School of Pharmaceutical Sciences, University of Geneva) is acknowledged for providing Figure 13.3. Dr. Jessica Litman is thanked for English correction.

References

1. Fiehn, O. (2002) *Plant Mol. Biol.*, **48**, 155–171.
2. Martin, C., Butelli, E., Petroni, K., and Tonelli, C. (2011) *Plant Cell*, **23**, 1685–1699.
3. Oksman-Caldentey, K.M. and Saito, K. (2005) *Curr. Opin. Biotechnol.*, **16**, 174–179.
4. Weckwerth, W. (2003) *Annu. Rev. Plant Biol.*, **54**, 669–689.
5. Verpoorte, R., Choi, Y., Mustafa, N., and Kim, H. (2008) *Phytochem. Rev.*, **7**, 525–537.
6. D'Auria, J.C. and Gershenzon, J. (2005) *Curr. Opin. Plant Biol.*, **8**, 308–316.
7. Dixon, R.A. (2001) *Nature*, **411**, 843–847.
8. Dixon, R.A. and Strack, D. (2003) *Phytochemistry*, **62**, 815–816.
9. Kaul, S., Koo, H.L., Jenkins, J., Rizzo, M., Rooney, T., Tallon, L.J., Feldblyum, T., Nierman, W., Benito, M.I., Lin, X.Y., Town, C.D., Venter, J.C., Fraser, C.M., Tabata, S., Nakamura, Y., Kaneko, T., Sato, S., Asamizu, E., Kato, T., Kotani, H., Sasamoto, S., Ecker, J.R., Theologis, A., Federspiel, N.A., Palm, C.J., Osborne, B.I., Shinn, P., Conway, A.B., Vysotskaia, V.S., Dewar, K., Conn, L., Lenz, C.A., Kim, C.J., Hansen, N.F., Liu, S.X., Buehler, E., Altafi, H., Sakano, H., Dunn, P., Lam, B., Pham, P.K., Chao, Q., Nguyen, M., Yu, G.X., Chen, H.M., Southwick, A., Lee, J.M., Miranda, M., Toriumi, M.J., Davis, R.W., Wambutt, R., Murphy, G., Dusterhoft, A., Stiekema, W., Pohl, T., Entian, K.D., Terryn, N., Volckaert, G., Salanoubat, M., Choisne, N., Rieger, M., Ansorge, W., Unseld, M., Fartmann, B., Valle, G., Artiguenave, F., Weissenbach, J., Quetier, F., Wilson, R.K., de la Bastide, M., Sekhon, M., Huang, E., Spiegel, L., Gnoj, L., Pepin, K., Murray, J., Johnson, D., Habermann, K., Dedhia, N., Parnell, L., Preston, R., Hillier, L., Chen, E., Marra, M., Martienssen, R., McCombie, W.R., Mayer, K., White, O., Bevan, M., Lemcke, K., Creasy, T.H., Bielke, C., Haas, B., Haase, D., Maiti, R., Rudd, S., Peterson, J., Schoof, H., Frishman, D., Morgenstern, B., Zaccaria, P., Ermolaeva, M., Pertea, M., Quackenbush, J., Volfovsky, N., Wu, D.Y., Lowe, T.M., Salzberg, S.L., Mewes, H.W., Rounsley, S., Bush, D., Subramaniam, S., Levin, I., Norris, S., Schmidt, R., Acarkan, A., Bancroft, I., Quetier, F., Brennicke, A.,

Eisen, J.A., Bureau, T., Legault, B.A., Le, Q.H., Agrawal, N., Yu, Z., Martienssen, R., Copenhaver, G.P., Luo, S., Pikaard, C.S., Preuss, D., Paulsen, I.T., Sussman, M., Britt, A.B., Selinger, D.A., Pandey, R., Mount, D.W., Chandler, V.L., Jorgensen, R.A., Pikaard, C., Juergens, G., Meyerowitz, E.M., Theologis, A., Dangl, J., Jones, J.D.G., Chen, M., Chory, J., and Somerville, M.C. (2000) *Nature*, **408**, 796–815.

10. Goff, S.A., Ricke, D., Lan, T.H., Presting, G., Wang, R.L., Dunn, M., Glazebrook, J., Sessions, A., Oeller, P., Varma, H., Hadley, D., Hutchinson, D., Martin, C., Katagiri, F., Lange, B.M., Moughamer, T., Xia, Y., Budworth, P., Zhong, J.P., Miguel, T., Paszkowski, U., Zhang, S.P., Colbert, M., Sun, W.L., Chen, L.L., Cooper, B., Park, S., Wood, T.C., Mao, L., Quail, P., Wing, R., Dean, R., Yu, Y.S., Zharkikh, A., Shen, R., Sahasrabudhe, S., Thomas, A., Cannings, R., Gutin, A., Pruss, D., Reid, J., Tavtigian, S., Mitchell, J., Eldredge, G., Scholl, T., Miller, R.M., Bhatnagar, S., Adey, N., Rubano, T., Tusneem, N., Robinson, R., Feldhaus, J., Macalma, T., Oliphant, A., and Briggs, S. (2002) *Science*, **296**, 92–100.

11. Fukusaki, E. and Kobayashi, A. (2005) *J. Biosci. Bioeng.*, **100**, 347–354.

12. Sumner, L.W., Mendes, P., and Dixon, R.A. (2003) *Phytochemistry*, **62**, 817–836.

13. Fiehn, O., Wohlgemuth, G., Scholz, M., Kind, T., Lee, D.Y., Lu, Y., Moon, S., and Nikolau, B. (2008) *Plant J.*, **53**, 691–704.

14. Glauser, G., Grata, E., Dubugnon, L., Rudaz, S., Farmer, E.E., and Wolfender, J.L. (2008) *J. Biol. Chem.*, **283**, 16400–16407.

15. Robert, C.A.M., Veyrat, N., Glauser, G., Marti, G., Doyen, G.A., Villard, N., Gaillard, M.D.P., Köllner, T.G., Giron, D., Body, M., Babst, B.A., Ferrieri, R.A., Turlings, T.C.J., and Erb, M. (2012) *Ecol. Lett.*, **15**, 55–64.

16. Leiss, K.A., Choi, Y.H., bdel-Farid, I.B., Verpoorte, R., and Klinkhamer, P.G.L. (2009) *J. Chem. Ecol.*, **35**, 219–229.

17. Shroff, R., Vergara, F., Muck, A., Svatos, A., and Gershenzon, J. (2008) *Proc. Natl. Acad. Sci. U.S.A.*, **105**, 6196–6201.

18. Gibon, Y., Usadel, B., Blaesing, O.E., Kamlage, B., Hoehne, M., Trethewey, R., and Stitt, M. (2006) *Genome Biol.*, **7**:R76.

19. Urbanczyk-Wochniak, E., Baxter, C., Kolbe, A., Kopka, J., Sweetlove, L.J., and Fernie, A.R. (2005) *Planta*, **221**, 891–903.

20. Harbourne, N., Marete, E., Jacquier, J.C., and O'Riordan, D. (2009) *LWT Food Sci. Technol.*, **42**, 1468–1473.

21. Schmitke, J.L., Wescott, C.R., and Klibanov, A.M. (1996) *J. Am. Chem. Soc.*, **118**, 3360–3365.

22. Kim, H.K. and Verpoorte, R. (2010) *Phytochem. Anal.*, **21**, 4–13.

23. Namiesnik, J. and Gorecki, T. (2000) *J. Planar Chromatogr.*, **13**, 404–413.

24. Sargenti, S.R. and Vichnewski, W. (2000) *Phytochem. Anal.*, **11**, 69–73.

25. Jonsson, P., Gullberg, J., Nordstrom, A., Kusano, M., Kowalczyk, M., Sjostrom, M., and Moritz, T. (2004) *Anal. Chem.*, **76**, 1738–1745.

26. Benthin, B., Danz, H., and Hamburger, M. (1999) *J. Chromatogr. A*, **837**, 211–219.

27. Castioni, P., Christen, P., and Veuthey, J.L. (1995) *Analusis*, **23**, 95–106.

28. Glauser, G. (2010) Etude de la réponse à la blessure mécanique chez les plantes: une approche métabolomique.

29. Yang, J. and Yen, H.E. (2002) *Plant Physiol.*, **130**, 1032–1042.

30. Choi, Y.H., Kim, H.K., Linthorst, H.J.M., Hollander, J.G., Lefeber, A.W.M., Erkelens, C., Nuzillard, J.M., and Verpoorte, R. (2006) *J. Nat. Prod.*, **69**, 742–748.

31. Colquhoun, I.J. (2007) *J. Pestic. Sci.*, **32**, 200–212.

32. Dunn, W.B. (2008) *Phys. Biol.*, **5**:011001.

33. Wishart, D.S. (2008) *Trends Anal. Chem.*, **27**, 228–237.

34. Kim, H.K., Choi, Y.H., and Verpoorte, R. (2010) *Nat. Protoc.*, **5**, 536–549.

35. van der Kooy, F., Maltese, F., Choi, Y.H., Kim, H.K., and Verpoorte, R. (2009) *Planta Med.*, **75**, 763–775.

36. Verpoorte, R., Choi, Y., and Kim, H. (2007) *Phytochem. Rev.*, **6**, 3–14.
37. Ward, J.L., Baker, J.M., and Beale, M.H. (2007) *FEBS J.*, **274**, 1126–1131.
38. Davies, H.V., Shepherd, L.V.T., Stewart, D., Frank, T., Rohlig, R.M., and Engel, K.H. (2010) *Regul. Toxicol. Pharm.*, **58**, S54–S61.
39. Manetti, C., Bianchetti, C., Bizzari, M., Casciani, L., Castro, C., D'Ascenzo, G., Delfini, M., Di Cocco, M.E., Lagana, A., Miccheli, A., Motto, M., and Conti, F. (2004) *Phytochemistry*, **65**, 3187–3198.
40. Baker, J.M., Hawkins, N.D., Ward, J.L., Lovegrove, A., Napier, J.A., Shewry, P.R., and Beale, M.H. (2006) *Plant Biotechnol. J.*, **4**, 381–392.
41. Fernie, A.R. and Schauer, N. (2009) *Trends Genet.*, **25**, 39–48.
42. Harrigan, G.G., Glenn, K.C., and Ridley, W.P. (2010) *Regul. Toxicol. Pharm.*, **58**, S13–S20.
43. Hegeman, A. (2008) *Hortscience*, **43**, 1057.
44. Urbain, A., Marston, A., Marsden-Edwards, E., and Hostettmanna, K. (2009) *Phytochem. Anal.*, **20**, 134–138.
45. Kim, H.K., Saifullah Khan, S., Wilson, E.G., Kricun, S.D.P., Meissner, A., Goraler, S., Deelder, A.M., Choi, Y.H., and Verpoorte, R. (2010) *Phytochemistry*, **71**, 773–784.
46. Fischedick, J.T., Hazekamp, A., Erkelens, T., Choi, Y.H., and Verpoorte, R. (2010) *Phytochemistry*, **71**, 2058–2073.
47. Palama, T.L., Khatib, A., Choi, Y.H., Payet, B., Fock, I., Verpoorte, R., and Kodja, H. (2009) *J. Agric. Food Chem.*, **57**, 7651–7658.
48. Ali, K., Maltese, F., Fortes, A.M., Pais, M.S., Choi, Y.H., and Verpoorte, R. (2011) *Food Chem.*, **124**, 1760–1769.
49. Abreu, I.N., Choi, Y.H., Sawaya, A.C.H.F., Eberlin, M.N., Mazzafera, P., and Verpoorte, R. (2011) *Planta Med.*, **77**, 293–300.
50. Lubbe, A., Choi, Y.H., Vreeburg, P., and Verpoorte, R. (2011) *J. Agric. Food Chem.*, **59**, 3155–3161.
51. Leiss, K.A., Choi, Y.H., Verpoorte, R., and Klinkhamer, P.G.L. (2011) *Phytochem. Rev.*, **10**, 205–216.
52. Widarto, H.T., Van der Meijden, E., Lefeber, A.W.M., Erkelens, C., Kim, H.K., Choi, Y.H., and Verpoorte, R. (2006) *J. Chem. Ecol.*, **32**, 2417–2428.
53. Jahangir, M., bdel-Farid, I.B., Choi, Y.H., and Verpoorte, R. (2008) *J. Plant Physiol.*, **165**, 1429–1437.
54. Abdel-Farid, I.B., Jahangir, M., Mustafa, N.R., van Dam, N.M., van den Hondel, C.A.M.J., Kim, H.K., Choi, Y.H., and Verpoorte, R. (2010) *Biochem. Syst. Ecol.*, **38**, 612–620.
55. Glauser, G., Marti, G., Villard, N., Doyen, G.A., Wolfender, J.L., Turlings, T.C.J., and Erb, M. (2011) *Plant J.*, **68**, 901–911.
56. Hostettmann, K. and Marston, A. (2007) *Chimia*, **61**, 322–326.
57. Verpoorte, R., Choi, Y.H., and Kim, H.K. (2005) *J. Ethnopharmacol.*, **100**, 53–56.
58. Urich-Merzenich, G., Zeitler, H., Jobst, D., Panek, D., Vetter, H., and Wagner, H. (2007) *Phytomedicine*, **14**, 70–82.
59. Yuliana, N.D., Khatib, A., Choi, Y.H., and Verpoorte, R. (2011) *Phytother. Res.*, **25**, 157–169.
60. Wang, Y.L., Tang, H.R., Nicholson, J.K., Hylands, P.J., Sampson, J., and Holmes, E. (2005) *J. Agric. Food Chem.*, **53**, 191–196.
61. Wu, B., Yan, S.K., Lin, Z.Y., Wang, Q., Yang, Y., Yang, G.J., Shen, Z.Y., and Zhang, W.D. (2008) *Mol. Biosyst.*, **4**, 855–861.
62. Cardoso-Taketa, A.T., Pereda-Miranda, R., Choi, Y.H., Verpoorte, R., and Villarreal, M.L. (2008) *Planta Med.*, **74**, 1295–1301.
63. Wolfender, J.L. (2009) *Planta Med.*, **75**, 719–734.
64. Holmes, E., Tang, H.R., Wang, Y.L., and Seger, C. (2006) *Planta Med.*, **72**, 771–785.
65. van der Kooy, F., Verpoorte, R., and Meyer, J.J.M. (2008) *S. Afr. J. Bot.*, **74**, 186–189.
66. Rasmussen, B., Cloarec, O., Tang, H.R., Staerk, D., and Jaroszewski, J.W. (2006) *Planta Med.*, **72**, 556–563.
67. Agnolet, S., Jaroszewski, J.W., Verpoorte, R., and Staerk, D. (2010) *Metabolomics*, **6**, 292–302.

68. Ioset, K.N., Nyberg, N.T., Van Diermen, D., Malnoe, P., Hostettmann, K., Shikov, A.N., and Jaroszewski, J.W. (2011) *Phytochem. Anal.*, **22**, 158–165.
69. Kang, J., Lee, S., Kang, S., Kwon, H.N., Park, J.H., Kwon, S.W., and Park, S. (2008) *Arch. Pharmacal Res.*, **31**, 330–336.
70. Yang, S.Y., Kim, H.K., Lefeber, A.W.M., Erkelens, C., Angelova, N., Choi, Y.H., and Verpoorte, R. (2006) *Planta Med.*, **72**, 364–369.
71. Shin, Y.S., Bang, K.H., In, D.S., Kim, O.T., Hyun, D.Y., Ahn, I.O., Ku, B.C., Kim, S.W., Seong, N.S., Cha, S.W., Lee, D., and Choi, H.K. (2007) *Arch. Pharmacal Res.*, **30**, 1625–1628.
72. Frederich, M., Wauters, J.N., Tits, M., Jason, C., de Tullio, P., Van der Heyden, Y., Fan, G.R., and Angenot, L. (2011) *Planta Med.*, **77**, 81–86.
73. Tarachiwin, L., Katoh, A., Ute, K., and Fukusaki, E. (2008) *J. Pharm. Biomed.*, **48**, 42–48.
74. Wolfender, J.L., Marti, G., and Queiroz, E.F. (2010) *Curr. Org. Chem.*, **14**, 1808–1832.
75. Webb, A.G. (2005) *J. Pharm. Biomed.*, **38**, 892–903.
76. Kim, H.K., Choi, Y.H., and Verpoorte, R. (2011) *Trends Biotechnol.*, **29**, 267–275.
77. Nicholson, J.K. and Lindon, J.C. (2008) *Nature*, **455**, 1054–1056.
78. Dwivedi, P., Wu, P., Klopsch, S.J., Puzon, G.J., Xun, L., and Hill, H.H. (2008) *Metabolomics*, **4**, 63–80.
79. Oikawa, A., Nakamura, Y., Ogura, T., Kimura, A., Suzuki, H., Sakurai, N., Shinbo, Y., Shibata, D., Kanaya, S., and Ohta, D. (2006) *Plant Physiol.*, **142**, 398–413.
80. McDougall, G., Martinussen, I., and Stewart, D. (2008) *J. Chromatogr. B*, **871**, 362–369.
81. Luthria, D.L., Lin, L.Z., Robbins, R.J., Finley, J.W., Banuelos, G.S., and Harnly, J.M. (2008) *J. Agric. Food Chem.*, **56**, 9819–9827.
82. Giavalisco, P., Hummel, J., Lisec, J., Inostroza, A.C., Catchpole, G., and Willmitzer, L. (2008) *Anal. Chem.*, **80**, 9417–9425.
83. Shroff, R., Rulisek, L., Doubsky, J., and Svatos, A. (2009) *Proc. Natl. Acad. Sci. U.S.A.*, **106**, 10092–10096.
84. Cha, S.W., Zhang, H., Ilarslan, H.I., Wurtele, E.S., Brachova, L., Nikolau, B.J., and Yeung, E.S. (2008) *Plant J.*, **55**, 348–360.
85. Holscher, D., Shroff, R., Knop, K., Gottschaldt, M., Crecelius, A., Schneider, B., Heckel, D.G., Schubert, U.S., and Svatos, A. (2009) *Plant J.*, **60**, 907–918.
86. Shrestha, B., Li, Y., and Vertes, A. (2008) *Metabolomics*, **4**, 297–311.
87. Horning, E.C. and Horning, M.G. (1971) *Clin. Chem.*, **17**, 802–809.
88. Sauter, H., Lauer, M., and Fritsch, H. (1991) *ACS Symp. Ser.*, **443**, 288–299.
89. Wolfender, J.L., Glauser, G., Boccard, J., and Rudaz, S. (2009) *Nat. Prod. Commun.*, **4**, 1417–1430.
90. Fiehn, O., Kopka, J., Dormann, P., Altmann, T., Trethewey, R.N., and Willmitzer, L. (2000) *Nat. Biotechnol.*, **18**, 1157–1161.
91. Roessner, U., Wagner, C., Kopka, J., Trethewey, R.N., and Willmitzer, L. (2000) *Plant J.*, **23**, 131–142.
92. von Roepenack-Lahaye, E., Degenkolb, T., Zerjeski, M., Franz, M., Roth, U., Wessjohann, L., Schmidt, J., Scheel, D., and Clemens, S. (2004) *Plant Physiol.*, **134**, 548–559.
93. Bottcher, C., von Roepenack-Lahaye, E., Schmidt, J., Schmotz, C., Neumann, S., Scheel, D., and Clemens, S. (2008) *Plant Physiol.*, **147**, 2107–2120.
94. Kusano, M., Fukushima, A., Arita, M., Jonsson, P., Moritz, T., Kobayashi, M., Hayashi, N., Tohge, T., and Saito, K. (2007) *BMC Syst. Biol.*, **1**:53
95. Levandi, T., Leon, C., Kaljurand, M., Garcia-Canas, V., and Cifuentes, A. (2008) *Anal. Chem.*, **80**, 6329–6335.
96. Tikunov, Y., Lommen, A., De Vos, C.H.R., Verhoeven, H.A., Bino, R.J., Hall, R.D., and Bovy, A.G. (2005) *Plant Physiol.*, **139**, 1125–1137.
97. Wu, W., Zhang, Q., Zhu, Y.M., Lam, H.M., Cai, Z.W., and Guo, D.J. (2008) *J. Agric. Food Chem.*, **56**, 11132–11138.

98. Dobson, G., Shepherd, T., Verrall, S.R., Conner, S., McNicol, J.W., Ramsay, G., Shepherd, L.V.T., Davies, H.V., and Stewart, D. (2008) *J. Agric. Food Chem.*, **56**, 10280–10291.
99. Garcia-Canas, V., Simo, C., Leon, C., Ibanez, E., and Cifuentes, A. (2011) *Mass Spectrom. Rev.*, **30**, 396–416.
100. Wu, L., Dixon, P.M., Nikolau, B.J., Kraus, G.A., Widrlechner, M.P., and Wurtele, E.S. (2009) *Planta Med.*, **75**, 178–183.
101. Stadler, M., Fournier, J., Laessoe, T., Chlebicki, A., Lechat, C., Flessa, F., Rambold, G., and Persoh, D. (2010) *Mycoscience*, **51**, 189–207.
102. Kang, D., Kim, J., Choi, J.N., Liu, K.H., and Lee, C.H. (2011) *J. Microbiol. Biotechnol.*, **21**, 5–13.
103. Fournier, J., Stadler, M., Hyde, K.D., and Duong, M.L. (2010) *Fungal Divers.*, **40**, 23–36.
104. Larsen, T.O., Smedsgaard, J., Nielsen, K.F., Hansen, M.E., and Frisvad, J.C. (2005) *Nat. Prod. Rep.*, **22**, 672–695.
105. Shu, X.L., Frank, T., Shu, Q.Y., and Engel, K.R. (2008) *J. Agric. Food Chem.*, **56**, 11612–11620.
106. Zhang, J.J., Wang, X., Yu, O., Tang, J.J., Gu, X.G., Wan, X.C., and Fang, C.B. (2011) *J. Exp. Bot.*, **62**, 1103–1118.
107. Ma, C.F., Wang, H.H., Lu, X., Xu, G.W., and Liu, B.Y. (2008) *J. Chromatogr. A*, **1186**, 412–419.
108. Hoefgen, R. and Nikiforova, V.J. (2008) *Physiol. Plant*, **132**, 190–198.
109. Kusano, M., Fukushima, A., Redestig, H., and Saito, K. (2011) *J. Exp. Bot.*, **62**, 1439–1453.
110. Hirai, M.Y., Yano, M., Goodenowe, D.B., Kanaya, S., Kimura, T., Awazuhara, M., Arita, M., Fujiwara, T., and Saito, K. (2004) *Proc. Natl. Acad. Sci. U.S.A.*, **101**, 10205–10210.
111. Nikiforova, V.J., Kopka, J., Tolstikov, V., Fiehn, O., Hopkins, L., Hawkesford, M.J., Hesse, H., and Hoefgen, R. (2005) *Plant Physiol.*, **138**, 304–318.
112. Urbanczyk-Wochniak, E. and Fernie, A.R. (2005) *J. Exp. Bot.*, **56**, 309–321.
113. Urano, K., Maruyama, K., Ogata, Y., Morishita, Y., Takeda, M., Sakurai, N., Suzuki, H., Saito, K., Shibata, D., Kobayashi, M., Yamaguchi-Shinozaki, K., and Shinozaki, K. (2009) *Plant J.*, **57**, 1065–1078.
114. Kim, J.K., Bamba, T., Harada, K., Fukusaki, E., and Kobayashi, A. (2007) *J. Exp. Bot.*, **58**, 415–424.
115. Kaplan, F., Kopka, J., Haskell, D.W., Zhao, W., Schiller, K.C., Gatzke, N., Sung, D.Y., and Guy, C.L. (2004) *Plant Physiol.*, **136**, 4159–4168.
116. Bino, R.J., De Vos, C.H.R., Lieberman, M., Hall, R.D., Bovy, A., Jonker, H.H., Tikunov, Y., Lommen, A., Moco, S., and Levin, I. (2005) *New Phytol.*, **166**, 427–438.
117. Kuzina, V., Ekstrom, C.T., Andersen, S.B., Nielsen, J.K., Olsen, C.E., and Bak, S. (2009) *Plant Physiol.*, **151**, 1977–1990.
118. Jansen, J.J., Allwood, J.W., Marsden-Edwards, E., van der Putten, W.H., Goodacre, R., and van Dam, N.M. (2009) *Metabolomics*, **5**, 150–161.
119. Jones, O.A.H., Maguire, M.L., Griffin, J.L., Jung, Y.H., Shibato, J., Rakwal, R., Agrawal, G.K., and Jwa, N.S. (2011) *Eur. J. Plant Pathol.*, **129**, 539–554.
120. Sana, T.R., Fischer, S., Wohlgemuth, G., Katrekar, A., Jung, K.H., Ronald, P.C., and Fiehn, O. (2010) *Metabolomics*, **6**, 451–465.
121. Bollina, V., Kumaraswamy, G.K., Kushalappa, A.C., Choo, T.M., Dion, Y., Rioux, S., Faubert, D., and Hamzehzarghani, H. (2010) *Mol. Plant Pathol.*, **11**, 769–782.
122. Allwood, J.W., Clarke, A., Goodacre, R., and Mur, L.A.J. (2010) *Phytochemistry*, **71**, 590–597.
123. Chan, E.C.Y., Yap, S.L., Lau, A.J., Leow, P.C., Toh, D.F., and Koh, H.L. (2007) *Rapid Commun. Mass Spectrom.*, **21**, 519–528.
124. Pongsuwan, W., Bamba, T., Harada, K., Yonetani, T., Kobayashi, A., and Fukusaki, E. (2008) *J. Agric. Food Chem.*, **56**, 10705–10708.
125. Tianniam, S., Tarachiwin, L., Bamba, T., Kobayashi, A., and Fukusaki, E. (2008) *J. Biosci. Bioeng.*, **105**, 655–659.
126. Liland, K.H. (2011) *TrAC Trends Anal. Chem.*, **30**, 827–841.

127. Ni, Y., Su, M.M., Qiu, Y.P., Chen, M.J., Liu, Y.M., Zhao, A.H., and Jia, W. (2007) *FEBS Lett.*, **581**, 707–711.

128. Moco, S., Forshed, J., De Vos, R.C.H., Bino, R.J., and Vervoort, J. (2008) *Metabolomics*, **4**, 202–215.

129. Moing, A., Aharoni, A., Biais, B., Rogachev, I., Meir, S., Brodsky, L., Allwood, J.W., Erban, A., Dunn, W.B., Kay, L., de Koning, S., De Vos, R.C.H., Jonker, H., Mumm, R., Deborde, C., Maucourt, M., Bernillon, S., Gibon, Y., Hansen, T.H., Husted, S., Goodacre, R., Kopka, J., Schjoerring, J.K., Rolin, D., and Hall, R.D. (2011) *New Phytol.*, **190**, 683–696.

130. Tikunov, Y.M., De Vos, R.C.H., Paramas, A.M.G., Hall, R.D., and Bovy, A.G. (2010) *Plant Physiol.*, **152**, 55–70.

131. Roussel, S., Bellon-Maurel, W., Roger, J.M., and Grenier, P. (2003) *J. Food Eng.*, **60**, 407–419.

132. Crockford, D.J., Holmes, E., Lindon, J.C., Plumb, R.S., Zirah, S., Bruce, S.J., Rainville, P., Stumpf, C.L., and Nicholson, J.K. (2005) *Anal. Chem.*, **78**, 363–371.

133. Brereton, R.G. (2006) *Trends Anal. Chem.*, **25**, 1103–1111.

134. Van Mechelen, I. and Smilde, A.K. (2010) *Chemom. Intell. Lab. Syst.*, **104**, 83–94.

135. Smilde, A.K., van der Werf, M.J., Bijlsma, S., van der Werff-van-der Vat, B.J., and Jellema, R.H. (2005) *Anal. Chem.*, **77**, 6729–6736.

136. Forshed, J., Idborg, H., and Jacobsson, S.P. (2007) *Chemom. Intell. Lab. Syst.*, **85**, 102–109.

137. Forshed, J., Stolt, R., Idborg, H., and Jacobsson, S.P. (2007) *Chemom. Intell. Lab. Syst.*, **85**, 179–185.

138. Biais, B., Allwood, J.W., Deborde, C., Xu, Y., Maucourt, M., Beauvoit, B., Dunn, W.B., Jacob, D., Goodacre, R., Rolin, D., and Moing, A. (2009) *Anal. Chem.*, **81**, 2884–2894.

139. Koch, B.P., Dittmar, T., Witt, M., and Kattner, G. (2007) *Anal. Chem.*, **79**, 1758–1763.

140. Kind, T. and Fiehn, O. (2007) *BMC Bioinformatics*, 8:105.

141. Kind, T. and Fiehn, O. (2006) *BMC Bioinformatics*, 7:234.

142. Hu, Q.Z., Noll, R.J., Li, H.Y., Makarov, A., Hardman, M., and Cooks, R.G. (2005) *J. Mass Spectrom.*, **40**, 430–443.

143. Makarov, A. (2000) *Anal. Chem.*, **72**, 1156–1162.

144. Meija, J. (2006) *Anal. Bioanal. Chem.*, **385**, 486–499.

145. Henry, K.D., Williams, E.R., Wang, B.H., McLafferty, F.W., Shabanowitz, J., and Hunt, D.F. (1989) *Proc. Natl. Acad. Sci. U.S.A.*, **86**, 9075–9078.

146. Werner, E., Heilier, J.F., Ducruix, C., Ezan, E., Junot, C., and Tabet, J.C. (2008) *J. Chromatogr. B*, **871**, 143–163.

147. Sheldon, M.T., Mistrik, R., and Croley, T.R. (2009) *J. Am. Soc. Mass Spectrom.*, **20**, 370–376.

148. Heinonen, M., Rantanen, A., Mielikainen, T., Kokkonen, J., Kiuru, J., Ketola, R.A., and Rousu, J. (2008) *Rapid Commun. Mass Spectrom.*, **22**, 3043–3052.

149. Hill, D.W., Kertesz, T.M., Fontaine, D., Friedman, R., and Grant, D.F. (2008) *Anal. Chem.*, **80**, 5574–5582.

150. Castillo, S., Gopalacharyulu, P., Yetukuri, L., and Oresic, M. (2011) *Chemom. Intell. Lab. Syst.*, **108**, 23–32.

151. Vivo-Truyols, G., Torres-Lapasio, J.R., van Nederkassel, A.M., Vander Heyden, Y., and Massart, D.L. (2005) *J. Chromatogr. A*, **1096**, 146–155.

152. (2011) NIST/EPA/NIH Mass Spectral Database.

153. Smith, C.A., O'Maille, G., Want, E.J., Qin, C., Trauger, S.A., Brandon, T.R., Custodio, D.E., Abagyan, R., and Siuzdak, G. (2005) *Ther. Drug Monit.*, **27**, 747–751.

154. Horai, H., Arita, M., Kanaya, S., Nihei, Y., Ikeda, T., Suwa, K., Ojima, Y., Tanaka, K., Tanaka, S., Aoshima, K., Oda, Y., Kakazu, Y., Kusano, M., Tohge, T., Matsuda, F., Sawada, Y., Hirai, M.Y., Nakanishi, H., Ikeda, K., Akimoto, N., Maoka, T., Takahashi, H., Ara, T., Sakurai, N., Suzuki, H., Shibata, D., Neumann, S., Iida, T., Tanaka, K., Funatsu, K., Matsuura, F., Soga, T., Taguchi, R., Saito, K., and

Nishioka, T. (2010) *J. Mass Spectrom.*, **45**, 703–714.
155. McLafferty, F.W. and Turecek, F. (1993) *Interpretation of Mass Spectra*, University Science Books, Mill Valley.
156. Neumann, S. and Bocker, S. (2010) *Anal. Bioanal. Chem.*, **398**, 2779–2788.
157. Kanehisa, M. and Goto, S. (2000) *Nucleic Acids Res.*, **28**, 27–30.
158. Caspi, R., Altman, T., Dale, J.M., Dreher, K., Fulcher, C.A., Gilham, F., Kaipa, P., Karthikeyan, A.S., Kothari, A., Krummenacker, M., Latendresse, M., Mueller, L.A., Paley, S., Popescu, L., Pujar, A., Shearer, A.G., Zhang, P., and Karp, P.D. (2010) *Nucleic Acids Res.*, **38**, D473–D479.
159. Croft, D., O'Kelly, G., Wu, G.M., Haw, R., Gillespie, M., Matthews, L., Caudy, M., Garapati, P., Gopinath, G., Jassal, B., Jupe, S., Kalatskaya, I., Mahajan, S., May, B., Ndegwa, N., Schmidt, E., Shamovsky, V., Yung, C., Birney, E., Hermjakob, H., D'Eustachio, P., and Stein, L. (2011) *Nucleic Acids Res.*, **39**, D691–D697.
160. Karp, P. and Caspi, R. (2011) *Arch. Toxicol.*, **85**, 1015–1033.
161. Zhang, P.F., Dreher, K., Karthikeyan, A., Chi, A., Pujar, A., Caspi, R., Karp, P., Kirkup, V., Latendresse, M., Lee, C., Mueller, L.A., Muller, R., and Rhee, S.Y. (2010) *Plant Physiol.*, **153**, 1479–1491.
162. Shinbo, Y., Nakamura, Y., taf-Ul-Amin, M., Asahi, H., Kurokawa, K., Arita, M., Saito, K., Ohta, D., Shibata, D., and Kanaya, S. (2006) in *Plant Metabolomics* (eds K. Saito, R.A. Dixon, and L. Willmitzer), Springer, Berlin Heidelberg, pp. 165–181.
163. Fiehn, O., Robertson, D., Griffin, J., van der Werf, M., Nikolau, B., Morrison, N., Sumner, L.W., Goodacre, R., Hardy, N.W., Taylor, C., Fostel, J., Kristal, B., Kaddurah-Daouk, R., Mendes, P., van Ommen, B., Lindon, J.C., and Sansone, S.A. (2007) *Metabolomics*, **3**, 175–178.
164. Grata, E., Boccard, J., Guillarme, D., Glauser, G., Carrupt, P.A., Farmer, E.E., Wolfender, J.L., and Rudaz, S. (2008) *J. Chromatogr. B*, **871**, 261–270.
165. Glauser, G., Guillarme, D., Grata, E., Boccard, J., Thiocone, A., Carrupt, P.A., Veuthey, J.L., Rudaz, S., and Wolfender, J.L. (2008) *J. Chromatogr. A*, **1180**, 90–98.
166. Glauser, G., Boccard, J., Rudaz, S., and Wolfender, J.L. (2010) *Phytochem. Anal.*, **21**, 95–101.

14
Metabolomics and Its Role in the Study of Mammalian Systems

Warwick B.Dunn, Mamas Mamas, and Alexander Heazell

14.1
Introduction – From Early Beginnings

In the last 15 years, metabolomics has emerged as an important research tool in the biological investigation of mammalian systems [1] with an emphasis on health, disease, lifestyle, and drug discovery. The emergence of metabolomics, especially in relation to the holistic study of large numbers of metabolites, has arisen from increases in computing power and technical advances in analytical instruments (e.g., new instrument designs [2]) and software [3]. Although metabolomics is considered as a postgenomic discipline, the first reported studies in the 1960s and 1970s focused on the application of gas chromatography-mass spectrometry (GC-MS) to the analysis of biological fluids (biofluids) [4, 5]. This was 30 years before the definitions of the metabolome [6, 7], metabolomics [8, 9], and metabonomics [10] were published and highlighted the foresight of the early pioneers.

In the last 15 years, a wide range of metabolomic applications have been observed that have focused on the investigation of human health (including the importance of nutrition [11, 12]) and disease [13, 14], aging (typically studied in other organisms [15]), lifestyle (e.g., exercise [16]), and drug discovery and development [17, 18]. The emergence of discovery-based strategies has been most prominent and has provided significant and novel insights related to molecular pathophysiological mechanisms [19] or have defined potential biomarkers associated with disease risk, onset, or progression [20].

Metabolites are implicated in a number of biological roles in human systems, as shown in Figure 14.1. Metabolomics is most frequently applied to define the phenotype, the biological determinant of the interactions between the genome and the environment (including lifestyle). The metabolome provides a dynamic and sensitive measure of the phenotype as it lies furthest downstream of the genome [1, 21]. Metabolites have other important roles derived from their interaction with other metabolites or biochemicals. The synthesis and catabolism of metabolites operate in a complex metabolic network (metabolism) through the interaction of metabolites and enzymes. Metabolites are the building blocks of other biochemicals (proteins, RNA, and DNA) and provide biological regulation by interactions with these other

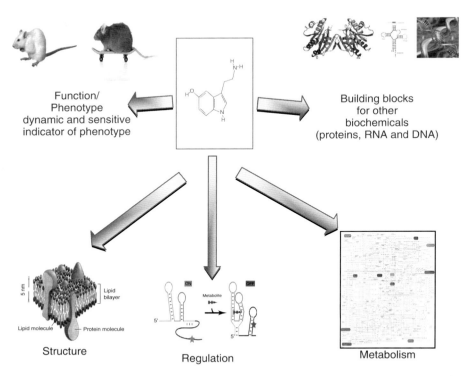

Figure 14.1 The roles which metabolites play in complex mammalian systems from acting as building blocks for other biochemicals to their complex interaction with other biochemicals and the environment to provide regulation and to define structures, networks, and phenotype.

biochemicals (e.g., allosterism [22] and riboswitches [23]). Metabolites are also important structural components (e.g., cell walls). As metabolites are involved in many different biological processes, the study of metabolites provides an overview of these many different processes in the mammalian body. Interactions at the biochemical level provide energy, growth, and reproduction which are all essential requirements for life. The human body acts and responds as a biochemical system and therefore studies at the systems level are often applied (defined as systems biology [24, 25] and discussed in Chapter 1).

All metabolomic studies follow a metabolome pipeline [26] from study and experimental design, through sample and data collection to data analysis and biological interpretation. In this chapter, an overview is presented for a number of general experimental questions to be considered when applying metabolomics to the study of mammalian systems. The experimental strategies that can be applied, the importance of experimental design and appropriate choice of sample type, sample collection/preparation, and analytical platform will all be discussed. The current difficulties associated with metabolite identification and the role of quality control (QC) samples in small-scale and large-scale metabolomic studies will also be discussed. Finally, applications will be described in two areas as examples of the general application of metabolomics: pregnancy-related complications and cardiovascular diseases (CVDs).

14.2
Hypothesis Generation or Hypothesis-Testing Studies

In all metabolomics studies, including the investigation of human systems, there are two generalized types of experiments: hypothesis-generating studies (otherwise defined as discovery or inductive studies) and hypothesis-testing studies [27].

In the 1990s and 2000s advances in analytical instruments, computing power, and data processing software provided the ability to simultaneously investigate hundreds or thousands of metabolites in a single analytical experiment. This enabled for the first time the holistic study of a wide diversity of metabolites in comparison to the targeted analysis of small numbers of metabolites as was applied in the preceding 50 years. At present, a large proportion of metabolomic studies apply the holistic data driven strategy in a hypothesis – generating approach to identify novel biological mechanisms or putative biomarkers not reported before the study. In these studies, an appropriate experimental design is constructed (see Section 14.4 for further details) without the presence of a detailed hypothesis to be tested. Data are acquired in semi-targeted or untargeted approaches (Section 14.3), followed by univariate and multivariate data analysis to define qualitative and quantitative differences in the metabolome of samples from two or more distinct classes. The observed differences are applied to construct a hypothesis. This type of experiment allows a system to be studied without the requirement for detailed biological knowledge and allows true discovery studies to be performed.

A hypothesis-testing experiment provides a hypothesis at the start of the study and the experiment is designed to test the hypothesis. Here, data for a limited number of metabolites (typically, 1–20 metabolites) are acquired in a targeted manner in relation to known biological mechanisms or networks. The data are applied to test whether a hypothesis is true or false. The hypothesis may have been constructed during a hypothesis-generating strategy and which requires validation through the hypothesis-testing approach.

Hypothesis-generating metabolomic studies have one of two general objectives: to discover a single metabolite biomarker or a group of metabolites acting as a biomarker panel *or* to discover new molecular pathophysiological mechanisms. Studies are typically designed to fulfill one of the two objectives, although the fulfillment of both objectives is not uncommon. However, the experimental design of a biomarker-focused study can be very different to the experimental design for a study to probe biological mechanisms. This distinction when designing a study or experiment is important. Knowing the study objective is an important first step so not to lead to disappointing or negative results for the researcher or collaborator.

The discovery of metabolite-based biomarkers in relation to health, disease, drug toxicity/efficacy, or lifestyle is an important scientific field of research. Glucose is an example of one metabolite applied as a biomarker, in this case to diagnose Diabetes Mellitus (DM). In these studies, the aim is to discover and then validate putative biomarkers and subsequently apply these in the appropriate environment. The route from discovery of a putative biomarker through validation to application is not easy and many putative biomarkers are shown not to be robust or applicable in validation studies. The level of validation required is dependent on the environment where they will be applied. For example, validation of biomarkers to be applied in the clinical environment require a longer path of validation before being applied compared to biomarkers of drug toxicity or efficacy applied in animal models during drug development studies.

The alternative objective is to define new molecular pathophysiological mechanisms related to the study of human disease, health, aging, lifestyle, and drug toxicity and efficacy. Here, metabolites showing a statistically valid change related to a biological or environmental perturbation are integrated with known biological networks or pathways (for which the metabolic network is one) to construct a hypothesis related to specific biological actions. These can be applied to enhance the current biological knowledge or to define specific targets for interventions in, for example, drug discovery.

The remaining sections of this chapter will focus on experimental considerations in hypothesis-generating strategies. Hypothesis-generating studies are a relatively new strategy with few published reviews and therefore will be the focus of this chapter. Hypothesis-testing studies have been performed routinely in the last 50 years and volumes of published research describe how this type of research has been performed.

14.3
Untargeted, Semi-Targeted, and Targeted Analytical Experiments

In Section 14.2, the two different types of experiments were described, specifically hypothesis-generating and hypothesis-testing experiments. To fulfill the objectives of these experiments, three different analytical strategies are applied: untargeted studies, semi-targeted studies, and targeted studies [1, 28]. Each of these strategies provides different advantages and limitations and none are universally applicable to all study objectives. The choice is dependent on the experimental objectives, experiment type (hypothesis-generating or hypothesis-testing) available resources (instruments, finances), available research expertise, and time available to perform each study.

Hypothesis-generating studies have the objective to acquire holistic metabolic profiling data on a diverse range of metabolites in a discovery study to define novel and previously unobserved changes in the metabolome which can be linked to biological function or mechanism or applied as a biomarker. To achieve this objective, appropriate sample preparation (discussed in Section 14.5) and advanced analytical instruments are required (Chapters 3–12) to allow data related to hundreds or thousands of metabolites to be reported. Only when a large fraction of the studied metabolome covering diverse areas of metabolism is detected can it be expected that previously unobserved changes would be identified. These are defined as semi-targeted studies or untargeted studies (otherwise known as *metabolic profiling* or *metabolite profiling*). Although the definition of metabolomics is the detection and identification of all metabolites [28], this is not yet routinely achievable because of the large ranges observed for physicochemical properties of metabolites (molecular weight, boiling point, and solubility as examples) and large concentrations range observed (millimolar to sub-picomolar). Instead large proportions of the metabolic network are studied, with the knowledge that all metabolites are not detected and the implications on discovery this provides.

There are major differences between semi-targeted and untargeted analyses [1]. Semi-targeted analyses define the metabolites to be assayed before the experiment and develop analytical methods to detect these metabolites with high accuracy, precision, sensitivity, and selectivity. Semiquantification is performed by comparison of raw analytical data to a calibration curve constructed with authentic chemical standards and, in some examples, with the presence of internal standards. The application of authentic chemical standards and internal standards provides some bias to the study as only metabolites which are commercially available or can be synthesized can be applied. A large fraction (>75%) of all known metabolites are not commercially available or the cost of purchasing is restrictive [29]. Typically, but not exclusively, liquid chromatography platforms coupled to triple quadrupole mass spectrometers are applied to provide high sensitivity and specificity. A single analytical method can typically detect 150–250 metabolites (e.g., see [30]). Multiple analytical methods, each being specific to one or a small number of metabolite classes (e.g., for nucleotides [31]), can be applied to provide greater coverage of the metabolome. This strategy of applying multiple analytical methods is highly appropriate as they combine the advantages of analytical methods of high sensitivity and specificity while providing good coverage of the metabolome. However, the limitation is that they are financially and technically intensive as one sample is analyzed on multiple occasions and sample throughput can be reduced if limited instruments are available to perform the work. These methods can be applied where the class of metabolites or area of metabolism expected to be perturbed is known, although metabolite-specific changes are not known. As the chemical identity of metabolites is defined before the experiment, the conversion of analytical data to biological knowledge is relatively simple, as has been shown [32, 33].

In untargeted studies, the metabolites to be identified are not known before the study. Instead analytical methods are developed, which provide detection of hundreds or thousands of metabolites in a single analytical experiment. The analytical methods developed do not provide as high a level of sensitivity or specificity as observed for semi-targeted studies but typically provide greater coverage of the metabolome in a single analytical experiment. Untargeted studies may be a more appropriate strategy where technical and financial resources may be limited or where high sample throughput is preferred. This strategy applies peak areas as parameters to define differences between metabolites, often referred to as *relative quantification*. There is no comparison of analytical raw data to calibration curves constructed with authentic chemical standards and typically internal standards are not applied. The application of calibration curves and internal standards is not observed because the metabolite identities are not known *a priori* and the selection of internal standards to apply to 100–1000s of metabolites present in a wide range of metabolite classes is technically difficult. Therefore, matrix effects should be expected to provide a source of variation in the data, especially when applying mass spectrometry instruments. Samples are prepared to minimize loss of any metabolites, although retention of all metabolites is technically difficult. The identity of metabolites is not known *a priori* and therefore identification of biologically interesting metabolites is required following data analysis and before biological interpretation is achievable. The process of metabolite identification can provides significant delays and limitations in applying untargeted studies, for mass spectrometry- and NMR spectroscopy-based studies (nuclear magnetic resonance) (Section 14.7). The decision on whether to apply semi-targeted or untargeted experiments can often be finalized on the choice of whether a separate metabolite identification step can be performed appropriately.

Targeted studies apply methods of high selectivity and sensitivity to target specific metabolites. A limited number of metabolites are assayed, typically <20, which have been defined as biologically relevant in previous studies. Sample preparation is more intensive compared to semi-targeted/untargeted studies and provides the separation of analytes from the sample matrix and other metabolites. Absolute quantitation is performed by the incorporation of appropriate internal standards and the comparison of data to calibration curves constructed with authentic chemical standards. The major differences between targeted and semi-targeted/untargeted studies are the number of metabolites assayed, the level of sample preparation, and the type of quantitation.

Semi-targeted/untargeted and targeted studies can operate in series. Semi-targeted/untargeted experiments can be performed to construct a hypothesis followed by targeted experiments in an independent sample set to define whether the hypothesis is true or false. This process of hypothesis-testing or validation is recommended to provide greater confidence in results constructed through hypothesis-generating studies. However, currently this is not frequently applied. For metabolomics to be successful and routinely applied in the clinical and drug development environments, validation following discovery studies is a necessity.

14.4
Study and Experimental Design

The study of mammalian systems typically involves two study environments [1]: those performed in a controlled laboratory environment (e.g., to determine the effect of a perturbation in an animal model or tissue culture) and those performed in the human population (e.g., to identify a biomarker associated with the risk of developing a disease). Many studies performed in a controlled environment applying animal, tissue, or cell models have the objective to mimic biological function and phenotype in humans.

Studies in the laboratory environment are closely controlled and the impact of variables present but not being directly studied (environment, genotype) can be minimized or eliminated. In culture-based systems, the variability in environmental conditions (e.g., growth medium and temperature) can be restricted and in animal models the environment, genotype and diet can all be controlled. Caution should always be taken as unexpected effects have been observed including differences in the gut microflora populations of rats as a result of differences in animal housing [34]. This highlights that expected and unexpected metabolic changes are present in metabolomic studies. As variability associated with variables not being directly studied is low it provides the ability to reduce or remove the impact of these variables not associated with the biological question while maintaining differences in the biological variable being investigated. The intraclass variability is low in comparison to studies performed in the general population. In addition, the perturbation applied can be extreme and therefore large quantitative differences in the metabolome can be observed [1]. For these reasons, low sample numbers (typically 6–20) are applied and can identify changes that are statistically robust. The use of small sample sizes can be important in animal studies where the cost per animal can prohibit the study size.

Studies in the general human population are more complex and difficult to perform. Significant biological variation is observed in the general population related to a multitude of variables including genotype, age, gender, BMI, lifestyle, and diet. The objective of this type of study is to define changes in the metabolome related to the onset or progression of biological or physical perturbations (e.g., presence of a disease). Interclass variability may be small compared to intraclass variability. To provide statistically robust results in this environment, large sample numbers should be studied to accurately define the variation associated with many different variables within a class and apply this variation to statistically test differences between sample classes. Medium- to large-scale studies applying a single sample collection point require hundreds or thousands of subjects.

There are a range of study types applied. In the general population, these studies can be controlled or observational [1]. Controlled studies provide a measurement of differences related to an induced perturbation with single or longitudinal studies performed where the researcher has control over class structure (e.g., a nutritional intervention study or clinical trial). Observational studies operate with no direct control by the researcher on the distribution of subjects between different classes. This could be a study of subjects from the general population, of which some are diagnosed as healthy and some are diagnosed with a specific disease. There are four generalized types of observational study applied in metabolomics: case-control, cross-sectional, cohort, and within-subject (for further details see [35]). In many applications, there will be a comparison at a single time point of two or more classes, each preferably matched in relation to potential confounding factors (biases) (e.g., see [36]). Other studies collect samples at multiple different time points (longitudinal studies). In longitudinal studies, especially when the time difference between samples is small, variability associated with many variables (e.g., genotype and lifestyle) are minimized. Further management of the environment can also be applied when studies are performed in study centers where random variables can be controlled to an even higher level. This can allow fewer subjects to be studied compared to single time point studies and is therefore advantageous. One example is a study where samples are collected before and after a specific perturbation, which as examples can be nutritional [37], exercise [16], or a combination of both [38].

Discovery-based studies applied in the general population have the potential to derive changes in the metabolome that are not biologically representative of the study objective. There are a number of issues to be considered when designing metabolomic experiments to provide a robust study which produces valid results including (i) sample size, (ii) presence of biases, (iii) high probability of type I false positives, and (iv) data analysis tools and overfitting of data. These issues have been discussed in detail previously [39]. Inappropriate choices related to any of these issues can invalidate the biological findings. For these reasons, the design of the metabolomic experiment is critical to produce reproducible data from which valid biological knowledge can be deduced [40]. During the experimental design, the different processes to be performed are assessed and planned in an appropriate manner to ensure a robust and valid conclusion to the study is acquired. In discovery studies, the design of the experiment is complex but essential to provide confidence in the biological conclusions. This problem is summarized in the following quotation, previously discussed from a metabolomics perspective [39].

> It can be proven that most claimed research findings are false [41].

Sample size is an important consideration. In controlled environments where variability associated with random variables is low, small sample sizes can be applied. However, in the study of the general population where single time points are collected, a range of different variables (age, gender, and BMI as examples) induce much higher levels of intraclass variability and therefore larger sample sizes are required to accurately quantify these changes and to ensure the study population is representative of the general population. This level of variability is reduced when applying longitudinal studies, where multiple time points are collected for a single subject, and is reduced further when applying a longitudinal study in a controlled environment such as a study center. Therefore, sample sizes of 6–20 are appropriate in a controlled environment and when applying longitudinal studies but are not appropriate in single time point studies applied in the general population.

Collection of metadata (e.g., data related to the study participants and including variables such as BMI, age, and gender) is an appropriate first process in any experimental design. This allows biases (or confounding factors) to be identified to ensure matching of variables between sample classes to remove the influence of these confounding factors. Removal of biases is not always possible but the ability to define confounding factors is important so to enable appropriate data analysis methods to be applied or the potential impact of biases on biological conclusions to be considered. For example, in a case-control study, subjects diagnosed with a disease may be prescribed drugs as a treatment, whereas it is not ethically appropriate for the control subjects to be prescribed these drugs. In this example, drug prescription is a bias that cannot be removed but can be considered when discussing biological conclusions. One solution would be the collection of samples from drug-naive subjects before treatment commences.

Bias is not just observed with variables associated with the study population but can be introduced at any stage along the experimental workflow [1]. Care should always be taken to ensure consistency in sample collection through the application of standard operating procedures (especially when samples are collected across multiple sites). Appropriate randomization is performed to ensure the order of sample preparation and order of sample analysis does not correlate with variables and class structure [42]. It has been reported that bias can be such an issue in observational research that a study should be presumed "guilty" until proven innocent [43].

In discovery metabolomic studies, a metabolite can be detected and reported as several different metabolic features. This colinearity can cause significant problems in univariate and multivariate data analysis. Many parallel tests are performed and the inappropriate choice of the data analysis method and critical p-value for univariate analysis can provide increased levels of type I false positives. Many published studies have applied a critical p-value of 0.05, which provides a higher probability of false positive identification where multiple testing is performed compared to lower critical p-values. Lower p-values or the application of tests, which take in to account multiple testing (e.g., Bonferroni [44] and false discovery rate, FDR [45]) can reduce the probability of false positives or quantify the probability of their presence and therefore provide greater confidence in the data. Another option to remove false biological hits is to classify metabolites shown to be of biological interest based on metabolite class or metabolic pathway. The grouping of statistically significant metabolites, which are biologically linked and show differences is an appropriate measure of their biological relevance. The probability of 2 or more metabolites being statistically significant is much lower compared to 1 metabolite. Finally, the appropriate choice of validation in supervised multivariate analysis is essential to provide robust multivariate models of high confidence [39].

One other point to consider during experimental design and related to confidence in biological discoveries is validation of discoveries in one or more independent studies. In discovery studies, the probability of reporting type I false positives is high, as discussed earlier. Statistical measures can be applied to minimize false positives, reported because of inappropriate critical p-values or because of multiple testing. Nonhomogenous sample populations applied in small studies of the general population may be another source of false positives. One strategy to eliminate or reduce the number of false positives reported is to perform validation studies in independent sample sets. This can apply a second discovery-style study or a targeted hypothesis-testing study [1]. When a second discovery study is applied, only metabolite-related changes observed in both studies should be reported as they provide greater confidence in their accuracy. Changes observed in one of two studies imply that they are applicable in only a specific sample population or are false positive discoveries. Many biomarker studies fail because a single discovery study has been applied without a validation study. Failure in these studies, especially when sample sizes are small, are caused by multiple factors including the choice of a nonhomogenous sample population which is not representative for the general study population. The application of two independent discovery studies where sample numbers are in the tens or low hundreds followed by a targeted validation study where $n = 100-1000$ s is an appropriate strategy which reduces risk associated with the large study costs of the collection and analysis of thousands of samples [1].

There are many other choices to be considered in the experimental design. The choice of an appropriate experimental system that can be applied to derive the data required is always difficult in relation to the increased complexity of population-based observational studies compared to studies performed in the controlled laboratory environment. The choice of sample to investigate is important and is discussed in Section 14.5. The choice of analytical platform applied can significantly influence the coverage of the metabolome acquired in untargeted studies. This choice is generally based on the metabolites of specific interest, the ease of data acquisition, data reproducibility, and the availability of specific analytical instruments. Three common analytical instruments are applied in untargeted metabolomic studies [1, 46]: NMR spectroscopy [47], liquid chromatography-mass spectrometry (LC-MS; and associated advances including ultra performance liquid chromatography) [42, 48] and GC-MS [42, 49]. For example, the study of lipoprotein fractions is appropriately performed applying NMR instruments. NMR instruments also provide robust and repeatable data acquisition across many months, especially when compared to chromatography-mass spectrometry platforms. However, NMR can suffer from a lack of sensitivity that allows detection of high abundance metabolites only. For more holistic approaches, coupling of chromatography and mass spectrometry platforms may be appropriate. The choice of gas or liquid chromatography is dependent on the types of metabolites to be studied: low molecular weight and volatile metabolites (e.g., amino and organic acids) are studied applying GC-MS, whereas high molecular weight metabolites (e.g., glycerophospholipids) are typically studied applying LC-MS. For true holistic studies, the application of both GC-MS and LC-MS is most appropriate as this will provide a greater coverage of the metabolome [42]. Sample throughput can also influence the decision of which analytical instrument to use. Experiments applying NMR spectroscopy and chromatography-mass spectrometry instruments typically apply analysis time of 5–60 min. To improve throughput (but potentially reduce biological information) direct infusion mass spectrometry (DIMS) [50] may be appropriate.

14.5
Sample Types

The human body is composed of a wide variety of unique cell types, tissues, organs, and biofluids. Each of these has a distinct and unique metabolome. Descriptions of human and tissue-specific metabolic networks have been reported [51, 52]. Work in the author's laboratories focused on the analysis of eight different mouse tissues show clear qualitative and quantitative differences in tissue metabolomes (Figure 14.2 shows UPLC-MS base peak chromatograms for four tissue types). Table 14.1 highlights a range of sample types that have been studied applying a hypothesis-generating metabolomic strategy. Examples of applications are also included.

The choice of sample to apply in a study is an important consideration and is based on the objectives of the study. The investigation of cell- or tissue-specific biological mechanisms are most appropriately performed in the target cells or tissues, either collected directly from humans or through the application of appropriate (i) cell or tissue culture systems or (ii) animals applied as model systems. This provides a

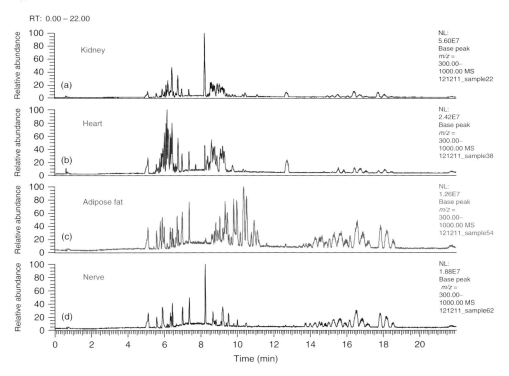

Figure 14.2 UPLC-MS base peak chromatograms for four different tissue types: (a) kidney, (b) heart, (c) adipose fat, and (d) nerve collected from a mouse. The chromatograms visually highlight qualitative and quantitative differences between the different tissue metabolomes.

Table 14.1 A review of different sample types investigated applying untargeted analysis methods with examples of their application.

Sample types	Published examples
Blood serum	[53, 54]
Blood plasma	[36, 55]
Urine	[50, 56]
Bile	[57, 58]
CSF	[59, 60]
Feces	[61]
Fecal water	[62]
Breathe	[63]
Sweat	[64]
Brain tissue	[65]
Kidney tissue	[66]
Lung tissue	[67]
Heart tissue	[68]
Liver tissue	[69]
Placental tissue and amniotic fluid	[70, 71]
Mammalian tissue culture	[72]
Mammalian cell culture	[73]
Mucosal colonic biopsies	[74]

sample-focused study where biases associated with biological processes elsewhere in the human body are eliminated or minimized. However, in applications focused on the development of biomarkers to be applied in the general human population or to define biological changes in a large human population, biofluids are most routinely applied. Biofluids are ethically, technically, and financially easier to collect than tissues and cells in observational studies comprised of 100–1000s of subjects. In some cases, integrated studies applying different sample types may be performed. For example, in a discovery study to identify a biomarker of tissue damage, the first study may focus on the specific tissue (human or animal/tissue/cell model) to identify potential biomarkers and then perform a second study applying a targeted assay in an appropriate biofluid to validate the applicability of these metabolites as biomarkers in the general population.

The collection of cells and tissues from large cohorts of the human population is ethically and technically difficult. Generally, this is only feasible when the tissue is being removed in a surgical procedure required as an intervention for a specific human disease. It is not ethically possible to remove tissue from healthy individuals and therefore control samples are difficult to acquire. Some tissues are more readily available and are easier to collect. Small skin biopsies acquired through surgery are one example as no internal surgical procedure is required [75]. One other example is placental tissue, which is naturally expelled following birth and can be applied to study complications associated with pregnancy [76].

In cases where collection of sufficient numbers and masses of tissue from human subjects is not achievable, the application of cell, tissue, or animal-based model systems is a preferred alternative. These systems mimic the surroundings and biological processes of the human body in an environment where controlled perturbations can be performed (genetic or environmental). This provides two advantages: (i) biological studies can be performed where direct sampling from humans is not possible and (ii) the controlled environment allows a reduction in intraclass variability and therefore smaller sample numbers are required. It is always recommended to validate discoveries, where possible, in human-based studies when applying this strategy.

The collection of a range of biofluids for large numbers of subjects in the general population is more readily achievable in comparison to the collection of cells and tissues. Biofluids can be collected in studies where large numbers of participants are required or where longitudinal collection of multiple samples from the same subjects is required. These include blood serum, blood plasma, urine, breath, and sputum that are collected routinely in the clinical environment through either noninvasive or minimally invasive procedures. These samples are integrative biofluids that represent the integration of many different biological processes operating in multiple cells, tissues, and organs with which the biofluid has direct or indirect interaction. The degree of tissue-specific changes may be diluted or constrained with other metabolite-related changes or the reduction in the concentration of the metabolite in the biofluid compared to the cell or tissue. Therefore, the discovery-based detection of tissue-specific changes in biofluids can at times be difficult to perform.

The collection of either serum or plasma from blood is an interesting choice. Plasma is acquired through collection of blood in to tubes containing anticoagulants (sodium or lithium heparin, sodium citrate, or potassium EDTA) followed by centrifugation to separate the aqueous plasma from white and red blood cells. Serum is acquired by collection in to tubes that do not contain an anticoagulant, followed by a period of blood clotting and a subsequent centrifugation step to separate the blood serum from the clot and blood cells. Collection of plasma is performed more routinely and the process is more reproducible as the extra step related to blood clotting can introduce variability. Plasma is applied in other omics-based biochemical studies (e.g., proteomics) although specific collection tubes are required. A number of studies have shown that the metabolome of serum and plasma are similar and any differences related to sample type or collection method are small compared to the intersubject differences observed [77–80]. Therefore, the choice of serum or plasma will have a minimal impact on most holistic studies although it is recommended that a single study apply collection of only one of these biofluids rather than the study of both [42]. In targeted studies, the stability of the metabolites in the chosen biofluid should be performed as part of any analytical validation study. This is because the studies defined above showed no general holistic change, although metabolite-specific changes for a limited number of metabolites were observed.

Other biofluids can also be collected although because of the difficulty of collection, the risk associated with collection, or the applicability of the sample are less frequently collected. Examples include cerebrospinal fluid, lymph fluid, and interstitial fluid.

Sample collection and preparation are two important aspects of the experimental workflow, which if performed applying inappropriate methods, can have large negative effects on the measured metabolome and the study. Many, but not all, metabolomes are dynamic in nature; they are present in a system where other biochemicals (e.g., enzymes) are present, which allow the continuation of metabolism following sample collection. Some metabolomes do not apply here including urine and breath as they do not contain enzymes, although metabolite-specific stability should always be considered. Metabolic flux can operate on timescales of seconds or minutes and this can result in the quantitative metabolic profile changing during sample collection and not representing the biological sample before sample collection [1]. Sample collection is therefore developed to minimize metabolism during and following collection. Sample collection normally applies a decrease (e.g., see [73]) in temperature, applied to inhibit metabolism through changes in enzyme activity or structure. In some cases this is simple, for example, the collection of blood and direct transfer on to ice during separation in to plasma or serum. Even during the collection of blood, care should be taken as there is generally a period between collection and processing, which can be anything from minutes to hours depending on the physical separation of collection point and processing laboratory. It is recommended that whole blood storage before processing is at $4\,°C$. Even on ice at $4\,°C$ where the temperature is significantly lower than the optimal for most enzymes ($37\,°C$) it may be expected that a small level of metabolic activity may be present. In other cases, it is not feasible to collect samples and rapidly inhibit metabolism. For blood collection, ice is not always available or cannot be located in specific blood collection areas for health and safety reasons. The same is true for operating theaters where for tissue collection, rapid washing in saline and freezing in liquid nitrogen would be recommended. Liquid nitrogen cannot be located in this environment based on health and safety grounds. Here, the tissue requires removal to a separate room where the process of washing and rapid freezing is performed. Again, this transfer of sample may not always be performed rapidly and the tissue may be present at room temperature for minutes or hours before sample processing is performed. These examples show the elevated difficulty of sample collection in the clinical environment in comparison to sample collection in the laboratory. It is not always feasible to collect samples in the most valid manner and compromises are sometimes required. It should always be considered whether the compromise will significantly affect any biological conclusions constructed and whether the study should be performed.

In cells and tissues collected following culture or from animal models applied in the laboratory, sample collection is somewhat easier to perform. All the appropriate instruments and chemicals are available and collection can be planned so that sample processing is rapid. The metabolism of non-adherent cell lines are quenched by addition of a cell culture volume to a cold quenching solution (typically containing methanol and/or water and potentially containing a buffer present at temperatures less than 4 °C [73]), followed by centrifugation and removal of the quenching solution. Washing of the cell pellet with water or saline to remove metabolites which were present in the culture medium and are trapped in the cell pellet is recommended [81].

For adherent cells the growth medium can be removed, the cells washed with water or saline and then a combined quenching and extraction solution applied to inhibit metabolism but also to perform metabolite extraction after cells are scrapped to release them in to the solution. Trypsin-based release of adherent cells is not recommended in metabolomic applications [82, 83].

For tissues cultured in a growth medium, collection is performed by removal from the growth medium, washing in water or saline, and then rapid freezing in liquid nitrogen [70]. It is not always easy to wash away all blood, especially in tissues perfused by blood and researchers should be aware that the extraction solution would contain metabolites from tissue and blood. For tissues collected from animal models, collection is performed following animal death and the mode of animal death should be considered as to whether it will impact detrimentally on the measured metabolome [84]. For example, asphyxiation could provide hypoxic stress and impact on specific cell and tissue metabolomes. Animal tissues are rapidly washed with water or saline and frozen in liquid nitrogen.

Metabolites present in cells and tissues from animals including humans are generally extracted applying either a lipophobic (e.g., water/methanol or methanol [73]) or combined lipophilic/lipophobic extraction procedure (chloroform/methanol/water [76, 85]). For cells, this requires cell wall disruption with freeze/thaw or physical disruption methods [73]. For tissues, the complete sample can be extracted after homogenization or alternatively selective study of specific regions of a tissue may be performed following specific dissection of the tissue on collection. However, further tissue dissection provides a longer time between collection and freezing and the potential for disruption of the metabolome. The extraction procedure has the objective to transport metabolites in to the extraction solution while ensuring other sample components (e.g., proteins, RNA, and DNA) are not soluble and are therefore removed as a pellet after centrifugation.

Sample preparation for biofluids is dependent on the sample and matrix. Urine normally contains no or little protein content and preparation typically involves centrifugation to remove particulates and either direct analysis (e.g., applying LC-MS [48]) or analysis after further procedures as observed for GC-MS. For GC-MS, further extraction and/or drying of samples is required because in the most common chemical derivatization step applied for GC-MS metabolomic studies, trimethylsilylation, the presence of water is detrimental to reproducible chemical derivatization and stable products. The removal of specific sample components may also be performed and urease treatment to enzymatically degrade the high levels of urea in urine is one common practice [86]. However, it has been reported that urease treatment can impact the metabolic profile [87].

Protein containing samples require a step to remove the protein complement, typically referred to as *deproteinization*. Here, an organic solvent is added to precipitate the protein, which can be pelleted with centrifugation to leave an extraction supernatant containing metabolites. A range of different solvents have been assessed and applied and no single and optimal solvent has yet been agreed on in the metabolomics community [42, 88–90]. Further sample fractionation can be performed to remove high levels of metabolites which may impact on analytical accuracy and precision. One example is the removal of phospholipids from plasma or serum, which are known to cause significant matrix effects in LC-MS analysis [91, 92]. However, in discovery studies the decision to remove these phospholipids will impact on the coverage of the metabolome detected. Phospholipids are related to many biological processes and their removal will potentially blind the researcher to these changes.

The time at which samples are collected can be important in animal or human studies, especially where a single time point sample is to be collected. It is always recommended to normalize the time of sampling where possible. Circadian rhythms can influence times when animals and humans eat and their associated metabolic profiles [93]. For example, rats generally eat at night and therefore samples collected during the day and night will be different. This is true for biofluid samples but also tissues where metabolism related to food intake are high (e.g., the liver). The potential effect of food consumption on the human metabolome is well known in health [50] and in relation to disease [94] and all samples should be collected under a similar regime, either all subjects are fasted or all are allowed to eat and drink as normal. A urine sample collected as the first urine of the day (representative of processes occurring during the evening before and during sleep) will be different to a sample collected after a day at work (as shown for metabolism of S-carboxymethyl-L-cysteine [95]). Twenty four hours urine collection is possible but technically is more difficult especially in large populations. The fraction of urine collected can be important. The first volume of passed urine will contain more cells and potential microbial contamination that can impact on sample stability and contaminate the urine metabolome with microbial metabolites. A mid-flow urine sample is recommended.

14.6
Quality Assurance and Quality Control

In studies applying chromatography and/or mass spectrometry instruments for targeted analysis of low numbers of metabolites, appropriate method development and validation is performed (e.g., following US Food and Drug Administration guidelines [96]) and quality assurance (QA) procedures are routinely applied. QA procedures are implemented to ensure that processes are generating data or products that meet required specifications. In analytical experiments, this relates to data acquisition with appropriate accuracy and precision. One aspect of the QA process is the analysis of QC samples composed of known concentrations (typically QC Low, QC Medium, and QC High) for all analytes to be assayed. QC samples are assayed to quantitatively define the accuracy and precision of the analytical method applied in a single analytical experiment. This process has been applied for many years in targeted assays.

In untargeted/semi-targeted studies, the application of QC samples as a process of QA is not routinely observed. A limited number of applications applying QC samples have been reported, initiated from the research of Professor Ian Wilson and his research team at AstraZeneca in the United Kingdom [97] and Professor van der Greef's research group in The Netherlands [98]. Some research groups have developed this strategy further and routinely apply QC samples in untargeted studies investigating biofluid samples [42, 48]. However, a more frequent application of QC samples is required for metabolomics to be accepted in many environments. This is especially important where large levels of variation or drift are expected to be present. For example, drift in response within and between analytical experiments applying NMR spectroscopy is small compared to biological variability allowing NMR spectroscopy to be applied in epidemiological studies [99]. However, drift in mass spectrometric and chromatographic properties (response, retention time, and mass accuracy as examples) are observed within and between analytical experiments [100]. Therefore, a measure of reproducibility (within experiment) and repeatability (between experiments) is recommended for chromatography-mass spectrometry focused discovery studies applying untargeted analyses.

Untargeted studies provide data on 100–1000s of metabolites in a single analysis applying a method developed for the holistic detection of a wide range of metabolites but which is not normally optimized for each specific metabolite. Generally, no internal standards are applied to detect and normalize for analytical variation introduced during sample preparation, data acquisition, and data preprocessing. In these studies, it is not generally feasible to develop methods which incorporate a large number of internal standards as the metabolites to be detected are not known, will differ across different studies applying the same sample type and would be technically and financially difficult. Therefore, a quantitative measure of any analytical variation within and between analytical experiments is highly recommended. The application of QC samples can provide this quantitative measure.

Two types of QC sample can be applied: pooled and commercially available [42]. Pooled QC samples take a part of all (or a large representative set of) samples in the study and mix thoroughly to provide a single pooled QC sample [97]. This sample is representative of the composition of all samples in the study, in effect an averaged composition. This is easy to perform for biofluids including serum, plasma, and urine where a small aliquot from each sample is easily collected and pooled. The pooled QC samples can then be sub-aliquoted, prepared, and analyzed as for all the study samples. This process of pooling is more difficult to perform for cell and tissue samples. Here, a further sample or samples can be cultured, collected, and applied as the QC sample after being processed through the identical sample preparation procedure. For tissue samples which are further dissected, the remaining waste tissue can be applied as a QC sample. Here, it is recommended to extract each QC sample separately and then pool the extract solutions in to a single pooled QC sample. Processes to construct a pooled QC sample before extraction are not always feasible. An alternative is to perform the extraction procedure and then take a small aliquot from each of the extraction solutions and combine to create a postextraction pooled QC sample. The level along the sample preparation step defines the processes for which the QC evaluation will be applicable. For samples pooled before sample extraction, QC evaluations will be a measure of process variation during sample preparation, data acquisition, and data preprocessing. Samples pooled after extraction will provide a measure of the variation associated with data acquisition and data preprocessing only but will not include variation introduced during sample extraction procedures.

Commercially available biofluids and tissues can be purchased to apply as QC samples. These samples may differ significantly to the study samples as they have not been collected and processed with the same procedures and in the same environment as for the study. However, in some cases where a pooled QC sample is not available they provide an alternative substitute. Biofluids are commercially available (e.g., Sigma-Aldrich supply serum and plasma). Human tissues can also be purchased (e.g., The National Disease Research Interchange [101]).

The pooled QC sample is always recommended, as its qualitative and quantitative composition will always be most similar to that of the study samples. Therefore, the QC sample will be highly representative of this sample and will provide the highest level of robust comparison. Commercially available QC samples can be applied as QC samples but their composition will differ to a greater level than samples collected as part of the study where they will be applied. For example, in the HUSERMET project [102], a commercially purchased serum sample was applied because data acquisition commenced before all samples (>3000 samples) were collected. The composition of the commercially available serum, although similar to subject samples, differed both qualitatively (some metabolites were present in QC sample but not subject samples and vice versa) and quantitatively (metabolites were present in QC sample and subject samples but at a different concentration). This provided a level of variability greater than the biological intersubject variability.

QC samples are applied to fulfill four objectives in a metabolomics study: (i) conditioning of the analytical system, (ii) to provide signal correction for drift, (iii) to assess within analytical experiment reproducibility, and (iv) to allow integration of data from multiple analytical experiments.

It has been shown that following routine maintenance of a chromatography and/or mass spectrometry instrument that significant drift in response (and sometimes retention time) can be observed for the first few injections. Injections of 5–10 QC samples provide conditioning of the system and a lower level of variability following conditioning [42, 48, 100, 103]. This is because routine cleaning has revealed active surface sites on to which metabolites can bind. QC samples allow this to be performed without using volume-limited or important samples.

Following conditioning, QC samples are intermittently analyzed throughout the analytical run (typically every 3–10th injection) [42, 48]. Random and systematic sources of variation will impact on the reproducibility of the data within an analytical experiment. Post data acquisition methods have been developed to correct for and minimize the impact of analytical variation. One developed method, which operates on a univariate basis, has applied Quality Control-Robust Loess Signal Correction (QC-RLSC) to reduce the level of technical variation in the data [42]. A low-order nonlinear locally estimated smoothing function (LOESS) is fitted to the QC data with respect to the order of injection. A correction curve for the complete data set is then interpolated, to which the total data set for that feature is normalized [42]. An alternative approach applies a multivariate strategy to reduce the influence of within- and between-batch variation [104]. Both of these methods reduce the technical variation observed while maintaining the same level of biological variation. An example of drift associated with analysis order of a high-glucose sample on a UPLC-MS platform and which is corrected for applying QC-RLSC is shown in Figure 14.3.

As the QC samples are analyzed intermittently during an analytical experiment and are qualitatively and quantitatively identical, indistinguishable results should be observed when comparing QC data. In reality there will be small levels of variation introduced during sample preparation, data acquisition, and data preprocessing and which remain following data preprocessing (e.g., QC-RLSC). The reproducibility of the sample preparation, data acquisition, and data preprocessing steps can be quantified applying the intermittently injected QC samples. If a metabolic feature detected in QC samples is not reproducibly detected then there is a large source of variation in the process. Calculating the percentage of times detected and the relative standard deviation (RSD) for the response for each metabolic feature for all of the QC samples across the run (but not including QCs applied for system conditioning) provides a quantitative measure of the variability present. This can be then applied for each metabolic feature to define whether the technical variability observed is acceptable [42]. Acceptance limits can be set which allow data to be removed for metabolites showing unacceptable variation. If reproducible data is not collected for replicate injections of the same sample (QC) then it cannot be expected to produce valid data for injection of different samples. In the HUSERMET project [102], RSD limits of 20% (UPLC-MS) and 30% (GC-MS) were applied along with the routine detection of the metabolic feature (which had to be detected in a minimum of 50% of all QC samples) [42].

QC samples also allow data to be integrated across different analytical experiments. In targeted and semi-targeted analyses, the response for each metabolite is compared to a calibration curve constructed with an authentic chemical standard and analyzed in the same analytical experiment. This allows changes in response within and between different analytical experiments to be normalized. However, in untargeted studies no comparison to calibration curves is applied and drifts or systematic changes between batches will impact on the technical variability of the integrated data set and biological interpretation. The use of the same pooled QC sample across multiple analytical experiments can allow the measure of reproducibility within batches but also allow the integration of data from multiple analytical experiments [42]. This is particularly important in human-based studies where because of inherent biological variability associated with genotype, environment, and lifestyle 100–1000s of subjects may require investigation. When an analytical experiment applying a chromatography-mass spectrometry instrument

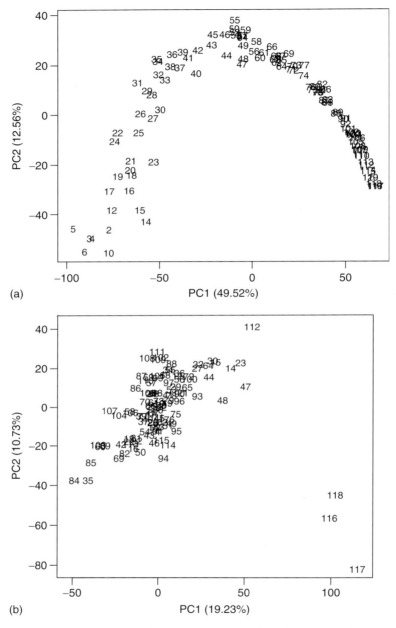

Figure 14.3 PCA scores plot showing (a) the effect of sample analysis order on the quality of data with the number defining the injection from 1 (first) to 120 (last) injection. The largest source of variation observed is related to analysis order and not to biological differences. (b) Following QC-RLSC the impact of technical variation is removed.

is constructed with >100–150 samples it is essential for the study to be separated in to different analytical experiments. When the number of injections between routine instrument maintenance is >100–150, significant drift can be observed which is not acceptable and cannot be corrected for applying QC-RLSC [100]. A recent example has shown how QC samples can be applied in a large study, the HUSERMET project, where >3000 subject samples were analyzed [102]. Here, QC-RLSC was applied to correct for drift within analytical experiments but also allow integration of data from multiple different analytical experiments (each composed of single serum samples from 120 subjects). This has allowed, for the first time, the collection of data in large-scale epidemiological studies applying untargeted chromatography-mass spectrometry methods across many months to be integrated in a robust manner and to derive biological knowledge from these data. This has been observed previously applying the more reproducible NMR instrument [99]. It is important in the experimental design of large-scale epidemiological studies to ensure that the diversity in the complete sample set is represented in each analytical experiment, for example, that a 60/40 gender ratio in the sample set is represented as a 60/40 gender ratio in each analytical experiment [42].

14.7
Metabolite Annotation and Identification

The chemical identity of metabolites is known before the analytical experiment when applying targeted and semi-targeted analyses. In untargeted studies, the chemical identity of metabolites is not known before the analytical experiment and must be determined applying data acquired during or after the analytical experiment. This is not a robust and automated process and typically does not provide a unique metabolite identification [29]. Metabolite annotation and identification is a significant bottleneck in untargeted studies and impacts on efficient biological interpretation [105].

There are four levels of identification as defined by the Metabolomics Standards Initiative in 2007 [106]. Definitive (level 1) identification is applied where comparison and matching of two or more orthogonal properties of the metabolite to data acquired applying the same analytical methodology (not necessarily in the researcher's laboratory) for an authentic chemical standard is performed. Putative identification (level 2 or 3) is applied where one or more properties are matched to experimental data acquired for authentic chemical standards applying different analytical methods or matched to theoretical data (e.g., matching accurate m/z measurements to the accurate monoisotopic mass of metabolites present in chemical or metabolite databases). This level of identification can provide metabolite-specific (level 2) or class-specific (level 3) identification.

In all untargeted studies some, and normally a significant number of, metabolic features are unidentified. The metabolite-specific search space is enormous. For example, the human metabolome database (HMDB) contains >8000 metabolites [107]. This search space is dependent on the metabolite complexity of different systems studied. The yeast metabolome [108], which contains 1168 metabolites, is less complex than the human metabolome as defined in HMDB. Incomplete qualitative descriptions of endogenous metabolomes associated with a large array of different sample types are currently not available. To provide an additional layer of complexity interaction with the environment will introduce further metabolites (e.g., derived from drugs or food [50]) and again these are not defined or catalogued well. A final layer of complexity is provided by the fact that a single metabolite can be detected as multiple different metabolic features. In NMR spectroscopy peak splitting is observed, in GC-MS multiple chemical derivatization products are formed, and in electrospray ionization-MS multiple different ion types are observed. Therefore, a greater number of metabolic features are detected compared to the number of metabolites present and which provides a greater level of complexity.

In traditional analytical chemistry, mass spectrometry and NMR spectroscopy are routinely applied for structural characterization of chemicals in single component solutions. However, these tools have not been sufficiently developed and validated to provide their routine applicability in metabolite identification in complex systems containing hundreds or thousands of metabolites. These studies provide a significant level of peak overlap in the data and for chromatography-mass spectrometry platforms detection of a single metabolite, compared to a mixture of metabolites, at any given time is not generally achievable. At present, the chemical characterization and identification of metabolites is a difficult and complex process for both NMR spectroscopy and mass spectrometry instruments and significant further developments are required.

In 1D-NMR applications, the chemical shift (^{1}H or ^{13}C) is commonly applied as the first parameter to match against chemical shift data in metabolite databases [107, 109]. Further 2D-NMR experiments (e.g., COSY, TOCSY, HSQC, and HMBC) can provide a NMR spectrum of greater resolution and also provides data related to coupling between ^{1}H and ^{13}C, depending on the application (as shown in Chapter 11). These all aid in chemical characterization of metabolites.

In GC-MS matching, the retention index and/or mass spectrum are commonly applied. A range of different mass spectral libraries are available, some contain data on a range of chemicals (some of which are not metabolites) [110] while some are metabolite specific [111, 112]. These libraries contain mass spectral fragmentation data acquired following electron impact (EI) ionization and some also contain retention index data. Retention indices [113] are calculated from nonmetabolite chemicals (e.g., n-alkanes) spiked in to a sample to allow normalization of the retention time to be performed, to correct for small variations in chromatographic performance and to allow data to be compared across different GC-MS instruments where a similar analytical method is applied.

Most GC-MS instruments employ EI ionization, which is a reproducible technique applied with an electron energy of 70 eV. The mechanism of ionization imparts significant internal energy to the positively charged molecular ion that cannot be lost by ion-molecule collisions in the high vacuum conditions applied. This energy is lost through the fission of covalent bonds to provide fragmentation of the molecular ion and subsequent detection of positively charged fragment ions. The mass spectrum is related to the chemical structure and can be applied for metabolite identification with high specificity. EI-derived mass spectral fragmentation data are available in mass spectral libraries and are reproducible and transferable between different instruments because of the reproducibility of the ionization and fragmentation processes.

The common identification workflow for GC-MS data is the comparison of EI mass spectra and/or retention indices against mass spectral libraries, which generally operate at unit mass resolution. More recently, mass spectrometers are becoming available, which can provide chemical ionization and high mass resolution and accuracy. In EI ionization, the fragmentation process typically produces a weak or nonexistent molecular ion. The determination of the molecular formula from accurate measurement of the m/z of the molecular ion is an appropriate method applied in the identification process. Chemical ionization provides a molecular ion with a high response and limited fragmentation. When combined with accurate m/z measurements this allows the deduction of the molecular formula or at least a reduction in the available search space containing all metabolites and molecular formula.

GC-MS mass spectral and NMR libraries are constructed with authentic chemical standards, but are not complete as authentic chemical standards for all known metabolites are not commercially available. Therefore comparison of data to mass spectral and NMR libraries can provide identification of metabolites where authentic chemical standards are available, but not identification of all metabolites. It is not possible to construct mass spectral libraries containing data for all known metabolites. *In-silico* methods to derive the chemical structure or to identify chemical substructures from GC-MS-derived mass spectral data are being developed [114]. *In-silico* prediction of retention indices is also being developed to assist in identification of metabolites not present in mass spectral libraries [115]. These provide further data to aid in metabolite identification where the authentic chemical standard is not readily available. For this reason level 2, but not level 1, identification is achievable.

In LC-MS, CE-MS, and DIMS, the accurate measurement of the m/z ratio of metabolites is the first tool commonly applied for metabolite identification. Many commercially available mass spectrometers applied in untargeted studies, in particular, Q-TOF, Orbitrap, and FT-ICR MS, operate at a high mass resolution (5000–200 000) and high mass accuracy (<5 ppm). The m/z is applied to calculate the molecular formula or formulae which are subsequently matched to metabolites present in metabolite-specific [116–118] or chemical-specific [119, 120] databases. However, electrospray ionization data can provide the observation of different ion types that produce a complex pattern. These include common ion types such as protonated ($[M + H]^+$) and deprotonated ($[M - H]^-$) ions but also ion types not necessarily expected including complex salt adducts containing multiple elements (e.g., sodium and chloride) [29]. These complex patterns of ion types arise from the sample matrix, which is not removed during sample preparation (e.g., high sodium content of biological fluids) and other factors including metabolite concentration (which can lead to the formation of dimers) and instrument (e.g., formation of

Fourier Transform artifact peaks) [29]. Incorrect identification of the ion type will lead to an inaccurate determination of the molecular formula of the metabolite and a false identification. However, freely available software is becoming available which apply multiple parameters to annotate metabolic features with the correct ion type and integrate data for several metabolic features of the same metabolite [121–123].

A second limitation of ESI-MS is that even with very high mass accuracy (sub 1 ppm), reduction of the search space to a single molecular formula is not always achievable, especially as m/z increases. Even if only one molecular formula is defined, this can relate to multiple different isomers (e.g., leucine/isoleucine or glucose/fructose). This is a significant issue in the analysis of lipids as a single molecular formula can be matched to many different lipids of the same lipid class where only the fatty acid chain length and double bond position of distinct fatty acids change (e.g., diacylglycerols (DGs) where DG(38 : 3) can relate to several different lipids of which two examples are DG(20 : 0/18 : 3) and DG (22 : 1/16 : 2)). For these reasons, the application of accurate measurements of m/z provides putative identifications with further experimental data being required to provide greater confidence of putative annotations or to provide definitive identifications. This can include traditional tools such as the seven golden rules applied to filter possible molecular formula [124] or the use of biological information related to metabolism to reduce the search space [125, 126]. The acquisition of MS/MS and MS^n data acquired during the analytical experiment and matched to MS/MS and MS^n libraries is also performed.

Mass spectral libraries containing retention time data can be constructed with commercially available metabolites [127, 128]. However, the wider range of column chemistries (C_8, C_{18}, HILIC, cyano, others) and chromatographic separation mechanisms result in a lower level of transferability between instruments or analytical methods compared to GC-MS. The collection of MS/MS data is applicable, although unlike EI, this requires further instrument complexity to acquire full scan accurate m/z data of metabolic features in parallel to the acquisition of good quality MS/MS data. The acquisition of full scan and MS/MS data is achievable to a certain level but generally MS/MS data for all metabolites is not acquired. Fraction collection and direct infusion experiments with nanoelectrospray ionization (to provide minutes of analysis time) can be applied to provide off-line identification of biologically interesting metabolites [129]. Methods to provide repetitive analyses of a single sample to acquire MS/MS data with appropriate use of inclusion and exclusion lists so to maximize the range of metabolites for which MS/MS data are collected while allowing collection of data for a single species only once have been applied in proteomics [130] and are being developed in metabolomics (Steffen Neumann, personal communication). Different ion activation mechanisms are available including collision-induced dissociation (CID) in an ion trap, CID in a quadrupole time-of-flight or triple quadrupole instrument and higher energy dissociation (HCD) in Orbitrap systems. These present limitations as these can produce different fragmentation mechanisms and therefore different mass spectral fragmentation patterns that are not comparable across different ion activation mechanisms. The level of reproducibility across the same manufacturer's instruments applying the same ion activation mechanism is currently being discussed and transferability of libraries is appropriate. However, the transferability of mass spectral libraries acquired applying different ion activation methods or acquired on different manufacturer's instruments is significantly less developed than for GC-MS. *In-silico* prediction of fragmentation patterns are available through commercial [131, 132] or freely available software [133, 134]. *In-silico* retention time prediction is also being developed [135].

The integration of data from two different analytical platforms can provide increased selectivity during the metabolite identification process. Methods to integrate data from different instruments in to a single data set for data analysis but also to aid in metabolite identification have been reported [136].

The limitations above define the current issues of metabolite identification in untargeted studies. There is a high probability of false identifications, especially where all potential metabolites are not present in databases or libraries applied for searching. It should always be considered whether the appropriate tools and expertise are available in research groups to undertake this process of metabolite identification in a valid and robust manner or whether a semi-targeted approach may be more suitable, metabolite identification after data acquisition is not required.

14.8
Applications

Metabolomics is applied in a diverse range of applications in mammalian-focused studies. Table 14.2 highlights some of the applications with a focus on diversity. There is too wide an application range to discuss and reviews are available which focus on specific areas of application (for the role of metabolomics in drug discovery and development see [137–140]). Below we will focus on the role of metabolomics in two specific application areas (pregnancy complications and CVD) and highlight important aspects discussed earlier.

14.8.1
Pregnancy Complications

Since the publication of the first metabolomic-focused paper in the field of reproductive biology in 2004 [153], there has been increasing interest in the utility of metabolomics with a year-on-year increase in the number of research papers, as shown in Figure 14.4. This highlights the growing importance of metabolomics in reproductive biology research and this trend is being observed in other areas of mammalian-based research. Forty seven manuscripts, of which 33 presented data from original research studies, were presented on PubMed. These can be divided into three areas of study: (i) oocyte and embryo variability, (ii) investigation and prediction of pregnancy complications, and (iii) understanding of pathophysiological mechanisms underlying pregnancy complications. Studies

Table 14.2 Review of different applications of metabolomics in the study of mammalian systems with examples of their application included.

Application area	Published examples
Cancer	[55, 141]
Diabetes	[14, 54]
Heart diseases	[53, 142]
Kidney diseases	[143, 144]
Cognitive diseases	[60, 145]
Liver diseases	[146, 147]
Drug discovery and development	[148, 149]
Nutrition	[150, 151]
Exercise	[16, 152]

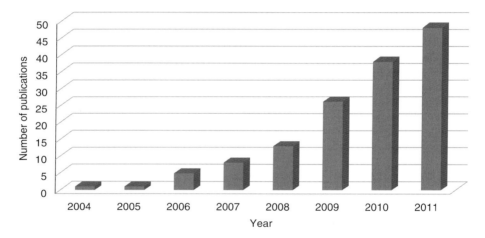

Figure 14.4 Cumulative number of papers published identified using PubMed/Medline with the MeSH search terms "metabolomics" and "pregnancy." Date of search 16 December 2011.

of oocyte and embryo viability have recently been comprehensively reviewed by Nel-Themaat and Nagy [154] and will not be discussed further.

Diagnosis or prediction of pregnancy outcome or complications during pregnancy is a highly desirable clinical goal. There were 210 million pregnancies worldwide in 2002 [155]. In the developed world, approximately 3–5% of pregnancies are complicated by preeclampsia [156], 4–8% by intrauterine growth restriction (IUGR), and 5–7% by preterm labor [157]. These complications may coexist and are responsible for the majority of perinatal mortality and morbidity causes. Although there are important prenatal risk factors for these conditions, the bulk of pregnancy complications occur in women classified as "low-risk" based on nonbiochemical data. Therefore, identifying patients most at risk of pregnancy complications before these complications are observed allows prenatal care to be focused. In economic terms, developing biochemical screening tests for pregnancy complications is a compelling argument; large numbers of women achieve a pregnancy each year and there is a relatively short time period between a screening time point and the development of disease pathology (typically a few months) compared to the slow onset of some other diseases measured over years or decades.

Metabolomic studies have adopted a hypothesis-generating approach to identify differences in metabolites between normal and complicated pregnancies and to provide the prediction of complications. Maternal blood [36, 158, 159], amniotic fluid [71, 160, 161], placental tissue [76], and conditioned culture medium (metabolic footprint) from placental tissue cultures [70, 72] have all been investigated.

Studies have defined putative biomarkers applicable for diagnosis (at time of disease) or to predict risk of developing a disease in asymptomatic women. Early pregnancy prediction of complications to occur in the third trimester is important to allow correct management of pregnancies. For example, by the study of 15 week blood samples collected routinely in the developed world this can be achieved [36]. To date metabolomic studies have investigated, at time of disease or before onset of symptoms; fetal malformations [160, 162], chromosal disorders [162], early rupture of membranes [71], preterm delivery with and without infection [161, 162], preeclampsia [159], IUGR [158], and gestational diabetes [162]. The study by Kenny *et al.* [36] has shown how validation studies can be performed in an independent sample set and that early metabolic disturbances in maternal plasma can be detected many weeks before symptoms are observed showing the sensitivity of metabolic changes to human disease.

Alternative applications have focused on the discovery of novel molecular pathophysiological mechanisms associated with pregnancy complications. These have studied the metabolic footprint of placental tissue after culturing in a serum-based media [70, 72] or analysis of placental tissue without culture [76]. This has been performed for normal pregnancies [70], small-for-gestational age pregnancies [163], and preeclamptic pregnancies [72], where hypoxia and oxidative stress have been implicated. Further evidence of hypoxia and oxidative stress metabolic markers detectable in the placenta have been reported from pregnancies occurring at sea level compated to pregnancies at an altitude of 3100 m [164].

14.8.2
Cardiovascular Diseases

CVDs are a group of disorders affecting the heart and blood vessels and which include coronary heart disease, cerebrovascular disease, hypertension, peripheral vascular disease, and heart failure. CVD is the commonest cause of mortality in the world. According to recent World Health Organization data, an estimated 17.3 million individuals died from CVD in 2008 representing 30% of all deaths [165]. One of the most important risk factors for the development of CVD is DM, which is characterized by dysregulation of glucose metabolism occurring secondarily to abnormalities in either insulin secretion and/or impaired insulin action at the level of the insulin receptor or pathways downstream. There are over 220 million individuals with DM worldwide and this is expected to double over the next 20 years predominantly because of the increasing burden of obesity and changes to less healthy lifestyles [166]. The study of CVDs and DM are therefore important clinical problems in relation to prevalence and economic burden.

The application of metabolomics to the study of CVD and DM has been reviewed [13]. Biomarkers are most commonly used in CVD and DM for diagnosis, prognosis, and risk stratification in clinical practice, most commonly in patients presenting with heart failure and acute coronary syndromes. Metabolomic strategies have been applied toward biomarker identification in CVD and biomarker discoveries associated with acute coronary syndromes [142, 167], cardiac ischemia [168–170], atherosclerosis [171], aortic aneurysm [172], and heart failure [53]. Biomarkers associated with the risk of developing diabetes [14] and early metabolic changes observed before diagnosis of diabetes with blood glucose measurements have also been reported [54, 173–175].

Other studies have investigated the mechanisms associated with CVD and DM. Atherosclerosis [176, 177], coronary artery disease [178], dilated cardiomyopathy [179], and the effect of gut microflora on CVD [180] have all been studied. PPARs [181], insulin resistance/sensitivity [182], and changes associated with the glucose tolerance test applied to diagnose DM [183] have also been studied. Finally, metabolomics has been applied in drug development related to CVD [184, 185] and DM [186, 187].

The application of metabolomics has increased the depth of understanding of the pathophysiological mechanisms that contribute to the development of CVD. Branched chain amino acids have been shown to be elevated in patients who go on to develop future DM in a number of patient populations, preceding the well described changes in glucoregulatory processes that were previously thought to characterize DM. These observations have led to the suggestion that these branched chain amino acids may have important roles in the pathogenesis of DM and therefore may not be "bystander" metabolites that merely predict the future development of DM. Branched chain amino acids may promote insulin resistance through mTOR activation leading to activation of p70 S6 kinase and increased serine phosphorylation of insulin receptor substrate 1 (IRS-1) [188]. IRS-1 plays a key role in transmitting signals from the insulin receptor to intracellular pathways leading to insulin resistance in skeletal muscle [189]. Such metabolomic studies not only identify potential biomarkers for predicting the development of DM many years before overt abnormalities in blood glucose levels are present, but define novel roles for branched chain amino acids in the pathophysiology of DM development. These branch chained amino acids and their associated metabolic pathways may represent novel targets for the development of future therapies that target the very earliest pathophysiological processes that lead to the eventual development of DM.

Another example is 2-oxoglutarate, which was discovered as a putative biomarker of heart failure which was at least as diagnostic for the presence of heart failure as the current gold standard biomarker brain natriuretic peptide (BNP) that is used in contemporary clinical practice [53]. Subsequent studies have revealed that serum levels of 2-oxoglutarate also correlate with the clinical severity of heart failure as defined by New York Heart Association class (unpublished data). 2-Oxoglutarate is a ligand of the GPR99 G-protein coupled receptor that regulates the renin–angiotensin system, one of the major pharmacological therapeutic targets in the treatment of heart failure [190].

The studies highlighted above have shown that in the last decade the application of metabolomics to "great obstetrical and cardiovascular syndromes" has facilitated advances in the understanding of the pathophysiology and etiology of these conditions and offered the potential of predictive tests. From the viewpoint of pregnancy complications, it is interesting to note that syndromes including preterm labor, SGA, and preeclampsia which were initially thought to be unrelated have a number of similar metabolic perturbations which mirror new understanding regarding their underlying causes [191]. In the future, metabolomics may further advance understanding and detection of pregnancy complications and CVDs and aid the movement to facilitate personalized medicine.

Acknowledgments

This work was supported by the NIHR Manchester Biomedical Research Centre. Warwick Dunn wishes to thank Dr Paul Begley, Dr Graham Mullard, and Dr Elizabeth Pawson for provision of samples and data discussed in this chapter.

References

1. Dunn, W.B. et al. (2011) *Chem. Soc. Rev.*, **40** (1), 387–426.
2. Hu, Q.Z. et al. (2005) *J. Mass Spectrom.*, **40** (4), 430–443.
3. Smith, C.A. et al. (2006) *Anal. Chem.*, **78** (3), 779–787.
4. Horning, E.C. (1968) *Clin. Chem.*, **14** (8), 777.
5. Pauling, L. et al. (1971) *Proc. Natl. Acad. Sci. U.S.A.*, **68** (10), 2374.
6. Oliver, S.G. et al. (1998) *Trends Biotechnol.*, **16** (9), 373–378.
7. Tweeddale, H., Notley-McRobb, L., and Ferenci, T. (1998) *J. Bacteriol.*, **180** (19), 5109–5116.
8. Fiehn, O. et al. (2000) *Nat. Biotechnol.*, **18** (11), 1157–1161.
9. Roessner, U. et al. (2000) *Plant J.*, **23** (1), 131–142.
10. Nicholson, J.K., Lindon, J.C., and Holmes, E. (1999) *Xenobiotica*, **29** (11), 1181–1189.
11. McNiven, E.M., German, J.B., and Slupsky, C.M. (2011) *J. Nutr. Biochem.*, **22** (11), 995–1002.
12. Zivkovic, A.M. and German, J.B. (2009) *Curr. Opin. Clin. Nutr. Metab. Care*, **12** (5), 501–507.
13. Dunn, W.B. et al. (2011) *Bioanalysis*, **3** (19), 2205–2222.
14. Wang, T.J. et al. (2011) *Nat. Med.*, **17** (4), 448–453.
15. Yoshida, R. et al. (2010) *Aging Cell*, **9** (4), 616–625.
16. Lewis, G.D. et al. (2010) *Sci. Transl. Med.*, **2** (33), 33ra37.
17. Wei, R. (2011) *Curr. Drug Metab.*, **12** (4), 345–358.
18. Waters, N.J. (2010) *Curr. Drug Metab.*, **11** (8), 686–692.
19. Sreekumar, A. et al. (2009) *Nature*, **457** (7231), 910–914.
20. Mamas, M. et al. (2010) *Arch. Toxicol.*, **85** (1), 5–17.

21. Goodacre, R. et al. (2004) *Trends Biotechnol.*, **22** (5), 245–252.
22. Liu, J. et al. (2005) *Biochem. J.*, **391** (Pt 2), 389–397.
23. Dambach, M.D. and Winkler, W.C. (2009) *Curr. Opin. Microbiol.*, **12** (2), 161–169.
24. Kell, D.B. (2006) *FEBS J.*, **273** (5), 873–894.
25. Naylor, S. and Chen, J.Y. (2010) *Pers. Med.*, **7** (3), 275–289.
26. Brown, M. et al. (2005) *Metabolomics*, **1** (1), 39–51.
27. Kell, D.B. and Oliver, S.G. (2004) *Bioessays*, **26**, 99–105.
28. Fiehn, O. (2002) *Plant Mol. Biol.*, **48** (1-2), 155–171.
29. Brown, M. et al. (2009) *Analyst*, **134** (7), 1322–1332.
30. Floegel, A. et al. (2011) *PLoS ONE*, **6** (6), e21103.
31. Klawitter, J. et al. (2007) *Anal. Biochem.*, **365** (2), 230–239.
32. Lawton, K.A. et al. (2008) *Pharmacogenomics*, **9** (4), 383–397.
33. Sabatine, M.S. et al. (2005) *Circulation*, **112** (25), 3868–3875.
34. Robosky, L.C. et al. (2005) *Toxicol. Sci.*, **87** (1), 277–284.
35. Lu, C.Y. (2009) *Int. J. Clin. Pract.*, **63** (5), 691–697.
36. Kenny, L.C. et al. (2010) *Hypertension*, **56** (4), 741–749.
37. Llorach, R. et al. (2010) *J. Proteome Res.*, **9** (11), 5859–5867.
38. Miccheli, A. et al. (2009) *J. Am. Coll. Nutr.*, **28** (5), 553–564.
39. Broadhurst, D.I. and Kell, D.B. (2006) *Metabolomics*, **2** (4), 171–196.
40. Dunn, W.B. (2008) *Phys. Biol.*, **5** (1), 011001.
41. Ioannidis, J.P.A. (2005) *Plos Med.*, **2** (8), 696–701.
42. Dunn, W.B. et al. (2011) *Nat. Protoc.*, **6** (7), 1060–1083.
43. Ransohoff, D.F. (2005) *Nat. Rev. Cancer*, **5** (2), 142–149.
44. Bland, J.M. and Altman, D.G. (1995) *Br. Med. J.*, **310** (6973), 170.
45. Benjamini, Y. and Hochberg, Y. (1995) *J. R. Stat. Soc. Ser. B*, **57** (1), 289–300.
46. Dieterle, F. et al. (2011) *Methods Mol. Biol.*, **691**, 385–415.
47. Beckonert, O. et al. (2007) *Nat. Protoc.*, **2** (11), 2692–2703.
48. Want, E.J. et al. (2010) *Nat. Protoc.*, **5** (6), 1005–1018.
49. Denkert, C. et al. (2006) *Cancer Res.*, **66** (22), 10795–10804.
50. Lloyd, A.J. et al. (2011) *Br. J. Nutr.*, **106** (6), 812–824.
51. Duarte, N.C. et al. (2007) *Proc. Natl. Acad. Sci. U.S.A.*, **104** (6), 1777–1782.
52. Hao, T. et al. (2011) *Mol. Biosyst.*, **8** (2), 663–670.
53. Dunn, W. et al. (2007) *Metabolomics*, **3** (3), 413–426.
54. Oresic, M. et al. (2008) *J. Exp. Med.*, **205** (13), 2975–2984.
55. Patterson, A.D. et al. (2011) *Cancer Res.*, **71** (21), 6590–6600.
56. Kuhara, T. et al. (2011) *Anal. Bioanal. Chem.*, **400** (7), 1881–1894.
57. Yang, L. et al. (2008) *Chem. Res. Toxicol.*, **21** (12), 2280–2288.
58. Plumb, R.S. et al. (2009) *J. Proteome Res.*, **8** (5), 2495–2500.
59. Wuolikainen, A. et al. (2011) *PLoS ONE*, **6** (4), e17947.
60. Verwaest, K.A. et al. (2011) *Biochim. Biophys. Acta*, **1812** (11), 1371–1379.
61. Martin, F.P. et al. (2010) *J. Proteome Res.*, **9** (10), 5284–5295.
62. Monleon, D. et al. (2009) *NMR Biomed.*, **22** (3), 342–348.
63. Basanta, M. et al. (2010) *Analyst*, **135** (2), 315–320.
64. Kutyshenko, V.P. et al. (2011) *PLoS ONE*, **6** (12), e28824.
65. Salek, R.M. et al. (2010) *Neurochem. Int.*, **56** (8), 937–947.
66. Boudonck, K.J. et al. (2009) *Toxicol. Pathol.*, **37** (3), 280–292.
67. Shin, J.H. et al. (2011) *J. Proteome Res.*, **10** (5), 2238–2247.
68. Andreadou, I. et al. (2009) *NMR Biomed.*, **22** (6), 585–592.
69. Wang, Y. et al. (2009) *Anal. Chim. Acta*, **633** (1), 65–70.
70. Heazell, A.E. et al. (2008) *Placenta*, **29** (8), 691–698.
71. Graca, G. et al. (2010) *J. Proteome Res.*, **9** (11), 6016–6024.
72. Dunn, W.B. et al. (2009) *Placenta*, **30** (11), 974–980.
73. Sellick, C.A. et al. (2011) *Nat. Protoc.*, **6** (8), 1241–1249.

74. Bjerrum, J.T. et al. (2010) *J. Proteome Res.*, **9** (2), 954–962.
75. Hollywood, K.A. et al. (2010) *Arch. Dermatol. Res.*, **302** (10), 705–715.
76. Dunn, W.B. et al. (2011) *Metabolomics*, doi: 10.1007/s11306-011-0348-6
77. Wedge, D.C. et al. (2011) *Anal. Chem.*, **83** (17), 6689–6697.
78. Denery, J.R., Nunes, A.A., and Dickerson, T.J. (2010) *Anal. Chem.*, **83** (3), 1040–1047.
79. Liu, L. et al. (2010) *Anal. Biochem.*, **406** (2), 105–112.
80. Dettmer, K. et al. (2010) *Electrophoresis*, **31** (14), 2365–2373.
81. Teng, Q. et al. (2009) *Metabolomics*, **5** (2), 199–208.
82. Martineau, E. et al. (2011) *Anal. Bioanal. Chem.*, **401** (7), 2133–2142.
83. Danielsson, A.P. et al. (2010) *Anal. Biochem.*, **404** (1), 30–39.
84. Petucci, C., Rojas-Betancourt, S., and Gardell, S.J. (2011) *Metabolomics*, doi: 10.1007/s11306-011-0370-8
85. Salek, R., Cheng, K.K., and Griffin, J. (2011) *Methods Enzymol.*, **500**, 337–351.
86. Matsumoto, I. and Kuhara, T. (1996) *Mass Spectrom. Rev.*, **15** (1), 43–57.
87. Kind, T. et al. (2007) *Anal. Biochem.*, **363** (2), 185–195.
88. Want, E.J. et al. (2006) *Anal. Chem.*, **78** (3), 743–752.
89. Jiye, A. et al. (2005) *Anal. Chem.*, **77** (24), 8086–8094.
90. Bruce, S.J. et al. (2008) *Anal. Biochem.*, **372** (2), 237–249.
91. Want, E.J. et al. (2006) *Metabolomics*, **2** (3), 145–154.
92. Michopoulos, F. et al. (2009) *J. Proteome Res.*, **8** (4), 2114–2121.
93. Park, Y. et al. (2009) *Am. J. Physiol. Regul. Integr. Comp. Physiol.*, **297** (1), R202–R209.
94. Lankinen, M. et al. (2009) *Nutr. Metab. Cardiovasc. Dis.*, **20** (4), 249–257.
95. Steventon, G.B. (1999) *Drug Metab. Dispos.*, **27** (9), 1092–1097.
96. (2001) http://www.fda.gov/downloads/Drugs/GuidanceComplianceRegulatoryInformation/Guidances/ucm070107.pdf (last accessed October 5th 2012).
97. Sangster, T. et al. (2006) *Analyst*, **131** (10), 1075–1078.
98. van der Greef, J., Hankemeier, T., and McBurney, R.N. (2006) *Pharmacogenomics*, **7** (7), 1087–1094.
99. Loo, R.L. et al. (2009) *Anal. Chem.*, **81** (13), 5119–5129.
100. Zelena, E. et al. (2009) *Anal. Chem.*, **81** (4), 1357–1364.
101. http://ndriresource.org/ (last accessed October 6th 2012).
102. http://www.husermet.org/ (last accessed October 6th 2012).
103. Begley, P. et al. (2009) *Anal. Chem.*, **81** (16), 7038–7046.
104. van der Kloet, F.M. et al. (2009) *J. Proteome Res.*, **8** (11), 5132–5141.
105. http://fiehnlab.ucdavis.edu/staff/kind/Metabolomics-Survey-2009/ (last accessed October 6th 2012).
106. Sumner, L.W. et al. (2007) *Metabolomics*, **3** (3), 211–221.
107. Wishart, D.S. et al. (2009) *Nucleic Acids Res.*, **37** (Database issue), D603–D610.
108. Herrgard, M.J. et al. (2008) *Nat. Biotechnol.*, **26** (10), 1155–1160.
109. Cui, Q. et al. (2008) *Nat. Biotechnol.*, **26** (2), 162–164.
110. http://chemdata.nist.gov/mass-spc/ms-search/ (last accessed October 6th 2012).
111. Kopka, J. et al. (2005) *Bioinformatics*, **21** (8), 1635–1638.
112. Kind, T. et al. (2009) *Anal. Chem.*, **81** (24), 10038–10048.
113. Haken, J.K. (1976) *Adv. Chromatogr.*, **14**, 367–407.
114. Hummel, J. et al. (2010) *Metabolomics*, **6** (2), 322–333.
115. Mihaleva, V.V. et al. (2009) *Bioinformatics*, **25** (6), 787–794.
116. http://www.lipidmaps.org/ (last accessed October 6th 2012).
117. http://www.genome.jp/kegg/ (last accessed October 6th 2012).
118. http://www.hmdb.ca/ (last accessed October 6th 2012).
119. http://pubchem.ncbi.nlm.nih.gov/ (last accessed October 6th 2012).
120. http://www.chemspider.com/ (last accessed October 6th 2012).
121. Brown, M. et al. (2011) *Bioinformatics*, **27** (8), 1108–1112.
122. Scheltema, R.A. et al. (2011) *Anal. Chem.*, **83** (7), 2786–2793.

123. Kuhl, C. et al. (2011) *Anal. Chem.*, **84** (1), 283–289.
124. Kind, T. and Fiehn, O. (2007) *BMC Bioinformatics*, **8**, 105.
125. Rogers, S. et al. (2009) *Bioinformatics*, **25** (4), 512–518.
126. Weber, R.J.M. and Viant, M.R. (2010) *Chemom. Intell. Lab. Syst.*, **104** (1), 75–82.
127. Horai, H. et al. (2010) *J. Mass Spectrom.*, **45** (7), 703–714.
128. Sana, T.R. et al. (2008) *J. Biomol. Tech.*, **19** (4), 258–266.
129. van der Hooft, J.J. et al. (2010) *Anal. Chem.*, **83** (1), 409–416.
130. Hoopmann, M.R. et al. (2009) *J. Proteome Res.*, **8** (4), 1870–1875.
131. http://www.acdlabs.com/products/adh/ms/ms_frag/ (last accessed October 6th 2012).
132. http://www.highchem.com/massfrontier/mass-frontier.html (last accessed October 6th 2012).
133. Wolf, S. et al. (2010) *BMC Bioinformatics*, **11**, 148.
134. Heinonen, M. et al. (2008) *Rapid Commun. Mass Spectrom.*, **22** (19), 3043–3052.
135. Creek, D.J. et al. (2011) *Anal. Chem.*, **83** (22), 8703–8710.
136. Crockford, D.J. et al. (2008) *Anal. Chem.*, **80** (18), 6835–6844.
137. D'Alessandro, A. and Zolla, L. (2012) *Drug Discov. Today*, **17** (1-2), 3–9.
138. Rabinowitz, J.D. et al. (2012) *Cold Spring Harb. Symp. Quant. Biol.*, doi:10.1101/sqb2011.76.010694.
139. Wishart, D.S. (2008) *Drugs R D*, **9** (5), 307–322.
140. Kell, D.B. (2006) *Drug Discov. Today*, **11** (23-24), 1085–1092.
141. Kim, K. et al. (2011) *OMICS*, **15** (5), 293–303.
142. Lewis, G.D. et al. (2008) *J. Clin. Invest.*, **118** (10), 3503–3512.
143. van der Kloet, F.M. et al. (2012) *Metabolomics*, **8** (1), 109–119.
144. Zhao, T. et al. (2012) *J. Pharm. Biomed. Anal.*, **60**, 32–43.
145. Han, X. et al. (2011) *PLoS ONE*, **6** (7), e21643.
146. Manna, S.K. et al. (2011) *J. Proteome Res.*, **10** (9), 4120–4133.
147. Legido-Quigley, C. et al. (2011) *Electrophoresis*, **32** (15), 2063–2070.
148. Beger, R.D., Sun, J., and Schnackenberg, L.K. (2009) *Toxicol. Appl. Pharmacol.*, **243** (2), 154–166.
149. Morvan, D. and Demidem, A. (2007) *Cancer Res.*, **67** (5), 2150–2159.
150. van Velzen, E.J. et al. (2009) *J. Proteome Res.*, **8** (7), 3317–3330.
151. van Duynhoven, J. et al. (2010) *Proc. Natl. Acad. Sci. U.S.A.*, **108** (Suppl 1), 4531–4538.
152. Chorell, E. et al. (2009) *J. Proteome Res.*, **8** (6), 2966–2977.
153. Cleves, M.A. and Hobbs, C.A. (2004) *J. Matern. Fetal Neonatal Med.*, **15** (1), 35–38.
154. Nel-Themaat, L. and Nagy, Z.P. (2011) *Placenta*, **32** (Suppl 3), S257–S263.
155. Neilson, J.P. et al. (2003) *Br. Med. Bull.*, **67**, 191–204.
156. Myers, J.E. and Brockelsby, J. (2004) The epidemiology of pre-eclampsia, in *Pre-eclampsia: Current Perspectives on Management* (eds P.N. Baker and J.C.P. Kingdom), Parthenon, London.
157. Tucker, J. and McGuire, W. (2004) *Br. Med. J.*, **329** (7467), 675–678.
158. Horgan, R.P. et al. (2011) *J. Proteome Res.*, **10** (8), 3660–3673.
159. Kenny, L.C. et al. (2008) *Reprod. Sci.*, **15** (6), 591–597.
160. Graca, G. et al. (2009) *J. Proteome Res.*, **8** (8), 4144–4150.
161. Romero, R. et al. (2010) *J. Matern. Fetal Neonatal Med.*, **23** (12), 1344–1359.
162. Diaz, S.O. et al. (2011) *J. Proteome Res.*, **10** (8), 3732–3742.
163. Horgan, R.P. et al. (2010) *Placenta*, **31** (10), 893–901.
164. Tissot van Patot, M.C. et al. (2010) *Am. J. Physiol. Regul. Integr. Comp. Physiol.*, **298** (1), R166–R172.
165. www.who.int/Mediacentre/Factsheets/Fs317/En/Index.html (last accessed October 6th 2012).
166. www.who.int/Mediacentre/Factsheets/Fs312/En/Index.html (last accessed October 6th 2012).
167. Bassand, J.P. et al. (2007) *Eur. Heart J.*, **28** (13), 1598–1660.
168. Sabatine, M.S. et al. (2005) *Circulation*, **112** (25), 3868–3875.

169. Barba, I. et al. (2008) *Magn. Reson. Med.*, **60** (1), 27–32.
170. Lin, H., Zhang, J., and Gao, P. (2009) *J. Clin. Lab. Anal.*, **23** (1), 45–50.
171. Chen, X. et al. (2010) *J. Sep. Sci.*, **33** (17-18), 2776–2783.
172. Ciborowski, M. et al. (2010) *J. Proteome Res.*, **10** (3), 1374–1382.
173. Rhee, E.P. et al. (2011) *J. Clin. Invest.*, **121** (4), 1402–1411.
174. Zhao, X. et al. (2010) *Metabolomics*, **6** (3), 362–374.
175. Lucio, M. et al. (2010) *PLoS ONE*, **5** (10), e13317.
176. Teul, J. et al. (2009) *J. Proteome Res.*, **8** (12), 5580–5589.
177. Mayr, M. et al. (2005) *Arterioscler. Thromb. Vasc. Biol.*, **25** (10), 2135–2142.
178. Turer, A.T. et al. (2009) *Circulation*, **119** (13), 1736–1746.
179. Alexander, D. et al. (2010) *Eur. J. Clin. Invest.*, **41** (5), 527–538.
180. Wang, Z. et al. (2011) *Nature*, **472** (7341), 57–63.
181. Atherton, H.J. et al. (2006) *Physiol. Genomics*, **27** (2), 178–186.
182. Koves, T.R. et al. (2008) *Cell Metab.*, **7** (1), 45–56.
183. Shaham, O. et al. (2008) *Mol. Syst. Biol.*, **4**, 214.
184. Kaddurah-Daouk, R. et al. (2010) *Metabolomics*, **6** (2), 191–201.
185. Li, N. et al. (2009) *J. Mol. Cell Cardiol.*, **47** (6), 835–845.
186. Gu, Y. et al. (2010) *Talanta*, **81** (3), 766–772.
187. Zhang, H. et al. (2008) *Am. J. Physiol. Renal Physiol.*, **295** (4), F1071–F1081.
188. Patti, M.E. et al. (1998) *J. Clin. Invest.*, **101** (7), 1519–1529.
189. Tremblay, F. and Marette, A. (2001) *J. Biol. Chem.*, **276** (41), 38052–38060.
190. He, M., Lu, L.M., and Yao, T. (2004) *Sheng Li Ke Xue Jin Zhan*, **35** (2), 188–192.
191. Brosens, I. et al. (2011) *Am. J. Obstet. Gynecol.*, **204** (3), 193–201.

15
Metabolomics in Biotechnology (Microbial Metabolomics)
Marco Oldiges, Stephan Noack, and Nicole Paczia

15.1
Introduction

The first documentation of the term *metabolomics* goes back to two independent publications in 1998 and was used in the context of describing the metabolic behavior of microbial systems, showing a distinct phenotype regarding their physiology and extracellular metabolite spectrum [1, 2]. For sure, these were not the first publications with such research activity, but these were the ones which obviously coined the term *metabolomics*. Later on, this term was picked up by the scientific community not only for the microbial field but also for many other scientific areas and paved its way to become one of the representatives of the omics era. In this sense, the suffix -ome is "used to direct attention to holistic abstractions, an eventual goal, of which only a few parts may be initially in hand" [3] and its origin goes back to the early twentieth century. Initially, the German botanist Hans Winkler is attributed to have coined the term *genome* by the combination of the terms *gene* and *chromosome* (Greek word *soma* = body) [4]. This keyword functioned as blueprint for the many other -omes that are currently investigated. Finally, they all go back to the Greek word *soma* and in opposite to the genome this term seems to be not precisely fitting for the metabolome, as the metabolites are not located or arranged in a body, but are distributed all over the system.

Since 1998, the term *metabolomics* was increasingly used in the literature showing significant increase (Figure 15.1) and indicating the fast development of the field.

The field of microbial metabolomics can be subdivided into the investigation of the endo- and exo-metabolome (intra- vs extracellular) as well as different approaches such as target analysis, metabolic profiling, and metabolic finger-/footprinting (Figure 15.2) [5, 6]. On the one hand, the target analysis aims at the quantitative analysis of substrate and/or product metabolites of a given metabolic conversion, whereas metabolic profiling focuses on the quantitative analysis of a set of predefined metabolites belonging to a class of compounds or members of particular pathways or a linked group of metabolites (e.g., sugars, sugar phosphates, lipids, organic acids). On the other hand, metabolic

Metabolomics in Practice: Successful Strategies to Generate and Analyze Metabolic Data, First Edition.
Edited by Michael Lämmerhofer and Wolfram Weckwerth.
© 2013 Wiley-VCH Verlag GmbH & Co. KGaA. Published 2013 by Wiley-VCH Verlag GmbH & Co. KGaA.

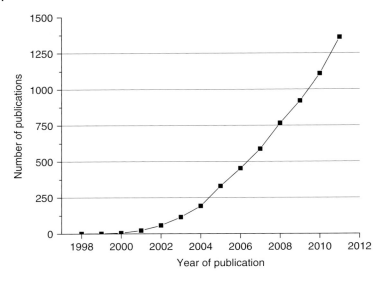

Figure 15.1 Usage of the term *metabolomics* since its coining in 1998, represented by the number of publications after search for the word "metabolom*" in the web of knowledge database (Thomson/Reuters).

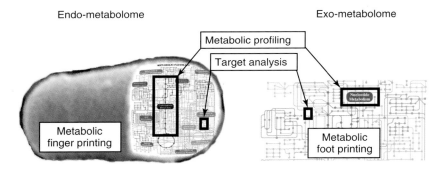

Figure 15.2 Classification of metabolomics into different fields. All approaches are used for microbial metabolomics. (With kind permission from Springer Science + Business Media: figure 2 from Ref. [5].)

fingerprinting and metabolic footprinting are concerned with the semiquantitative analysis of a comprehensive number of intracellular and extracellular metabolites, respectively.

These approaches differ regarding the number of metabolites in the research focus (i.e., comprehensiveness of the study), their localization (endo- or exo-metabolome), and aimed information content of the data (qualitative, semiquantitative, quantitative). Qualitative approaches only aim at identification of certain metabolites to verify their presence, whereas the semiquantitative approach uses the relative change of the detector signal, to estimate the relative increase or decrease of certain metabolites during comparison of a set of samples from different strain phenotypes. The quantitative analysis aims to get the accurate concentration value employing methods for correction of analytical and experimental errors. In this context, the sampling and sample processing procedure is very critical to generate samples which are a valid representation of the physiological state of the microbial culture in the cultivation vessel before sampling [7, 8].

It is well known that turnover rates of metabolites from central carbon metabolism have been measured to be at least in the range of seconds, but are sometimes even faster at the subsecond level. This strictly requires the application of reliable procedures to stop metabolic activity in the cells immediately and to prevent further metabolic conversion [9]. Such approaches are typically referred to as *metabolic quenching methods*, employing harsh acidic or basic conditions, liquid nitrogen or deep frozen organic solvents (e.g., cold methanol quenching) to name only a few [10]. The application of such conditions can negatively influence the cell integrity during sample processing causing the so-called metabolite leakage or complete mixing of intracellular and extracellular metabolites. Thus, the choice of the sampling method can have a pivotal effect on the final metabolome result (Figure 15.3). The sample preparation is the first important step to generate a valid sample which is suitable for quantitative metabolomics (for further details see Chapter 1 of this book). The second step is the analytical determination of the correct amount of a single metabolite or a more comprehensive set of metabolites. To achieve this, isotope-labeled reference compounds are typically used for isotope-dilution mass spectrometry [11] and other internal standards can

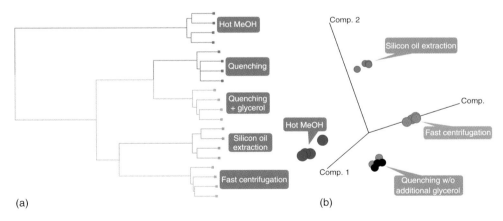

Figure 15.3 Comparison of different protocols for sampling and sample processing illustrating their pivotal impact on the final metabolome result. Visualization of GC-TOF-MS metabolome data analyzed by simple statistical analysis and displayed by (a) hierarchical cluster analysis (HCA) and (b) principal component analysis (PCA). Samples for all processing procedures were taken from a common bioreactor cultivation of *Corynebacterium glutamicum*.

be used to correct analytical and experimental errors originating from sample processing and analytical measurement (for further details see Chapters 3 and 12 of this book).

It seems that microbial metabolomics investigations are often done in a more qualitative and semiquantitative manner. This is not surprising as the correction of almost all experimental and analytical errors is typically difficult, laborous, and time-consuming. Hence, the throughput in quantitative investigations must be lower as more hardware measurement time and personnel resources are required to perform it. For this reason, it looks like that the qualitative or semiquantitative approach seems to produce more results in less amount of time. Of course this is not true, but it points to the necessity to define the required type of metabolomics data and their quality to serve the intended application or research.

For a rough microbial phenotyping, it could be sufficient to use qualitative and semiquantitative data if the expected difference in the metabolome of a mutant strain in reference to the parental strain is significant enough [12, 13]. Nevertheless, the same information quality is usually not sufficient to serve for mathematical modeling and validation of a mechanistic metabolic model. Such an application demands for strictly quantitative data with sufficient accuracy and precision.

To conclude, performing a microbial metabolomics study requires the initial decision about the aim of the study, because this can lead to restrictions regarding experimental and analytical constraints for sampling, sample processing, and analytical procedures (cf. Chapters 1, 3, and 12 of this book). Strictly speaking, an approach providing an extremely comprehensive data set for a large mutant strain collection with several hundreds of metabolite entries of relative peak intensity changes for each strain will be hardly a benefit for mechanistic kinetic metabolic modeling, but of high relevance to understand genotype–phenotype relation of a mutant strain collection. A very detailed and focused data set for only a few strains with limited number of metabolites precisely quantified could be of high relevance for thorough mechanistic metabolic modeling [14, 15].

15.2
Analytical Methods Applied for Microbial Metabolomics

The emerging field of the instrumental analytics made significant technological improvements, which also directly influences the field of microbial metabolomics. A long time before sophisticated analytical hardware became available, enzymatic assays were used to determine microbial metabolite concentrations [16]. They make use of the substrate specificity of enzymes to overcome selectivity issues because of the biological matrix and presence of all other metabolites in the same sample. By coupling of enzymes to reaction cascades or nonnatural reaction partners and conditions, thermodynamic limitations of reaction equilibria were overcome to allow stoichiometric coupling of distinct metabolite concentration to the consumption or generation of, for example, reduced or oxidized redox cofactors [16]. While offering the determination of a broad range of metabolites with standard laboratory equipment the major drawback is the necessity for pure enzymes ideally from commercial sources and the low throughput and parallelization option, as each metabolite requires the conduction of a unique enzyme assay.

The chromatography based methods, that is, liquid chromatography (LC) and gas chromatography (GC) as well as capillary electrophoresis (CE), principally best suited for sample automation, suffer from insufficient selectivity in the metabolite detection, and strong matrix interference when using standard optical or electrochemical detector setups. Combination with mass spectrometry is nowadays a key to success for many research activities in the field of microbial metabolomics [5, 10]. Among the multiple options to couple chromatography to available ionization sources and mass spectrometric detector types, it seems like almost every combination have been applied for looking at microbial metabolites. Among them are GC-MS and LC-MS approaches employing various types of MS detectors [5, 6, 17]. For LC-MS, the connection to detectors with superior resolution and mass accuracy, that is, Orbitrap-MS and FT-ICR-MS detector technology have been described providing much more structural information about the metabolites and improved peak identification opportunities, although such detectors suffer from reduced sensitivity [18]. Also for CE many successful applications show that various MS detector types can be used efficiently for this separation technology [19]. To analyze the metabolites present in the gas phase above microbial cultures ion-mobility MS has been successfully employed to statistically analyze characteristic metabolite pattern as well as targeted analysis of metabolic compounds [20].

The significantly reduced sensitivity is also a weak point for the application of NMR technology for microbial metabolomics. Although it is already in use for a long time for investigating changes of metabolite concentrations in microbial samples, its application is hampered by the low concentration of metabolites and the necessity to monitor NMR-active nuclei. This is done for phosphorylated compounds as ^{31}P is the exclusive naturally occurring isotope [21, 22], while ^{13}C needs to be incorporated from stable isotope labeling experiments using cost intensive substrates. In contrast to all other mentioned measurement techniques, the NMR offers the possibility for noninvasive metabolite determination providing the opportunity to use it as online tool during cultivation experiments [23, 24]. In contrast to MS tools, the NMR technique can provide the full spectrum of positional labeling information which is a substantial benefit compared to MS data delivering usually mass isotopomer data or only limited access to the positional labeling information [25, 26].

The application of enzyme sensors based on fluorescence resonance energy transfer (FRET) technology has gained increasing interest also in the microbial field [27]. FRET sensors which are sensitive against intracellular metabolites have been widely used mainly for qualitative analysis of presence or time course of metabolite concentrations because of metabolic, regulatory, or genetic changes. More and more enzyme FRET sensors have become available specifically targeted against metabolites from many parts of metabolism. This FRET-based approach for metabolomics can offer advantage over standard analytical approaches, because the sensor is directly placed inside the biological system and could allow noninvasive optical measurement of metabolites [28, 29]. Nevertheless, for developing FRET to become a valid quantitative tool for microbial metabolomics, some important basic issues of analytics regarding calibration, accuracy, and precision of the sensor under the intracellular conditions seem still to be investigated.

15.3
Custom-Made Separation for Microbial Metabolomics

Nowadays, it is common sense that ion suppression is a general problem of LC-MS techniques [30]. In order to reduce the underlying matrix interferences several strategies have been proposed concentrating mainly on the MS part [31–33]. However, if high sensitivity is needed, as true for intracellular metabolite analysis, the only way of minimizing ion suppression is to improve the sample quality by better sample preparation or higher selectivity of the LC part. Especially in the latter case, a straightforward approach is to adjust the chromatographic conditions for a given column in such a way, that the analyte peaks are not eluting in regions of ion suppression [34]. This can be realized by modifying the mobile phase strength or gradient conditions as well as the addition of chromatographic reagents such as tributylamine [35]. However, for the same reason as for biological matrix components, such additives can cause ion suppression [36, 37].

Interestingly, selectivity optimization for a certain LC method is usually based on a single stationary phase material, which is commercially available and hence more or less specific for the separation problem of interest. Clearly, this is mainly due to the fact that the mobile phase and all operating parameters can be modified in a continuous manner generating "intermediate" separation and selectivity [38]. In contrast, adapting the stationary phase is much more demanding as it requires to figure out the right composition of different column materials or serial coupling of single columns. A standard approach for stationary phase optimization is the use of a so-called method development kit containing different very short columns to investigate the retention behavior of analytes on different stationary phase materials. Such approach is usually taken to select the most suitable stationary phase material, followed by mobile phase optimization, although best performance would have been possible using a mixture of more than one of the stationary phase materials.

Recently, a more convenient technique called "phase optimized liquid chromatography" (POP-LC) was proposed, which allows a speedup of the development and optimization of stationary phase compositions [39]. In general, the method utilizes a set of column modules with typical stationary materials with alternative phase chemistry covering a broad range of retention mechanisms. The column modules can be coupled via connectors in a dead volume free manner to generate diverse combinations of column materials and lengths. The technical part is complemented by the POP-LC software tool that allows calculation of the optimal stationary phase composition of the column on the basis of predefined experiments. The approach clearly aims at the development and application of custom-made chromatographic columns tailored to specific metabolomics applications. Besides the many established and newly developed stationary phase materials, the tailor-made combination of stationary phase materials is expected to show a significant impact for analytical method development and application to the field of microbial metabolomics.

15.4
Microbial Metabolomics with Higher Throughput

In the previous years, the typical motivation for further method development was to increase data quality or number of metabolites that could be measured using a specific method. With increasing investigation of cellular metabolism on the metabolome level in the context of systems biology and necessity for very detailed look inside metabolic pathways of microbial cells, there is a strong pull effect from application side to improve the data quality as well as the throughput in metabolomics investigations.

The time necessary for chromatographic metabolite separation is the time critical step of the overall analysis and, thus, is the primary target for optimization. To process more samples in a given time can be achieved by at least two options: (i) chromatographic separation performance could be increased with concomitant reduction in run times or (ii) chromatographic separation could be completely eliminated from the analysis. Besides typical steps in optimization of chromatographic performance especially the application of Ultra-HPLC (UHPLC) [40–42] and tailor-made columns for specialized shortened applications seem to provide substantial potential to speedup analysis. To enhance the duty cycle of the chromatographic system in front of the detector, also multiplexing is an option to enhance overall throughput [43].

Avoidance of any chromatographic separation will provide the maximum sample throughput, as measurement time is limited by the data acquisition of the detector only. For nontargeted metabolomics, flow injection analysis (FIA) coupled to TOF-MS [44] have been applied for rapid microbial phenotyping of strain collections. Although such FIA approaches will most likely increase the measurement error of the data obtained the acceleration of the analysis is tremendous and also quantitative metabolomics applications using MS-based measurements to investigate *in vitro* pathways have been described [45]. Although MALDI-TOF is usually used for peptide and protein analysis, it lacks chromatography and showed principal applicability for metabolite measurements [46].

The increase of sample throughput is accompanied with a proportional increase of amount of data, which is further elevated if metabolites can be present in different labeling states representing a further set of metabolites. Although multicomponent analytical methods are state-of-the-art technology, the complex microbial samples often show peaks not completely resolved or with nonideal peak symmetry interfering with automated peak detection and integration algorithms. Although this can be compensated by tuning integration parameters in the software, column performance changes between sets of samples and cumulative loss in chromatographic performance because of matrix interference can cause slowdown of raw data processing. In fact, such issues can be easily handled manually if only a few samples need to be processed, but for a comprehensive set of microbial metabolomics samples this can be a time-consuming effort, diminishing the advantage of higher throughput analytics.

15.5
Application of Microbial Metabolomics

In the post genome era, where genome sequencing has become sometimes more a service than a research issue, the most important questions of how microbial metabolism is operating seem still to be answered. Nevertheless, the fundamental knowledge of the genome sequence is the essential basis for microbial research activities. Compared to other information levels accessible through analytical omics technologies there is one special issue in the application of metabolomics which shows a clear difference. Compared to other omics data, the metabolites show the fastest response time to changes in the environment [47]. Metabolite concentrations can change in the subsecond time frame, so that efficient and smart methodologies have to be applied, to determine the metabolite pattern or concentrations as accurate as possible [9]. Such methods are often called *metabolic quenching methods* and have been described in various ways, although the "gold standard" technology is still to be developed. The necessity to use metabolic quenching steps might be reduced in cases where high throughput approaches are used to characterize a comprehensive microbial mutant collection [44]. In such application, very strong changes in the metabolite pattern are most likely still detectable independent of the sample processing. Nevertheless, data for smaller changes in the metabolome pattern might get lost, because of inadequate sample processing leading to data sets with reduced data quality and information content.

The application of metabolomics for microbial systems follows the sections as is shown in Figure 15.2. Metabolic fingerprinting [10] as well as footprinting [48, 49] are usually associated with high throughput technologies and therefore requires the measurement with MS instrumentation often without any separation technology to maintain sufficient throughput [44]. Please note that such approaches are hardly able to deliver valid and correct absolute values. For the sake of short analysis time, the quantitative aspect has to be subordinated.

For application of metabolomics in the field of metabolic profiling or distinct analysis of a target metabolite (Figure 15.2), the demand for sufficient data quality automatically leads to a drastically reduced throughput, due to necessity of doing replicate measurements and correction of analytical and experimental errors. Such quantitative data can serve as a basis for metabolic modeling [50] and thermodynamic considerations of quantitative metabolome data [51, 52]. In order to be able to monitor concentration-time-courses, people started to construct fast sampling devices which allowed sampling time below 1 s [53]. Such devices were applied for stimulus–response experiments where the microbial culture was disturbed by, for example, a substrate pulse and the rapid concentration changes of the metabolites was detected by fast sampling and simultaneous metabolic quenching [9]. For the first data sets, extensive enzymatic assays were performed to determine the concentration levels, which limited analysis to only few metabolites because of the required large sample volume [15, 54, 55]. Later on, LC-MS [35, 56, 57] and GC-MS [58] were used for multiple metabolite determination and allowed miniaturization of the sampling devices [14, 59] because of minimal sample requirement for the MS analysis.

Improved knowledge about necessary analytical [11] and experimental [10] corrections in recent years further improved the quantitative metabolome measurements and generation of quantitative multiomics data sets [60], allowing a detailed model-based description of microbial metabolism and derived useful model-based prediction. In combination with other data from transcriptomics, proteomics, fluxomics and so on, the metabolome data sets can be regarded as extremely useful in the context of systems biology.

With the technology switch from enzyme assays to MS devices not only access to the metabolite concentration is possible but also to the isotopic labeling state. Owing to the natural presence of the ^{13}C isotope of about 1.1% all carbon containing metabolites show a natural isotope labeling distribution. For determination of the intracellular reaction rates, specifically ^{13}C-labeled glucose substrate mixtures are applied for ^{13}C metabolic flux analysis [61]. Instead of analyzing the ^{13}C-labeling in the amino acids of hydrolyzed biomass, the measurement of the labeling state of the free intracellular metabolites gives a direct insight to the metabolite labeling state [62–64]. Moreover, access to labeling information of primary metabolites was also combined with stimulus-response experiments with ^{13}C-substrate resulting in data providing dynamic labeling information at short time scale [65] enabling isotopically nonstationary ^{13}C metabolic flux analysis [66]. Combination of high throughput together with ^{13}C-labeling technology resulted in a series of fluxome data [67, 68].

Besides the above-mentioned application areas, metabolomics has found widespread application in the field of metabolic and bioprocess engineering, microbial physiology and functionality of metabolic networks [5, 6] and also synthetic biology [69]. Successful applications cover a broad range of research topics and microbial systems, for example, *in vivo* NMR [23], optimization of amino acid formation [70, 71], investigation of ethanol formation in yeast [72] and beta-lactam antibiotics [73–75].

Despite its importance in pharmaceutical industry concerning drug target and drug metabolism [76], the application of metabolomics in industrial biotechnology seems to be not as widespread as, for example, genome sequencing or transcriptome technology. Compared to metabolome analysis such technologies are much more matured and hence easier applicable in the industrial context. Nevertheless, there are examples clearly showing how metabolomics and other omics technologies can successfully contribute [77–80].

An emerging field is the development of single cell metabolomics. The current sample processing and analytical technologies only allow monitoring an averaged metabolite level representing the mean value of the whole cell population rather than information of individual cells. With developments in the field of microfluidics, tools become available for cultivation and manipulation (e.g., sorting) of cells at single cell level [81–85]. Combined with miniaturized analytics the assessment of metabolites at single cell level proves to offer perspective for successful future application [86–88].

15.6
Conclusion

There is no doubt that microbial metabolomics has become an established tool for a diversified application to investigate microbial metabolism and physiology, metabolic engineering and bioprocess optimization, systems biology, industrial biotechnology, and many more topics. Although many success stories and applications are found in literature it is obvious that metabolomics is not as matured as other omics technologies and most applications still belong to the field of academic research. One reason for this can be seen in the lack of a "gold standard" for sample processing, unifying integrity of the microbial cell and metabolic quenching. Such challenge does affect the detailed quantitative metabolite analysis more severely than the semiquantitative approaches with higher throughput. Nevertheless, there is a very strong demand for high quality quantitative metabolome data for metabolic modeling application. In combination with other omics data such data will be used for improved description of microbial cells and their small metabolic ingredients as well as the function of these parts in the concert of the whole cell.

References

1. Oliver, S.G., Winson, M.K., Kell, D.B., and Baganz, F. (1998) *Trends Biotechnol.*, **16** (9), 373–378.
2. Tweeddale, H., Notley-McRobb, L., and Ferenci, T. (1998) *J. Bacteriol.*, **180** (19), 5109–5116.
3. Lederberg, J. and McCray, A.T. (2001) *Scientist*, **15** (7), 8–8.
4. Winkler, H. (1920) *Verbreitung und Ursache der Pathogenesis im Pflanzen- und Tierreich*, Fischer Verlag, Jena.
5. Oldiges, M., Lutz, S., Pflug, S., Schroer, K., Stein, N., and Wiendahl, C. (2007) *Appl. Microbiol. Biotechnol.*, **76** (3), 495–511.
6. Mashego, M.R., Rumbold, K., De Mey, M., Vandamme, E., Soetaert, W., and Heijnen, J.J. (2007) *Biotechnol. Lett.*, **29** (1), 1–16.
7. Canelas, A.B., Ras, C., ten Pierick, A., van Dam, J.C., Heijnen, J.J., and Van Gulik, W.M. (2008) *Metabolomics*, **4** (3), 226–239.
8. Wellerdiek, M., Winterhoff, D., Reule, W., Brandner, J., and Oldiges, M. (2009) *Bioprocess Biosyst. Eng.*, **32** (5), 581–592.
9. van Gulik, W.M. (2010) *Curr. Opin. Biotechnol.*, **21** (1), 27–34.
10. Villas-Boas, S.G., Mas, S., Akesson, M., Smedsgaard, J., and Nielsen, J. (2005) *Mass Spectrom. Rev.*, **24** (5), 613–646.
11. Wu, L., Mashego, M.R., van Dam, J.C., Proell, A.M., Vinke, J.L., Ras, C., van Winden, W.A., van Gulik, W.M., and

Heijnen, J.J. (2005) *Anal. Biochem.*, **336** (2), 164–171.

12. Strelkov, S., von Elstermann, M., and Schomburg, D. (2004) *Biol. Chem.*, **385** (9), 853–861.
13. Buchinger, S., Strosser, J., Rehm, N., Hanssler, E., Hans, S., Bathe, B., Schomburg, D., Kramer, R., and Burkovski, A. (2009) *J. Biotechnol.*, **140** (1–2), 68–74.
14. Mashego, M.R., van Gulik, W.M., Vinke, J.L., Visser, D., and Heijnen, J.J. (2006) *Metab. Eng.*, **8** (4), 370–383.
15. Chassagnole, C., Noisommit-Rizzi, N., Schmid, J.W., Mauch, K., and Reuss, M. (2002) *Biotechnol. Bioeng.*, **79** (1), 53–73.
16. Bergmeyer, H. (1984) *Methods of Enzymatic Analysis*, vols. 6 and 7, 3rd edn, Wiley-VCH Verlag GmbH, Weinheim.
17. Koek, M.M., Jellema, R.H., van der Greef, J., Tas, A.C., and Hankemeier, T. (2011) *Metabolomics*, **7** (3), 307–328.
18. Breitling, R., Pitt, A.R., and Barrett, M.P. (2006) *Trends Biotechnol.*, **24** (12), 543–548.
19. Monton, M.R.N. and Soga, T. (2007) *J. Chromatogr., A*, **1168** (1–2), 237–246.
20. Juenger, M., Vautz, W., Kuhns, M., Hofmann, L., Ulbricht, S., Baumbach, J.I., Quintel, M., and Perl, T. (2012) *Appl. Microbiol. Biotechnol.*, **93** (6), 2603–2614.
21. Gonzalez, B., de Graaf, A., Renaud, M., and Sahm, H. (2000) *Yeast*, **16** (6), 483–497.
22. De Graaf, A.A., Striegel, K., Wittig, R.M., Laufer, B., Schmitz, G., Wiechert, W., Sprenger, G.A., and Sahm, H. (1999) *Arch. Microbiol.*, **171** (6), 371–385.
23. Andersen, A.Z., Carvalho, A.L., Neves, A.R., Santos, H., Kummer, U., and Olsen, L.F. (2009) *Comput. Biol. Chem.*, **33** (1), 71–83.
24. Fonseca, C., Neves, A.R., Antunes, A.M.M., Noronha, J.P., Hahn-Hagerdal, B., Santos, H., and Spencer-Martins, I. (2008) *Appl. Environ. Microbiol.*, **74** (6), 1845–1855.
25. de Graaf, A.A., Mahle, M., Mollney, M., Wiechert, W., Stahmann, P., and Sahm, H. (2000) *J. Biotechnol.*, **77** (1), 25–35.
26. Petersen, S., de Graaf, A.A., Eggeling, L., Mollney, M., Wiechert, W., and Sahm, H. (2000) *J. Biol. Chem.*, **275** (46), 35932–35941.
27. Frommer, W.B., Davidson, M.W., and Campbell, R.E. (2009) *Chem. Soc. Rev.*, **38** (10), 2833–2841.
28. Bermejo, C., Haerizadeh, F., Takanaga, H., Chermak, D., and Frommer, W.B. (2011) *Nat. Protoc.*, **6** (11), 1806–1817.
29. Bermejo, C., Haerizadeh, F., Takanaga, H., Chermak, D., and Frommer, W.B. (2010) *Biochem. J.*, **432**, 399–406.
30. Jessome, L.L. and Volmer, D.A. (2006) *LCGC*, **24** (5), 498–511.
31. Souverain, S., Rudaz, S., and Veuthey, J.L. (2004) *J. Chromatogr., A*, **1058** (1–2), 61–66.
32. Holcapek, M., Volná, K., Jandera, P., Kolárová, L., Lemr, K., Exner, M., and Církva, A. (2004) *J. Mass Spectrom.*, **39** (1), 43–50.
33. Gangl, E.T., Annan, M.M., Spooner, N., and Vouros, P. (2001) *Anal. Chem.*, **73** (23), 5635–5644.
34. Nelson, M.D. and Dolan, J.W. (2002) *LCGC*, **15**, 73–79.
35. Luo, B., Groenke, K., Takors, R., Wandrey, C., and Oldiges, M. (2007) *J. Chromatogr., A*, **1147** (2), 153–164.
36. Temesi, D. and Law, B. (1999) *LCGC*, **17** (7), 627–632.
37. Enke, C.G. (1997) *Anal. Chem.*, **69** (23), 4885–4893.
38. Nyiredy, S., Szucs, Z., and Szepesy, L. (2007) *J. Chromatogr., A*, **1157** (1–2), 122–130.
39. Lamotte, S. (2006) *Nachr. Chem.*, **54** (4), 439–440.
40. Wang, X.J., Sun, H., Zhang, A.H., Wang, P., and Han, Y. (2011) *J. Sep. Sci.*, **34** (24), 3451–3459.
41. Guillarme, D., Ruta, J., Rudaz, S., and Veuthey, J.L. (2010) *Anal. Bioanal. Chem.*, **397** (3), 1069–1082.
42. Buescher, J.M., Moco, S., Sauer, U., and Zamboni, N. (2010) *Anal. Chem.*, **82** (11), 4403–4412.
43. Trapp, O. (2007) *Angew. Chem. Int. Ed.*, **46** (29), 5609–5613.
44. Fuhrer, T., Heer, D., Begemann, B., and Zamboni, N. (2011) *Anal. Chem.*, **83** (18), 7074–7080.
45. Bujara, M., Schumperli, M., Pellaux, R., Heinemann, M., and Panke, S. (2011) *Nat. Chem. Biol.*, **7** (5), 271–277.

46. Wittmann, C. and Heinzle, E. (2001) *Biotechnol. Bioeng.*, **72** (6), 642–647.
47. Oldiges, M. and Takors, R. (2005) *Technology Transfer in Biotechnology: From Lab to Industry to Production*, Springer-Verlag, Berlin, pp. 173–196.
48. Mapelli, V., Olsson, L., and Nielsen, J. (2008) *Trends Biotechnol.*, **26** (9), 490–497.
49. Pope, G.A., MacKenzie, D.A., Defemez, M., Aroso, M., Fuller, L.J., Mellon, F.A., Dunn, W.B., Brown, M., Goodacre, R., Kell, D.B., Marvin, M.E., Louis, E.J., and Roberts, I.N. (2007) *Yeast*, **24** (8), 667–679.
50. Wiechert, W. and Noack, S. (2011) *Curr. Opin. Biotechnol.*, **22** (5), 604–610.
51. Zamboni, N., Kuemmel, A., and Heinemann, M. (2008) *BMC Bioinformatics*, **9**, 199.
52. Kuemmel, A., Panke, S., and Heinemann, M. (2006) *Mol. Syst. Biol.*, **2**, 1–10.
53. Schaedel, F. and Franco-Lara, E. (2009) *Appl. Microbiol. Biotechnol.*, **83** (2), 199–208.
54. Theobald, U., Mailinger, W., Baltes, M., Rizzi, M., and Reuss, M. (1997) *Biotechnol. Bioeng.*, **55** (2), 305–316.
55. Schaefer, U., Boos, W., Takors, R., and Weuster-Botz, D. (1999) *Anal. Biochem.*, **270** (1), 88–96.
56. van Dam, J.C., Eman, M.R., Frank, J., Lange, H.C., van Dedem, G.W.K., and Heijnen, S.J. (2002) *Anal. Chim. Acta*, **460** (2), 209–218.
57. Seifar, R.M., Zhao, Z., van Dam, J., van Winden, W., van Gulik, W., and Heijnen, J.J. (2008) *J. Chromatogr., A*, **1187** (1–2), 103–110.
58. Hofmann, U., Maier, K., Nicbel, A., Vacun, G., Reuss, M., and Mauch, K. (2008) *Biotechnol. Bioeng.*, **100** (2), 344–354.
59. Visser, D., van Zuylen, G.A., van Dam, J.C., Oudshoorn, A., Eman, M.R., Ras, C., van Gulik, W.M., Frank, J., van Dedem, G.W.K., and Heijnen, J.J. (2002) *Biotechnol. Bioeng.*, **79** (6), 674–681.
60. Palsson, B. and Zengler, K. (2010) *Nat. Chem. Biol.*, **6** (11), 787–789.
61. Wiechert, W. (2001) *Metab. Eng.*, **3** (3), 195–206.
62. Iwatani, S., Van Dien, S., Shimbo, K., Kubota, K., Kageyama, N., Iwahata, D., Miyano, H., Hirayama, K., Usuda, Y., Shimizu, K., and Matsui, K. (2007) *J. Biotechnol.*, **128** (1), 93–111.
63. van Winden, W.A., van Dam, J.C., Ras, C., Kleijn, R.J., Vinke, J.L., van Gulik, W.M., and Heijnen, J.J. (2005) *FEMS Yeast Res.*, **5** (6–7), 559–568.
64. Bartek, T., Blombach, B., Lang, S., Eikmanns, B.J., Wiechert, W., Oldiges, M., Noh, K., and Noack, S. (2011) *Appl. Environ. Microbiol.*, **77** (18), 6644–6652.
65. Noh, K., Gronke, K., Luo, B., Takors, R., Oldiges, M., and Wiechert, W. (2007) *J. Biotechnol.*, **129** (2), 249–267.
66. Noh, K. and Wiechert, W. (2011) *Appl. Microbiol. Biotechnol.*, **91** (5), 1247–1265.
67. Fischer, E. and Sauer, U. (2005) *Nat. Genet.*, **37** (6), 636–640.
68. Blank, L.M., Kuepfer, L., and Sauer, U. (2005) *Genome Biol.*, **6** (6).
69. Ellis, D.I. and Goodacre, R. (2012) *Curr. Opin. Biotechnol.*, **23** (1), 22–28.
70. Oldiges, M., Kunze, M., Degenring, D., Sprenger, G.A., and Takors, R. (2004) *Biotechnol. Prog.*, **20** (6), 1623–1633.
71. Magnus, J.B., Hollwedel, D., Oldiges, M., and Takors, R. (2006) *Biotechnol. Prog.*, **22** (4), 1071–1083.
72. Ding, M.Z., Cheng, J.S., Xiao, W.H., Qiao, B., and Yuan, Y.J. (2009) *Metabolomics*, **5** (2), 229–238.
73. Nasution, U., van Gulik, W.M., Ras, C., Proell, A., and Heijnen, J.J. (2008) *Metab. Eng.*, **10** (1), 10–23.
74. Schiesel, S., Laemmerhofer, M., and Lindner, W. (2010) *Anal. Bioanal. Chem.*, **396** (5), 1655–1679.
75. Preinerstorfer, B., Schiesel, S., Lammerhofer, M.L., and Lindner, W. (2010) *J. Chromatogr., A*, **1217** (3), 312–328.
76. Wei, R. (2011) *Curr. Drug Metab.*, **12** (4), 345–358.
77. Papini, M., Salazar, M., and Nielsen, J. (2012) Systems biology of industrial microorganisms, *Biosystems Engineering I: Creating Superior Biocatalysts*, Springer-Verlag Berlin, Berlin, Germany, pp. 51–99.
78. Otero, J.M. and Nielsen, J. (2010) *Biotechnol. Bioeng.*, **105** (3), 439–460.

79. Demain, A.L. and Adrio, J.L. (2008) *Mol. Biotechnol.*, **38** (1), 41–55.
80. Takors, R., Bathe, B., Rieping, M., Hans, S., Kelle, R., and Huthmacher, K. (2007) *J. Biotechnol.*, **129** (2), 181–190.
81. El-Ali, J., Sorger, P.K., and Jensen, K.F. (2006) *Nature*, **442** (7101), 403–411.
82. Yin, H. and Marshall, D. (2012) *Curr. Opin. Biotechnol.*, **23** (1), 110–119.
83. Kovarik, M.L. and Allbritton, N.L. (2011) *Trends Biotechnol.*, **29** (5), 222–230.
84. Kortmann, H., Blank, L.M., and Schmid, A. (2011) Single cell analytics: an overview, *High Resolution Microbial Single Cell Analytics*, Springer-Verlag Berlin, Berlin, Germany, pp. 99–122.
85. Grunberger, A., Paczia, N., Probst, C., Schendzielorz, G., Eggeling, L., Noack, S., Wiechert, W., and Kohlheyer, D. (2012) *Lab Chip.*, doi: 10.1039/c2lc40156h
86. Amantonico, A., Urban, P.L., and Zenobi, R. (2010) *Anal. Bioanal. Chem.*, **398** (6), 2493–2504.
87. Heinemann, M. and Zenobi, R. (2011) *Curr. Opin. Biotechnol.*, **22** (1), 26–31.
88. Borland, L.M., Kottegoda, S., Phillips, K.S., and Allbritton, N.L. (2008) *Annu. Rev. Anal. Chem.*, **1**, 191–227.

16
Nutritional Metabolomics
Hannelore Daniel

16.1
Introduction

Metabolomics applications in the context of nutrition seem unlimited when performed in cell cultures or model organisms including rodent models. However, dietary components usually affect a biological system less pronounced than, for example, drugs or toxic compounds. Nutritional metabolomics therefore has to deal with rather small effects on the metabolome. In addition, applications in human studies face special problems. Although tissue biopsies may be collected after ethical approval, the sample size is usually small and samples are heterogeneous containing different cell populations and blood. Moreover, already the sampling procedure itself may cause acute stress responses in the tissue. Consequently, most of the human studies rely on the use of blood, urine, or other body fluids that can be obtained noninvasively or by minimal invasive techniques. That metabolite profiling in blood or urine is able to separate healthy states from disease states is well documented by the success of the newborn screening programs for identification of inherited diseases of metabolism [1–3] and a variety of more recent studies in different target populations. Yet, to determine human metabolome responses to environmental conditions in otherwise healthy subjects is still a huge challenge.

16.2
The Metabolome of Human Plasma and Urine: General Considerations

Data from nutritional studies are multidimensional and complex and treatment effects are usually much smaller than the biological variations between individuals. Properly designed experiments, well-selected and phenotyped study populations, and an adjusted data analysis strategy are therefore essential for successful nutritional metabolomics studies. It has to be emphasized that timescales and effect sizes for alterations of the metabolome differ for acute dietary challenges or chronic treatments. Acute responses to food intake can cause major changes in body fluid

Metabolomics in Practice: Successful Strategies to Generate and Analyze Metabolic Data, First Edition.
Edited by Michael Lämmerhofer and Wolfram Weckwerth.
© 2013 Wiley-VCH Verlag GmbH & Co. KGaA. Published 2013 by Wiley-VCH Verlag GmbH & Co. KGaA.

metabolomes by the invasion of nutrients and metabolites after digestion and absorption of the food components. These diet-induced changes usually last a couple of hours and are controlled by the hormones (mainly insulin) secreted following meal intake. Changes in the urinary output of metabolites from an acute diet test can last for more than 24 h. In contrast, effects of long-lasting dietary interventions are mostly small and need studies of large cohorts of well phenotyped individuals to detect changes in the metabolome that exceed the natural biological variation. Nutritional intervention trials are usually designed as crossover studies with the benefit that each individual serves as his own control. This allows a direct comparison of treatments and is particularly efficient to identify differences in the presence of large interindividual variation. Then, multilevel data analysis (MLDA) tools are applied permitting a separate analysis of the interindividual and intraindividual variation in the data sets obtained [4, 5]. The MLDA is therefore preferred over classical megavariate data analysis methods such as principal component analysis (PCA) or partial least squares discriminant analysis (PLS-DA).

A typical feature of the metabolome of human body fluids is its continuous change. As an "integrated metabolome" it is composed of the nutrients and metabolites provided by the diet, metabolites produced endogenously in the interorgan metabolism and by the microbiota in the large intestine (Figure 16.1).

16.2.1
The Plasma Metabolome

For practical but not necessarily scientific reasons, plasma samples are usually taken after overnight fasting. Although sufficient in most studies such as for a comparison of cohorts, an overnight fasting state may not represent the best reference point in all cases. Human metabolism is a continuum driven by the need to continuously provide ATP as a fuel for metabolism and for anabolic synthetic metabolism. After overnight fasting, the body mainly utilizes fatty acids for β-oxidation and ATP production. Therefore, as fasting continues, plasma-free fatty acid concentrations steadily increase by enhanced lipolysis because of the action of glucagon and catecholamines on adipose tissue. Maintenance of plasma glucose is achieved by breakdown of liver glycogen or by hepatic gluconeogenesis to feed cells that rely on glucose as a sole energy substrate (for example, red blood cells). In addition, a varying amount of amino acids is released into the plasma from accelerated protein degradation in muscle. The amino acids are partially used for oxidation while glucogenic amino acids serve also as precursors in hepatic *de novo* glucose production. The catabolic state induced by overnight fasting therefore produces plasma metabolite profiles that are characterized by elevated levels of free fatty acids and selected (branched chain and aromatic) amino acids. The absolute plasma levels of those metabolites during fasting, however, will depend on the metabolic status of the volunteer (lean, overweight, and obese), on its nutritional condition before fasting and on the amount of stored glycogen in liver. It may therefore be advised to control and standardize food intake as good as possible also for the day prior to collecting the fasting blood or urine samples.

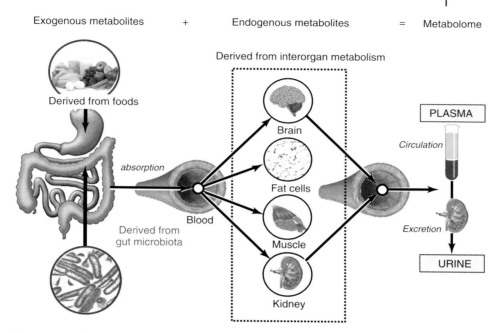

Figure 16.1 The surrogate character of the human metabolome. Metabolomes derived from human body fluids (plasma, urine) are a surrogate of nutrients and metabolites provided by the diet, their intermediate products as generated in interorgan metabolism, and those produced by bacteria (microbiota) in the large intestine from either nondigested/nonabsorbed food components or released by the host into the intestine.

16.2.2
The Urinary Metabolome

For urine analysis either by ^{1}H-NMR or GC-/LC-MS, 24 h urine sampling with a representative aliquot analyzed is superior to a spontaneously collected urine sample. As absolute metabolite concentrations in urine depend on water homeostasis to maintain tissue osmolarity and blood pressure, urinary metabolites should be expressed relative to an internal reference compound such as creatinine. It is produced from creatine phosphate in muscle with a more or less constant rate and its level in urine therefore reflect mainly the constant muscle mass. There are, however, a variety of conditions in which creatinine levels in urine also change and thus corrections of urinary metabolites may be better based on the osmolarity or on the total volume of urine collected over 24 h to account for differences in fluid intake.

The excreted products in urine mainly derive from nitrogen metabolism (amino acids and their metabolic intermediates) such as urea and ammonium for nitrogen elimination. Urine samples obtained from fasting volunteers display only low levels of metabolites from interorgan metabolism, whereas other metabolites originate from the gut microbiome. They enter circulation after absorption from the large intestine and usually undergo secondary or tertiary metabolism in the host's liver before being also excreted via the kidneys. The interpretation of the urine metabolome is therefore a difficult task. Among the metabolites of the gut microbiota hippuric acid and derivatives thereof are identified. Hippuric acid can originate from plant foods as a key microbial degradation product of polyphenolic secondary metabolites of plants. In addition, hippuric acid is produced endogenously in the liver from ingested benzoic acid, contained in larger quantities in plant foods or used as a food preservative and, in addition, it can derive from phenylalanine degradation. Thus, not all hippuric acid found in urine comes from the bacterial degradation of polyphenols. Numerous conjugation products of dietary polyphenols, however, are also found in urine and those appear to reflect the intake of particular foods containing high levels of individual polyphenols.

Trimethylamine N-oxide (TMAO), recently proposed as a marker metabolite for cardiovascular diseases [6], may serve as another example to demonstrate the difficulties in interpretation of urinary metabolomes. TMAO is a characteristic product found in urine and derives mainly from the microbiome. Its precursor trimethylamine is produced by gut bacteria from dietary choline or phosphatidylcholines (food derived or secreted into the gut), which then is oxidized in the host's liver to TMAO. However, TMAO in urine is also increased after sea fish consumption as it is a prominent osmolyte in such fish.

When identical metabolites are quantified in plasma and urine of the same volunteers at the same sampling time, the metabolomes do not correlate (Figure 16.2). Despite a high level of cross-correlation among the metabolites in the plasma samples (Figure 16.2a) or among other metabolites in the urine samples (Figure 16.2b) essentially no correlation is obtained when identical metabolites from plasma and urine are plotted against each other (Figure 16.2c). This is the consequence of kidney function with the reabsorption of all nutrients, while allowing other metabolites to be cleared. The kidney tubular system is equipped with literally hundreds of transport systems for allowing efficient reabsorption of nutrients and metabolites from the glomerular filtrate to prevent renal loss of energy substrates as well as other essential nutrients (minerals, trace elements, and vitamins). Consequently, especially renal diseases show prominent changes in the urinary metabolomes [7–10].

16.3
The Food Metabolome and Its Signature in Human Samples

Nutrition studies including those dedicated to metabolomics generally face the problem to accurately assess and quantify the food intake of the volunteers. Usually the volunteers have to report the kind and quantity of the individual food items consumed in food frequency questionnaires as good as possible. From these, experts can calculate mean daily nutrient intake levels based on databases containing reference values of nutrients of hundreds of representative food items.

16.3 The Food Metabolome and Its Signature in Human Samples | 397

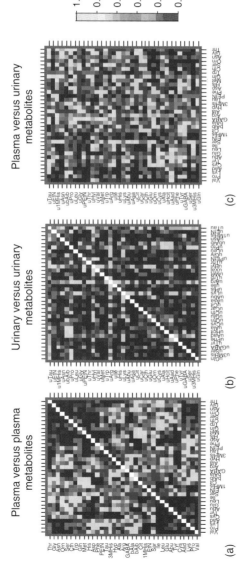

Figure 16.2 Plasma and urinary metabolite profiles show minimal overlap. Plasma and urine samples from humans kept under controlled dietary conditions were collected at the same time and were analyzed via LC-MS/MS for 35 amino acids and derivatives. In plasma, cross-correlation among subsets of the metabolites was obtained and urinary amino acids and metabolites (denoted with u in axis labeling) revealed again for subsets cross-correlations. However, for identical metabolites, correlation analysis between plasma and urine samples failed to deliver any interrelationship. This demonstrates the role of the kidney in efficiently reabsorbing the metabolites from the primary filtrate of plasma leading to a urinary signature of metabolites that shows a distinct individual pattern but that does not refer to changes in plasma.

These methods are, however, prone to problems such as underreporting of actual food intake by the volunteer or inaccuracy of the calculated nutrient intake because of the huge variability in the composition of the food items. To overcome some of these problems metabolomics approaches are thought to yield more reliable information on food intake by the metabolite signatures produced in human intermediary metabolism.

The metabolite composition of foods of animal origin match largely with the metabolites found in humans, whereas plant foods contain a large spectrum of compounds of the plant secondary metabolism in addition to the contained nutrients. Most of these secondary plant components possess two or three phenolic rings in the core structure to which multiple substituents can be attached. There exist around 10 000 different polyphenolic entities in the edible plant kingdom. As the patterns and concentrations of these compounds depend on the plant species they are considered to serve as a fingerprint of a given plant food.

Unlike the macronutrients carbohydrates, proteins, and lipids that can be completely oxidized in the human system, the plant polyphenols cannot be utilized. They are xenobiotics and, like other foreign compounds, undergo substantial modification in phase I and phase II metabolism. They are methylated or conjugated with glucuronic acid or sulfate moieties after intestinal absorption already in intestinal epithelial cells or in liver with numerous metabolites formed. Those conjugates are either secreted back into the intestinal lumen via the bile or are delivered to the kidneys for excretion. The site of intestinal absorption of the plant compounds and their absorption rate can vary considerably – although bioavailability is generally low. Some of the polyphenols are not absorbed in the upper small intestine and reach the distal large intestine, where the gut microbiota converts the compounds to new derivatives which then can be absorbed, appear in the blood, get modified in the liver and are excreted in the urine. Depending on the composition of the microbiota, the patterns of metabolites, production rates, and concentrations in plasma or urine can differ enormously. In some cases, cohorts of volunteers can be easily subcategorized in producers and nonproducers for distinct metabolites derived from ingested plant polyphenols because of their differing gut mircobiota. However, various studies have succeeded in identifying characteristic compounds in plasma or urine, which originate directly from the ingestion of individual food items [11–18]. Although it can be expected that food metabolomics will provide more such surrogate markers of food intake, it remains to be seen whether they allow a quantitative intake assessment of individual food items in human volunteers.

16.4
The Variability of the Human Metabolome in Health and Disease States

Besides the contribution of the gut microbiome to the appearance of metabolites in plasma and urine, the variability of the human metabolome itself is intriguing and deserves more research addressing its origin. As a first source, genetic heterogeneity is considered and several recent studies have employed metabolomics in combination with genetic screening including genome-wide associations. Some 40 individual single nucleotide polymorphisms (SNP's) have been identified which correspond to changes in plasma metabolite patterns or concentrations mostly in the lipidome [19–21]. Although the absolute concentration changes of the metabolites in plasma or urine that associate with a given SNP are rather small [21–24], the power of the large cohorts allows significant correlations to be established. Accordingly, a recent study analyzed the "familiality" (as the sum of genetic and common environmental factors) of metabolite profiles in plasma and urine by ^{1}H-NMR and estimated, that sample sizes of a few thousand volunteers should provide sufficient statistical power to quantify robust biomarkers with a disease predisposing character [25]. However, this also questions whether metabolomics for diagnostic purposes when taken down to the individual level, will be able to deliver data of reliable prognostic value.

A second source, significantly affecting the variability in the metabolome, are differences in body weight, body total fat mass, and other anthropometric measures in human cohorts. Without doubt, genetics also determines body mass to a large extent, as studies in monozygotic twins reared apart or reared together have demonstrated. Various genome-wide association studies on genetic determinants of body fat mass revealed that individual SNP's constitute only with minor effects to total body fat mass. The key drivers in overweight and obesity and their associated diseases such as type 2 diabetes, cardiovascular diseases, or different cancer entities are environmental factors such as sedentary life styles with lack of sufficient exercise and hypercaloric nutrition. A variety of recent studies revealed early alterations in the metabolome in obese individuals that can be detected even before insulin resistance (IR) established or type 2 diabetes was diagnosed. In a very elegant study in monozygotic twins discordant for body mass, changes in plasma lipidomic profiles indicated that obesity *per se* caused an increase in plasma lysophosphatidylcholines, and a decline in some phospholipids [26]. Obesity, already in its early stages and seemingly independent of genetic influences, seems therefore to be associated with changes in lipid metabolism that also are known to be associated with atherosclerosis and IR.

When humans develop an IR, metabolome changes are becoming even more pronounced. Among the metabolites that most strongly associate with IR and type 2 diabetes are amino acids [27], their keto acid derivatives as well as free fatty acids and their carnitine derivatives as well as some lysophosphatides. A number of studies in humans with IR and type 2 diabetes consistently identified elevated plasma levels of branched chain amino acids (BCAAs), aromatic amino acids (AAAs) but also keto acids such as α-OH-butyrate and α-ketobutyrate and related carnitine derivatives and the classical ketone bodies as markers [28–30]. The increase of these metabolites may all relate to a common mechanism that originates form the impaired insulin action in muscle, adipose tissue, and liver. Despite increased insulin levels in IR, the reduced action of insulin on its target tissues causes a disbalance in the anabolic to catabolic hormone profiles with the latter causing a mild catabolic metabolic condition. The resulting changes in the plasma metabolite profile in IR therefore resemble more or less the metabolite changes found in healthy individuals during prolonged fasting. Such a mild catabolic state is characterized by elevated plasma levels of free fatty acids, corresponding acyl-carnitines, elevated BCAA, and AAA as well as hydroxy- and keto acids derived from the degradation of amino acids and fatty acids. This "metabolomic signature" of IR and type 2 diabetes in humans is displayed in Figure 16.3 showing the key metabolites and their interrelationships as revealed by recent metabolomics studies. Most surprisingly, the elevated levels in the plasma amino acids in still healthy individuals had a very high power in predicting the individual's diabetes development long before IR established [30]. Although a proper mechanistic explanation is still lacking for this finding, it may be speculated that metabolomics here identified metabolites derived from an already known epidemiological risk factor for diabetes: the dietary protein intake [31, 32]. A high protein intake – usually based on high meat consumption – causes an increase in plasma amino acids, mainly of BCAA and AAA, as those bypass liver metabolism they appear in peripheral blood in higher concentrations. Moreover, their uptake into tissues is dependent on insulin and if insulin action is reduced they may show reduced clearance and elevated plasma levels. Further studies are necessary to clarify the underlying causal relationships between the altered plasma levels of those amino acids and IR and diabetes risk. When urinary analysis via ^{1}H-NMR was performed in rodent models of diabetes and in human samples, again amino acids and their derivatives were among the most discriminating metabolites separating healthy controls from diseased individuals [33–35].

16.5
The Dynamic Nature of the Metabolome

A major problem in most studies assessing the human metabolome is that only one single measurement per volunteer (and mostly after overnight fasting) is performed. It is, however, a key feature of metabolism and thus of the metabolome that it is highly dynamic in time and space and changes every minute. As mammals, we constantly shift between anabolic conditions after food intake and catabolic states between meals or in extending fasting periods. Both conditions are marked by characteristic changes in hormone levels that control the metabolite fluxes. It is thus rewarding to analyze the responses of a metabolic system to a certain dietary challenge such as a simple oral glucose tolerance test (OGTT). This standard assay has been performed millions of times in clinical practice for diagnosis of IR and diabetes by measuring plasma glucose and insulin levels over time after oral administration of 75 g of D-glucose. Recently, metabolomics approaches have highlighted that the known time-dependent changes in plasma glucose and insulin levels are associated with major alterations in a large variety of other metabolites. Among those are increases in the levels of bile acids and marked decreases in BCAA and AAA levels next to known changes in free fatty acids and ketone bodies [36–38]. The changes observed for those metabolites in plasma after an OGTT

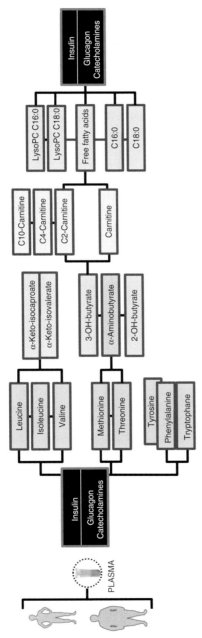

Figure 16.3 Plasma metabolomic signature of humans with insulin resistance or type 2 diabetes. Schema compiled from reported plasma metabolite changes and their putative interrelationship as derived from human cohort studies targeted to identify biomarkers of insulin resistance and type 2 diabetes by metabolomics approaches. It may be speculated that the changes in lipid and amino acid biomarker in plasma originate from the impaired sensitivity of peripheral tissues such as muscle and adipose tissue and of the liver to respond properly to the anabolic action of insulin. This simultaneously leads to a slightly increased activity of the antagonistic hormones glucagon and catecholamines that promote lipolysis and gluconeogenesis and reduced protein synthesis.

last up to 4 h while plasma glucose levels usually return to baseline levels within 2 h. What can also be observed is that the metabolite profiles of the volunteer's exhibit an increased variation (SD or SEM) after the OGTT when compared to a much smaller variation of the same metabolites in the fasting state. This suggest that such defined challenges of a metabolic system combined with comprehensive metabolite profiling may be suited to identify individual phenotypic differences or "metabotypes" that may be missed otherwise when only fasting samples are analyzed.

16.6
The Future of Nutritional Metabolomics and Research Needs

Like in most other areas, metabolomics applications were introduced only very recently into nutritional studies that assess the effects of diets or individual dietary constituents on human metabolism in health and disease conditions. Therefore, the current state is mainly of descriptive character. Most of the existent studies provide just qualitative or semiquantitative measures of metabolites combined with component analysis. The future of metabolomics should be based on quantitative analysis by which metabolomics could become a more objective and meaningful scientific area. There is also a need to expand the analysis platforms to cover more "low abundant" metabolites. Both, ^{1}H-NMR as well as MS-based methods currently cover only the metabolome "tip of the iceberg." It may therefore not be surprising that identical "biomarkers" with similar changes appear in such different diseases as type 2 diabetes, atherosclerosis, and cardiovascular diseases or even depression [39–42]. This finding questions the specificity and selectivity of current metabolomics approaches when analyzing only at the "tip of the iceberg."

Another major problem is the lack of knowledge on the physiological origin of the changes observed. This applies to metabolites that arise from human metabolism as well as to those produced by the gut microbiota. It can be a rather frustrating scouting exercise to identify the most plausible cause for a change in plasma or urinary metabolites in human studies. Although data bases such as the human metabolome database (HMDB), at present containing about 8000 metabolite entries, are very helpful in linking metabolites and functional context, for some of the metabolites there are not even normal plasma reference values available. Most of our knowledge on the human plasma metabolome and on the reference values originates from studies of rare monogenetic diseases and newborn screening programs, which because of their nature cover usually only a narrow spectrum of metabolites. In mammalian samples, there are metabolites found for which no reference at all can be found and no information is available of whether mammals are capable to produce them or not. Here, more studies in germ-free rodents (without a gut microbiota) may be helpful in defining of whether the metabolites are indeed derived from the microbiota or produced by the host itself.

Future metabolomics studies in humans should more often apply stable isotope labeled precursors to determine the fate of the compounds and their metabolites. Stable isotope labeled compounds should also be employed as reference standards in studies that targeted to better define the "food metabolome." Furthermore, research strategies have to be improved by more well-controlled studies in the same volunteers – with thus identical host and microbiome metabolic phenotypes. In combination with genotyping, such data could better reveal the underlying dietary effects on host metabolism and microbiota and their relationships. The science community should define the minimal descriptors of such studies and develop highly standardized study protocols. Only on the basis of such controlled approaches can nutritional metabolomics deliver robust and meaningful data and improve the quality of the databases.

Finally, the author is convinced that it is important to assess the dynamics of changes in the metabolome and the resilience of the biological system. The capability to cope with a large variety of environmental conditions with a high build in buffering capacity is a prime feature of the mammalian system. In human evolution, it was most essential to adapt rapidly to changes in food supply and to quite different food sources. If metabolomics applications in human studies are targeted toward the characterization of metabolic phenotypes, it may be wise to push the system to its limits to overcome its intrinsic buffering capacity. This may be accomplished for instance, by an extended fasting period or an acute dietary glucose or lipid loading test or combinations thereof, followed by the analysis of metabolite profiles in a time-dependent manner. On the basis of the dynamic changes in metabolites we may as well be able to apply kinetic models to the data like in pharmacokinetics allowing characteristic time constants to be derived. When using stable isotope labeled food components, such a "nutrikinetic" approach could reveal discrete human metabotypes characterized by quantified differences and time constants in their response to a given metabolic condition.

References

1. Garg, U. and Dasouki, M. (2006) *Clin. Biochem.*, **39** (4), 315–332.
2. McCabe, L.L. and McCabe, E.R. (2008) *Annu. Rev. Med.*, **59**, 163–175.
3. Lehotay, D.C., Hall, P., Lepage, J., Eichhorst, J.C., Etter, M.L., and Greenberg, C.R. (2011) *Clin. Biochem.*, **44** (1), 21–31.
4. van Velzen, E.J., Westerhuis, J.A., van Duynhoven, J.P., van Dorsten, F.A., Hoefsloot, H.C., Jacobs, D.M., Smit, S., Draijer, R., Kroner, C.I., and Smilde, A.K. (2008) *J. Proteome Res.*, **7** (10), 4483–4491.
5. Westerhuis, J.A., van Velzen, E.J., Hoefsloot, H.C., and Smilde, A.K. (2010) *Metabolomics*, **6** (1), 119–128.
6. Wang, Z., Klipfell, E., Bennett, B.J., Koeth, R., Levison, B.S., Dugar, B., Feldstein, A.E., Britt, E.B., Fu, X., Chung, Y.M., Wu, Y., Schauer, P., Smith, J.D., Allayee, H., Tang, W.H., DiDonato, J.A., Lusis, A.J., and Hazen, S.L. (2011) *Nature*, **472** (7341), 57–63.
7. Weiss, R.H. and Kim, K. (2011) *Nat. Rev. Nephrol.*, **8** (1), 22–33.
8. Kim, K., Taylor, S.L., Ganti, S., Guo, L., Osier, M.V., and Weiss, R.H. (2011) *OMICS*, **15** (5), 293–303.

9. Lin, L., Huang, Z., Gao, Y., Yan, X., Xing, J., and Hang, W. (2011) *J. Proteome Res.*, **10** (3), 1396–1405.
10. Rhee, E.P. and Thadhani, R. (2011) *Curr. Opin. Nephrol. Hypertens.*, **20** (6), 593–598.
11. Edmands, W.M., Beckonert, O.P., Stella, C., Campbell, A., Lake, B.G., Lindon, J.C., Holmes, E., and Gooderham, N.J. (2011) *J. Proteome Res.*, **10** (10), 4513–4521.
12. Manach, C., Hubert, J., Llorach, R., and Scalbert, A. (2009) *Mol. Nutr. Food Res.*, **53** (10), 1303–1315.
13. McGhie, T.K. and Rowan, D.D. (2011) *Mol. Nutr. Food Res.*, doi: 10.1002/mnfr.201100545
14. Favé, G., Beckmann, M., Lloyd, A.J., Zhou, S., Harold, G., Lin, W., Tailliart, K., Xie, L., Draper, J., and Mathers, J.C. (2011) *Metabolomics*, **7** (4), 469–484.
15. Heinzmann, S.S., Brown, I.J., Chan, Q., Bictash, M., Dumas, M.E., Kochhar, S., Stamler, J., Holmes, E., Elliott, P., and Nicholson, J.K. (2010) *Am. J. Clin. Nutr.*, **92** (2), 436–443.
16. Llorach, R., Urpi-Sarda, M., Jauregui, O., Monagas, M., and Andres-Lacueva, C. (2009) *J. Proteome Res.*, **8** (11), 5060–5068.
17. O'Sullivan, A., Gibney, M.J., and Brennan, L. (2011) *Am. J. Clin. Nutr.*, **93** (2), 314–321.
18. Stella, C., Beckwith-Hall, B., Cloarec, O., Holmes, E., Lindon, J.C., Powell, J., van der Ouderaa, F., Bingham, S., Cross, A.J., and Nicholson, J.K. (2006) *J. Proteome Res.*, **5**, 2780–2788.
19. Tukiainen, T., Kettunen, J., Kangas, A.J., Lyytikäinen, L.P., Soininen, P., Sarin, A.P., Tikkanen, E., O'Reilly, P.F., Savolainen, M.J., Kaski, K., Pouta, A., Jula, A., Lehtimäki, T., Kähönen, M., Viikari, J., Taskinen, M.R., Jauhiainen, M., Eriksson, J.G., Raitakari, O., Salomaa, V., Järvelin, M.R., Perola, M., Palotie, A., Ala-Korpela, M., and Ripatti, S. (2012) *Hum. Mol. Genet.*, **21** (6), 1444–1455.
20. Chasman, D.I., Paré, G., Mora, S., Hopewell, J.C., Peloso, G., Clarke, R., Cupples, L.A., Hamsten, A., Kathiresan, S., Mälarstig, A., Ordovas, J.M., Ripatti, S., Parker, A.N., Miletich, J.P., and Ridker, P.M. (2009) *PLoS Genet.*, **5** (11), e1000730.
21. Gieger, C., Geistlinger, L., Altmaier, E., Hrabé de Angelis, M., Kronenberg, F., Meitinger, T., Mewes, H.W., Wichmann, H.E., Weinberger, K.M., Adamski, J., Illig, T., and Suhre, K. (2008) *PLoS Genet.*, **4** (11), e1000282.
22. Suhre, K., Shin, S.Y., Petersen, A.K., Mohney, R.P., Meredith, D., Wägele, B., Altmaier, E. CARDIOGRAM, Deloukas, P., Erdmann, J., Grundberg, E., Hammond, C.J., de Angelis, M.H., Kastenmüller, G., Köttgen, A., Kronenberg, F., Mangino, M., Meisinger, C., Meitinger, T., Mewes, H.W., Milburn, M.V., Prehn, C., Raffler, J., Ried, J.S., Römisch-Margl, W., Samani, N.J., Small, K.S., Wichmann, H.E., Zhai, G., Illig, T., Spector, T.D., Adamski, J., Soranzo, N., and Gieger, C. (2011) *Nature*, **477** (7362), 54–60.
23. Suhre, K., Wallaschofski, H., Raffler, J., Friedrich, N., Haring, R., Michael, K., Wasner, C., Krebs, A., Kronenberg, F., Chang, D., Meisinger, C., Wichmann, H.E., Hoffmann, W., Völzke, H., Völker, U., Teumer, A., Biffar, R., Kocher, T., Felix, S.B., Illig, T., Kroemer, H.K., Gieger, C., Römisch-Margl, W., and Nauck, M. (2010) *Nat Genet.*, **43** (6), 565–569.
24. Illig, T., Gieger, C., Zhai, G. et al. (2010) *Nat Genet.*, **42** (2), 137–141.
25. Nicholson, G., Rantalainen, M., Maher, A.D., Li, J.V., Malmodin, D., Ahmadi, K.R., Faber, J.H., Hallgrímsdóttir, I.B., Barrett, A., Toft, H., Krestyaninova, M., Viksna, J., Neogi, S.G., Dumas, M.E., Sarkans, U., Silverman, B.W., Donnelly, P., Nicholson, J.K., Allen, M., Zondervan, K.T., Lindon, J.C., Spector, T.D., McCarthy, M.I., Holmes, E., Baunsgaard, D., and Holmes, C.C. (2011) *Mol. Syst. Biol.*, **7**, 525. (The Molpage Consortium). doi: 10.1038/msb.2011.57
26. Pietiläinen, K.H., Sysi-Aho, M., Rissanen, A., Seppänen-Laakso, T., Yki-Järvinen, H., Kaprio, J., and Oresic, M. (2007) *PLoS ONE*, **2** (2), e218.
27. Rhee, E.P., Cheng, S., Larson, M.G., Walford, G.A., Lewis, G.D., McCabe, E., Yang, E., Farrell, L., Fox, C.S.,

O'Donnell, C.J., Carr, S.A., Vasan, R.S., Florez, J.C., Clish, C.B., Wang, T.J., and Gerszten, R.E. (2011) *J. Clin. Invest.*, **121** (4), 1402–1411.
28. Fiehn, O., Garvey, W.T., Newman, J.W., Lok, K.H., Hoppel, C.L., and Adams, S.H. (2010) *PLoS ONE*, **5** (12), e15234.
29. Gall, W.E., Beebe, K., and Lawton, K.A. (2010) *PLoS ONE*, **5** (5), e10883.
30. Wang, T.J., Larson, M.G., Vasan, R.S., Cheng, S., Rhee, E.P., McCabe, E., Lewis, G.D., Fox, C.S., Jacques, P.F., Fernandez, C., O'Donnell, C.J., Carr, S.A., Mootha, V.K., Florez, J.C., Souza, A., Melander, O., Clish, C.B., and Gerszten, R.E. (2011) *Nat. Med.*, **17** (4), 448–453.
31. Sluijs, I., Beulens, J.W., van der A, D.L., Spijkerman, A.M., Grobbee, D.E., and van der Schouw, Y.T. (2010) *Diabetes Care*, **33** (1), 43–48.
32. Tinker, L.F., Sarto, G.E., Howard, B.V., Huang, Y., Neuhouser, M.L., Mossavar-Rahmani, Y., Beasley, J.M., Margolis, K.L., Eaton, C.B., Phillips, L.S., and Prentice, R.L. (2011) *Am. J. Clin. Nutr.*, **94** (6), 1600–1606.
33. Griffin, J.L., Atherton, H.J., Steinbeck, C., and Salek, R.M. (2011) *BMC Res Notes*, **4**, 272.
34. Salek, R.M., Maguire, M.L., Bentley, E., Rubtsov, D.V., Hough, T., Cheeseman, M., Nunez, D., Sweatman, B.C., Haselden, J.N., Cox, R.D., Connor, S.C., and Griffin, J.L. (2007) *Physiol. Genomics*, **29** (2), 99–108.
35. van Doorn, M., Vogels, J., Tas, A., van Hoogdalem, E.J., Burggraaf, J., Cohen, A., and van der Greef, J. (2007) *Br. J. Clin. Pharmacol.*, **63** (5), 562–574.
36. Shaham, O., Wei, R., Wang, T.J., Ricciardi, C., Lewis, G.D., Vasan, R.S., Carr, S.A., Thadhani, R., Gerstzten, R.E., and Mootha, V.K. (2008) *Mol. Syst. Biol.*, **4**, 214.
37. Zhoa, X., Peter, A., Fritsche, J., Elcernova, M., Fritsche, A., Häring, H.U., Schleicher, E.D., Xu, G., and Lehmann, R. (2009) *Am. J. Physiol. Endocrinol. Metab.*, **296**, E384–E393.
38. Skurk, T., Rubio-Aliaga, I., Stamfort, A., Hauner, H., and Daniel, H. (2010) *Metabolomics*, 1–12, doi: 10.1007/s11306-010-0255-2
39. Herder, C., Karakas, M., and Koenig, W. (2011) *Clin. Pharmacol. Ther.*, **90** (1), 52–66.
40. Barderas, M.G., Laborde, C.M., Posada, M., de la Cuesta, F., Zubiri, I., Vivanco, F., and Alvarez-Llamas, G. (2011) *J. Biomed. Biotechnol.*, **2011**, 790132.
41. Zhang, F., Jia, Z., Gao, P., Kong, H., Li, X., Lu, X., Wu, Y., and Xu, G. (2010) *Mol. Biosyst.*, **6** (5), 852–861.
42. Zhang, F., Jia, Z., Gao, P., Kong, H., Li, X., Chen, J., Yang, Q., Yin, P., Wang, J., Lu, X., Li, F., Wu, Y., and Xu, G. (2009) *Talanta*, **79** (3), 836–844.

Index

a

absorbance 246
accuracy 41
accurate mass and retention time (AMRT) library 164
acidic acetonitrile-methanol (AANM) 13
advanced exploratory analysis 275–277
airway lining fluid (ALF) 128
analyte 41
– detection 245
– – electrochemical 246–247
– – optical 245–246
ANOVA and univariate methods 273–275
aqueous normal phase chromatography (ANP) 24, 25
aromatic amino acids (AAAs) 399, 400
ASCA 274, 275
atmospheric pressure chemical ionization (APCI) 40, 97, 141, 156, 324
atmospheric pressure photoionization (APPI) 40, 156, 326
Automatic Liquid Sampler (ALS) 165
automatic mass spectral deconvolution and identification system (AMDIS) 170–172
axial trapping frequency 122–123

b

background electrolyte (BGE) 178, 244, 245
bacterial fatty acid methyl esters (BAMEs) 145
baseline offsets and corrections 83–85
bidirectional orthogonal-PLS (O2PLS) 227
binning. See bucketing
biofluids 345, 355, 356–359
bioinformatics 126
biological fluids and cellular material 146–148
biological questions
– data processing and
– – biomarkers 264
– – methods following questions 264
– – networks and mechanistic insight 265
– – treatment effects 264–265
– microfluidics
– – metabolic response to stimulation and cell-to-cell signaling, monitoring 249–251
– pharmacokinetics and pharmacodynamics 252–254
– – clinical diagnostics 254
biomarkers 32–33, 264, 323, 371, 372, 402
biomass density 2, 4
biosynthetic pathways, for stress-induced metabolites 229
biotechnology 379–382
– analytical methods for microbial metabolomics 382–384
– microbial metabolomics
– – applications 386–388
– – custom-made separation for 384–385
– – higher throughput and 385–386
blood 190, *201*
boiling ethanol (BE) 12, 13
Bonferroni correction 274
bottom-up approach 42
– calculation, according to bottom-up approach 42–48
brain natriuretic peptide (BNP) 372
branched chain amino acids 372, 399, 400
bucketing 225

c

capillary electrochromatography (CEC) 211, 245

Metabolomics in Practice: Successful Strategies to Generate and Analyze Metabolic Data, First Edition.
Edited by Michael Lämmerhofer and Wolfram Weckwerth.
© 2013 Wiley-VCH Verlag GmbH & Co. KGaA. Published 2013 by Wiley-VCH Verlag GmbH & Co. KGaA.

capillary electrophoresis-mass spectrometry (CE-MS) 177–179, 211, 326
– applications 190–202
– – nontargeted approaches 196–200, 202–203
– – targeted approaches 190, 191–195, 202
– coupling
– – interfacing 180, 182–184
– – mass analyzers 184–185
– data analysis 187–190
– sample pretreatment 185–187
– separation conditions 179–180
capillary zone electrophoresis (CZE) 178, 244–245
carbon transitions modeling 298
cause-and-effect diagram (CaED) 44
cell extracts LC-MS measurement uncertainty calculation 63–65
ceramide (CERs) 149, 152, 159, 164, 165, 166
charged lipids 156
chemometrics 77, 126, 231
chemotaxonomy 319
Chemspider 109
ChenomX NMR Suite 222
chloroform methanol (CM) 12
cholesterol esters (CEs) 141
Circadian rhythms 360
cold methanol quenching 5, 7
collision-induced dissociation (CID) 39, 157, 326, 368
combined standard uncertainty 42–43
– calculation 45
– relative contribution to 58
combined two-dimensional methods 217, 219
complex regional pain syndrome (CRPS) 202
conjugated linoleic acids (CLAs) 168
Corneofix® 148
correlation spectroscopy (COSY) 215–216
cross-talk 28
cross-validation 227–228
– and permutations 267–271

d

data-dependent acquisition (DDA) 157
data processing 261
– biological questions
– – biomarkers 264
– – methods following questions 264
– – networks and mechanistic insight 265
– – treatment effects 264–265
– characteristics

– – correlation structure 261–262
– – dynamics 263
– – informative versus noninformative variation 262–263
– – low samples-to-variables ratio 263
– – measurement error 263
– – nonlinear relations 264
– methods overview
– – advanced exploratory analysis 275–277
– – ANOVA and univariate methods 273–275
– – discriminant analysis 280–281
– – exploratory analysis 272–273
– – multilevel approaches 281–282
– – network inference 282–283
– – regression methods 277–280
– validation
– – cross-validation and permutations 267–271
– – dimensionality curse 266–267
– – levels 265
DC amperometry 246, 247
deconvolution 86–87
dendogramms 127
deproteinization 359
desorption electrospray ionization (DESI) 145
diabetes 139
direct infusion mass spectrometry 354, 367, 368
direct injection mass spectrometry 324
discriminant analysis 280–281
diurnal variation 99
duty cycle 28

e

electron capture dissociation (ECD) 124, 246, 247
electron impact (EI) ionization 39, 62, 79, 325, 366–367
electroosmotic flow (EOF) 178, 180, 181, 243, 245
electrospray ionization (ESI) 21, 40, 60, 97, 141, 155, 157, 179, 182
Entner–Doudoroff pathway 285, 293, 294, 300
enzymatic assays 13
Eurachem/CiTAC guide 44
Exactive Orbitrap 29
exhaled breath condensate, application example using 128
– biochemical mass difference networking and statistical results synthesis 131–133

– C–H–N–O–S–P formula annotation 130
– data preprocessing 129–130
– experiment 128–129
– FT-ICR/MS measurement 129
– statistical preprocessing 130–131
expanded uncertainty 45, 57
extracted compound chromatograms (ECCs) 153
extracted ion chromatograms (EICS) 153
extraction efficiency 54, 57
extractive electrospray ionization (EESI) 97

f

false discovery rate (FDR) 274, 353
fast scan cyclic voltammetry (FCSV) 247
fatty acid methyl esters (FAMEs) 145, 165, 166, *167*, 168, 170, 172
fatty acids (FAs) 149, 151, 158, 160
fecal metabolome analysis 110
fermentations, sampling and sample preparation in metabolite profiling in 49–53
field amplified sample injection (FASI) 244
flame ionization detection (FID) 75
flow channel systems 249
flow injection analysis 385
flow–injection-based (FIA-MS) metabolomics 22
fluidic pumping for on-chip mixing 242
fluorescence microscopy 245
fluorescence resonance energy transfer (FRET) 383
flux dynamics 307–308
fluxome profiling 307
fluxomics 70, 285, 302, 308
Fourier Transform-ion cyclotron resonance (FT-ICR) 119–120, 212
– application example using exhaled breath condensate 128
– – biochemical mass difference networking and statistical results synthesis 131–133
– – C–H–N–O–S–P formula annotation 130
– – data preprocessing 129–130
– – experiment 128–129
– – FT-ICR/MS measurement 129
– – statistical preprocessing 130–131
– network analysis and NetCalc composition assignment 126–127
– – dataset statistics 127–128
– principles
– – applied physical techniques 123–124
– – natural ion movement inside ICR cell subjected to magnetic and electric fields 121–123
– – practical advantages 124–126
Fourier transform mass spectrometry (FT-MS) 29, 40, 124
free induction decay (FID) 212
freezing-thawing in methanol (FTM) 13
fructose bisphosphatase 303, *305*
functional genomics 263, 273
functional networks 126

g

galphimine series 322
gas chromatography (GC) 39, 40
– based metabolite profiling, sample preparation for 71–74
– data analysis strategies and software 82–87
– illustrative examples 88–89
– MS 61–63, 210–211, 325, 367, 383
– – GC-TOFMS instrumentation and 74–82
– MS-based lipidomics 165
– – data processing and analysis 170–172
– – GC-MS 166–170
– – sample preparation 165–166
Gaussian graphical models (GGM) 282
genetically modified crops 318–319
genome 379, 386, 387
genomics 139
glutathione (GSH) 58
glutathione disulphide (GSSG) 58
glycerolipids (GLs) 151
glycerophospholipids (GPs) 141, 149, 152
glyoxylate 285, 296, 302, 303
graph theory 126
Guide for Measurement Uncertainty (GUM) 41, 42, 45

h

heatmaps 127
heteronuclear multiple bond coherence (HMBC) 217
heteronuclear multiple quantum coherence (HMQC) 217
heteronuclear single quantum coherence (HSQC) 217
heteronuclear two-dimensional methods 217
hierarchical cluster analysis (HCA) *381*
hierarchical principal component analysis (HPCA) 87

high-performance liquid chromatography (HPLC) 22, 24, 25, 27, 40, 59, 124, 322
high-throughput flux screening 307
hippuric acid 396
Homeostasis Model Assessment Insulin Resistance (HOMA-IR) 112
host plant resistance and plant–environment interactions 320–321
hot water method (hw) 12
Human Metabolome Database (HMDB) 109, 130, 189, 365–366, 402
hydrophilic interaction liquid chromatography (HILIC) 24, 25, 96, 106, 112, 144, 325
hyphenated methods 1, 13

i

information-dependent acquisition (IDA) 34
injection system 59–60
insulin resistance 399–400, *401*
internal standards (IS) 73, 103
– ^{13}C-labeled 13, 15–17
International Lipid Classification and Nomenclature Committee (ILCNC) 141
intracellular metabolite quantitation, in yeast
– sample preparation uncertainty calculation for 53
– – general sample workflow 53–54
– – sample preparation protocol for uncertainty estimation 54–59
intracellular metabolites 2, 3, 4, 5, *10–11*
determination, quenching procedure for 6–7
ion cyclotron resonance (ICR) 121, 123
ion distribution 27
ionization performance 94
ionization techniques 27–28
ion mobility mass spectrometry (IMS) 29–30
ion pairing chromatography (IP-LC) 24
ion suppression 95, 101
Ishikawa diagram. See cause-and-effect diagram (CaED)
isotope dilution mass spectroscopy (IDMS) 5–6, 15
isotope labeling 292–293
– pattern definition 289–290
– radiolabeled isotopes 293–295
– stable isotopes 295–296
isotope ratio (ir) 301
isotopologs 15, 43
isotopomer 43
– mapping 299

– mass 289, 290
– positional 289

j

J-resolved spectroscopy (JRES) 215

k

KEGG 130
Kragten spreadsheet, for uncertainty propagation 46, 47, *57*

l

labeling fingerprint 293
laminar diffusion 241–243
laser-induced fluorescence (LIF) 245
LASSO 275
latent vectors 278
limit of quantification (LOQ) 28, 31
limits of detection (LODs) 81, 184, 245, 246
linear discriminant analysis (LDA) 280
LipidMaps 130
LIPID MAPS Structure Database (LMSD) 140
lipidomics 139–140
– GC-MS-based 165
– – data processing and analysis 170–172
– – GC-MS 166–170
– – sample preparation 165–166
– LC-MS based 146
– – data processing and analysis 161–162, 164–165
– – ionization characteristics 155–156
– – lipid extraction 146–150
– – lipid identification 156–161
– – retention time characteristics 151–155
– lipid diversity 140–141
– state-of-the-art tackling 141–146
liquid chromatography (LC) 22, 24–27. See also nontargeted metabolomics, LC-MS based
– MS 59–61, 211, 325–326, 367, 383
– – cell extracts measurement uncertainty calculation 63–65
– – lipidomics 146–165
– – targeted metabolomics 22
liquid chromatography/electrospray ionization mass spectrometry (LC-ESI-MS) 13, 15
liquid–liquid extraction 316
loading plot 225, 226
Lorentz force 121
low-order nonlinear locally estimated smoothing function (LOESS) 363

m

Madin Darby Canine Kidney (MDCK) cells 253
magnetron frequency 123
mammalian systems study 345–347
- analytical experiments 348–350
- applications 369
- – cardiovascular diseases 371–373
- – pregnancy complications 369–371
- hypothesis generation and testing studies 347–348
- metabolite identification 365–369
- quality assurance and quality control 360–365
- sample types 355–360
- study and experimental design 350–354
Mann–Whitney test 274
mass analyzers 28–30
MassHunter Qualitative Analysis software 164
mass isotopomers 289, 290
mass–mass difference networks 126
mass spectral overlap 60
MassTRIX 130
MATHCAD 291
MATLAB 104, 291
matrix-assisted laser desorption ionization (MALDI) 97, 145, 256, 324
- time-of-flight (TOF) 301, 385
Matrix Generator algorithm 129
maximum likelihood methods 263
mean centering 225
measurand 41
- definition 43, 63
- model 43
- model equation 56, 64
- value calculation and associated expanded uncertainty 45, 47
measurement error 263
measurement uncertainty, in quantitative metabolomics
- calculation, according to bottom-up approach 42–48
- definition 41–42
- MS-based techniques 39–40
- MS experiments 48
- – cell extracts LC-MS measurement uncertainty calculation 63–65
- – GC-MS 61–63
- – uncertainties in sample preparation 48–59
- reporting and documentation 47–48
- uncertainty sources
- – identification 44

- – quantification 44–45
metabolic fingerprinting 119, 209, 228, 246
metabolic flux analysis 285–286
- application 303
- – industrial production strains improvement 303–305
- – integration into systems biology approaches 306–307
- labeling studies using isotopes 292–293
- – radiolabeled isotopes 293–295
- – stable isotopes 295–296
- prerequisites
- – isotope labeling pattern definition 289–290
- – metabolic and isotopic steady state 288–289
- – network formulation and condensation 287–288
- – network topology and cellular composition 287
- recent advances
- – flux dynamics 307–308
- – high-throughput flux screening 307
- state-of-art 13C flux analysis 296–298
- – carbon transitions modeling 299
- – experimental design 299
- – flux calculation and flux data statistical evaluation 300–301
- – labeling analysis by mass spectrometry 301–302
- – labeling analysis by nuclear magnetic resonance spectroscopy 302
- stoichiometric flux analysis 290–292
metabolic profiling 48, 93, 97, 98, 101, 103, 109, *111*, 112, 118, 209, 228, 247, 349
- sampling and sample preparation in 49–53
metabolic quenching methods 381, 386
metabolite cartography, in metabolomics 118
Metabolite Mass Spectrometry Analysis Tool (MMSAT) 34
metabolome 21, 22, 32, 33, 35, 117. *See also individual entries*
MetAlign 104
methyl-*tert*-butylether (MTBE) 148
METLIN 109
micellar electrokinetic chromatography (MEKC) 178, 245
microchip capillary electrophoresis 243
- sample injection 244
- separations 244–245
- systems 243–244

microfluidics 239–240
– cellular analysis and
– – biological questions 249–254
– – instrumentation types 248–249
– – single cell metabolomics requirements 247–248
– sample processing 240
– – fluidic pumping for on-chip mixing 242
– – laminar diffusion 241–243
– – solid phase extraction 240–241
– separations for metabolic analysis 243
– – analyte detection 245–247
– – microchip capillary electrophoresis 243–245
modulation period 76
molar enrichment 290
Molecular Biometrics 112
MSDChem software (Agilent Technologies) 172
multilevel approaches 281–282
multilevel component analysis (MSCA) 282
multilevel data analysis (MLDA) 394
multiple linear regression (MLR) 277–278
multiple reactions monitoring (MRM) 28, 103, 142, 187
multivariate data analysis (MVA) 319
multivariate statistical analysis 94
mutants and metabolic pathways 326–327
MZmine 104
mzXML 104

n

natural product discovery 329
– phytopreparation effect and 321–322
negative ionization 97
netCDF 104
network analysis and NetCalc composition assignment 126–127
– FTICR-MS dataset statistics 127–128
network inference 282–283
neutral lipids 156
nontargeted metabolomics, LC-MS based
– analytical strategies 103–104
– applications 109–112
– data analysis 104–107
– LC issues 94–97
– mass spectrometry 97–98
– metabolite identification 107, 109
– sample preparation 100–102
– study design 98–100
normal-phase liquid chromatography (NP-LC) 144, 148
nuclear magnetic resonance (NMR) 118, 119, 209–210, 317–318, 323
– applications 228
– – agricultural applications 233
– – application to bioactivity screening 230–231
– – chemotaxonomy 232
– – herbal medicines quality control 231–232
– – stress response understanding 228–230
– metabolomics and 219
– – bidirectional orthogonal-PLS (O2PLS) 227
– – data preprocessing 223–225
– – metabolite identification 221–223
– – partial least squares (PLS) projections to latent structures 226–227
– – principal component analysis 225–226
– – sample preparation 220–221
– – validation 227–228
– metabolomics platforms 210
– – Fourier Transform–Infrared spectroscopy (FT–IR) 212
– – mass spectrometry (MS) 210–212
– principles and techniques 212
– – combined two-dimensional methods 217, 219
– – correlation spectroscopy (COSY) 215–216
– – heteronuclear two-dimensional methods 217
– – J-resolved spectroscopy (JRES) 215
– – one-dimensional 213–215
– – total correlation spectroscopy (TOCSY) 217
– labeling analysis by 302
nutrition, of plant 320, 327
nutritional metabolomics 393–394
– dynamic nature 400, 402
– food metabolome and human samples 396, 398
– future and research needs 402–410
– health and disease states, human metabolome variability in 398–400
– plasma 394
– urine 395–396

o

oral glucose tolerance test (OGTT) 400, 402
Orbitrap 29, 40, 97, 112, 368
orthogonal partial least-squares (OPLS) 127, 281
– discriminant analysis (OPLS-DA) 87, 110, *111*, 281

– O2PLS-DA models 131
orthogonal signal correction (OSC) 130
– filter 227
– PLS model 131
2-oxoglutarate 372–373
ozone-induced dissociation (OzID) 160

p

Packed column supercritical fluid chromatography (pSFC) 144
PARAFAC 276, 277, *279*
PARAFASCA 276, *278*, 280
Pareto scaling 225
partial least square-discriminant analysis (PLS-DA) 127, 187, 227, 229, 330, 394
partial least squares (PLS) 246
– projections to latent structures 226–227
partition and adsorption chromatography 59
peak alignment 225
peak capacity 77–82, 89, *120*
Penicillium chrysogenum 6, 7, 15
pentose phosphate pathway (PPP) 285, 288, 293, *294*, 296, 300, 303, 305
phase optimized liquid chromatography (POP-LC) 384
phosphoinositol (PI) lipids 155
phytopharmaceuticals, quality control of 322–323, 328
Pichia pastoris 6
pipeline, metabolomics 262
plant physiology 230
plant science applications 313–314
– analytical methods 316–317
– – applications 318–323, 326–329
– – combined approaches 329–330
– – direct MS methods 324
– – direct NMR fingerprinting 317–318
– – future trends 323
– – MS hyphenation to separation techniques 324–326
– – NMR hyphenation to separating techniques 323
– metabolite identification
– – databases 332–334
– – mass spectra interpretation 330–332
– novel metabolites and model validation 334–336
– sample preparation 314
– – culture and harvesting 314–315
– – extraction 315–316
– – storage and drying 315
pneumatic valving 242

polar lipids 156
positional isomers 145, 168
positional isotopomers 289
principal component analysis (PCA) 106, 127, 187, 225–226, 229, 246, 272, 273, 275, 276, 278, *381*, 394
principal component discriminant analysis (PCDA) 280
principal component regression (PCR) 278
projected one-dimensional spectrum (p-JRES) 215
pulsed electrochemical detection (PED) 246

q

QTRAP 98
quadrupole mass spectrometry (qMS) 78, 79
quadrupoles 24, 28, 29, 35, 141
– ion trap (QIT) 182, *183*
Quality Control-Robust Loess Signal Correction (QC-RLSC) 363
quality control samples 103–104
quenching
– bacteria 7
– procedure validation and metabolic leakage minimization 5–6
– procedures and properties 4–5
– procedures for intracellular metabolites determination in extracellular abundance presence 6–7

r

radiolabeled isotopes 293–295
random forest 127, 246, 275
receiver operating characteristic (ROC) curve 110, *111*
regression methods 277–280
relative quantification 350
relative standard deviations (RSDs) 190, 203, 363
retention time locked (RTL) 166, 171, 172
reversed-phase (RP) liquid chromatography (RP-LC) 96, 144, 146, 148
reversed-phase pentafluorophenylpropyl (PFPP) 24, 25

s

samples-to-variables ratio, low 263
sampling and sample preparation
  problem 1
– ^{13}C-labeled internal standards 13, 15–17
– metabolite extraction

sampling and sample preparation
problem (contd.)
– – methods and properties   9
– – validation of methods for yeast
metabolomics   9, 12–13
– microorganisms and properties   1–2
– quenching
– – bacteria   7
– – procedures and properties   4–5
– – procedures for intracellular metabolites
determination in extracellular
abundance presence   6–7
– – procedure validation and metabolic
leakage minimization   5–6
– sampling methods
– – rapid sampling need   2–3
– – sampling systems   3–4
sampling methods
– rapid sampling need   2–3
– sampling systems   3–4
scale freeness   131
score plot   225, 226
secondary ion (SI)   145
selected reaction monitoring (SRM)   21, 28, 31, 34
selection bias   268
shotgun lipidomics   142
SIMCA-P11.5   131
single nucleotide polymorphisms (SNP)   399
SMART   276
solid-phase extraction (SPE)   148, 186, 187, 316
solid phase microextraction (SPME)   30, 71, 240–241
sphingoid base (SB)   159
stable isotope labeled internal standard (SIL-IS)   31
stable isotopes   295–296
statistical heterospectroscopy (SHY) method   329–330
statistical total correlation spectroscopy (STOCSY)   221
SteroIDQ   34
stoichiometric flux analysis   290–292
stratum corneum (SC)   148
structural networks   126
summed fractional labeling (SFL)   290
support vector machine (SVM)   127
surface-enhanced laser desorption ionization (SELDI)   97
surrogate analyte   31–32
systems biology   209

t

tandem mass spectrometry   1
– applications   32–34
– LC-MS targeted metabolomics   22
– liquid chromatography   22, 24–27
– mass spectrometry
– – ionization techniques   27–28
– – mass analyzers   28–30
– relative and absolute quantification   31–32
– sample preparation   30
– synopsis   34–35
target analysis   118
targeted metabolomics   21
– LC-MS   22
taxonomy and phylogeny   327
technologies, metabolomics   118–121
tert-butyldimethylsilyl (TBDMS)   73, 74
time-of-flight (TOF) geometry   29, 40, 74–82, 97, 112, 124, 146, 185, 189
time resolved data   263, 282
top-down approach   42
total combined uncertainty   45, 64
total correlation spectroscopy (TOCSY)   217
traceability   41–42
transition matrix approach   299
triacylglycerols (TGs)   141
tricarboxylic acid (TCA) cycle   285, 293, 294, 303
trimethylamine N-oxide (TMAO)   396
trimethylsilyl (TMS)   73, 74
trimethylsilylation   61
t-test   273–274

u

ultra (high) performance liquid chromatography (U(H)PLC)   28, 94, 96, 97, 110, 111, 112, 124, 151, 326, 385
uncertainty budget   64
unit variance (UV)   225
urine   180, 181, 187, 189, 202, 205

v

variable selection   274–275, 279
vertical drop method   60
Vocabulary of Basic and General Terms in Metrology (VIM)   41

w

water suppression enhanced through T1 effects (WET) technique   213

weak anion exchange (WAX) 149
Wilcoxon test. See Mann–Whitney test

x
XCMS 104, 106

y
yeast metabolomics, validation of extraction methods for 9, 12–13